Forest Pathology

The American Forestry Series

HENRY J. VAUX, *Consulting Editor*

ALLEN AND SHARPE · An Introduction to American Forestry

BAKER · Principles of Silviculture

BOYCE · Forest Pathology

BROCKMAN · Recreational Use of Wild Lands

BROWN, PANSHIN, AND FORSAITH · Textbook of Wood Technology

Volume II—The Physical, Mechanical, and Chemical Properties of the Commercial Woods of the United States

BRUCE AND SCHUMACHER · Forest Mensuration

CHAPMAN AND MEYER · Forest Mensuration

CHAPMAN AND MEYER · Forest Valuation

DANA · Forest and Range Policy

DAVIS · American Forest Management

DAVIS · Forest Fire: Control and Use

DUERR · Fundamentals of Forestry Economics

GRAHAM AND KNIGHT · Principles of Forest Entomology

GUISE · The Management of Farm Woodlands

HARLOW AND HARRAR · Textbook of Dendrology

HUNT AND GARRATT · Wood Preservation

PANSHIN, DE ZEEUW, AND BROWN · Textbook of Wood Technology

Volume I—Structure, Identification, Uses, and Properties of the Commercial Woods of the United States

PANSHIN, HARRAR, AND BETHEL · Forest Products

PRESTON · Farm Wood Crops

SHIRLEY · Forestry and Its Career Opportunities

STODDART AND SMITH · Range Management

TRIPPENSEE · Wildlife Management

Volume I—Upland Game and General Principles

Volume II—Fur Bearers, Waterfowl, and Fish

WACKERMAN · Harvesting Timber Crops

Walter Mulford was Consulting Editor of this series from its inception in 1931 until January 1, 1952.

FOREST PATHOLOGY

John Shaw Boyce

PROFESSOR EMERITUS OF FOREST PATHOLOGY
YALE UNIVERSITY

THIRD EDITION

McGRAW-HILL BOOK COMPANY

New York Toronto London

1961

FOREST PATHOLOGY

 Library of Congress Catalog Card Number 60-15282.

9-MAMM-7 6 5 4 3 2 1

06898

To
The Many Investigators
Whose Work Has Made This Book Possible

Preface

This volume is a combined text and reference book, in that the large number of selected literature citations that are closely correlated with the text should make the work a useful reference for specialists, who generally have direct or indirect access to adequate libraries. In selecting references no attempt has been made to cite the work of the principal investigator of the particular topic discussed; instead, later papers that contain adequate bibliographies have been given preference. Space limitations which made it impossible to develop a complete reference book also forbade a discussion of culture methods and other pathological technique, but these topics have been adequately treated by others, and, furthermore, they are of value largely to specialists.

Effort has been made to correlate the lengths of the discussions on the various diseases with their relative economic importance, but there is always a tendency on the part of an author to emphasize those with which he is personally experienced. Furthermore, no two authorities will quite agree on the relative importance of many different diseases. Then, too, conditions change, so that a disease which now seems important may in time be relegated to a minor position.

Years of experience in forest pathology as an investigator and teacher have convinced the author that a book on tree diseases based on the parts of the tree affected as well as on the tree species, with the sequence of diseases caused by pathogens roughly following the life of the tree from the seedling stage to maturity, is the most practical. For those who have received no formal instruction in forest pathology, it is relatively simple to find information in a book on tree diseases based on the part of the tree affected, but if it is necessary to know first the name, or at least the relative taxonomic position of an organism, the search often becomes difficult.

The author is indebted to a number of sources for illustrations that are credited in the legends. Particular acknowledgment is made to

J. S. Boyce, Jr., of the Southeastern Forest Experiment Station, United States Forest Service, for revising Chaps. 7, 8, 14, and 19. This work was performed on his own time and not as part of his official duties for the Forest Service. For suggestions on or help with various portions of the manuscript grateful appreciation is given to the following: W. V. Benedict, T. W. Bretz, T. W. Childs, R. W. Davidson, C. L. Fergus, L. S. Gill, F. G. Hawksworth, G. H. Hepting, P. D. Keener, J. W. Kimmey, J. L. Mielke, and W. W. Wagener.

John Shaw Boyce

Contents

CHAPTER 1

Introduction

Forest pathology is the branch of botanical science that deals with diseases of forest trees for the purpose of preventing or controlling such diseases. The deterioration of wood in use, through organic agencies or chemical reactions, is usually considered as a part of forest pathology, although the names wood pathology and products pathology have also been applied to this specialty. Shade-tree pathology is concerned with diseases of shade and ornamental trees. These trees have such a high value as compared with forest trees that specialized and intensive practices can be used to combat their diseases that cannot be economically applied against diseases of forest trees.

Just as forestry is a highly specialized branch of agriculture, forest pathology is a specialized branch of plant pathology. The forest pathologist must not only have a knowledge of the methods employed by the plant pathologist in investigating and controlling diseases of the usual agricultural crops; he must also thoroughly understand the methods and economics of producing forest crops in order that he may develop economically applicable procedures for the prevention and control of forest-tree diseases. Thus forest pathology is also a branch of forest protection. In general the low value of the forest crop sharply circumscribes control measures.

Historical. The science of forest pathology did not originate in the United States. Robert Hartig (1839–1901), the last in a family line of German foresters, was the father of forest pathology. The published accounts of his investigations of many tree diseases are classics today, and he was the first investigator (1874) to diagnose properly the relation of fungous hyphae in decayed wood to the conks or fructifications on the tree, although the presence of hyphae in decayed wood had previously been recognized by other workers.

In the United States, the credit for the development of forest pathology belongs to the Department of Agriculture of the Federal government (Hartley 1950). Organized work began in 1899 when the Mississippi Valley Laboratory at St. Louis was created, but this laboratory was discontinued in 1907 when the Division of Forest Pathology was started in the Bureau of Plant Industry with headquarters at Washington, D.C. In 1953 work on forest pathology was transferred to the Forest Service.

The development of forest pathology in the United States has been necessitated by the realization that losses resulting from tree diseases must be reduced in order to achieve full success in the practice of forestry. It has been particularly stimulated by the ravages of chestnut blight and white pine blister rust, both caused by introduced pathogens.

Losses from Disease. Diseases, particularly, and insects work slowly and insidiously, whereas fire is sudden and spectacular. Yet loss from fire is much less than loss from diseases and insects. Of the growth impact (mortality plus growth loss) on sawtimber in the United States including Coastal Alaska in 1952, 45 per cent was estimated as due to diseases, 20 per cent to insects, 17 per cent to fire, and 18 per cent to all other agencies (Hepting and Jemison 1958). This was a loss of almost 20 billion board feet of sawtimber from diseases alone.

Furthermore, after timber is converted, much of it, in storage or in service, is lowered in value or destroyed by stain and decay, thus necessitating an increased annual cut for replacement. Although the overcut necessitated is a significant loss, the major loss is the cost of replacement for wood in service. This cost is greater than the value of the material destroyed even though wood in the form of forest products is worth several times as much as the stumpage of the timber from which it came. A considerable proportion of this damage to forest products is preventable.

For stands not yet of merchantable size, reduction of increment, causing lengthening of the rotation, and actual killing of trees, causing understocking, result in heavy losses. Root, foliage, and stem diseases are widespread, all contributing to a reduction in the future capital of timber. Dwarf mistletoes, so prevalent in the West, alone take a heavy toll, and the less conspicuous foliage diseases also cause considerable reduction in increment. Outright killing of immature trees does not necessarily cause loss unless understocking results or unless the quality of the stand is lowered by disproportionate killing of the more valuable species or individual trees. Outright killing may even function as a beneficial thinning, although such killing is likely to be irregular in distribution, resulting in openings in the stand. Killing of nursery stock results in the easily figured direct loss of the cost of growing the stock and the potentially more serious indirect loss of the higher cost of growing the remaining plants and the possible disruption of the planting program.

Loss is determined by the type of land use. The standards for judging loss in forests for watershed protection or for recreation including game production are quite different from those applied in forests for wood production. The effect of native diseases on watershed protection is apparently slight. So far as our present knowledge goes, native diseases do not materially lower the efficiency of forest cover either in delaying runoff of water or in preventing erosion. In areas used primarily for recreation, tree diseases, generally speaking, are of importance only when they detract from the beauty of the landscape or upset the natural balance of plant associations which should be maintained as far as possible. Heart rots and growth-reducing diseases that might make the difference between profit and loss in a timber-growing project have much less effect on aesthetic values; in fact to some people, hollow trunks and gnarly or irregular crowns make trees more picturesque. However, killing diseases often materially hurt the appearance of the forest. Trees with trunk and butt rots are predisposed to windbreak and wind throw and are dangerous in the immediate vicinity of roads, trails, and camps. Diseases which cause a conspicuous amount of unsightly yellow and brown dying foliage are sometimes very objectionable. There is no obvious way in which ordinary diseases impair the value of the forest for game purposes, although a virulent killing disease, such as chestnut blight, must have some effect on the food habits and possibly the number of certain small animals, particularly squirrels.

Since the degree of loss through disease is determined by land use, the major purpose of any forest area should be defined before disease control is attempted. The principles of forest-disease control are discussed in the concluding chapter.

REFERENCES

Hartig, R.: "Wichtige Krankheiten der Waldbäume." Berlin: Springer-Verlag (1874).

Hartley, C.: The Division of Forest Pathology. *U.S. Dept. Agr., Plant Disease Reptr. Suppl.*, 195:445–462 (1950).

Hepting, G. H., and G. M. Jemison: Forest Protection. Timber Resources for America's Future. *U.S. Dept. Agr., Forest Serv., Forest Resource Rept.*, 14:185–220 (1958).

USEFUL BOOKS AND PAPERS ON TREE DISEASES

Baldwin, H. I., et al. (eds.): "Important Tree Pests of the Northeast," 2d ed. Concord, N.H.: Evans Printing Company (1952).

Baxter, D. V.: "Pathology in Forest Practice," 2d ed. New York: John Wiley & Sons, Inc. (1952).

Beckwith, L. C., and R. L. Anderson: The Forest Insect and Disease Situation, Lake States, 1956. *Lake States Forest Expt. Sta., Sta. Paper*, 42:1–26 (1956).

Bier, J. E.: Some Common Tree Diseases of British Columbia. *Can. Dept. Agr., Sci. Serv., Div. Botany and Plant Pathol.* (1949).

Biraghi, A.: "Patalogia delle piante forestali. Parte speciale. Malattia prodotte da virus, batteri e funghi." Firenze: Editrice Universitaria (no date).

Campbell, W. A., and P. Spaulding: Stand Improvement of Northern Hardwoods in Relation to Diseases in the Northeast. *Allegheny Forest Expt. Sta., Occas. Paper*, 5:1–25 (1942). (Mimeographed.)

Carter, J. C.: Illinois Trees: Their Diseases. *Illinois Nat. Hist. Survey Circ.*, 46:1–99 (1955).

Christensen, C. M., R. L. Anderson, and A. C. Hodson: Enemies of Aspen. *Lake States Forest Expt. Sta., Lake States Aspen Rept.*, 22:1–16 (1951).

Edlin, H. L., and M. Nimmo: "Tree Injuries." London: Thames & Hudson (1956).

Felt, E. P., and W. H. Rankin: "Insects and Diseases of Ornamental Trees and Shrubs." New York: The Macmillan Company (1932).

Ferdinandsen, C., and C. A. Jørgensen: "Skovtraernes sygdomme." Copenhagen: Gyldendalske Boghandel—Nordisk Forlag (1938–1939).

Hartig, R. (transl. from the German by W. Somerville): "Text-book of the Diseases of Trees." London: Geo. Newnes, Ltd. (1894).

Hartig, R. "Lehrbuch der Pflanzenkrankheiten," 3d ed. Berlin: Springer-Verlag (1900).

Hepting, G. H.: Reducing Losses in Tree Diseases in Eastern Forests and Farm Woodlands. *U.S. Dept. Agr., Farmers' Bull.*, 1887:1–22 (1942).

Hubert, E. E.: "An Outline of Forest Pathology." New York: John Wiley & Sons, Inc. (1931).

Jackson, L. W. R., G. E. Thompson, and H. O. Lund: Forest Diseases and Insects of Georgia's Trees. Georgia Forestry Commission (1954).

Jemison, G. M., and G. H. Hepting: Timber Stand Improvement in the Southern Appalachian Region. *U.S. Dept. Agr., Forest Serv., Misc. Publ.*, 693:1–80 (1949).

Lorenz, R. C., and C. M. Christensen: A Survey of Forest Tree Diseases and Their Relation to Stand Improvement in the Lake and Central States, U.S. Dept. Agr., Bur. Plant Ind. (1937). (Mimeographed.)

Marchal, E.: "Éléments de pathologie végétale appliquée à l'agronomie et la sylviculture." Gembloux: Jules Ducolot (1925).

Marshall, R. P., and Alma M. Waterman: Common Diseases of Important Shade Trees. *U.S. Dept. Agr., Farmers' Bull.*, 1987:1–53 (1948).

Möller, A.: "Der Waldbau." Berlin: Springer-Verlag (1929), vol. I.

Moore, Agnes Ellis (ed.): Bibliography of Forest Disease Research in the Department of Agriculture. *U.S. Dept. Agr., Misc. Publ.*, 725:I-II, 1–186 (1957).

Neger, F. W.: "Die Krankheiten unserer Waldbäume und wichtigsten Gartengehölze," 2d ed. Stuttgart: Ferdinand Enke (1924).

Orlos, H.: Przewodnik do oznaczania chorób drzew i zgnilizny drewna [A Guide to the Determination of Tree Diseases and Wood Rots]. Warsaw: State Agricultural and Forestry Publisher (1951).

Pomerleau, R.: The Problems of Forest Pathology in Quebec. *Forestry Chronicle*, 21:267–280 (1945).

Rankin, W. H.: "Manual of Tree Diseases." New York: The Macmillan Company (1918).

Roll-Hansen, F.: "Forstpatalogi." Norges Landbrukshögskole (1958).

Rostrup, E.: "Afbildning og beskrivelse de farligste snyltesvampe i Denmarks skove." Copenhagen: Philipsens Boghandel (1889).

Schrenk, H. von: Some Diseases of New England Conifers: A Preliminary Report. *U.S. Dept. Agr., Div. Veg. Physiol. and Pathol. Bull.*, 25:1–56 (1900).

Schrenk, H. von, and P. Spaulding: Diseases of Deciduous Forest Trees. *U.S. Dept. Agr., Bur. Plant Ind. Bull.*, 149:1–85 (1909).

Schwerdtfeger, F.: "Die Waldkrankheiten. Ein Lehrbuch der Forstpathologie und des Forstschutzes," 2d ed. Hamburg and Berlin: P. Parey (1957).

Tubeuf, K. von (transl. from the German by W. G. Smith): "Diseases of Plants Induced by Cryptogamic Parasites." London: Longmans, Green & Co., Ltd. (1897).

Ward, H. M.: "Timber and Some of Its Diseases." London: Macmillan & Co., Ltd. (1909).

CHAPTER 2

Disease

Disease in plants may be defined as "Sustained physiological and resulting structural disturbances of living tissues and organs, ending sometimes in death" (Ehrlich 1941). The entire plant or only a portion of it may be affected, and disease is not necessarily significantly injurious.

CAUSES OF DISEASE

Plant diseases may be caused by the nonliving environment, such as unfavorable soil or atmospheric conditions. They may be caused by viruses whose nature is not yet understood, or they may be caused by an unknown agent. The majority of diseases are caused by the activities of living organisms, viz., slime molds, bacteria, fungi, algae, seed plants, and animals including insects. Two or more agencies may combine to cause disease, and it is frequently necessary for a plant to be weakened by unfavorable environmental conditions before it can be attacked by living organisms. Among forest trees, diseases caused by unfavorable environmental conditions are common.

Viruses, common causers of disease in plants and animals, are protein particles so small that they can pass through the finest porcelain filter and can be seen only through an electron microscope. The largest virus is about one-tenth the size of the smallest bacterium. Viruses multiply readily in living cells, but they do not respire or have a metabolism, nor do they have a cellular structure. Many biologists consider viruses as intermediate between nonliving matter and living microorganisms. Virus diseases are frequent and destructive on many plants (Bawden 1956; Smith 1957). The symptoms vary; commonly the leaves of the host are malformed, stunted, wrinkled, or mottled. The entire plant may be stunted or blighted, the number of leaves may be greatly increased, or the leaves may be abnormal in size and color. Witches'-brooms may result. Virus diseases occur to some extent on trees (Atanasoff 1935), the best known being the destructive spike disease of sandal (*Santalum album* L.) in India. In the United States the brooming disease of black locust (page 303) and the phloem necrosis of American elm (page 114) are caused by viruses. There are unexplained abnormalities of little consequence on forest trees that may be indications of virus diseases (Hartley 1933).

Slime molds do not cause any disease of forest trees in North America, although these organisms are abundant as saprophytes in forest stands. Bacteria cause many serious plant diseases (Elliott 1951; Dowson 1957), severely attacking certain fruit and nut trees, but they have been of little consequence on forest trees. Fungi are prolific causers of forest-tree diseases; algae alone are not known to induce any tree disease; and certain seed plants, notably mistletoes and dwarf mistletoes, are widespread on and confined to trees and shrubs, causing severe injury. Animals, particularly insects, cause many and severe tree diseases, but these organisms, with the exception of nematodes, eelworms, or roundworms, are left to the province of zoology and entomology. Serious diseases caused by nematodes occur on a wide variety of plants (Chitwood and Birchfield 1956), but they occur infrequently on trees and so far have been reported only on seedlings.

To determine that an organism is the actual cause of a disease, rules of proof, known as "Koch's postulates," must be fulfilled if possible: (1) establish constant association of the organism with the disease; (2) isolate the organism from the diseased plant and grow it in pure culture; (3) inoculate healthy plants of the same species or variety with the organism from the pure culture and produce the same disease in them; and (4) reisolate the same organism from the inoculated plants as was present in the pure culture. These postulates cannot be fulfilled completely with organisms that will not grow on artificial media.

INDICATIONS OF DISEASE

The indications or apparent expressions of disease processes in the plant, i.e., the visible evidences of an unhealthy condition, are known as the symptoms and signs of disease, although generally both types of evidence are simply termed symptoms. Whetzel (1929:1209) considers symptoms to be expressions by the plant itself of the pathologic condition by which a diseased individual is distinguished from a healthy one, and signs to be other evidences such as structures of the causal organism produced in or on the lesions, emanations, and so forth. It is essential to be thoroughly familiar with the normal appearance and function of plants in order to recognize the symptoms and signs of disease.

SYMPTOMS

The symptoms of plant diseases fall naturally into three types—necrotic, atrophic, and hypertrophic—although a disease caused by one organism may have more than one type of symptom. For example, dwarf mistletoe on Douglas fir causes witches'-broom, a hypertrophic symptom, although the brooms have needles shorter than normal, an atrophic symptom. The symptoms of white pine blister rust may be classed as hypertrophic because of the spindle-shaped swelling produced on small stems, or as necrotic because of killing of the bark that often causes larger stems to be constricted. Furthermore it is not always possible to determine by the macroscopic appearance alone to which category a symptom belongs, so that study of the microscopic anatomy of the diseased tissues is then necessary.

Necrotic Symptoms. Here are included those symptoms which are evidences of necrosis or death of the affected tissues. The best known of these is death of green tissue, which is shown by a marked change in color, usually first yellowing or icterus followed by browning, reddening, or occasionally graying when the tissue finally dies. The entire organ may be killed, to which the names "leaf blight," "bud blight," and so forth, are applied, or only a portion of it may be affected as exemplified by the well-known leaf spots of hardwoods. Wilt implies a wilting of the plant because the outgo of water by transpiration has exceeded the intake. Dieback means the dying of shoots from the tip back, canker the death of areas of the cortex or bark, and rot or decay the breaking down of tissues. Included here is rot of the heartwood of living trees, of slash, and of forest products caused by wood-destroying fungi, although, of course, the tissues are already dead when attacked. Some wood-destroying fungi do, however, kill and then break down living sapwood.

Atrophic Symptoms. These symptoms are a slowing down in development of the affected plant or parts thereof, resulting from subnormal cell division (hypoplasia) or from degeneration of the cells. Dwarfing or nanism is the best-known symptom in this category, and among its causes are unfavorable environmental conditions, such as repeated frosts, unfavorable soil conditions, viruses, and pathogenic organisms. Chlorosis, a yellowing of green tissues, expresses underdevelopment of the chlorophyll apparatus and is caused by various agencies, such as viruses, bacteria, fungi, and excess or deficiency of certain mineral elements in the soil. Leaves that are both chlorotic and dwarfed are symptomatic of certain virus diseases and of diseases caused by parasitic organisms. When yellowing is caused by lack of light, the term "etiolation" is used. Some individual plants may be entirely without chlorophyll, and these are known as albinos. Unless chlorophyll is developed they cannot live, since they cannot produce carbohydrate food.

Hypertrophic Symptoms. These symptoms are overgrowths of all kinds resulting from an abnormal increase in the number of cells, i.e., excessive cell division (numerical hypertrophy or hyperplasia), or from an abnormal increase in the size of cells (simple hypertrophy), or from both. They are striking symptoms of disease expressed by various prominent malformations and excrescences, such as galls, burls or tumors, erinoses, intumescences, hypertrophied lenticels, fasciations, witches'-brooms, hairy root, leaf curls, leaf blisters, and deformation of fruit and flowers. A general rather than localized overdevelopment of cells so that an extensive enlargement of organs results is termed edema or dropsy. Erinoses, which are an abnormal development of hairs on leaves forming feltlike patches, sometimes brilliantly colored, were once thought to be caused by fungi but are now known to result from the work of parasitic mites. A bright red erinose of dwarf maple is common in the Pacific Northwest.

SIGNS

Most signs of disease are either vegetative or fruiting (reproductive) structures of the organism causing the disease, which become apparent either on or near the affected plant. Consequently it is necessary to know the various stages in the development of disease-causing organisms, particularly fungi, in order to recognize and properly evaluate the various signs. The different fructifications of fungi are common and valuable signs. In some instances there may be no symptoms of the abnormal condition of the plant but only signs. For example, usually the only external evidence of decay of the heartwood of living trees by wood-destroying fungi is the development of conks (fructifications) of the fungi. In addition to structures of the causal organism, another common sign is emanations from the diseased plant.

Exudations. Certain exudations are the result of normal physiological processes, such as forcing out of water or cell sap upon the free surface of leaves (known as "guttation") and bleeding from the cut ends of branches that so frequently occurs on hardwoods, particularly in the spring. Other exudations are definitely signs of disease.

Bacterial Exudates. The first sign of a common canker disease of poplars in Europe is an exudation of bacteria from cracks in the bark.

covering the stem with a shiny layer (Koning 1938:14). Fire blight of apple, pear, and other trees is characterized by a sticky, fluid mass composed of fire-blight bacteria mixed with decomposition products from the affected tissue oozing from cracks in the bark of active lesions. On drying, the exudate may form a firm mass.

Slime Flux. This name indicates a liquid or semiliquid exudation from the inner bark or wood of hardwoods, preceded in all cases by injury (Guba 1934; Ogilvie 1924; Stautz 1931). It occurs primarily on shade trees, less often on forest trees. The phenomenon is actually one of excessive bleeding, but the sap, beginning as a clear watery fluid, is invaded by various microorganisms including bacteria, fungi, and nematodes so that it is changed into a viscid, malodorous, variously colored slime. These exudations do not appear from small stems but are common from wounds on large stems where the sap pressure is great. Slime flux delays or prevents the closing of wounds from which it occurs and encourages the growth of organisms causing decay. Furthermore the exudate is somewhat toxic so that its persistent flow down a trunk or branch may kill the bark.

Gummosis. The production, on the surface of the affected part, of a clear or amber-colored exudate that sets into solid masses insoluble in water is termed "gummosis." It is peculiar to certain hardwoods, notably cherries and plums. It can result from a number of widely different causes, such as poorly aerated soils, wounding, bacteria, and fungi, so that it is not a sign of one specific disease. Wounding or fungous invasion stimulates in many hardwoods the development of gum that is deposited in the tissues (Higgins 1919; Rankin 1933) but not exuded in quantity as in gummosis.

Resinosis. Since production of resin in the bark, wood, and other tissues is a normal process in many conifers, a small amount of resin exudate on the surface is natural, but an excessive outflow of resin, known as "resinosis," is frequently a sign of disease. Resin secretion is stimulated by wounds, and resin completely covering an open wound is an excellent antiseptic dressing. However, excessive resin flow from the affected portion of a stem frequently characterizes white pine blister rust, larch canker, and similar diseases. Red spruce with the heartwood affected by red ring rot, caused by *Fomes pini*, sometimes shows heavy resin flow along the trunk from the base of branch stubs. Resinosis from the root collar of pines with shoestring root rot, caused by *Armillaria mellea*, is often so profuse that the surrounding soil and duff become caked into a solid mass.

TERMINOLOGY OF DISEASE

For the discussion of food habits of plants and plants as causal agents of disease, a special and at times confusing terminology has developed, the confusion arising because authorities differ in their definition of certain terms. All plants may be divided into two broad classes, viz., autophytes, autotrophic plants, or independent plants that secure all necessary food materials from inorganic sources, subsequently converting these food materials into foods; and heterophytes, heterotrophic plants, or dependent plants, whose existence is dependent upon the development of other organic forms since they must obtain part of their food at least from organic sources. Heterophytes in turn may be divided into parasites and saprophytes.

The material on which a living organism becomes established may be termed the "substratum." When the substratum is a living plant or part thereof, it is termed the "host," and the organism living upon it the "parasite." The term parasite connotes a one-sided food relationship that is never beneficial and is often harmful to the host. Stevens and Young (1927:410) define parasite as follows: "A parasite is an organism which lives in or is attached to some other species of living organism from the living matter of which it secures part or all of its food materials." According to Link (1933:857), "The essence of the concept of parasitism, in so far as nutrition is concerned, is dependence of *one individual* upon another living individual for food, irrespective of whether the materials used by the dependent individual or parasite are actually derived from the living or dead parts of the host." Consequently Link classifies a wood-destroying fungus that feeds upon the dead heartwood of a living tree as a parasite, whereas following the concept of Stevens and Young it would be a saprophyte. The latter authorities define a saprophyte as "An organism which secures its food from dead organic matter," and in the words of Link, "An organism growing on a dead body or an artificial medium is currently designated as a saprophyte." The definitions of parasitism and saprophytism given by Link are more useful in forest pathology, where so many fungi attack the dead heartwood of living trees and sometimes encroach on the living sapwood.

A parasite does not necessarily cause any serious disturbance in the life of its host, and, although these terms are commonly used in connection with disease, a more exact terminology is to call the causal organism the "pathogen" and the plant that it affects the "suscept" (Whetzel 1929:1208). A pathogen then is any organism capable of causing disease, and a suscept is any plant liable to infection by a given pathogen.

Although a pathogen is always parasitic, a parasite is not always pathogenic, i.e., it does not always cause disease. Consequently there are pathogenic and nonpathogenic parasites. A given pathogen is often variable in its pathogenicity, i.e., in its ability to produce disease, without being influenced by any external factors. To these different strains or races within a species, the term "pathogenic strains" is applicable, and the broader terms "biologic" or "physiologic races" or strains have also been much used. An organism causing decay, e.g., a fungus decaying dead wood, may be termed a "saprogen" and one imparting a stain by reason of its own pigment, a "chromogen" (Ehrlich 1941).

Parasitic organisms are variable in their habits. Many fungi regularly spend part of their life cycle as saprophytes and the remainder of it as parasites. Obligate parasites, also termed "holoparasites," "true parasites," "complete parasites," or "strict parasites," are unable to live except parasitically, i.e., they must obtain all their organic food from another living organism. Among fungi, the rusts are the outstanding obligate parasites. Facultative parasites, also termed "hemisaprophytes" or "occasional parasites," usually live as saprophytes but become parasitic when a susceptible host is debilitated or when other conditions temporarily become extremely favorable to their development. Facultative saprophytes or hemiparasites live usually as parasites but are capable of becoming saprophytic for a period if necessary. Plants such as mistletoes, which although always parasitic do contain chlorophyll so that they can elaborate some of their own food from inorganic food materials taken from their hosts, have also been termed "hemiparasites," as well as "semiparasites," "partial parasites," and "water parasites." Obligate saprophytes, also known as "holosaprophytes" or "true saprophytes," never have any parasitic existence. Because several categories of parasitism do not accord precisely with those given above, other terms for different kinds and degrees of parasitism have been suggested (Link 1933:858; Stevens and Young 1927).

Infection has been defined as the condition of being contaminated by a minute foreign organism, i.e., a microorganism, and infestation as contamination by a large foreign organism, although in common usage this distinction is rarely observed. Plant pathologists are prone to refer to the activities of infinitesimally tiny pathogenic bacteria and of large mistletoes as infections; entomologists commonly refer to infestation by insects, no matter how small. A disease in which infection is an essential antecedent is referred to as an infectious disease as opposed to a noninfectious disease where no contaminating agent is necessary. Of course, a disease, of itself, is not actually infectious, since only the pathogenic agent is actually transmissible. A contagious disease is one in which the pathogenic agent is transmitted by contact only, whereas the agent of an infectious disease is transmitted by soil, water, or air, or by other organisms, usually insects, acting as carriers (vectors). Commonly, however, the terms "infectious" and "contagious" are used interchangeably. With an infectious or contagious disease, for the time elapsing between inoculation and the first observed evidences of disease, the term "incubation period" or "stage" is commonly used, whereas, for the period from the first visible evidences of disease through the final reaction of the suscept, the term "infection period" or "stage" is used. Whetzel (1929:1212) has limited these terms and proposed several new ones to designate various stages in disease from the moment when the inoculum of the causal agent leaves its source through the final reaction of the affected plant. The infection court, also known as the "point" or "place of infection," is the place where the pathogen enters or penetrates the suscept. For example, open wounds and dead branch stubs are infection courts for many wood-destroying fungi attacking the heartwood of living trees.

A disease is said to be "epidemic" when it suddenly becomes abundant in a locality or region. Strictly speaking, epidemic refers to the appearance of a disease among people, epiphytotic having the same meaning for plants and epizootic for animals other than human beings. However, epidemic is commonly used by plant pathologists when referring to outbreaks of plant diseases. The opposite of the foregoing terms are endemic, enphytotic, and enzootic. These terms indicate a disease occurring regularly in a locality or region but in moderate severity only. An endemic disease may become epidemic for a time and then return to its normal status. A disease endemic in one region may become epidemic when the pathogen causing it is introduced into a new region and may increase in severity until the organisms affected by it are almost completely destroyed. Chestnut blight is endemic in Asia but epidemic in North America, the ultimate result being the destruction of American chestnut as a commercial tree. Conditions necessary for an epidemic are a large population of susceptible trees, a favorable environment for disease development, large amounts of inoculum, effective method of spread of the inoculum, and high virulence of the pathogen.

A special terminology has developed for tree diseases (Buckland, Redmond, and Pomerleau 1957).

CLASSIFICATION OF DISEASES

Diseases may be classified in various ways, the classification depending on the purpose to be served. On the basis of cause, they may be divided into (1) noninfectious diseases, also known as "nonparasitic," "physiological," or "physiogenic" diseases; (2) infectious diseases, also known as "parasitic" diseases; and (3) diseases of unknown origin. Noninfectious diseases are those caused by the nonliving environment, whereas infectious diseases are largely caused by other living organisms and may be wholly so caused if virus diseases are finally determined to be caused by pathogens. On the basis of symptoms, they may be classified as necrotic diseases, atrophic diseases, and hypertrophic diseases, or more elaborate classifications can be made.

On the basis of hosts, they may be classified strictly by the type of host affected, as pine diseases, oak diseases, and maple diseases, or by hosts as they are grouped for commercial purposes, as fruit-tree diseases, shade-tree diseases, and forest-tree diseases. The manner in which the affected plants are grown or various stages in their development may determine the classification, as diseases affecting nurseries, plantations, immature forests, and merchantable forests. Again they may be divided according to the part of the plant affected, as root diseases, stem diseases, and foliage diseases. The latter is the classification largely followed in this book.

REFERENCES

Atanasoff, D.: Old and New Virus Diseases of Trees and Shrubs. *Phytopathol. Z.*, 8:197–223 (1935).

Bawden, F. C.: "Plant Viruses and Virus Diseases," 3d ed. Waltham, Mass.: Chronica Botanica Company (1956).

Buckland, D. C., D. R. Redmond, and R. Pomerleau: Definitions of Terms in Forest and Shade Tree Diseases. *Can. J. Botany*, 35:675–679 (1957).

Dowson, W. J.: "Plant Diseases Due to Bacteria," 2d ed. New York: Cambridge University Press (1957).

Chitwood, B. G., and W. Birchfield: Nematodes, Their Kinds and Characteristics, vol. 2. *Florida State Plant Bd. Bull.*, 9:1–49 (1956).

Ehrlich, J.: Etiological Terminology. *Chron. Botanica*, 6:248–249 (1941).

Elliott, Charlotte: "Manual of Bacterial Plant Pathogens," 2d rev. ed. Waltham, Mass.: Chronica Botanica Company (1951).

Guba, E. F.: Slime Flux. *Natl. Shade Tree Conf. Proc.*, 10:56–60 (1934).

Hartley, C.: Unexplained Abnormalities in Forest Trees May Be Due to Virus Diseases. *Forest Worker*, 9:13–14 (1933).

Higgins, B. B.: Gum Formation with Special Reference to Cankers and Decays of Woody Plants. *Georgia Agr. Expt. Sta. Bull.*, 127:23–59 (1919).

Koning, Henriëtte C.: The Bacterial Canker of Poplars. *Phytopathol. Lab. "Willie Commelin Scholten," Mededeel.*, 14:.3–42 (1938).

Link, G. K. K.: Etiological Phytopathology. *Phytopathology*, 23:843–862 (1933).

Ogilvie, L.: Observations on the "Slime-fluxes" of Trees. *Brit. Mycol. Soc. Trans.*, 9:167–182 (1924).

Rankin, W. H.: Wound Gums and Their Relation to Fungi. *Natl. Shade Tree Conf. Proc.*, 9:111–115 (1933).

Smith, K. M.: "A Textbook of Plant Virus Diseases," 2d ed. London: J. & A. Churchill, Ltd. (1957)

Stautz, W.: Beitrage zur Schleimflussfrage. *Phytopathol. Z.*, 3:163–229 (1931).

Stevens, F. L., and P. A. Young: On the Use of the Terms Saprophyte and Parasite. *Phytopathology*, 17:409–411 (1927).

Whetzel, H. H.: The Terminology of Phytopathology. *Proc. Intern. Congr. Plant Sci.*, 2:1204–1215 (1929).

CHAPTER 3

The Fungi

Since fungi cause so many tree diseases, it is necessary to know something about the structure, habit of life, and classification of these plants. However, it is impossible here to give more than a cursory discussion so that for detailed treatment of them the reader must consult the references given at the end of this chapter.

The principal nonflowering vegetable parasites that cause plant diseases belong to three subdivisions in the plant kingdom, viz., (1) Myxomycetes or slime molds, (2) Schizomycetes or bacteria, and (3) Eumycetes or fungi. The term "fungi" (singular fungus) in its broadest sense has been used to include all three subdivisions, but the present tendency is to consider slime molds and bacteria as distinct from fungi. The members of all three subdivisions are lacking in chlorophyll. Transpiration, respiration, and true assimilation are the same as for green plants in all but a few forms, but photosynthesis or starch manufacture cannot be carried on by them. Sunlight then being useless to them directly, they can live in the dark as well as in the light.

Not being able to elaborate their own foods from inorganic matter, these organisms are limited to such nutriment as they can obtain from living or dead organic matter. Consequently they have acquired various food habits. Some are nearly omnivorous and can subsist upon almost any dead tissue or upon soils or solutions rich in organic debris. Others thrive only upon special substances, e.g., some particular plant or animal, or perhaps only upon some special part of it. Many of these organisms, being parasitic, cause disease.

MYXOMYCETES OR SLIME MOLDS

Slime molds do not cause any diseases of trees, but they are commonly found on the ground or on wood debris in dark, moist places in the forest. The slime molds are primitive organisms with characteristics of both plants and animals, a condition that was recognized by the old name Mycetozoa (fungous animals or animal fungi) applied to them. They resemble plants only in their reproductive stages.

The vegetative or plasmodial stage is a multinucleate mass of naked protoplasm, usually colorless, termed a "plasmodium." In consistency this protoplasm is about the same as the white of an egg. Plasmodia vary in size from the minute to a foot or more in diameter. Plasmodia are capable of a limited amoebic or flowing motion. The plasmodial stage is passed in somewhat dark and damp situations.

The reproductive or sporangial stage develops when the protoplasm of the plasmodium collects into definite heaps of one form or another and these heaps become transformed into sporangia or spore cases, often in beautiful forms and colors (Crowder 1926). The protoplasm is divided into numerous spores, among which may be mixed numbers of distinct threads, collectively forming what is known as the "capillitium."

SCHIZOMYCETES OR BACTERIA

Although bacteria severely attack certain fruit and nut trees, they are of minor importance on forest trees. Bacteria are the smallest of living plants, many of them being scarcely visible with the highest power of the compound microscope. It is possible that forms exist which are too minute for microscopical detection. They are one-celled plants with a definite cell wall, composed of nitrogenous compounds, that in many species is surrounded by a thin gelatinous sheath. The cell wall is protective, but it is also thin enough for the passage of food materials through it. There is no well-organized nucleus.

Many bacteria have a limited power of motion owing to the lashing movements of one or more cilia or flagella, which are delicate whiplike threads attached to the cell wall. In most forms reproduction is by simple fission or cell division, in which the original cell becomes split or divided into two daughter cells of equal size, the separation of a single cell requiring from 20 to 30 minutes under suitable conditions. The name Schizomycetes means splitting fungi.

Some bacteria form spores that are able to resist prolonged periods of drought, high or low temperatures, and disinfecting chemicals (Fig. 1*d*). The common type, the endospore, is formed by the cell content compacting into one or more bodies of regular shape, usually globular or oval, and surrounded by a firm, solid membrane. At first the spore is retained within the wall of the parent cell, but this ultimately disappears. Since usually only a single spore is formed in each cell, spore formation is not a method of reproduction or multiplication but is merely a resting stage to tide the plant over unfavorable conditions.

The primary classification of bacteria is based on the form of the vegetative cell. On this basis there are three types: (1) globular or spherical forms, the coccus type; (2) short or long straight cylindrical or rod-shaped forms, the bacillus type; and (3) short or long spiral cylindrical forms, the spirillum type. The individual cells may be held to-

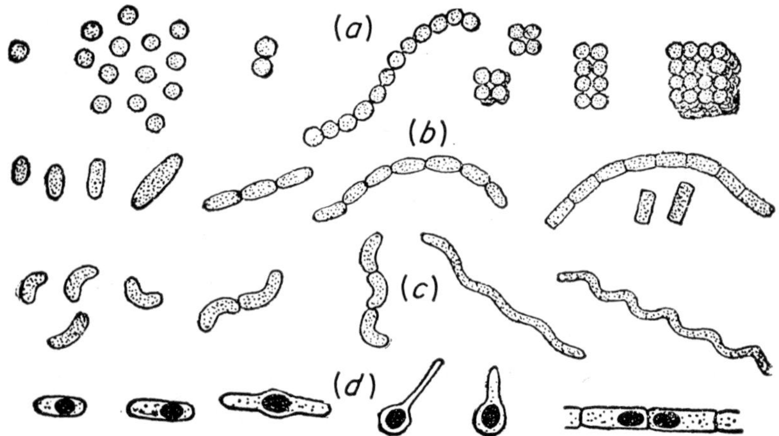

Fig. 1. Diagram illustrating the forms of bacteria. (*a*), spherical forms, the coccus type; (*b*), straight cylindrical or rod-shaped forms, the bacillus type; (*c*), spiral cylindrical forms, the spirillum type; (*d*), types of spore formation. (*After F. D. Heald.*)

gether in pairs, chains, filaments, long spirals, or, in the coccus type, in masses, but these groupings are merely aggregates of individuals. Only bacteria of the bacillus type are known to cause plant diseases.

EUMYCETES OR FUNGI

Fungi are thallophytes without chlorophyll that reproduce by means of spores. A thallophyte is a plant without differentiation into stem, leaves, and roots; consequently it has a very simple structure and is devoid of any special vascular system. Fungi are either saprophytes or parasites, the latter causing many and varied diseases of forest trees. Fungi generally have two reasonably distinct phases in their development, the vegetative and the reproductive stage, the latter usually being the more conspicuous. For example, the microscopically fine mycelium hidden from view in the cells of the heartwood is the vegetative stage of a wood-destroying fungus causing decay in a living tree, whereas the fructification or conk plainly visible on the trunk is the reproductive stage.

THE VEGETATIVE STAGE

All fungi, with the exception of a few unicellular species, have an extensive mycelium in or on the substratum, which is the medium in or on which the mycelium grows. The mycelium originates by the germination of a spore, a small tubelike or filamentlike process, known as a germ tube, growing out from the spore. The germ tube rapidly elongates and branches to form the mycelium, growth being largely apical and nourish-

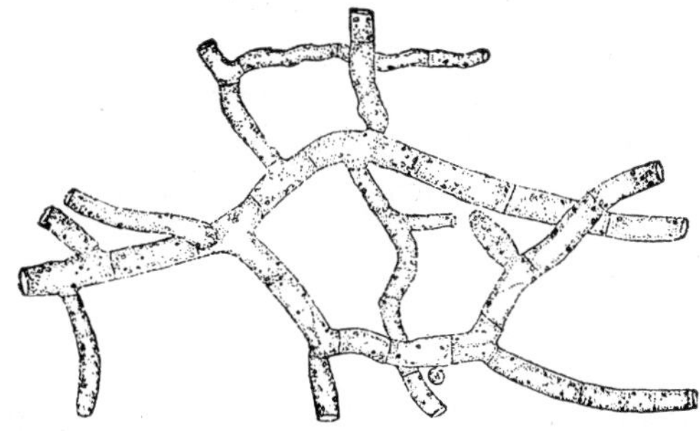

Fig. 2. Small portion of a typical mycelium showing branching, septation, and the granular protoplasm. (*After J. B. Hill, L. O. Overholts, and H. W. Popp.*)

ment being obtained by direct absorption from the substratum through the cell walls. The mycelium then consists of a complicated mass of interwoven, microscopically fine, delicate, branched, and often anastomosed, tubular filaments. Each filament is known as a hypha (plural hyphae); collectively all the hyphae that form the vegetative stage of a single fungus plant are the mycelium.

The hyphae may or may not have cross walls (septa). When no septa are present, the mycelium is a single cell, this condition being known as a "coenocytic" type of plant body. Among the fungi the Phycomycetes are coenocytes. When septa are formed, the hyphae are divided into numerous more or less elongated cells. In some cases the cell wall is apparently pure cellulose; in others it is largely composed of a fatty acid complex with a chitin base. The cell contents consist of vacuolated protoplasm with one or more nuclei and stored food materials. Hyphae are usually colorless, but they may be colored through pigmentation of the cell wall and sometimes of the cell content. The hyphae, particularly at their tips, secrete enzymes that have the power of dissolving many organic substances, enabling delicate hyphae to penetrate hard, lignified cell walls. Furthermore, these enzymes convert organic materials into forms that can be assimilated by the fungus.

A parasitic mycelium may be either internal, developing in the substratum, or external, developing on it. An internal mycelium can usually be seen only by a microscopic examination except when it aggregates into dense masses in fissures or cavities in the substratum. An internal mycelium may be either intercellular, growing between the cells of the substratum, or intracellular, growing inside the cells, or both. From intercellular hyphae, special branches, known as "haustoria," commonly penetrate into the cell cavity or lumen to absorb food. Haustoria (singular haustorium) have been defined as "Intracellular, sac- and clublike, bluntly lobed, coarsely branched or coralloid, feeding organs of parasites which do not immediately injure or kill the host cell but live in a somewhat balanced relationship with it." An external mycelium appears either as whitish cobweblike threads or as sooty brown or black threads on the surface of the substratum, as exemplified by the powdery mildews and sooty molds.

Mycelia sometimes assume special forms. Mycelial felts or plates are developed when hyphae aggregate into dense, compact, tough, often thick, white, feltlike masses occupying cracks or cavities in the host tissue. These are commonly formed by certain wood-destroying fungi such as *Fomes laricis.* Mycelial fans are usually thinner than felts with the mycelium spread out in a fanlike form. Tawny fans formed between the bark and wood characterize the chestnut blight fungus. Mycelial strands or rhizomorphs are aggregations of mycelium into cordlike or tapelike strands, usually more or less branched and sometimes fused to form a network. They vary in color from white to tawny or dark brown and in size from threadlike to an inch or more in thickness. The shoestringlike rhizomorphs of *Armillaria mellea* that ramify through the soil have a tough dark-brown to almost black outer rind with a white interior. These strands carry the fungus to new hosts or to new parts of the same host, while, as in *Poria incrassata,* they may also serve to conduct water.

Sclerotia (singular sclerotium) are densely compacted masses of hyphae varying in size from less than a pinhead to several inches or more in diameter and in shape from elongated-cylindrical, globular, or ellipsoid masses to more or less flattened and irregular forms. Commonly they are dark in color, either with a dark rind and a white interior or dark-colored throughout. Sclerotia are rich in reserve food and are able to withstand extreme desiccation as well as high and low temperatures so that their function is to tide the fungus over unfavorable periods. When conditions again become favorable, the sclerotia may produce new vegetative hyphae or fructifications bearing spores.

THE REPRODUCTIVE STAGE

Fungi reproduce by means of spores. A spore is a cell or a small number of united cells set apart for reproducing the plant. Spores are microscopically small, varied in shape, and of simple structure quite different from the complex structure of the seeds of higher plants, although in function the two are similar. When a spore is more than one-celled, each cell acts independently and is usually capable of producing a new fungus. Spores are produced by specialized structures known as "sporophores," "fructifications," or "fruit bodies," which are usually well-differentiated from the vegetative structures. Sporophores range from microscopically minute structures to such large ones as mushrooms, puffballs, and the conks on decaying wood. Usually they are annual, some existing for only a few hours, but a few are perennial, an age of 75 to 80 years having been reported in some instances. A single fructification may produce from a few to several billion spores.

Spores are produced endogenously or exogenously. Endogenous spores are produced inside normal hyphal cells or inside large specialized cells developed for the purpose. Exogenous spores are produced externally, usually on the ends or sides of specialized hyphal branches. Spores are either sexual or asexual; the latter are also occasionally termed "accessory." Sexual spores result from the union of two separate cells or elements that represent male and female. Asexual spores are formed directly from the hyphae without a preceding conjugation. Asexual spores are by far the most common. In addition to their reproductive function certain thick-walled resting spores serve to tide a fungus over periods unfavorable for growth and are the regular means by which certain fungi live through the winter. Many fungi have more than one kind of spore, and some have produced as many as five different types during various stages in their life history. Of the many kinds of spores some are characteristic of different groups of fungi and are discussed under the respective groups.

Highly specialized types of spores may be produced. Hyphae may break up into chains of thin-walled, elongated or barrel-shaped, nonresting spores known as "oïdia." Again, thick-walled resting spores, known as "chlamydospores," are developed intercalarily by hyphae. Conidia (singular conidium) or conidiospores are spores cut off in succession or cut off singly from the ends of specialized hyphae that are termed "conidiophores."

Most spores germinate by means of a germ tube and require certain atmospheric conditions for germination. They must have moist air or in some cases an actual film of water. Extremes of temperature inhibit

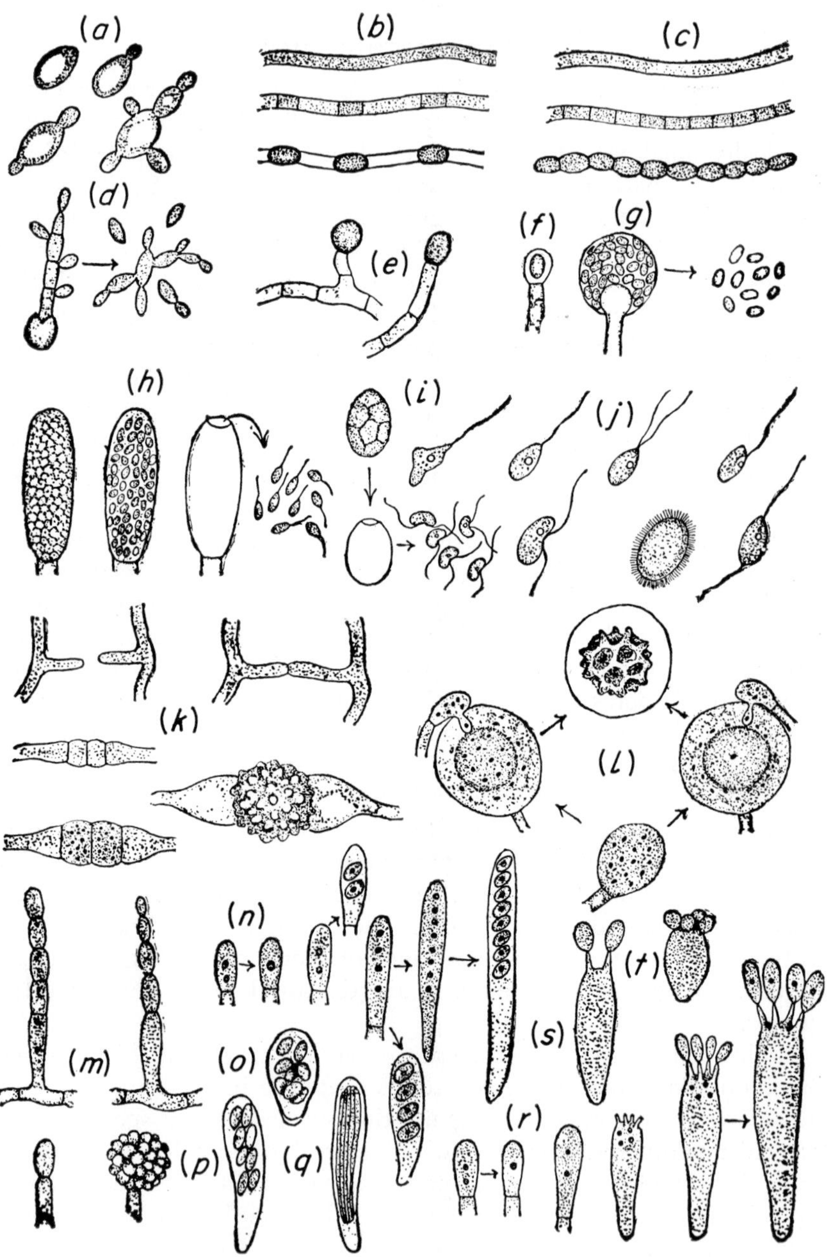

For caption see page 23.

or retard germination, the optimum for most spores being between 15 and 20°C.

DISSEMINATION

Most fungi are disseminated by spores that are generally carried by wind. Certain more or less aquatic fungi have motile spores that can swim in free water. Nonmotile spores may be carried by running water or washed about by rain. Insects and animals including man also carry spores. The tools used in various agricultural and commercial practices are often contaminated when used on diseased plants and should be sterilized before use again on healthy individuals. Sclerotia are rarely carried by wind but are sometimes disseminated by man. Mycelium is often spread by transportation of the substratum on which it occurs. The white pine blister rust fungus was introduced into North America from Europe as mycelium in the bark of living trees.

CLASSIFICATION

The fungi are usually divided into four classes, Phycomycetes, Ascomycetes, Basidiomycetes, and Deuteromycetes, which in turn are subdivided into subclasses, orders, families, genera, and species. No universally accepted classification of the fungi has yet been established, so that there are several systems in current use.

PHYCOMYCETES OR ALGAL FUNGI

The Phycomycetes have certain relationships with the green algae from which they probably originated, so the common name "algal fungi" is applied to them. The group contains a number of obligate parasites, and several of the various fungi causing damping-off of tree seedlings are

---

Fig. 3. Semidiagrammatic drawings showing types of spores and their manner of formation. (*a*), yeast cells, showing primitive method of propagation by budding; (*b*), formation of chlamydospores; (*c*), chlamydospores formed in a continuous chain; (*d*), yeastlike method of production of secondary spores characteristic of some higher fungi, e.g., smuts; (*e*), chlamydospores produced at the ends of hyphae; (*f*), a sporangium bearing a single nonmotile spore; (*g*), a globular sporangium (*Mucor* type) with numerous nonmotile spores; (*h*), stages in the formation of uniciliate swarm spores from a sporangium; (*i*), a zoosporangium from which biciliate swarm spores have been formed; (*j*), types of swarm spores; (*k*), stages in the development of a zygospore from the union of equal and similar cells (gametes); (*l*), two types of öospore formation characteristic of white rusts and downy mildews; (*m*), four types of origin of conidia; (*n*), stages in the development of asci and ascospores; (*o*), (*p*), (*q*), three asci showing different arrangement of the ascospores; (*r*), stages in the development of a basidium and basidiospores; (*s*), (*t*), two other types of basidia. (*After F. D. Heald.*)

Phycomycetes. A few members of the group have a vegetative stage in which a mycelium is rudimentary or lacking, the plant body then being a simple protoplasmic mass, but the majority have a well-developed mycelium, the hyphae of which are without cross walls until the reproductive stage develops. This results in long, branched hyphae containing many nuclei.

Sexual reproduction is of two types: isogamous and heterogamous. Isogamous reproduction consists of a union of two sexual protoplasmic bodies (gametes), with their nuclei, borne in cells (gametangia) that to all appearances are alike. A gametangium may contain many gametes. The result of this union in the higher Phycomycetes is a thick-walled resting spore known as a "zygospore." In heterogamous reproduction the gametangium bearing the male gametes (sperms) is termed an "antheridium." It is usually smaller and different in shape from the gametangium bearing the female gametes, which is termed an "oögonium." Each oögonium contains one or more oöspheres or eggs. The nuclei and protoplasm from the antheridium pass into the oögonium, where they fuse with the female nuclei, resulting in the oöspheres' developing into resting spores known as "oöspores."

Asexual reproduction is by the production of spores of three general types: (1) zoospores, motile by means of cilia and borne in indefinite (usually large) numbers in a sac called a "sporangium"; (2) sporangiospores, nonmotile and borne in indefinite (usually large) numbers in a sac called a "sporangium"; and (3) conidia or conidiospores.

The Phycomycetes are divided into three subclasses: (1) Archimycetes with the mycelium rudimentary or obsolete; (2) Oömycetes with a well-developed mycelium, sexual reproduction by oöspores (heterogamous), and asexual spores often motile; and (3) Zygomycetes with a well-developed mycelium, sexual reproduction by zygospores (isogamous), and asexual spores nonmotile.

ASCOMYCETES OR SAC FUNGI

The Ascomycetes comprise a large group of fungi, many of which cause diseases of forest trees. Most members of the class have a well-developed, septate mycelium. The characteristic structure of the Ascomycetes is the ascus with its ascospores, the ascus being a modified terminal cell of a hypha. The ascus is usually a cylindrical or club-shaped sac, somewhat similar to the sporangium of the Phycomycetes, but with a definite number of endogenous spores, usually eight, although two spores or other multiples thereof do occur. Occasionally the primary ascospores divide by budding to form many secondary ascospores within the ascus. Furthermore, in many species the ascus is a product of a heterogamous sexual

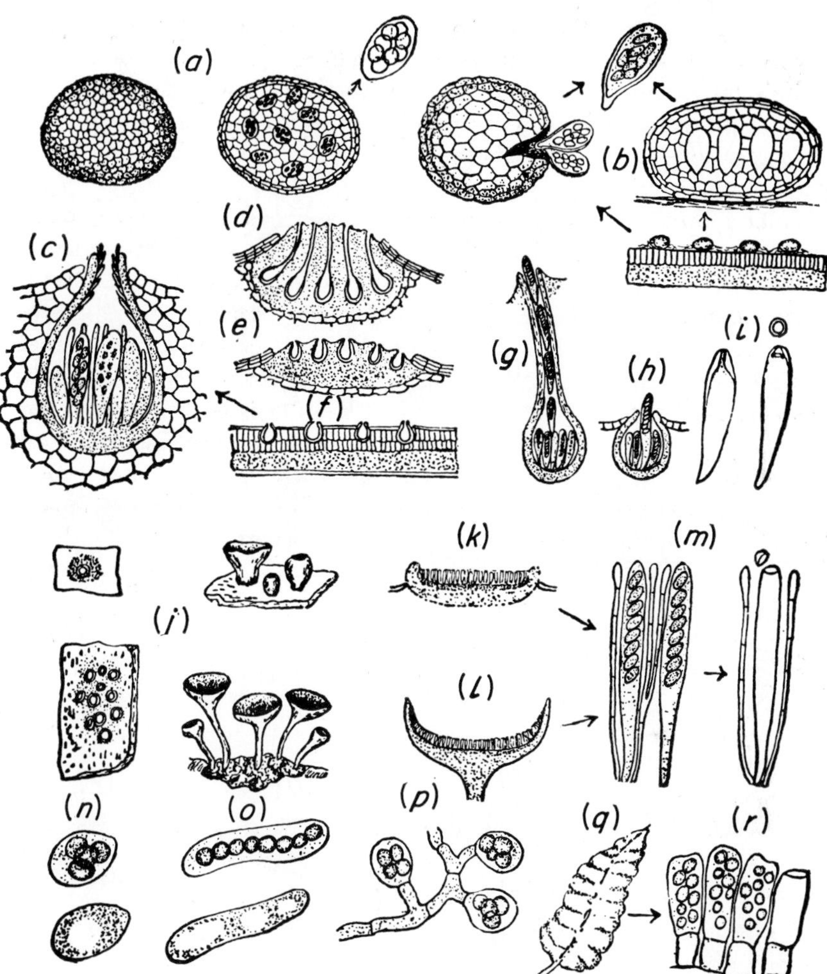

Fig. 4. Semidiagrammatic drawings of ascigerous fruits and simple ascus-forming fungi. (*a*), surface and sectional view of a closed ascocarp of *Aspergillus,* with a single enlarged ascus; (*b*), habit sketch, surface view with escaping asci, enlarged ascus, and section of a perithecium of a powdery mildew; (*c*), vertical section of a typical ostiolate perithecium; (*d*), (*e*), sections of stromata with immersed perithecia; (*f*), habit sketch of typical perithecia; (*g*), section of a perithecium with long neck showing how the asci are released and forced up to the ostiole for the discharge of spores; (*h*), section of a perithecium showing the manner of elongation of asci through the ostiole for the expulsion of the ascospore; (*i*), two types of asci, one with terminal sphincter, the other terminal canal, structures used in spore discharge; (*j*), four types of apothecia or fruits of cup fungi; (*k*), section of a sessile apothecium; (*l*), section of a stalked or stipitate apothecium; (*m*), asci and paraphyses, or sterile filaments from a typical apothecium, showing one empty ascus with the lid separated; (*n*), (*o*), vegetative yeast cells and sporulating cells or simple asci; (*p*), asci of a simple filamentous fungus, *Endomyces;* (*q*), leaf of peach affected by leaf curl, or *Taphrina deformans;* (*r*), asci from the surface of the same leaf. (*After F. D. Heald.*)

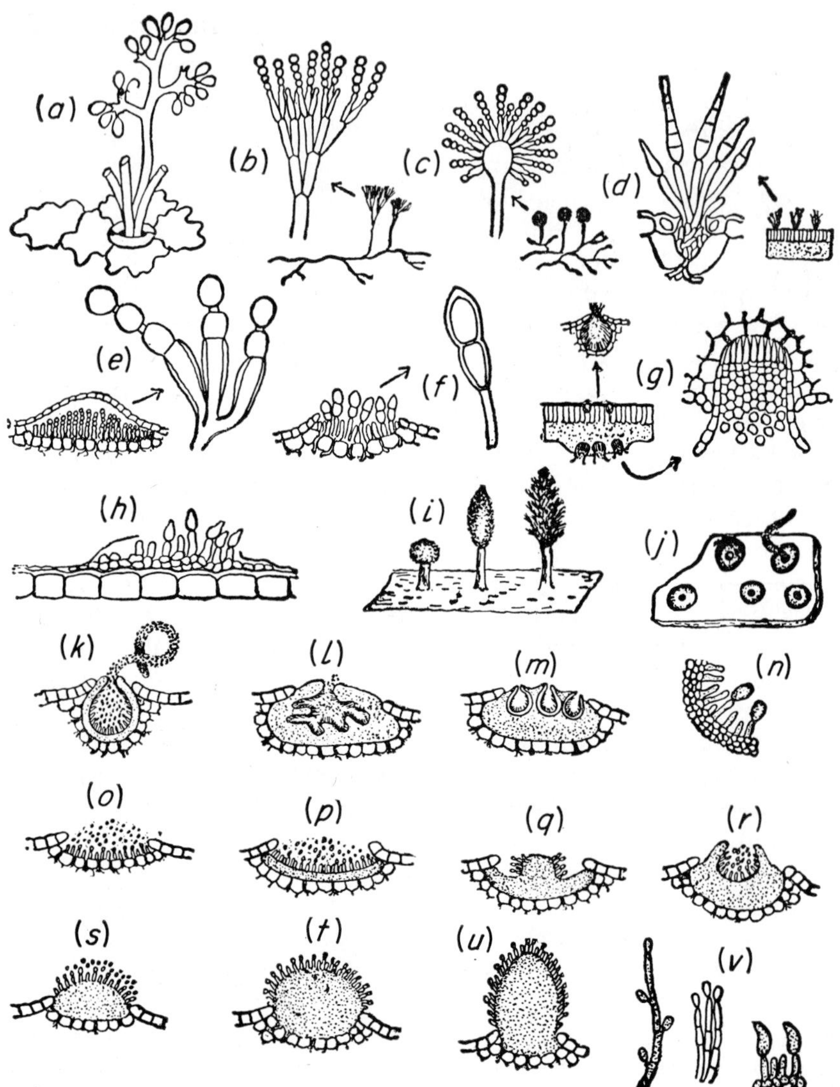

Fig. 5. Semidiagrammatic drawings of types of conidiophores and spore fruits. (*a*), a downy mildew, *Peronospora;* (*b*), blue mold or *Penicillium;* (*c*), *Aspergillus;* (*d*), *Cercospora;* (*e*), section of a sorus of white rust (*Albugo*) with enlarged conidiophores and conidia; (*f*), section of a telium or teleutosorus of a true rust with a single enlarged teliospore; (*g*), section of cluster cups or aecia and pycnia (above) with each spore fruit enlarged; (*h*), conidiophores and conidia of the apple-scab fungus; (*i*), three types of coremia; (*j*), habit sketch of pycnidia; (*k*), section of a pycnidium with a spore tendril protruding from the ostiole; (*l*), section of a stromatic pycnidium; (*m*), section of a stroma with several immersed pycnidia; (*n*), detail of small portion of the wall of a pycnidium showing conidiospores and conidia or pycnidiospores; (*o*) to (*r*), types of acervuli; (*s*) to (*u*), types of sporodochia; (*v*), three different types of conidiophores from spore fruits. (*After F. D. Heald.*)

fusion so that the ascospores are sexual spores, the only sexual spores developed in this group. The stage during which the asci are produced is known as the "perfect stage" of the fungus.

The asci may be scattered irregularly over the substratum as in *Taphrina*, but usually they stand side by side in a definite hymenium or spore-producing layer. Sterile threadlike bodies known as "paraphyses," which are usually much narrower and slightly longer than the asci, are commonly intermingled with the asci in the hymenium. Their chief function apparently is to protect the asci. The asci are generally contained in a fructification known as an "ascocarp." Most ascocarps are small so that they cannot be seen readily except under a hand magnifier, but some attain a size approximating the smaller mushrooms and toadstools of the Basidiomycetes. When the hymenium is exposed at maturity as the inner lining of a cup-shaped or saucer-shaped fructification, the ascocarp is termed an "apothecium." If the apothecium has a stalk, stem, or stipe, it is said to be stipitate; otherwise it is sessile. A perithecium is a more or less globose or flask-shaped ascocarp enclosing the hymenium in a central cavity. There may be a small opening termed an "ostiole" in the top of the perithecium for the escape of the ascospores; otherwise the perithecium merely disintegrates to free the spores. Perithecia are nearly always sessile and may stand on or be buried in the substratum or may be on or in a dense and compact mass of hyphae known as a "stroma," which in turn may be on or in the substratum. Sunken perithecia have a long neck reaching to the surface of the substratum or the stroma.

The Ascomycetes also reproduce freely by means of asexual conidia or conidiospores, and this reproductive stage is known as the "conidial" or "imperfect stage." Several types of conidia in different kinds of fructifications may be produced by a single fungus. The conidiophores that bear the conidia may be scattered irregularly over the substratum. If the conidiophores are gathered into a fascicle somewhat resembling a shock of wheat, the fructification is a coremium. When the conidiophores are gathered into a flat or a saucer-shaped structure without any protective covering of fungous tissue, the structure is known as an "acervulus." If the conidiophores are unprotected but arise from a well-defined stroma, the fructification is called a "sporodochium." A pycnidium is a structure similar to a perithecium, but it contains conidiophores with conidia; the spores are sometimes termed "pycnidiospores." Pycnidiospores are sometimes exuded in long, more or less sticky tendrils known as "spore horns" or "cirri."

The Ascomycetes can be divided into three subclasses: (1) Plectomycetes with asci irregularly arranged as represented by *Taphrina* or the ascocarp with no definite ostiole as represented by the powdery mildews;

(2) Discomycetes with the ascocarp wide open when mature, typical fructification an apothecium as represented by the larch canker fungus; and (3) Pyrenomycetes with the ascocarp flask-shaped when mature, typical fructification a perithecium as represented by the chestnut blight fungus.

BASIDIOMYCETES OR BASIDIA FUNGI

This class contains a number of virulent pathogens on forest trees and nearly all the wood-destroying fungi. All have a well-developed septate mycelium. The Basidiomycetes are characterized by the development somewhere in their life history of a basidium on which the basidiospores are produced exogenously. Since in many species the basidia and basidiospores are known to be the product of isogamous sexual fusion, the basidiospores are regarded as sexual spores. Basidia are simple in structure, generally club-shaped, usually one-celled, but in some forms the basidia may be two- or four-celled by transverse or longitudinal septa. In any case the basidium usually produces four external stalks called "sterigmata" (singular sterigma), each of which bears a single basidiospore. The basidium is the modified terminal cell of a hypha.

The basidia may be scattered over the substratum, but usually they are aggregated into a hymenium and may be accompanied by sterile structures of two forms: (1) paraphyses similar to those in the Ascomycetes, and (2) cystidia, which are much larger than paraphyses and commonly project considerably beyond the basidia. Both paraphyses and cystidia are modified terminal cells of sterile hyphae. The hymenium is borne by a fructification termed a "basidiocarp." Basidiocarps vary from insignificant structures to large elaborate fructifications such as mushrooms, toadstools, and conks.

Except in the Uredinales or rust fungi, asexual spores are not developed in the abundance that characterizes the Ascomycetes. Most asexual spores occurring in the group are conidia. Although perfect stage can be used to designate the reproductive stage during which basidiospores develop and imperfect stage the stage characterized by asexual spores, these terms are not so generally applied to Basidiomycetes as they are to Ascomycetes.

The Basidiomycetes are divided into three subclasses: (1) Hemibasidiomycetes with the number of basidiospores indefinite; (2) Protobasidiomycetes with the number of basidiospores definite, usually four, and the basidia septate; and (3) Autobasidiomycetes with the number of basidiospores definite, usually four, and the basidia without septa.

Hemibasidiomycetes

This subclass includes a single order, the Ustilaginales or smuts. Although the smuts cause serious plant diseases, none of them attack trees or shrubs, so they will not be discussed.

Protobasidiomycetes

This subclass includes three orders, the Uredinales, the Auriculariales, and the Tremellales. The two last comprise the jelly fungi, so named because of their gelatinous basidiocarps so often found on badly decayed wood on the forest floor. Since none of them cause diseases of trees or even primary decay of wood, they will not be considered further.

Uredinales or Rust Fungi. All the Uredinales are obligate parasites, and many of them cause serious diseases of forest trees, particularly conifers. They are a highly specialized group of fungi readily distinguished by the form and arrangement of their spores. The different species may have from one to five spore forms in their life cycle. If the spore forms are produced on one host, the rust is termed "autoecious," but if the spore forms are produced on two unrelated hosts, the rust is "heteroecious." Practically all the rust fungi on trees are heteroecious. Four stages are recognized in the life cycle of a rust having the full complement of spore forms. These stages are designated as the pycnial, aecial, uredial, and telial stages or for convenience 0, I, II, and III. When all stages are present, they develop in the order given, and in a heteroecious rust the pycnial and aecial stages occur on one host and the uredial and telial stages on the other.

Pycnial Stage. This stage is characterized by the development of the pycnium or spermogonium, which is either (1) flat and indefinite, subcuticular, or subcortical; or (2) globular to flask-shaped and definite, subcuticular, or subepidermal. The tiny pycniospores or spermatia are produced on the tips of hyphae in the pycnial cavity, where they are mingled with a sweetish liquid and are exuded through an ostiole or through a longitudinal slit in the wall of the pycnium. The pycniospores cannot cause new infections, but they have a sexual function in a number of species. Pycnia never appear alone but are either accompanied or followed by one of the other stages.

Aecial Stage. This stage, also known as the "aecidial" or "cluster-cup" stage, closely follows the pycnial stage. It is characterized by the development of the aecium, aecidium, or cluster cup, a more or less cup-shaped structure sunken in the host tissue. In the bottom of the aecium is a mass of upright hyphae from which the aeciospores are cut off in chains. Surrounding and covering the sporiferous hyphae and aeciospores is a cellular membrane or wall known as the "peridium" that often protrudes above the surface of the host tissue in the form of a blister, a short column, or a horn. The peridium is white to dingy white in color but often appears orange or light brown because of the color of the densely packed aeciospores within. The aeciospores are released when

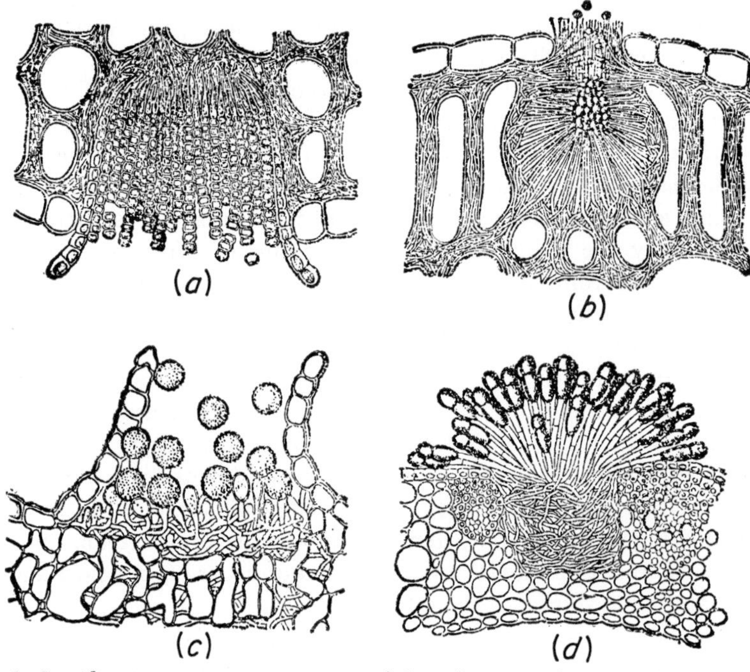

Fig. 6. Semidiagrammatic representation of fructifications of a typical rust. (*a*), an aecium with aeciospores; (*b*), pycnium with pycniospores; (*c*), uredium with urediospores; (*d*), telium with teliospores. (*After Wettstein, Handbuch der systematischen Botanik.*)

the peridium ruptures in various ways. Some aecia are without a peridium, merely being covered by the epidermis of the host; these are termed "caeoma." Not all aecia are cup-shaped. The aeciospores are always one-celled, usually thick-walled, the walls are variously sculptured, and the cell contents are orange, yellow, or colorless. Aeciospores are largely wind-disseminated and germinate by a germ tube.

Uredial Stage. This stage, also known as the "uredinial stage," develops from infections by aeciospores. It is characterized by the production of a uredium, uredinium, or uredosorus, which when mature is usually an aggregation of hyphae standing erect on the surface of the host and bearing urediospores at their apex. These sporiferous hyphae may occupy a slightly saucer-shaped depression. A peridium is usually absent, the hyphae and spores being covered by the host epidermis and released by its rupture. Paraphyses may be present forming a protective border around the sporiferous hyphae. The urediospores, uredospores, or urediniospores are always one-celled and thin- to thick-walled, the walls almost always spiny or warty, and the spores are usually orange or reddish brown in color either because of a colored cell content or a colored wall. The urediospores are usually produced singly on short stalks from which they separate readily when mature. Sometimes they are produced in chains. Urediospores are wind-disseminated and germinate by a germ tube.

Telial Stage. This stage, also known as the "teleuto stage" or "perfect stage," arises later from the same mycelium that produced the uredial stage or from new infections. The telia or teleutosori that characterize this stage vary greatly so that a general characterization is difficult. Some may resemble the uredia, and spores of both stages sometimes occur in the same fructification. Others may appear as waxy layers (*Coleosporium*), as dark-colored crusts (*Melampsora*), as slender projecting hairs or bristles (*Cronartium*), as relatively thick columns that gelatinize when moist (*Gymnosporangium*), and as erumpent, exposed cushionlike masses (*Puccinia*). In a few cases the teliospores are borne singly within the host tissues (*Uredinopsis*). The teliospores or teleutospores may be one- to several-celled and formed in chains (*Cronartium*) or singly (*Puccinia*); if formed singly, they usually have a permanent stalk. Their wall is frequently much thickened and generally smooth, although some have roughened walls, but very rarely are the walls genuinely spiny. In the main the teliospores vary in color from some shade of brown to almost black. In many rusts the teliospores are resting spores germinating only after a shorter or longer period of dormancy that frequently coincides with the winter season, enabling the fungus to "overwinter" as a spore, so that the name "winter spores" is sometimes applied

The important characteristic of the teliospores is their method of germination. They are not wind-disseminated but remain in the telium. On germination a short germ tube or promycelium is produced from each of one or more cells of the spore. This promycelium is the basidium, and it becomes four-celled at maturity. From each cell of the promycelium, a tiny stalk called a "sterigma" is developed on the end of which a basidiospore (also known as a "sporidium") is formed. The sporidia are small, thin-walled, readily disseminated by wind, and so delicate that they quickly lose their viability if conditions are not suitable for germination. The germ tube gives rise to a mycelium forming the pycnial and aecial stage, thus again starting the life cycle.

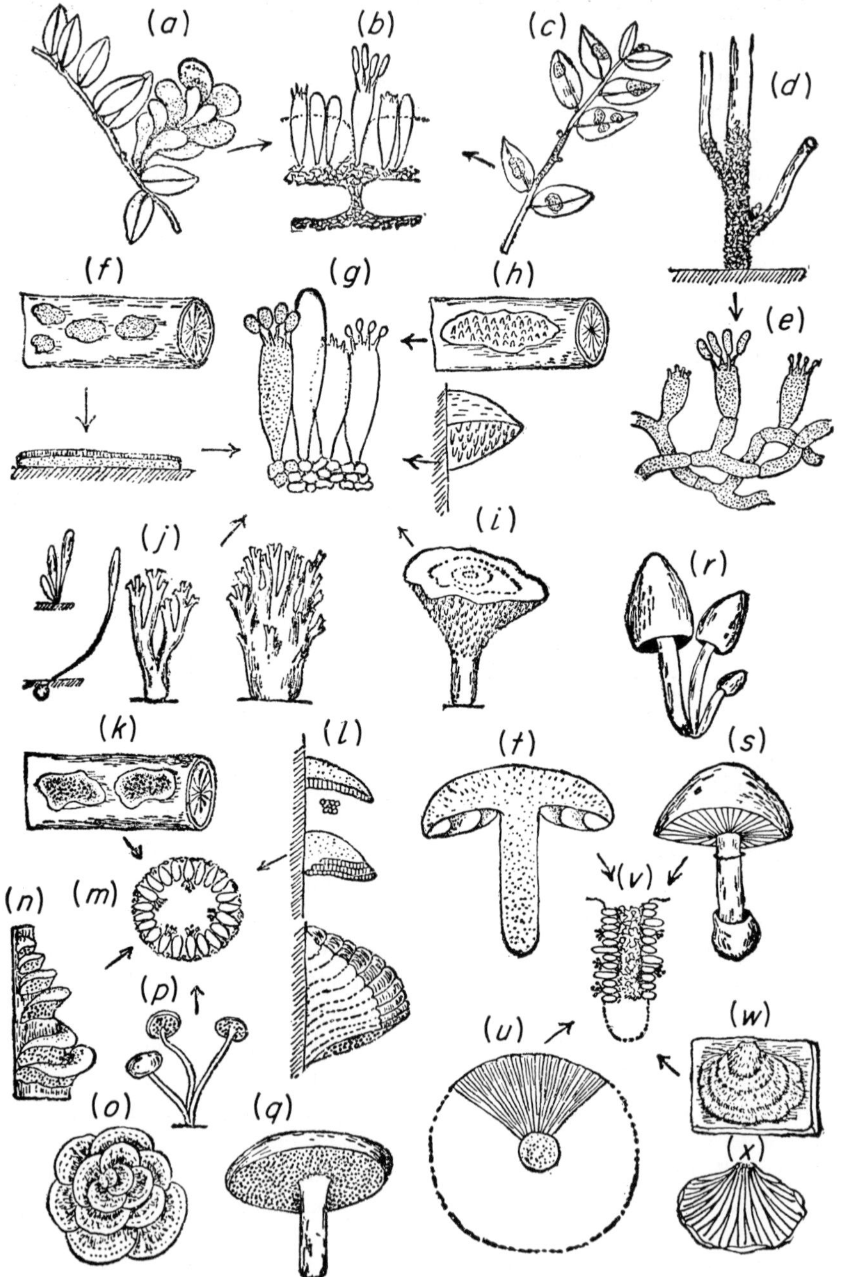

For caption see page 33.

Classification. The Uredinales are divided into two or more families on the basis of varying characters and arrangement of the teliospores. The various genera are also based on the teliospores, so that, until the telial stage of a rust is known, it cannot be permanently named. However, study of the life of any species requires time, particularly for the heteroecious forms. Consequently a series of form genera based on the aecia and uredia are recognized so that a species can be provisionally named when the telia are not known:

Aecia with catenulate spores* and peridia
- On angiospermous hosts .... *Aecidium*
- On gymnospermous hosts .... *Peridermium*

Aecia with catenulate spores and no peridia .... *Caeoma*
Aecia with pedicellate spores .... *Uraecium*
Uredia with or without peridia .... *Uredo*

* Catenulate spores, i.e., individual spores borne in a series or chain.

Another form genus, *Roestelia,* with the aecia cornute (hornlike) in form and the long axis perpendicular to the host tissue, may be recognized, or it may be included in *Aecidium.* A species of *Rostelia* is always the aecial stage of a *Gymnosporangium.*

To illustrate the value of these form genera, the white pine blister rust fungus may be cited. The uredia and telia were discovered on *Ribes,* and the fungus named *Cronartium ribicola.* The aecial stage when found on *Pinus strobus* was named *Peridermium strobi* and carried this name for some time, since it was not realized that the rusts on the two hosts were different stages of the same fungus. Later when the connection was proved *C. ribicola* became the correct name for all stages of the fungus.

Fig. 7. Semidiagrammatic drawings of basidium fruits. (*a*), cranberry affected with rose bloom (*Exobasidium oxycocci*), showing enlarged flowerlike lateral shoot; (*b*), basidia from the surface of one of the hypertrophied leaves shown in (*a*); (*c*), red leaf spot of the cranberry, with leaf and stem lesions that produce basidia similar to those shown in (*b*); (*d*), base of stem showing a simple basidial felt (*Hypochnus* or *Corticium* type); (*e*), hyphae and basidia from (*d*); (*f*), habit sketch and section of a plain resupinate sporophore (*Stereum* type); (*g*), typical basidia with one sterile cell or cystidium; (*h*), (*i*), resupinate, shelving, and stalked sporophores with toothed basidial surfaces (*Hydnum* type); (*j*), sporophores of fairy clubs and coral fungi (*Clavaria* type); (*k*), habit sketch of a resupinate, poroid sporophore; (*l*), sections of annual and perennial bracket sporophores (*Polyporus* and *Fomes* types); (*m*), section through a single pore showing arrangement of the basidia; (*n*), (*o*), imbricated bracket fruits of the pore type; (*p*), (*q*), stipitate or stalked sporophores, the basidia in tubes; (*r*), (*s*), sporophores of gill fungi or toadstool forms (*Agaricus* type); (*t*), section of a toadstool form through the middle of the cap or pileus and the stipe or stalk showing the varying lengths of the basidium-bearing plates or lamellae; (*u*), arrangement of the gills as viewed from the undersurface of the pileus; (*v*), section of a portion of a gill showing arrangement of the basidia; (*w*), upper surface of the sporophore of a common gill fungus (*Schizophyllum alneum*); (*x*), undersurface of the same sporophore showing the arrangement of the gills or lamellae. (*After F. D. Heald.*)

Autobasidiomycetes

This subclass is characterized by most of its members having well-developed and prominent basidiocarps, such as mushrooms, toadstools, the shelf- or bracketlike conks on trees, and puffballs. The basidia are one-celled and club-shaped, and nearly always bear four basidiospores.

The parts of the fructification are (1) the stipe, stem, or stalk; (2) the pileus or cap; (3) the hymenium or fruiting surface; and (4) the context or flesh. A form with a stem is designated as "stipitate," but, if the stem is absent and the pileus is attached directly to the substratum as in most bracketlike conks on trees, the fructification is termed "sessile." If it is spread out flat on the substratum without a pileus or stipe, i.e., forming a crust, it is termed "resupinate." All gradations between the three forms occur. When in the bracketlike forms a series of more or less overlapping brackets occur, the fruit body is said to be "imbricate."

The Autobasidiomycetes may be divided into two orders: the Hymenomycetales and the Gasteromycetales.

Hymenomycetales. The members of this order are characterized by having the hymenium exposed at maturity and often throughout development. The order includes five important families that contain most of the wood-destroying fungi: (1) the Thelephoraceae or leather fungi with the hymenium spread over a smooth or corrugated surface as represented by the genus *Stereum;* (2) the Clavariaceae or coral or club fungi with the hymenium covering all sides of the smooth surface of simple or branched clubs as represented by *Clavaria;* (3) the Hydnaceae or spine, tooth, or hedgehog fungi with the hymenium spread over spines or teeth as represented by *Hydnum* and *Echinodontium;* (4) the Agaricaceae or gill fungi with the hymenium spread over gills (lamellae) as represented by *Agaricus* and *Armillaria;* and (5) the Polyporaceae or tube or pore fungi with the hymenium lining tubes as represented by *Fomes* and *Polyporus.*

Gasteromycetales. The members of this order are characterized by having the hymenium enclosed until after the spores are ripe. The best known representatives are the puffballs. Although some species form mycorrhiza on tree roots, none cause any significant disease of forest trees or primary decay of wood in service, so they will not be considered further.

DEUTEROMYCETES OR IMPERFECT FUNGI

This class, more commonly known as Fungi Imperfecti, is not a natural one but comprises those fungi whose life histories are still in doubt in that no spore stage has been found which enables them to be allocated to one of the other three classes. Certain ones may be reduced forms that permanently reproduce asexually. Some have only sterile mycelium, but the majority have asexual spores, usually conidia. In general the fructifications are the same as those described for the imperfect or conidial stages of the Ascomycetes. Ultimately most Fungi Imperfecti will be found to be asexual stages of Ascomycetes, as many already have been. A few will connect with Basidiomycetes and occasional ones with Phycomycetes, but, since the mycelium of fungi in this last class is not septate, it is more possible to determine whether a fungus is a Phycomycete in the absence of any characteristic spore stage.

REFERENCES

Ainsworth, G. C., and G. R. Bisby: "A Dictionary of the Fungi," 4th ed. Kew, Surrey: The Commonwealth Mycological Institute (1954).

Alexopoulos, C. G.: "Introductory Mycology." New York: John Wiley & Sons, Inc. (1952).

Arthur, J. C.: "The Plant Rusts (Uredinales)." New York: John Wiley & Sons, Inc. (1929).

Bessey, E. A.: "Morphology and Taxonomy of Fungi." New York: McGraw-Hill Book Company, Inc., Blakiston Division (1950).

Clements, F. E., and C. L. Shear: "The Genera of Fungi." New York: Hafner Publishing Company (1954).

Crowder, W.: Marvels of Mycetozoa. *Natl. Geograph. Mag.*, 54:421–443 (1926).

Frobisher, M., Jr.: "Fundamentals of Microbiology," 5th ed. Philadelphia: W. B. Saunders Company (1953).

Gäumann, E.: "Die Pilze." Basel: Verlag Birkhäuser (1949).

Gäumann, E. (transl. from the German by F. L. Wynd): "The Fungi." New York: Hafner Publishing Company (1952).

Hagelstein, R.: "The Mycetozoa of North America." Mineola, N.Y.: Published by the author (1944).

Hawker, Lillian E.: "Physiology of Fungi." London: University of London Press, Ltd. (1950).

Hill, J. B., L. O. Overholts, and H. W. Popp: "Botany," 2d ed. New York: McGraw-Hill Book Company, Inc. (1950). (Chap. 15, Thallophyta—Bacteria, Slime Molds, Fungi, pp. 410–485.)

Lancaster, Margaret E.: "Forest Fungi." Wellington, New Zealand: Government Printer (1955).

Lilly, V. G., and H. L. Barnett: "Physiology of the Fungi." New York: McGraw-Hill Book Company, Inc. (1951).

MacBride, T. H., and G. W. Martin: "The Myxomycetes." New York: The Macmillan Company (1934).

Salle, A. J.: "Fundamental Principles of Bacteriology," 5th ed. New York: McGraw-Hill Book Company, Inc. (1961).

Seymour, A. B.: "Host Index of the Fungi of North America." Cambridge, Mass.: Harvard University Press (1929).

Snell, W. H., and Esther A. Dick: "A Glossary of Mycology." Cambridge, Mass.: Harvard University Press (1957).

Wolf, F. A., and F. T. Wolf: "The Fungi." New York: John Wiley & Sons, Inc. (1947), vols. I and II.

CHAPTER 4

Noninfectious Diseases

Noninfectious diseases of plants, also called "nonparasitic" and "physiological" diseases, have been extensively investigated (Delacroix 1926; Graebner 1920; Sorauer 1922). They are diseases having specific symptoms, although such symptoms are not known to be produced by pathogenic organisms, and the causes of such symptoms may be known or as yet undetermined. Of the many noninfectious diseases of forest trees (Pomerleau 1944), few have been carefully studied. Compared to diseases caused by parasites, the aggregate damage they cause to the forests of this country is not great, but intermittently they do occasion severe losses over limited areas. Certain noninfectious leaf diseases are heritable (Smith and Cochran 1943).

Although the symptoms and effects of certain diseases are well known, the factor or combination of factors producing them is still questionable. Among these are needle blight of eastern white pine, cause unknown (Campana 1954; Linzon 1960), which may be separated into red needle blight, commonly characterized by the death of the ends of the leaves of the current season, giving affected trees a rust-brown color, but occasionally characterized by the death of entire leaves followed by premature leaf fall and the death of trees affected annually over a period of years (Spaulding and Hansbrough 1943; Quirke 1954), with fertilizer applications giving a promise of control (Ibberson and Streater 1952) and yellow dwarf blight or chlorotic dwarf, characterized by stunted roots and tops and yellow-green needles (Swingle 1944); pole blight of western white pine causing death of pole-size trees between 40 and 100 years old, preceded by yellow, thin, and stunted foliage, reduced leader growth, and flat faces along the trunk with abundant resin flow from some of them (Leapheart, Copeland, and Graham 1957; Leapheart 1958), and a similar disease of lodgepole pine (Parker 1959); sweetgum blight characterized by premature fall coloration of the leaves on individual branches or throughout the crown, reduction in size of leaves, amber-brown to deep-brown streaks in the white sapwood, and death of individual branches or of the entire crown, possibly brought on by soil moisture deficiency (Toole 1959); and extensive dying of species of the red (or black) oak group in Pennsylvania and West Virginia, probably caused by drought (Tryon and True 1958).

Sudden changes in environmental conditions can bring about disease and death of forest trees, although the exact factors primarily responsible are not known. When a dense stand is removed, leaving only scattered seed trees, there must be profound changes in site conditions, not only in humidity, temperature, and light above the ground, but also in moisture, microflora, microfauna, and probably other qualities of the soil. Many of the scattered trees left standing after the heavy wind throw of January, 1921, on the Olympic Peninsula in Washington died within a year or two, without any parasite being responsible. When mature Douglas fir stands in western Oregon and Washington are removed by cutting, some of the reserved, scattered seed trees often die for no apparent reason. The primary cause of the death of yellow and paper birches, as well as other northern hardwoods left standing on cutover areas in New England and the Lake states, is the exposure to which they are suddenly subjected. Insects and fungi are merely contributory factors (Spaulding and MacAloney 1931; Hall 1933). Mortality of eastern hemlock also follows undue exposure (Graham 1944).

Soil changes profound enough to cause the death of forest trees also result from excessive use by man of limited forest areas for recreational or other purposes (Lutz 1945). Fine large conifers die on public camp grounds in the forests of the West, and the cumulative effects of continued use will result in steadily increasing losses, unless the trees are protected by diverting the people to other areas.

Unfavorable weather conditions are the most common cause of noninfectious diseases (Hubert 1930; 1932), and the symptoms of some weather-caused or climatic diseases are spectacular and alarming out of proportion to the actual damage resulting, as, for example, with the form of winter drying known as "red belt." Adverse weather conditions may act in combination; e.g., the slight drought in the Adirondacks during the growing season of 1933 coupled with the severe cold of the following winter resulted in considerable damage to beech, which became apparent in the spring of 1934 (Spaulding and Hansbrough 1935). Frost, drought, and heat resistance are basically the same, that is, dehydration resistance (Levitt 1951).

HIGH TEMPERATURES

High temperatures are a cause of diseases in trees, since the various parts of the tree are adjusted to certain maximum temperatures, beyond which injury will occur. However, high temperatures are generally more important when acting in combination with other factors than as a single direct agency in causing disease to trees beyond the seedling stage. For example, most drought injury is accompanied by high temperature.

Heat Defoliation. Premature casting of leaves may occur in both deciduous and evergreen trees when the temperature rises above the maximum to which their tissues are adjusted (Heald 1933:142). Leaves of the inner crown are first affected, probably because they are more sensitive to heat and because heat loss from them is checked by the foliage of the outer crown. Extensive browning of tender foliage of oaks in California has followed hot weather (Mielke and Kimmey 1942).

Sugar Exudation. In the dry interior valleys of British Columbia and eastern Washington, the leaves and twigs of Douglas fir are sometimes covered with white sugar. Although a similar sugar excreted by aphids and known as "manna" is found on European larch (Henry 1924), it is thought that the Douglas fir sugar is exuded in solution directly from the leaf tips, the leaves having developed an excess supply during the hot, dry, cloudless days of summer. The water then evaporates, leaving the sugar encrusted on the leaves (Davidson 1919). The effect on the trees is negligible.

Sunscald. An actual overheating and drying out of the bark, resulting in open wounds, occurs on certain young trees with smooth, thin bark when they are suddenly exposed to the summer sun by removal of their neighbors. The injury is also known as "bark scorch." It is rare on forest trees.

An exactly similar injury termed "winter sunscald" (Heald 1933:182) is caused by a combination of high and low temperatures during the late winter or early spring. The living bark tissue and cambium on the south and southwest sides of the tree are warmed to a temperature above freezing by the sun's rays on relatively warm days and frozen again by the rapid drop in temperature during the night. This alternate freezing and thawing causes a rending of the tissue and results in the formation of cankers. Trees with smooth bark are most liable to injury. Aspen in Colorado is occasionally injured (Hartley and Hahn 1920:143), eastern white pine is susceptible (Huberman 1943), yellow birch can be affected (Godman 1959), and sugar maple may be severely damaged (Pomerleau 1944:176). A method of control, applicable only to ornamental or orchard trees, is to spray or paint the trunks with whitewash in the fall or early winter (Mix 1916*b*:283). Sudden exposure of forest trees with thin bark should be avoided.

Birch Dieback. This disease of yellow birch, and to a lesser degree of paper birch, has been widespread for the past 25 years or more in the northeastern United States and eastern Canada, where it has caused such heavy mortality that merchantable birch has almost disappeared from some stands. According to Redmond (1955),

> The first symptom of decline is a thinning of the foliage accompanied by small, sometimes chlorotic or curled, leaves at the tip of shoots in the periphery of the crown. Later, twigs become bare because buds die or fail to open and the foliage grows thinner in the interior of the crown. Still later, branches and whole portions of the crown die, the foliage remaining is bunched because of shortening of internodes, and it is usually confined to lower branches and adventitious shoots. Death of trees, when it occurs, takes place within three to five years after the initial appearance of symptoms.

It has not been proved that fungi (Hahn and Eno 1956), viruses, or insects cause birch dieback, although some authorities insist that it is an infectious disease (Clark and Barter 1958). Another possibility is that increased soil temperature following an increase in summer air temperature during a 30-year period prior to 1950 increased the mortality of birch rootlets. Birches are much more sensitive to environmental changes than are their associates.

LOW TEMPERATURES

Trees, depending on their natural range, are adjusted to varying degrees of normal low temperature but may be injured by seasonable but exceptionally low temperatures or by unseasonable low temperatures. Sudden fluctuations from high to low temperatures are more unfavorable than a gradual drop. The tops of thrifty pines from 15 to 100 feet tall have been killed by a sudden drop in the autumn temperature, even though the temperature was well above that normally considered to be damaging (Wagener 1949). Species outside their natural range are most likely to suffer, especially those which have been moved from a mild to a more severe climate.

Frost. Injury to trees by low temperatures may occur at any time throughout the year (Day and Peace 1934), but it is during the growing season that living tissues are most susceptible, and it is then that most of the serious frost damage occurs. Frosts are commonly classified as winter frosts, spring or late frosts, and autumn or early frosts (Zon 1903).

Winter frosts, which occur during the dormant season for trees, usually cause little injury, unless the temperature is excessively low. Then extensive damage may occur, particularly to hardwoods, resulting in the death of limbs and leading shoots, or the death of the entire tree. Lesions on balsam fir in Quebec similar to those caused by sunscald, but on the northeast side of the trees, have been attributed to severely cold winds or to coatings of ice that persist after severe storms (Pomerleau 1944: 175). Even roots may be killed by low winter temperatures, resulting in subsequent disease or dying of trees, as occurred to beech in New York during the winter of 1917–1918 (Graves 1919:117). The formation of dark-colored false heartwood or frost heartwood is induced in red beech in Europe by unusually low temperatures (Raunecker 1956; Rennerfelt and Thunell 1950), although this phenomenon has not been recorded for beech in North America. Late and early frosts that occur during the growing season are more serious, since living tissues are then active and are more susceptible to low temperatures. The greatest damage is caused by late frosts (Day 1928*a*; Day and Peace 1946), since trees are most active and have the largest amount of tender, succulent tissue in the spring and early summer; although species such as honey locust and Siberian elm, which continue growth until late in the season, may be seriously injured by early frost. Hardwoods, because they begin growth earlier in the spring, are more liable to late frost injury than conifers. Young trees close to the ground are more susceptible than taller, older trees, although Beal (1926) reports extensive killing of mature white oaks in North Carolina by a late frost on May 26 and 27. In a given locality a late frost may injure hardwoods but not affect conifers, which have not yet developed new shoots. Hardwoods also suffer more from early frosts than conifers, coppice forests being especially sensitive because the sprouts do not have time to harden.

The principal types of injury by late and early frosts, adapted from the classification by Day (1928*b*), may be listed as follows:

Injury by late frost
- 1. Before shoot development
  - *a*. Reddening of the leaves of conifers
  - *b*. Death of buds
  - *c*. Injury to the cambium
- 2. After shoot development
  - *a*. Injury to the cambium
  - *b*. Injury to the bark
  - *c*. Death or injury of new shoots and leaves

Injury by early frost
  - *a*. Injury to the cambium
  - *b*. Death of tips of unripened shoots

Reddening and then dropping of the leaves of conifers result, as shown by Neger (1915:206) for Norway spruce, when a frost occurs just at the time growth is starting but before the buds have opened. The needles on shoots of the previous season are most injured and may be the only ones involved. Extensive damage of this type has been reported on conifers in the Southwest (Phillips 1907). Frost injury of this kind is readily con-

Fig. 8. Cross section of a branch of western hemlock, showing frost-ring formation, either partially or completely around the stem, in every growth ring but the first one. (*Reproduced from U.S. Dept. Agr. Bull.* 1131.)

fused with winter drying, and only an exact knowledge of weather conditions will enable a correct diagnosis.

At the time when buds are swelling, they are liable to injury or death by late frost. When the entire bud whorl of the leading shoot of the main stem of planted Scotch pine and other conifers was killed, the leader died back to the previous year's growth of laterals, leaving a dead stub that could be mistaken for a leader killed by weevils except for the absence of exit holes (Belyea and MacAloney 1926). The buds on the outer portion of sycamore branches have been killed over extensive areas by late frosts, leaving only the buds on the innermost portion of the branches alive, so that by midsummer the trees were characterized by long dead twigs around the bases of which bushy masses of leaves had developed (Schrenk 1907). Unless this injury constantly recurs, the trees recover, although they make irregular growth. Killing of young shoots, particularly leaders, by late frost results in crooked leaders, forking of the main stem, and retarded growth in red pine (Kienholz 1933). Injury by late frost to young shoots of pines, not severe enough to cause death,

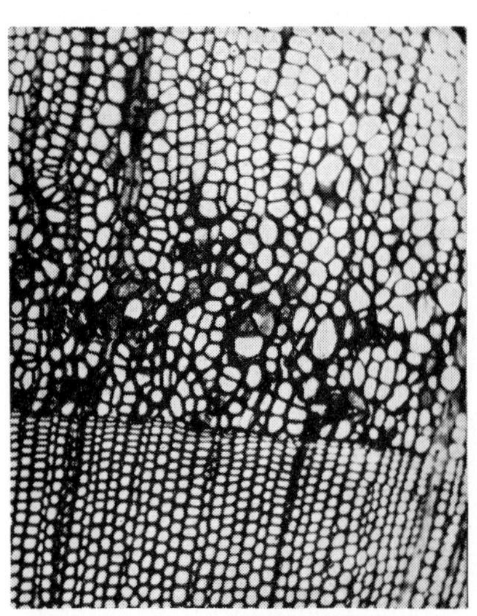

Fig 9. Photomicrograph of a cross section through a frost ring in red pine showing broadening and displacement of the rays. The injury occurred at the beginning of a growing season. (*Photograph by L. S. Gill.*)

Fig. 10. Leader of a balsam fir sapling killed by a late frost.

can result in crooking or bending (Stone 1952). A needle droop of young, planted slash pine is caused by cold (Weber 1957).

Injury to the cambium is a common form of frost damage (Day and Peace 1934:29). It may occur at any time during the growing season or even when the tree is dormant. Both conifers and hardwoods are affected. With severe cold the cambium may be killed outright, or injured beyond recovery; or with a less severe temperature it may develop abnormal tissue, the so-called "frost ring" (Figs. 8, 9). In conifers the abnormal tissue of a frost ring varies with the severity of the injury, being characterized by these symptoms: gummosis, crumpling of the wood cells that were but slightly lignified at the time of injury, development of parenchymatous tissue instead of tracheids, and marked broadening and strong lateral displacement of the rays (Rhoads 1923:13). These features will occur in various combinations. Frost rings are also formed in hardwoods (Hemenway 1926; Mix 1916*a*), corresponding in all important particulars to those in conifers (Harris 1934). Frost injury to the cambium is usually confined to young stems with thin bark. Rhoads (1923:6) found no frost-ring formation in coniferous stems larger than 2 inches in diameter. The cambium of stems with thick bark is sometimes killed or injured by excessively low temperatures in winter. Injury to young trees similar to frost rings may be caused occasionally by drought, lightning, and probably other factors.

If the temperature is low enough, the cambium may be killed entirely around the young stem, thus girdling the stem, or it may be killed over localized areas, causing cankers. Douglas fir and Sitka spruce are susceptible to girdling of the main stem near the ground by frost (Day 1928*b*:182), and frost cankers are also formed on these species. European larch is quite subject to this injury (Day 1931). Frost canker is accompanied by a frost ring, which is a valuable diagnostic character. Hardwoods with smooth bark are liable to frost canker.

Injury to the bark of conifers may occur at the same time as injury to the cambium. Abnormally rough and furrowed bark at the base of the main stem of young Douglas fir has been correlated with frost rings in the wood (Day 1928*b*:187). A slighter bark injury in Sitka spruce is characterized by the development of unusually large resin cavities. American elm (Humphrey 1913) and sycamore discard their outer bark following injury to the cork cambium by low temperature during the winter, but this does not affect the health of the trees (Stone 1916:207).

The most conspicuous injury by late frost is the discoloration, wilting, and death of the tender, succulent leaves and shoots of the current season. In conifers this usually takes the form of reddening or browning and death of the needles with a characteristic curling of the shoot (Fig. 10). The injury may be confused with gray mold blight caused by *Botrytis cinerea*, but the presence of the mycelium and fruit bodies of the fungus serve to distinguish the two injuries. Conifers that have their young shoots repeatedly killed by late frosts become stunted and bushy (Fig. 11). The leaves of hardwoods shrivel and turn brown or even black when severely frosted. In horse chestnut a characteristic slitting or tearing of the leaves between the veins is caused by frost (Tubeuf 1904). Such slitting may also occur in other species. The injury is common in this country during certain years. The periodic failures of pine seed crops may be caused by frost injury during flowering (Campbell 1955).

Injury by early frosts in the autumn occurs occasionally. Trees that continue growth late in the season are liable to this injury. Leaves, shoots, and buds are injured or killed, and the symptoms are similar to but not so striking as those described for injury by late frosts. Early frosts can damage the cambium and living bark of pines, causing a dieback of the tops (Day 1945). Locust normally continues growth each year until

Fig. 11. Young tree of lowland white fir repeatedly injured by late frosts, showing the abnormally compact and bushy form of growth resulting from the ready development of numerous compensatory shoots. (*Reproduced from U.S. Dept. Agr. Bull.* 1131.)

its young shoots are frozen back in the autumn. Basswood, sycamore, and elm are more often injured by early frosts than other hardwoods. As demonstrated experimentally by Harvey (1930) with American elm, temperatures at or close to freezing alternating with temperatures sufficiently high for active growth serve to harden seedlings against cold. This is the normal climatic condition encountered by trees in the autumn.

Frost injury can occur when the temperature drops below 32°F during the growing season, although experimental evidence indicates that it must drop at least several degrees below freezing before forest trees are affected (Day and Peace 1934:13). Injury is extremely variable because of modifying conditions preceding or accompanying it such as rapidity of freezing or thawing, the duration of freezing, the repetition of freezing, wind, and the presence of water on the plants (Day and Peace 1936–1937). Injury to plant tissues by freezing to death has been extensively investigated (Levitt 1956; 1957). Before plant tissue will freeze to death, ice must form in it. There are three main types of frost injury and hardening changes that oppose them: (1) intracellular freezing, which is nearly always fatal, is usually prevented by increased cell impermeability to water so that the ice forms outside the cells; (2) the mechanical effect of freezing and thawing when the ice forms outside the cells, and from which recovery is frequent, is prevented by reduced structural viscosity of the cytoplasm; and (3) the physicochemical effect of dehydration is prevented by reduced coagulability of the protoplasm (Scarth 1944). Sugars play an important part in frost resistance. Soluble sugars are at a maximum during the coldest part of the winter. Frost hardening produces the same changes within plants as drought hardening. There are marked differences in resistance to frost between different tree species, between individuals, strains, or races within a species, and in individuals at different times during the year.

Control of frost injury is dependent on sound silviculture. Damage is most severe on trees in the seedling and sapling stage. The value of an overstory to protect coniferous reproduction from frost is recognized in Europe (Amann 1930; Day and Peace 1946:98). Plantations of species susceptible to frost should not be established on grasslands because the number and intensity of frosts immediately over the ground is greater than normal (Day 1928*a*; Staubacher 1924). Frost intensity is closely correlated to topography (Day 1939). In frost hollows or pockets (Hough 1945; Byram 1948; Pomerleau and Ray 1957), i.e., low areas in which cold air settles and cannot drain off, only frost-resistant trees should be grown. Areas where frost pockets may form should not be clear-cut. In a Douglas fir plantation in Oregon where the trees had attained an average height of 12 feet, many trees in a small frost hollow within this plantation had been killed, and the remainder were only 2 to 3 feet high, stunted and bushy. Where frosts are severe, hardwoods should not be coppiced. In reforestation, racial differences in frost hardiness within a species must not be overlooked. Bates (1930) considers it practically certain that red pine will show racial differences in frost hardiness and considers it unwise to grow trees in a region whose summer temperatures are normally lower than the region occupied by their parents. For example, trees grown in northern Minnesota from seed collected in lower Michigan would not be frost-hardy.

Frost Cracks. Both conifers and hardwoods are subject to radial cracking or splitting of the trunk (Lamprecht 1951; Rol 1953). Frost cracks are usually formed only during the dormant period, when there is a sudden and pronounced drop in temperature, so that the inner wood remains

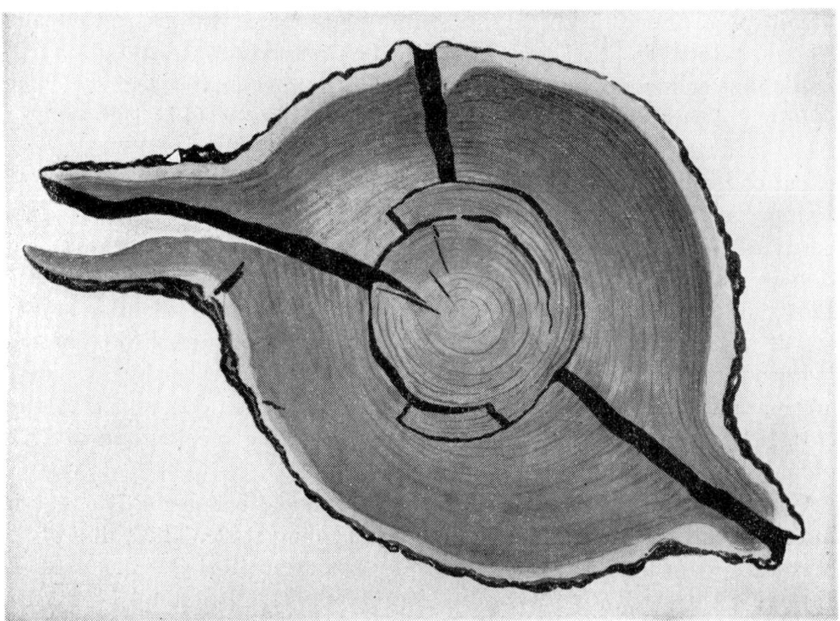

Fig 12. Cross section of an oak trunk showing frost cracks (radial), frost shakes (peripheral), and frost ribs. (*After F. W. Neger.*)

comparatively warm, while the outer wood becomes cold and contracts rapidly. Shrinkage in wood is always greater tangentially than radially. The cracks usually originate in the base of the trunk, extend upwards from a few to many feet, extend deeply into the wood, and split the bark (Fig. 12). Healing of the wound produces considerable callous growth, whereas repeated opening of the crack by cold or strains induced by wind will result in a very pronounced, protruding callous growth, particularly in hardwoods, known as a "frost rib." Small internal radial cracks in the wood, confined to a single growth ring but not evident in the bark, have been caused in Japanese larch by frost (Orr 1925), and more extensive internal cracks, which did not split the bark, in silver fir and oak (Hartig 1896).

Frost cracks damage such species as oaks (Fergus 1956), hybrid black poplars (Poskin and Flon 1955), lodgepole pine (Mason 1915:174), and red and white fir, in the latter species also forming infection courts for decay (Meinecke 1916:3). Frost cracks are a common phenomenon in the northern hardwood forests.

Peripheral cracks or frost shakes are formed by the sudden warming of the outer layers of the wood while the inner are still cold (Bruce 1902), but these ring shakes or cup shakes are also caused by other agencies, so that it is usually difficult to determine their origin. Blister shake, found in the butt of 30-year-old yellowpoplars and caused by frost, is a tangential cavity between growth rings extending partly or completely around the bole that results from the separation of bark from wood and subsequent healing (Tryon and True 1952).

WATER DEFICIENCY

A shortage of water so that trees cannot carry on normal physiological functions results in diseases that present different symptoms even though their cause is fundamentally the same. There is an underlying similarity between drought resistance and cold or frost hardiness (Levitt 1951; Iljin 1957). Although they may differ in extent, the effects of drying out, freezing, and plasmolysis are considered by Harvey (1918) to be similar, all these phenomena causing changes in the acidity and salt concentration of the cell sap. Drought may cause zones of abnormal tissue (drought rings) in stems of young trees similar to frost rings caused by freezing (Day and Peace 1934:40), even to the lateral displacement of the rays (Somerville 1916), which is considered a criterion of frost injury (Steinmetz and Hilborn 1937:24). Cankers may be initiated by drought (True and Tryon 1956).

Drought. Here is included injury caused by an actual deficiency of water in the soil for the needs of trees (Parker 1956). This is brought about by periods of subnormal precipitation and drying winds. Injury may result from a season of extremely low precipitation or from a series of subnormal years, the cumulative effect of which finally reduces soil moisture below tree requirements. This slow drought injury is often difficult to diagnose, the affected trees frequently succumbing to weakly parasitic fungi or insects (Secrest, MacAloney, and Lorenz 1941; MacAloney 1944). Even though the trees survive, their growth is retarded. Evidences of drought injury may appear promptly, or their appearance may be delayed until the next season, so that records of precipitation are often essential to diagnose this injury correctly. With some conifers reduced diameter and height growth tends to appear one year after the drought occurs (McIntyre and Schnur 1936). The severity of the injury depends on the severity of the drought and will vary from slight injury to the foliage to complete death of the tree. Under the same conditions, young trees are more seriously affected than older ones that have their roots deeper in the soil. Trees in shallow soil are more readily affected by drought than the same species on deep soil.

Trees die from the top down and from the outside in. Injury is most severe on south and southwest slopes. The most striking symptom is discoloration of the foliage, which may be preceded by wilting. In hardwoods the leaves turn yellowish to reddish, beginning usually at the tip or at the margins, but occasionally midway between the main veins, until finally the entire leaf is involved. In hardwoods, premature leaf fall commonly results, and the growing season for all trees is shortened. In conifers, the needles turn reddish brown beginning with the youngest, and the discoloration takes place in late summer, autumn, or early winter. As a result of the summer and fall drought of 1929 on the Pacific Coast, particularly in California, reproduction of Douglas fir was extensively killed or injured on the poorer, drier sites. Ponderosa pine, sugar pine, and white fir suffered also but in a lesser degree. Douglas fir showed various degrees of injury. Some trees were completely dead above ground; in others the crown had died back for varying distances from the top; in some only the outer part of the crown, i.e., the ends of the branches, had succumbed, leaving the inner crown green; and in others all the branches had died, but the main stem remained alive and had short shoots developing on it (Boyce 1933:671). It is characteristic of drought injury for the roots still to be alive after the entire portion of the tree above ground appears dead.

Timber being worked for turpentine is easily injured. During the protracted drought of 1931–1932, slash and longleaf pines in turpentine stands died in enormous numbers (Cary 1932). Few unbled trees died, those being worked for the first year suffered slightly, but those being worked for the fourth year suffered heavily. Furthermore there was a great increase in "dry facing" on trees remaining alive, and the yield of resin from all trees was lower.

Trees vary considerably in their capacity to resist drought. Cedars (*Thuja* spp.) quickly show the effects of drought by premature browning of the foliage. In the Great Plains region, jack pine and Scotch pine suffered severe losses in 1934, while ponderosa pine and Austrian pine were less affected. Mature eastern hemlock was killed extensively in southern Connecticut following several dry years, although the species associated with it suffered much less (Stickel 1933), and the drought of 1930 caused greater mortality to conifers than to hardwoods in Pennsylvania, with the exception of scarlet and black oak (McIntyre and Schnur 1936). Eastern hemlock proved quite susceptible, whereas chestnut oak, red oak, and white oak were the most resistant. Maximov (1929) considers drought resistance in plants to be based on conditions within the cell, such as an increase in sap density and modification in the protoplasm itself that increases its waterholding capacity. Korstian (1924:900) found generally greater sap densities in the more drought-resistant species.

Needle Droop. This disease of young planted pines, particularly red pine, characterized by abnormal drooping and dying of the current season's needles, is caused by sudden and excessively rapid transpiration resulting in an absorption lag and loss of turgor in highly succulent young tissue at the needle base (Patton and Riker 1954).

Scorch. This injury, also known as "leaf scorch" or "sun scorch," may occur during periods of severe, warm, dry winds because of a sudden and rapid loss of water from the leaves which cannot be replaced by the roots, since the soil moisture is low (Stone 1916:210). In fruit trees it has been related to a deficiency of potash in the leaves, which disturbs the proper water balance (Warne 1934). Hardwoods are usually affected, but coniferous seedlings in nurseries are sometimes damaged (page 78). Maples are particularly liable to scorch. Affected leaves show dead, brownish areas on their margins or similar areas between the veins, but the leaves remain alive and do not drop so that the tree is little injured. The symptoms are most pronounced on the side of the tree facing the wind. Bronzing of leaves is a milder manifestation of scorch.

Drought Cracks. Pronounced cracks, similar to frost cracks, sometimes extending completely through the trunk and occasionally entirely internal, occur in conifers, particularly spruces (Day 1954; Lutz 1952). These cracks, first termed "heat cracks" (*Hitzerisse*), are evidently the result of collapse caused by unbalance between loss and replenishment of water when the soil becomes unusually dry.

Winter Drying. This injury, also known as "winterkilling" and "winter injury," is prevalent on conifers, although evergreen and deciduous broadleaf trees are also affected. It occurs during midwinter or early spring, and the generally accepted explanation for conifers is that, after a period of cold weather, sudden great increases in temperature often accompanied by drying winds cause excessive loss of water from the leaves, which cannot be replaced because the soil moisture, being either frozen or too cold, is not available to the roots, or because the wood of the stems is frozen and water cannot pass through it. The injury then results from a form of drought. It is a common phenomenon in the mountainous regions of the West, after warm weather with a "chinook" wind has succeeded a cold period. Foliage under snow is not affected. However, Gail (1926) believes that the extensive injury to ponderosa pine in Montana, Idaho, Oregon, Washington, and British Columbia during the winter of 1924–1925, which has been considered typical winter drying, was actually caused by such a sudden drop in temperature that the cell sap could not concentrate quickly enough to prevent freezing of the cell protoplasm; hence the needles died. On the other hand Gail and Cone (1929) find that concentration of cell sap in ponderosa pine increases with the age of the needles for any given time. If this is so, the older needles should be most resistant to low temperature, since cold resistance in plants increases with increased concentration of the cell sap. Actually the older needles are the first and sometimes the only ones to suffer from this injury to which the name winter drying is applied.

Although all conifers retaining their leaves throughout the year are liable to winter drying, ponderosa pine, lodgepole pine, Douglas fir, and blue spruce are frequently injured, one reason being that these species commonly occupy sites subject to sudden extremes of winter temperature. Eastern white pine is occasionally affected (Morse 1909:21). When winter drying is severe, damage may be extensive, as evidenced by the killing of 40,000 acres of coniferous timber in Montana during the winter of 1908–1909 (Hedgcock 1912) and the considerable loss of ponderosa pine in the Black Hills of South Dakota in January, 1909 (Hartley 1912:40). Bates (1916) concludes that the poor quality of coniferous forests, particularly Douglas fir and ponderosa pine in the Central Rocky Mountain region east of the Continental Divide is caused "not so much by a general shortage of precipitation as by the desiccating effect of wind at times when the air becomes unusually dry and when soil moisture is cut off by freezing."

The disease manifests itself by browning of the foliage upon the advent of warm weather in the spring. In cases of severe injury all the leaves and buds are killed and the trees die. Generally, even though all the leaves are killed, the buds escape and the trees survive. Finally the dead needles drop off. These symptoms may be most pronounced on the sides of the trees facing the warm winds. During the summer, injured Douglas fir may be completely bare of foliage except for the current season's shoots. While the discolored foliage is still evident and before the buds have opened and the new shoots formed, winter drying often appears catastrophic. However, by late summer the trees appear more or less normal and, unless the injury recurs, apparently suffer no permanent ill effects. No attempt should be made to determine the amount of damage by winter drying until well along in the summer at the earliest.

The name "red belt" is applied to winter drying that occurs in zones or belts more or less following contour lines (Hartley 1912; Hubert 1918). The explanation advanced by Mason as reported by Hedgcock (1913:113), viz., that this is brought about by a combination of meteorological conditions resulting in frozen soil occurring at certain elevations along mountainous slopes while the soil is not frozen at other elevations, although reasonable, is not the only possible explanation. Ward (1923:163) points out that warm, dry, chinook winds themselves may follow belts. This would explain the condition recorded by Melrose (1919) in British Columbia on Douglas fir and ponderosa pine in which red belt occurred at a very distinct level, sharply defined: "As the bottom of the strip was reached the injury got further into the tops of the trees leaving the needles on the lower limbs unharmed. The trees just reaching the edge of the belt had the tips of the crowns just touched with red. In the same way the lower limbs of the trees on the upper edge of the belt were injured, the unharmed tops becoming longer and longer the further up the hill the trees grew."

Douglas fir in western Oregon and Washington is subject to winter drying to which the name "parch blight" has been applied (Munger 1916). This occurs during brief periods in the winter when dry winds from the east sweep across the Cascade Mountains. The dead brown needles are most prevalent on the eastern side of the trees, particularly on exposed trees or those on the eastern edge of groves. Sometimes the 1-year-old twigs are also killed. Affected trees are usually not permanently injured. The soil is not frozen when parch blight occurs, but the soil is probably too cold for water to be absorbed and conducted rapidly enough to offset the loss by transpiration. Water absorption by roots is reduced as soil cools (Kramer 1942), and water conduction in stems practically stops at 0°C (Fries 1943; Handley 1939). Actual killing of the Douglas fir leaves by low temperatures that always occur during these periods is a lesser possibility.

WATER EXCESS

An excess of water can be unfavorable to trees by stimulating them to excessive development of tender, succulent tissue, which is more readily invaded by fungous parasites and more susceptible to extremes of heat and cold. Intumescences are produced on the leaves of aspen and largetooth aspen when all or portions of the leaves are surrounded by stagnant, moist air, such as occurs in nature when the leaves are rolled or fastened together by insects, thus forming small moist chambers (La Rue 1933). Tree roots must get oxygen necessary for their development from the soil air, and, since most trees cannot obtain sufficient oxygen from water, a saturated soil will result in death of the roots by asphyxiation.

Fig. 13. Lodgepole pines killed by high water around a reservoir.

Stunting and dying of red pine plantations has occurred in soils that have a long period of saturation in the spring (Stone, Morrow, and Welch 1954).

Dropsy. Young eastern white pines, both planted and naturally regenerated, occasionally exhibit a marked thickening of a portion or all of the main stems or branches. In cross section the wood appears unchanged, but the bark, very soft and spongy, is several times thicker than normal. The cortical cells are unusually large and thin-walled. Affected trees have a generally unhealthy appearance with many yellowish to brown needles. Although a similar disease in other plants has been ascribed to an excess of water in the tissues (Atkinson 1893), its cause in white pine is undetermined.

High-water Injury. Trees inundated by floods or by backing-up of water behind dams are injured, or, if the ground is flooded permanently or for long periods, they are killed (Hall and Smith 1955). Species vary in their reaction to flooding, and standing or stagnant water is more serious than flowing water. Klose (1927), for example, found in one flood area that jack pine and white ash together with several other species were not injured by standing or flowing floodwater. On the other hand Douglas fir was severely injured, and Norway spruce was killed by both standing and flowing water. Alder, maple, elm, and birch were killed by standing water but not by running water.

The 1927 Mississippi flood caused considerable damage to young hardwoods in the hardwood bottomland type in Louisiana (Lentz 1928). The floodwater remained for about 3 months at depths of 8 to 20 feet. All trees entirely covered by water were killed, and many of the less thrifty trees that were partially covered died. Ash, gum, elm, oak, and hickory suffered, but as would be expected willow and cottonwood were little affected.

In the West where many reservoirs have been created to furnish water for irrigation and hydroelectric power, not only are the trees standing in water killed (Fig. 13) but conifers on unsubmerged land fringing the reservoir commonly die because of the sudden and excessive increase in soil moisture. Flooding by sea water is especially injurious (Little, Mohr, and Spicer 1955).

NUTRITION

Disease in plants may be caused by an excess or a deficiency of food materials in the soil, by toxic substances, or by other unfavorable soil conditions that upset nutritive processes. The fused needle disease of exotic pines in Australia and Great Britain is an example (Young 1940; Neilson-Jones 1945). Deficiency diseases caused by a lack of one or more essential elements in the soil are common and difficult to diagnose because the symptoms vary not only between elements but between species when the same missing element is responsible (Scott 1947; Stone and Baird 1956; Walker 1956). Potassium deficiency in planted pines and spruces was expressed by a general chlorosis followed by browning and dying of the needles, decreased height and diameter growth, decrease in number of years the needles persisted, and shortening of the needles (Heiberg and White 1951). The pale-green foliage of ponderosa pine reproduction on the rocky, mineral soil of old placer mines in California, as contrasted with the darker green color of reproduction of the same species on adjacent areas having soil with a normal humus content, seems to be a direct expression of nitrogen deficiency. Chlorosis (page 80) in some instances results from a deficiency of iron in the soil. It has been successfully treated in cottonwoods and other broadleaf trees by injecting into the trunk a ¼ per cent solution of ferrous sulfate (Burke 1932:330) or dry ferric phosphate powder (Starr 1942), or injecting under pressure into the soil 6 ounces of chelated iron in 10 gallons of water around each tree (Crum 1954). Fasciation (page 304) has been ascribed to local overnutrition.

Proliferation. The development of an abnormal number of stems, flowers, or fruits in the position normally occupied by a single organ may also be the result of local overnutrition. On certain hard pines, clusters of cones are sometimes found. Witches'-brooms of unknown origin occur on conifers and hardwoods, although many such malformations are caused by parasitic fungi and insects.

INDUSTRIAL PROCESSES

Orchard and shade trees are commonly injured by unfavorable conditions of the air or soil arising from industrial processes. Forest trees escape most of these adverse factors, because timber-producing stands are not often located close to dense industrial areas in this country. Death or injury by the escape of artificial illuminating gas into the soil (Deuber 1936), or by the escape of steam overheating the soil, although common enough among shade trees, is unknown among trees in forest stands. The presence of natural gas by itself is less injurious. Air pollution is a nuisance (McCabe et al. 1952; Magill, Holden, and Ackley 1956), and trees suffer along with other living things (Scurfield 1960). The gases most frequently injuring plants are sulfur dioxide, hydrofluoric acid gas, chlorine, hydrogen sulfide, ammonia, growth-regulating substances, and constituents of manufactured illuminating gas (Thomas 1951; Wienhaus 1955; Zimmerman 1955). Species vary in their susceptibility to different gases (Zimmerman and Hitchcock 1956), and ponderosa pine affected by fluorine shows differences in susceptibility from one individual to another (Adams, Shaw, and Yerkes 1956).

Smoke Injury. Trees are injured by gases in the air arising from the incomplete combustion of coal during manufacturing processes, or from the smelting of ores containing sulfur. Dust and metallic fume cause minor damage only, but gases, particularly sulfur dioxide (SO_2), frequently cause severe damage. In forests serious injury occurs to stands in the vicinity of smelters, where large quantities of sulfur dioxide are released into the air, so the following discussion is principally concerned with the effects of this gas (Jahnel 1954–55; Katz 1949; Scheffer and Hedgcock 1955; Zieger 1955).

Evergreen coniferous trees in general are more severely injured by smelter fumes than trees that do not retain their foliage in winter. Consequently, larch is usually less affected than other conifers. The order of susceptibility of tree species shows little agreement between different authorities because the relative degrees of injury within a species tend to differ with location, year, and distance from the source of sulfur dioxide. Silver fir affected by smoke is said to be less resistant to winter cold (Meyer 1957).

Three stages of injury from smelter smoke are recognized: the acute stage, due to high concentrations of sulfur dioxide in the air, which is characterized by rapid discoloration of the foliage followed by defoliation and in extreme cases by the death of the plant; the chronic stage, due to small quantities of gas generally present, which is characterized by an unhealthy condition of the plant expressed in stunted growth; and the invisible stage, due to the action of sulfur dioxide in very dilute quantity, characterized by reduction in growth not visible to the eye but determinable by careful measurements or by chemical analysis of the foliage. Actually the invisible stage is a mild form of the chronic stage so that the distinction between them is scarcely warranted and difficult to draw.

Acute injury is first indicated by color changes in the foliage. Coniferous needles become wine red in color sometimes throughout their entire length, but more commonly first at the base, at the tip, or in the middle. The needles finally turn brown and drop if the action of the gas is protracted or sufficiently severe. Conifers close to a smoke source may retain only the needles of the current year. The older needles suffer first. In broadleaf trees, yellowish-brown to dark-brown dead areas appear in the leaf tissue between the veins, while the tissue next to the larger veins remains green (Fig. 14). Occasionally, instead of being intercostal, the dead areas may border the leaf or be confined to the tip or base. If the injury is sufficiently severe, the entire leaf is discolored. The symptoms of chronic injury are less striking. Discoloration of the foliage is not so apparent, and in conifers defoliation is greatly lessened. Affected trees, however, have an unhealthy and a stunted appearance. Evergreen broadleaf trees tend to become deciduous. Invisible injury is difficult to detect and is easily confused with other factors causing small reductions in growth.

The diagnosis of acute smoke injury is relatively simple because of the characteristic symptoms. Chronic injury, and particularly invisible injury, are easily confused with diseases caused by other noninfectious agencies such as winter drying or red belt, sun scorch, and summer drought. In the West, separation of smoke injury from winter drying of conifers

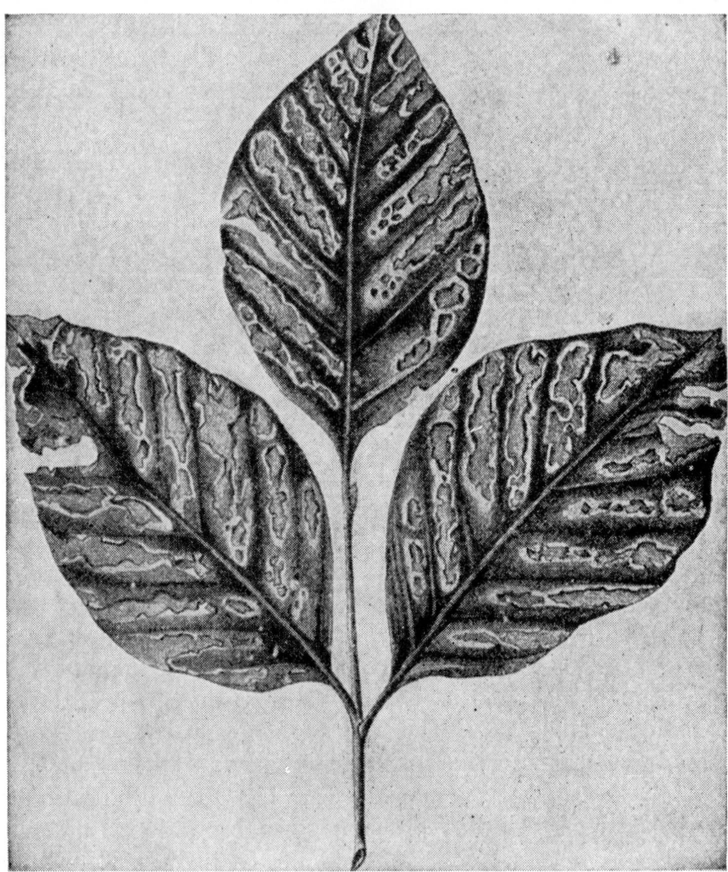

Fig. 14. Beech leaves injured by sulfur dioxide. (*After J. V. Schroeder and C. Reuss.*)

has proved difficult at times, but according to Hedgcock (1912) smoke injury causes a brighter red discoloration of pine needles than does winter drying and does not so often kill the entire leaf. Killing by winter drying is sudden; by smoke more gradual. Smoke injury causes a gradual reduction in the width of the annual rings, whereas winter drying results in little annual ring growth for the growing season after the injury, followed by practically normal growth for succeeding seasons. The symptoms of smoke injury will gradually decrease in intensity as the distance from the source of smoke increases. Smoke injury will naturally follow the drift of the air currents during daylight hours, since injury does not occur in the dark. Where the topography is rough, the heavy smoke will concentrate in and be carried along natural air drainages, such as valleys or gorges, so that injury will be irregularly distributed around the source. Under such conditions when large quantities of sulfur dioxide were being released, symptoms of injury were found on hardwood trees around a small lake 40 miles from the source of smoke, although on intervening areas for some distance there were no evidences of injury. Slopes facing the drift of smoke frequently show more injury than those facing in the opposite direction. The use of indicator plants known to be especially susceptible to smoke injury is helpful in diagnosis. Such plants can be determined by field study of the native vegetation in a locality subject to smoke injury. Other methods useful for diagnosis are analysis of the air for its sulfur dioxide content and microscopic study of sensitive plants known to develop an area of brown dead tissue under their lenticels when affected by sulfur dioxide (Neger 1919). Where sulfur dioxide is present, lichens are absent from the flora, and metals corrode rapidly. It is impossible to predict from analyses of the poison content of the air, of the soil, or of the plant tissues just what damage will occur to plants from smoke (Mühlsteph 1942), because so many factors influence the susceptibility of plants to injury by sulfur dioxide (Setterstrom and Zimmerman 1939).

The most important causal agency in smoke injury is sulfur dioxide, a colorless gas with a characteristic suffocating odor. It is 2.21 times as heavy as air. It is extremely toxic to plants, injuring pines in the low concentration of 1 part of sulfur dioxide to 500,000 parts of air. The odor cannot be detected when the concentration is less than 3 parts to 1,000,000 parts of air. In fully formed leaves the gas enters nearly exclusively through stomata, but in immature tender leaves some penetration occurs directly through the epidermal wall. After entering the leaf, sulfur dioxide penetrates the living cells, and in chemical combination with substances in the cells it retards or completely checks physiological functions. Injured leaves contain few or no starch grains; hence the food supply of affected plants is materially reduced. Sulfur dioxide in small quantities also affects transpiration, which increases quickly when this gas penetrates the leaves and then decreases to below normal. Bright light, high humidity, and high temperature increase injury; thus the greater the photosynthetic activity of the plant, the greater the degree of injury. Plants in the dark exposed to toxic quantities of sulfur dioxide are not injured. The gas confines its effects to the chlorophyll-bearing or green parts of plants, woody tissues not being directly affected. The theory that conifers are particularly affected by smoke because the stomata are clogged with soot and tars is not acceptable (Rhine 1924).

In addition to direct injury to foliage indirect injury is claimed because of soil changes, such as the acidity of the smoke depleting the soil of calcium carbonate (Wieler 1922) and thus reducing the number and lessening the activity of nitrifying bacteria (Ruston 1921). However, McCool and Mehlich (1938) found that soils about centers of considerable sulfur dioxide production had not been noticeably altered after many years of exposure, and timber stands recovered satisfactorily after sulfur dioxide was abated (Scheffer and Hedgcock 1955:44).

The control of smoke injury to forest trees can be accomplished only at the source, either by eliminating the injurious substances in the smoke or by reducing their concentration below toxic limits. Methods used are electrically filtering out the solids, diluting the gases with high smokestacks or with numerous small stacks, superheating gases in the stacks so they will diffuse more widely in the air when released, washing with water, neutralizing the acid gases with basic materials, and the manufacture of sulfur or sulfuric acid from sulfur dioxide.

SALT SPRAY

Trees along the seacoast are injured by salt spray, and, after winds of high velocity, injury may be found many miles inland (Little, Mohr, and Spicer 1958). Conifers are more affected than broadleaf trees. Injured needles are bright orange yellow or orange red for varying distances from the tip down. In mass, affected conifers appear as though damaged by fire. Eastern white pine is highly susceptible to injury, but blue spruce, Austrian pine, Japanese black pine (Wyman 1939), and live oak (Wells 1939) are highly resistant.

MECHANICAL INJURIES

Few forest trees attain maturity without mechanical injury. Injuries causing open wounds, i.e., wounds exposing the wood, are of most importance because such wounds offer entrance points for wood-destroying and other fungi (page 373). Falling trees frequently break off branches or tops or knock off large patches of bark from the trunks of their neighbors. Logging operations, particularly where high-speed machinery is used, are a prolific source of injury to the valuable trees in the reserved stand. In some regions ground fires cause many open wounds, but the formation of these wounds may be a slow process, so that it is not for some years after fire occurs that the full measure of injury is apparent (Lachmund

1921, 1923). The relation between decay and fire scars is well understood, and the bottom-land hardwood stands of the South would be largely free from disease except that ground fires have led to heavy losses from decay. Even small wounds, such as blazes made directly by man, afford access to fungi. Along seacoasts or at high elevations, wind acting on exposed trees deforms and stunts them and may even cause shrinkage in diameter of the lower bole for a period of several years (Haasis 1933). Wind is also injurious in breaking branches and tops or overthrowing trees. However, the prevention of wind throw is largely the province of the silviculturist.

Ice and Snow. Trees are frequently damaged or destroyed by ice or wet snowstorms. The most exhaustive accounts of snow injury, par-

Fig. 15. Douglas firs with the tops broken off by a severe ice storm. Most of this young stand had been killed by fire before the picture was taken. (*Reproduced from U.S. Dept. Agr. Tech. Bull.* 286.)

ticularly to conifers, are found in Swedish and Finnish forestry literature. The unusual weight of snow or ice breaks off branches, sometimes stripping the trunk almost bare, or the top may be broken off (Fig. 15). Tall wide-crowned trees and those with soft, weak wood, such as poplar, suffer most. Injuries caused by the severe ice storm of November, 1921, in New England were still in evidence 10 years later. Not only are trees deformed by crooks in the trunk, where the top is broken off and later replaced by a branch, but the open wounds are infection courts for fungi that decay the heartwood (page 373). Destruction of many trees, with the consequent sudden exposure of the remainder, may result in much sunscald in the residual stand with decay entering through the lesions (Spaulding and Bratton 1946). Coniferous reproduction may be bent or

Fig. 16. Hail wounds on branches. New wounds on Carolina poplar below and old wounds completely healed over on eastern white pine above. (*Courtesy of Conn. Agr. Expt. Sta.*)

broken by heavy snow (Boerker 1914). Recurring ice storms may explain the poor quality of many mountain hardwood stands on slopes and ridges, which has been attributed to other factors (Abell 1934).

Prominent narrow but not significantly injurious scars extending from one-quarter to one-half way around the circumference of the stem often follow the formation of ice crusts on trunks and branches of hardwoods, being most common on stems 1 inch or less in diameter (Lutz 1936).

Hail. Severe hailstorms cause defoliation in varying degrees, laceration of leaves of broadleaf trees, wounding of the bark and cambium, and occasionally the death of young trees or sprouts (Crug 1928; Riley 1953; Tanner 1953; Thomas 1956). Cook (1925) reported the breaking off of all the twigs and branches on one side of the trees in a 10-year-old plantation of eastern white pine. Hail wounds (Fig. 16) are confined to the upper side of the branch and to the side of the main stem facing the storm. Only young stems with tender bark are wounded. Wounds 12 to 15 inches long and 1 inch broad have been found on hardy catalpa, but usually the scars on most species are only 1 or 2 inches long. When partially healed they might be confused with injuries caused by the cicada, except for the torn wood projecting from the insect wounds. Fungi and insects may enter through hail wounds. Decay in hardy catalpa caused by *Polystictus versicolor* is sometimes associated with hail wounds.

Lightning. Although it has been considered that tree species differ in their liability to lightning stroke (Stahl 1912), it is concluded by Plum-

Fig. 17. A sugar maple shattered by lightning. (*Photograph by G. E. Nichols.*)

mer (1912:36) that trees of the dominant species will be most frequently struck in any locality, and that no species is immune to lightning stroke. Although many trees are struck without death or mechanical injury of consequence occurring (Stone 1914:12), others are killed or injured. The complete shattering of a tree or tearing off of the bark and a shallow layer of wood in a narrow strip, sometimes extending spirally around the trunk from high in the crown to the butt, is a familiar sight (Figs. 17, 18). The cambium, being the line of least resistance, is particularly susceptible to electrical discharge, and the classical investigations of Hartig (1897) have demonstrated the formation of abnormal tissue in trees showing no outward indications of injury. In the bark, small vertically elongated areas of hard tissue (*Blitzspuren*) lighter in color than normal are developed (Fig. 19), and in the wood lightning rings are formed that may extend entirely around the trunk circumferentially and for many feet vertically. These rings are formed in the annual growth rings, and in some conifers are commonly characterized by an abnormal number of resin ducts (Fig. 20), although Tubeuf (1906) describes and illustrates in Norway spruce a zone of abnormal tissue caused by lightning that is without traumatic resin ducts and is exactly similar to a frost ring. Lightning rings of similar structure occur in incense cedar (Fig. 21). Lightning rings are a defect in wood used in small sizes and requiring great strength, such as airplane members, since the lightning ring forms a plane of cleavage along which checking may occur (Boyce 1923:17).

Fig. 18. A lightning wound on a white ash. (*Photograph by R. P. Marshall.*)

Small, roughly circular groups of conifers, but not hardwoods, may be killed by a diffuse electrical discharge without the trees showing any outward visible lightning wounds (Tubeuf 1905, 1906; Hoepffner 1910; Peace 1940). This possibly may explain the small groups of dead trees which often occur in the mature Douglas fir forests of western Oregon and Washington, for which no definite cause has been found. The tops of conifers may be killed in the same way without any outward signs of

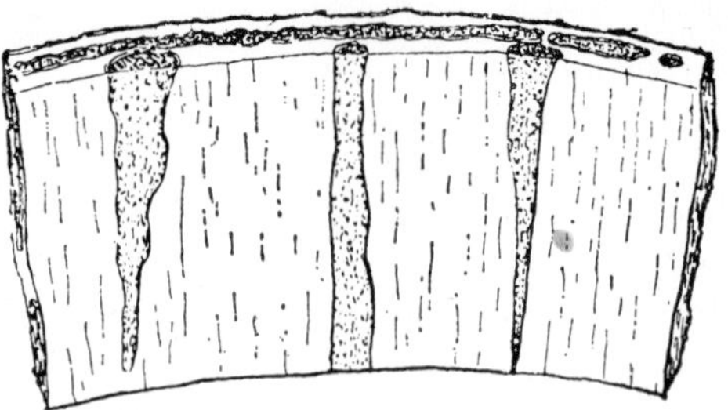

Fig. 19. Areas of hard tissue in the bark of Norway spruce caused by lightning. (*After R. Hartig.*)

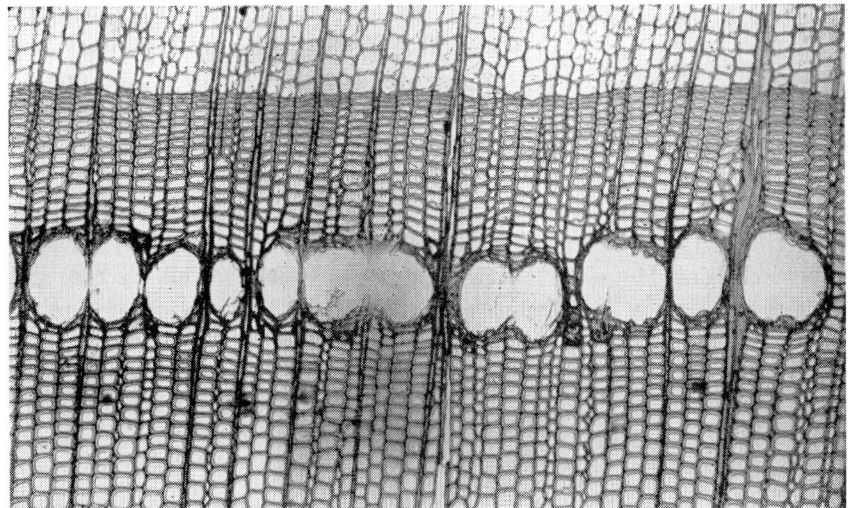

Fig. 20. Photomicrograph of a cross section through a lightning ring in Sitka spruce showing the traumatic resin ducts.

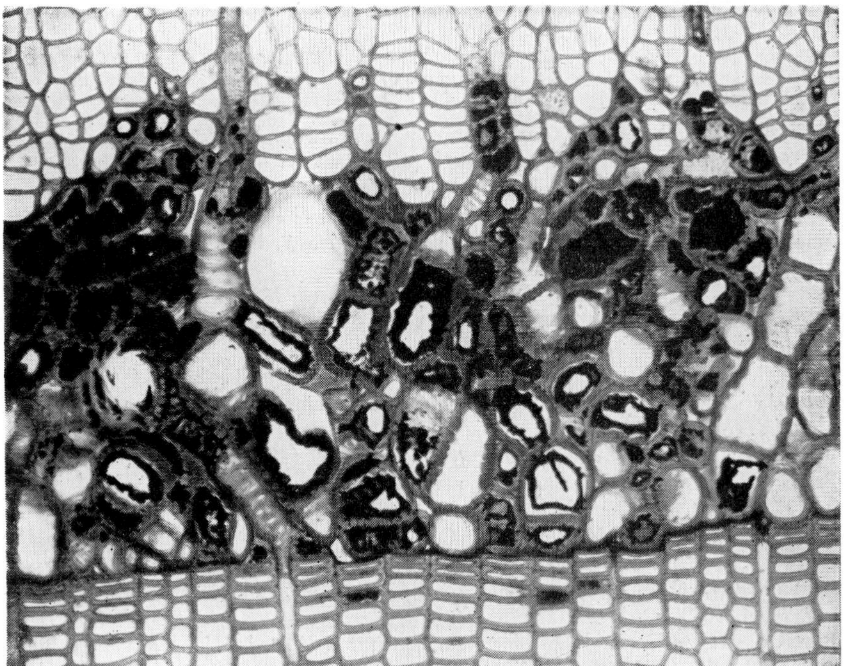

Fig. 21. Photomicrograph of a cross section through a lightning ring in incense cedar showing lateral displacement and broadening of the rays.

lightning being apparent (Tubeuf 1903:445). Although large trees are most likely to be affected by lightning, groups of nursery seedlings and young trees may also be killed or injured (Jackson 1940; Rhoads 1943).

In June, 1923, in the Columbia National Forest in Washington a mixed coniferous stand was curiously affected during a severe electrical storm. On a strip of timber about 3 miles long by ¼ mile wide, the needles were partially or completely killed and browned on the trees from the top to within 3 to 50 feet from the ground, and the stand on the strip appeared dead. However the twigs and buds were not affected, so the trees resumed growth and developed new shoots as usual. Several other similar, though smaller, areas of injury occurred in the same locality. A July storm scorched about twenty-five acres of timber in the same way and killed about three hundred trees outright. In no case were there any outwardly visible lightning wounds on the trees.

Lightning can cause significant loss. It accounts for about one-third of the total timber mortality in the ponderosa pine forests of northern Arizona and in addition causes a considerable amount of defect (Wadsworth 1943). In an Arkansas locality, lightning was directly or indirectly responsible for 70 per cent, on a volume basis, of the mortality in southern pines (Reynolds 1940).

REFERENCES

Abell, C. A.: Influence of Glaze Storms upon Hardwood Forests in the Southern Appalachians. *J. Forestry,* 32:35–37 (1934).

Adams, D. F., C. G. Shaw, and W. D. Yerkes, Jr.: Relationship of Injury Indexes and Fumigation Fluoride Levels. *Phytopathology,* 46:587–591 (1956).

Amann, H.: Birkenvorwald als Schutz gegen Spätfroste. *Forstwiss. Centr.,* 52:492–503, 581–592 (1930).

Atkinson, G. F.: Dropsical Diseases of Plants. *Science,* 22:323–324 (1893).

Baldwin, H. J.: Needle Blight in Eastern White Pine. *U.S. Dept. Agr., Plant Disease Reptr.,* 38:725–727 (1954).

Bates, C. G.: The Effect of Wind. *Soc. Am. Foresters Proc.,* 11:443–444 (1916).

Bates, C. G.: The Frost Hardiness of Geographic Strains of Norway Pine. *J. Forestry,* 28:327–333 (1930).

Beal, J. A.: Frost Killed Oak. *J. Forestry,* 24:949–950 (1926).

Belyea, H. C., and H. J. MacAloney: Weather Injury to Terminal Ends of Scotch Pine and Other Conifers. *J. Forestry,* 24:685–690 (1926).

Berry, F. H.: Winter Injury to Asiatic Chestnut Trees in the South during November 1950. *U.S. Dept. Agr., Plant Disease Reptr.,* 35:504–505 (1951).

Boerker, R. H.: Damage to Reproduction by Snow. *Soc. Am. Foresters Proc.,* 9:267–268 (1914).

Boyce, J. S.: Decays and Discolorations in Airplane Woods. *U.S. Dept. Agr. Bull.,* 1128:1–51 (1923).

Boyce, J. S.: A Canker of Douglas Fir Associated with *Phomopsis lokoyae. J. Forestry,* 31:664–672 (1933).

Bruce, E. S.: Frost Checks and Wind Shakes. *Forestry and Irrig.*, 8:159–164 (1902).

Burke, E.: Chlorosis of Trees. *Plant Physiol.*, 7:329–334 (1932).

Byram, G. M.: Terrestrial Radiation and Its Importance in Some Forestry Problems. *J. Forestry*, 46:653–658 (1948).

Campana, R. J.: *Corticium galactinum* Does Not Cause White Pine Needle Blight. *U.S. Dept. Agr., Plant Disease Reptr.*, 38:297–303 (1954).

Campbell, T. E.: Freeze Damages Shortleaf Pine Flowers. *J. Forestry*, 53:452 (1955).

Cary, A.: On the Recent Drought and Its Effects. *Naval Stores Rev.*, 42(17):14–15; (18):14–15, 20; (19):14–15, 18–19 (1932).

Clark, J., and G. W. Barter: Growth and Climate in Relation to Dieback of Yellow Birch. *Forest Sci.*, 4:343–364 (1958).

Cook, H. O.: Hail Storm Injury to Young Trees. *J. Forestry*, 23:849–850 (1925).

Crug, K.: Der Hagelschlag vom 27 Juli 1829 in den oberbayerischen Staatsforsten. *Forstwiss. Centr.*, 50:542–556, 561–575, 620–627, 687–694 (1928).

Crum, P.: Chelating Agents for the Control of Lime-induced Chlorosis in Southern Magnolia. *Natl. Shade Tree Conf. Proc.*, 30:267–270 (1954).

Davidson, J.: Douglas Fir Sugar. *Can. Field Nat.*, 33:6–9 (1919).

Day, W. R.: Damage by Late Frost on Douglas Fir, Sitka Spruce and Other Conifers. *Forestry*, 2:19–30 (1928*a*).

Day, W. R.: Frost as a Cause of Disease in Trees. *Quart. J. Forestry*, 22:179–191 (1928*b*).

Day, W. R.: The Relationship Between Frost Damage and Larch Canker. *Forestry*, 5:41–56 (1931).

Day, W. R., Local Climate and the Growth of Trees with Special Reference to Frost. *Quart. J. Roy. Meteorol. Soc.*, 65:195–209 (1939).

Day, W. R.: A Discussion of Causes of Dying-back of Corsican Pine, with Special Reference to Frost. *Forestry*, 19:4–26 (1945).

Day, W. R.: Drought Crack of Conifers. *Gt. Brit., Forestry Comm., Forest Rec.*, 26:1–40 (1945).

Day, W. R., and T. R. Peace: The Experimental Production and the Diagnosis of Frost Injury on Forest Trees. *Oxford Forestry Mem.*, 16:1–60 (1934).

Day, W. R., and T. R. Peace: The Influence of Certain Accessory Factors on Frost Injury to Forest Trees. *Forestry*, 10:124–132; 11:13–29, 92–103 (1936–1937).

Day, W. R., and T. R. Peace: Spring Frosts, with Special Reference to the Frosts of May 1935. *Gt. Brit., Forestry Comm. Bull.*, 18 (2d ed.): 1–111 (1946).

Delacroix, G.: Maladies non-parasitaires. In G. Delacroix and A. Maublanc: "Maladies des plantes cultivées," 3d rev. ed. Paris: J. B. Ballière (1926) vol. I.

Deuber, C. G.: Effects on Trees of an Illuminating Gas in the Soil. *Plant Physiol.*, 11:401–412 (1936).

Fergus, C. L.: Frost Cracks on Oak. *Phytopathology*, 46:297 (1956).

Fries, N.: Zur Kenntnis des winterlichen Wasserhaushalts der Laubbäume. *Svensk Botan. Tidskr.*, 37:241–265 (1943).

Gail, F. W.: Osmotic Pressure of Cell Sap and Its Possible Relation to Winter Killing and Leaf Fall. *Botan. Gaz.*, 81:434–445 (1926).

Gail, F. W., and W. H. Cone: Osmotic Pressure and pH Measurements on Cell Sap of *Pinus ponderosa*. *Botan. Gaz.*, 88:437–441 (1929).

Godman, R. M.: Winter Sunscald of Yellow Birch. *J. Forestry*, 57:368–369 (1959).

Graebner, P.: "Lehrbuch der nicht-parasitären Pflanzenkrankheiten." Berlin: Paul Parey (1920).

Graham, S. A.: Causes of Hemlock Mortality in Northern Michigan. *Univ. Mich. School Forestry and Conserv. Bull.*, 10:1–61 (1944).

Graves, A. H.: Some Diseases of Trees in Greater New York. *Mycologia,* 11:111–124 (1919).

Haasis, F. W.: Shrinkage in a Wind-dwarfed Redwood. *J. Forestry,* 31:407–412 (1933).

Hahn, G. G., and H. G. Eno: Fungus Association with Birch "Dieback" and Its Significance. *U.S. Dept. Agr., Plant Disease Reptr.,* 40:71–79 (1956).

Hall, R. C.: Post-logging Decadence in Northern Hardwoods. *Univ. Mich. School Forestry and Conserv. Bull.,* 3:1–66 (1933).

Hall, T. F., and G. E. Smith: Effects of Flooding on Woody Plants, West Sandy Dewatering Project, Kentucky Reservoir. *J. Forestry,* 53:281–285 (1955).

Handley, W. R. C.: The Effect of Prolonged Chilling on Water Movement and Radial Growth in Trees. *Ann. Botany,* 3:803–813 (1939).

Harris, H. A.: Frost Ring Formation in Some Winter-injured Deciduous Trees and Shrubs. *Am. J. Botany,* 21:285–298 (1934).

Hartig, R.: Innere Frostspalten. *Forstl. Naturw. Z.,* 5:483–488 (1896).

Hartig, R.: Untersuchungen über Blitzschläge in Waldbäumen. *Forstl. Naturw. Z.,* 6:97–120, 145–165, 193–206 (1897).

Hartley, C.: Notes on Winterkilling of Forest Trees. *Nebraska Univ. Forest Club Ann.,* 4:39–50 (1912).

Hartley, C., and G. G. Hahn: Notes on Some Diseases of Aspen. *Phytopathology,* 10:141–147 (1920).

Harvey, R. B.: Hardening Process in Plants and Developments from Frost Injury. *J. Agr. Research,* 15:83–112 (1918).

Harvey, R. B.: Length of Exposure to Low Temperature as a Factor in the Hardening Process in Tree Seedlings. *J. Forestry,* 28:50–53 (1930).

Heald, F. D.: "Manual of Plant Diseases," 2d ed. New York: McGraw-Hill Book Company, Inc. (1933).

Hedgcock, G. G.: Winter-killing and Smelter Injury in the Forests of Montana. *Torreya,* 12:25–30 (1912).

Hedgcock, G. G.: Notes on Some Diseases of Trees in Our National Forests. III. *Phytopathology,* 3:111–114 (1913).

Heiberg, S. O., and D. P. White: Potassium Deficiency of Reforested Pine and Spruce Stands in Northern New York. *Soil Sci. Soc. Am. Proc.,* 15(1950):369–376 (1951).

Hemenway, A. F.: Late Frost Injury to Some Trees in Central Kentucky. *Am. J. Botany,* 13:364–366 (1926).

Henry, A.: Manna of Larch and of Douglas Fir, Melezitose and Lethal Honey. *Pharm. J. & Pharm.,* 112:387–390 (1924).

Hepting, G. H., J. H. Miller, and W. A. Campbell: Winter of 1950–51 Damaging to Southeastern Woody Vegetation. *U.S. Dept. Agr., Plant Disease Reptr.,* 35:502–503 (1951).

Hoepffner, A.: Beobachtungen über elektrische Erscheinungen im Walde. Ein weiterer Beitrag zum Kapital "Blitzlöcher" im Walde. *Naturw. Z. Forst- u. Landwirtsch.,* 8:411–416 (1910).

Hough, A. F.: Frost Pocket and Other Microclimates in Forests of the Northern Allegheny Plateau. *Ecology,* 26:235–250 (1945).

Huberman, M. A.: Sunscald of Eastern White Pine, *Pinus strobus* L. *Ecology,* 24:456–471 (1943).

Hubert, E. E.: A Report on the Red Belt Injury of Forest Trees Occurring in the

Vicinity of Helena, Montana. *Montana State Forester Bien. Rept.*, 5:33–38 (1918).

Hubert, E. E.: Forest-tree Diseases Caused by Meteorological Conditions. *U.S. Monthly Weather Rev.*, 58:455–459 (1930).

Hubert, E. E.: How Weather Causes Tree Diseases. *Northwest Sci.*, 6:10 (1932).

Humphrey, C. J.: Winter Injury to White Elm. *Phytopathology*, 3:62–63 (1913).

Ibberson, J. E., and H. Streater: White Pine Blight Responds to Fertilizer Applications. *Penna. Forests and Waters*, 4:30–31, 46 (1952).

Iljin, W. S.: Drought Resistance in Plants and Physiological Processes. *Ann. Rev. Plant Physiol.*, 8:257–274 (1957).

Jackson, L. W. R.: Lightning Injury of Black Locust Seedlings. *Phytopathology*, 30:183–184 (1940).

Jahnel, H.: Physiologisches über Einwirkung von Schwefeldioxid auf die Pflanzen. *Wiss. Z. tech. Hochschule Dresden*, 4(3):3–7 (1954–1955).

Kahl, A.: Der Winterfrost 1928–1929 und seine Auswirkungen auf Baum und Strauch. *Deut. Dendrol. Ges. Mitt.*, 42:222–245 (1930).

Katz, M.: Sulfur Dioxide in the Atmosphere and Its Relation to Plant Life. *Ind. Eng. Chem.*, 41:2450–2465 (1949).

Kienholz, R.: Frost Damage to Red Pine. *J. Forestry*, 31:392–399 (1933).

Klose: Die Hochwasserschäden 1926 in den schlesischen Forsten. *Schles. Forstver. Jahrb.*, 1927:134–177 (1927).

Korstian, C. F.: Density of Cell Sap in Relation to Environmental Conditions in the Wasatch Mountains of Utah. *J. Agr. Research*, 28:845–907 (1924).

Kramer, P. J.: Species Differences with Respect to Water Absorption at Low Soil Temperatures. *Am. J. Botany*, 29:828–832 (1942).

Lachmund, H. G.: Some Phases in the Formation of Fire Scars. *J. Forestry*, 19:638–640 (1921).

Lachmund, H. G.: Bole Injury in Forest Fires. *J. Forestry*, 21:723–731 (1923).

Lamprecht, H.: Ueber den Einfluss von Umweltsfaktoren auf die Frostrissbildung bei Stiel- und Traubeneiche im nordostschweizerischen Mittelland. *Schweiz. Anst. forstl. Versuchsw. Mitt.*, 26:359–418 (1951).

La Rue, C. D.: Intumescences on Poplar Leaves. I. Structure and Development. II. Physiological Considerations. *Am. J. Botany*, 20:1–17, 159–175 (1933).

Leapheart, C. D.: Pole Blight—How It May Influence Western White Pine Management in Light of Current Knowledge. *J. Forestry*, 56:746–751 (1958).

Leapheart, C. D., O. L. Copeland, and D. P. Graham: Pole Blight of Western White Pine. *U.S. Dept. Agr., Forest Serv., Forest Pest Leaflet*, 16:1–4 (1957).

Lentz, G. H.: The 1927 Flood Damage to Young Hardwoods. *Forest Worker*, 4:14 (1928).

Levitt, J.: Frost, Drought, and Heat Resistance. *Ann. Rev. Plant Physiol.*, 2:245–268 (1951).

Levitt, J.: "The Hardiness of Plants." New York: Academic Press, Inc. (1956).

Levitt, J.: The Moment of Frost Injury. *Protoplasma*, 48:289–302 (1957).

Linzon, S. N.: The Development of Foliar Symptoms and the Possible Cause and Origin of White Pine Needle Blight. *Can. J. Botany*, 38:153–161 (1960).

Little, S., J. J. Mohr, and L. L. Spicer: Salt-water Storm Damage to Loblolly Pine Forests. *J. Forestry*, 56:27–28 (1958).

Lutz, H. J.: Scars Resulting from Glaze on Woody Stems. *J. Forestry*, 34:1039–1041 (1936).

Lutz, H. J.: Soil Conditions of Picnic Grounds in Public Forest Parks. *J. Forestry,* 43:121–127 (1945).

Lutz, H. J.: Occurrence of Clefts in the Wood of Living White Spruce in Alaska. *J. Forestry,* 50:99–102 (1952).

MacAloney, H. J.: Relation of Root Condition, Weather, and Insects to the Management of Jack Pine. *J. Forestry,* 42:124–129 (1944).

McCabe, L. C., et al.: Proc. U.S. Tech. Conf. on Air Pollution. New York: McGraw-Hill Book Company, Inc. (1952).

McCool, M. M., and A. Mehlich: Soil Characteristics in Relation to Distance from Industrial Centers. *Contribs. Boyce Thompson Inst.,* 9:353–369 (1938).

McIntyre, A. C., and G. L. Schnur: Effect of Drought on Oak Forests. *Penna. Agr. Expt. Sta. Bull.,* 325:1–43 (1936).

Magill, P. L., F. R. Holden, and C. Ackley (eds.): "Air Pollution Handbook." New York: McGraw-Hill Book Company, Inc. (1956).

Mason, D. T.: The Management of Lodgepole Pine. *Forestry Quart.,* 13:171–182 (1915).

Maximov, N. A.: The Physiological Nature of Drought-resistance of Plants. *Proc. Intern. Congr. Plant Sci., 1st Congr., Ithaca, 1926,* 2:1169–1175 (1929).

Meinecke, E. P.: Forest Pathology in Forest Regulation. *U.S. Dept. Agr. Bull.,* 275:1–62 (1916).

Melrose, G. P.: Red-belt Injury in British Columbia. *Can. Forestry J.,* 15:164 (1919).

Meyer, H.: Die Bedeutung von Klimaextremen für die Frostresistenz der Weisstanne (*Abies alba* Mill.) in ihren nördlichen Randgebiet. *Forst u. Jagd,* 7:11–14 (1957).

Mielke, J. L., and J. W. Kimmey: Heat Injury to the Leaves of California Black Oak and Some Other Broadleaves. *U.S. Dept. Agr., Plant Disease Reptr.,* 26:116–119 (1942).

Mix, A. J.: The Formation of Parenchyma Wood Following Winter Injury to the Cambium. *Phytopathology,* 6:279–283 (1916*a*).

Mix, A. J.: Sun-scald of Fruit Trees: A Type of Winter Injury. *N.Y. State Agr. Expt. Sta. (Cornell) Bull.,* 382:235–284 (1916*b*).

Morse, W. J.: Notes on Plant Diseases, 1908. *Maine Agr. Expt. Sta. Bull.,* 164:1–28 (1909).

Mühlsteph, W.: Wege zum chemischen Nachweis von Abgas- (Rauch-) Schäden mit besonderer Rücksicht auf den Wald. *Tharandt. forstl. Jahrb.,* 93:631–658 (1942).

Munger, T. T.: Parch Blight on Douglas Fir in the Pacific Northwest. *Plant World,* 19:46–47 (1916).

Neger, F. W.: Rauchwirkung, Spätfrost und Frosttrocknis und ihre Diagnostik. *Tharandt. forstl. Jahrb.,* 66:95–212 (1915).

Neger, F. W.: Ein neues untrügliches Merkmal für Rauchschäden bei Laubholzern. *Angew. Botan.,* 1:129–138 (1919).

Neilson-Jones, W.: Further Field Observations on Fused Needle Disease of Pines. *Empire Forestry J.,* 24:235–239 (1945).

Orr, M. Y.: The Effect of Frost on Wood of the Larch. *Roy. Scot. Arbor. Soc. Trans.,* 39:38–41 (1925).

Parker, A. K.: An Unexplained Decline in Vigor of Lodgepole Pine. *Forestry Chronicle,* 31:298–302 (1959).

Parker, J.: Drought Resistance in Woody Plants. *Botan. Rev.,* 22:241–289 (1956).

Patton, R. F., and A. J. Riker: Needle Droop and Needle Blight of Red Pine. *J. Forestry,* 52:412–418 (1954).

Peace, T. R.: An Interesting Case of Lightning Damage to a Group of Trees. *Quart. J. Forestry,* 34:61–63 (1940).

Phillips, F. J.: Effect of a Late Spring Frost in the Southwest. *Forestry and Irrig.,* 13:484–492 (1907).

Plummer, F. G.: Lightning in Relation to Forest Fires. *U.S. Dept. Agr., Forest Serv. Bull.,* 111:1–39 (1912).

Pomerleau, R.: Observations sur quelques maladies non parasitaires des arbres dans le Quebec. *Can. J. Research. C. Botan. Sci.,* 24:171–189 (1944).

Pomerleau, R., and R. G. Ray: Occurrence and Effects of Summer Frost in a Conifer Plantation. *Can. Dept. Northern Affairs Natl. Resources, Forestry Branch, Research Div., Tech. Note,* 51:1–15 (1957).

Poskin, A., and P. Flon: Contribution à l'étude de la gélivure des peupliers euraméricains en Belgique. *Intern. Poplar Comn. 8th Sess. (Madrid) Spec. Rept. No. FAO/CIP/75-D Add.* 1(1954):3–5 (1955).

Quirke, D. A.: White Pine Needle Blight. *Can. Dept. Agr., Forest Biol. Div., Sci. Serv., Ann. Rept. Forest Insect and Disease Survey,* 1953:80 (1954).

Raunecker, H.: Der Buchenrotkern nur eine Alterserscheinung? *Allgem. Forst- u. Jagdztg.,* 127:16–31 (1956).

Redmond, D. R.: Studies in Forest Pathology XV. Rootlets, Mycorrhiza, and Soil Temperatures in Relation to Birch Dieback. *Can. J. Botany,* 33:595–627 (1955).

Rennerfelt, E., and B. Thunell: Undersökningar över bokens rödkärna. [German summary.] *Statens Skogsforskningsinst. (Sweden), Medd.,* 39:1–36 (1950). [*Forestry Abstr.* 12:482–483 (1951).]

Reynolds, R. R.: Lightning as a Cause of Timber Mortality. *Southern Forest Expt. Sta., South. Forestry Notes,* 31:1 (1940).

Rhine, J. B.: Clogging of Stomata of Conifers in Relation to Smoke Injury and Distribution. *Botan. Gaz.,* 78:226–231 (1924).

Rhoads, A. S.: The Formation and Pathological Anatomy of Frost Rings in Conifers Injured by Late Frosts. *U.S. Dept. Agr. Dept. Bull.,* 1131:1–15 (1923).

Rhoads, A. S.: Lightning Injury to Pine and Oak Trees in Florida. *U.S. Dept. Agr., Plant Disease Reptr.,* 27:556–557 (1943).

Riley, C. G.: Hail Damage in Forest Stands. *Forestry Chronicle,* 29:139–143 (1953).

Rol, R.: Les gélivures. *Soc. forest. de Franche-Comté et Provs. de l'Est, bull. trimestr.,* 26:641–651 (1953).

Ruston, A. G.: The Plant as an Index of Smoke Polution. *Ann. Appl. Biol.,* 7:390–402 (1921).

Scarth, G. W.: Cell Physiological Studies of Frost Resistance: A Review. *New Phytologist,* 43:1–12 (1944).

Scheffer, T. C., and G. G. Hedgcock: Injury to Northwestern Trees by Sulfur Dioxide from Smelters. *U.S. Dept. Agr. Tech. Bull.,* 1117:1–49 (1955).

Schrenk, H. von: On Frost Injuries to Sycamore Buds. *Missouri Botan. Garden, Ann. Rept.,* 18:81–83 (1907).

Scott, C. E.: Defficiency Diseases in Trees. *Natl. Shade Tree Conf. Proc.,* 23:363–370 (1947).

Scurfield, G.: Air Pollution and Tree Growth. *Forestry Abstr.,* 21:339–347, 517–528 (1960).

Secrest, H. C., H. J. MacAloney, and R. C. Lorenz: Causes of the Decadence of

Hemlock at the Menominee Indian Reservation, Wisconsin. *J. Forestry,* 39:3–12 (1941).

Setterstrom, C., and P. W. Zimmerman: Factors Influencing Susceptibility of Plants to Sulphur Dioxide Injury. I. *Contribs. Boyce Thompson Inst.,* 10:155–181 (1939).

Smith, C. O., and L. C. Cochran: A Non-infectious Heritable Leaf-spot and Shothole Disease of the Beaty Plum. *Phytopathology,* 33:1101–1103 (1943).

Somerville, W.: Abnormal Wood in Conifers. *Quart. J. Forestry,* 10:132–136 (1916).

Sorauer, P. (transl. from the German by Frances Dorrance): "Manual of Plant Diseases." Vol. I. Non-parasitic Diseases. (1922).

Spaulding, P.: Susceptibility of Beech to Drought and Adverse Winter Conditions. *J. Forestry,* 44:377 (1946).

Spaulding, P., and A. W. Bratton: Decay Following Glaze Storm Damage in Woodlands of Central New York. *J. Forestry,* 44:515–519 (1946).

Spaulding, P., and J. R. Hansbrough: The Dying of Beech in 1934. *Northeast. Forest Expt. Sta. Tech. Note,* 20:1–2 (1935).

Spaulding, P., and J. R. Hansbrough: The Needle Blight of Eastern White Pine. *U.S. Dept. Agr., Div. Forest Pathol.* (1943). (Mimeographed.)

Spaulding, P., and H. J. MacAloney: A Study of Organic Factors Concerned in the Decadence of Birch Cut-over Lands in Northern New England. *J. Forestry,* 29:1134–1149 (1931).

Stahl, E.: "Die Blitzgefährdung der verschiedenen Baumarten." Jena, Germany: Gustav Fischer Verlagsbuchhandlung (1912).

Starr, G. H.: The Control of Chlorosis in Cottonwood Trees and Oher Plants. *Wyoming Agr. Expt. Sta. Bull.,* 252:1–16 (1942).

Staubacher: Die Frostschäden im Forstbetrieb, deren Ursachen und Bekämpfung. *Forstwiss. Centr.,* 46:1–13, 54–66, 98–111 (1924).

Steinmetz, H. F., and M. T. Hilborn: A Histological Evaluation of Low Temperature Injury to Apple Trees. *Maine Agr. Expt. Sta. Bull.,* 388:1–32 (1937).

Stickel, P. W.: Drought Injury in Hemlock-hardwood Stands in Connecticut. *J. Forestry,* 31:573–577 (1933).

Stone, E. L., Jr.: An Unusual Type of Frost Injury in Pine. *J. Forestry,* 50:560 (1952).

Stone, E. L., and G. Baird: Boron Level and Boron Toxicity in Red and White Pine. *J. Forestry,* 54:11–12 (1956).

Stone, E. L., R. R. Morrow, and D. S. Welch: A Malady of Red Pine on Poorly Drained Sites. *J. Forestry,* 52:104-114 (1954).

Stone, G. E.: Electrical Injuries to Trees. *Mass. Agr. Expt. Sta. Bull.,* 156:1–19 (1914).

Stone, G. E.: Shade Trees, Characteristics, Adaptation, Diseases and Care. *Mass. Agr. Expt. Sta. Bull.,* 170:123–264 (1916).

Swingle, R. U.: Chlorotic Dwarf of Eastern White Pine. *U.S. Dept. Agr., Plant Disease Reptr.,* 28:824–825 (1944).

Tanner, H.: Hagelschäden an Waldbaumen und ihre Folgen. *Schweiz. Z. Forstw.,* 104:232–237 (1953).

Thomas, J. B.: Hail Damage in Lake Nipigon Area. *Can. Dept. Agr., Sci. Serv., Forest Biol. Div., Bi-monthly Progr. Rept.,* 12(3):2 (1956).

Thomas, M. D.: Gas Damage to Plants. *Ann. Rev. Plant Physiol.,* 2:293–322 (1951).

Toole, E. R.: Sweetgum Blight. *U.S. Dept. Agr., Forest Serv., Forest Pest Leaflet,* 37:1–4 (1959).

True, R. P., and E. H. Tryon: Oak Stem Cankers Initiated in the Drought Year 1953. *Phytopathology*, 46:617–622 (1956).

Tryon, E. H., and R. P. True: Blister-shake of Yellow Poplar. *West Va. Univ. Agr. Expt. Sta. Bull.*, 350T:1–15 (1952).

Tryon, E. H., and R. P. True: Recent Reductions in Annual Radial Increments in Dying Scarlet Oaks Related to Rainfall Deficiencies. *Forest Sci.*, 4:219–228 (1958).

Tubeuf, C. von: Über den anatomischpathologischen Befund bei gipfeldürren Nadelhözern. III. Die Gipfeldürre der Fichte. *Naturw. Z. Land- u. Forstw.*, 1:417–447 (1903).

Tubeuf, C. von: Frostwirkungen auf Laubblätter. *Naturw. Z. Land- u. Forstw.*, 2:293–295 (1904).

Tubeuf, C. von: Beobachtungen über elektrische Erscheinungen im Walde. VI. Absterben ganzer Baumgruppen durch den Blitz. *Naturw. Z. Land- u. Forstw.*, 3:493–507 (1905).

Tubeuf, C. von: Beobachtungen über elektrische Erscheinungen im Walde. VII. Über sogenannte Blitzlöcher im Walde. *Naturw. Z. Land- u. Forstw.*, 4:344–351 (1906).

Wadsworth, F. H.: Lightning Damage in Ponderosa Pine Stands of Northern Arizona. *J. Forestry*, 41:684–685 (1943).

Wagener, W. W.: Top Dying of Conifers from Sudden Cold. *J. Forestry*, 47:49–53 (1949).

Walker, L. C.: Foliage Symptoms as Indicators of Potassium Deficient Soils. *Forest Sci.*, 2:113–120 (1956).

Ward, R. DeC.: Hot Waves, Hot Winds and Chinook Winds in the United States. *Sci. Monthly*, 17:146–167 (1923).

Warne, L. G. G.: Distribution of Potassium in Normal and Scorched Foliage. *Ann. Botany*, 48:57–67 (1934).

Weber, G. F.: Cold Injury to Young Pine Trees. *U.S. Dept. Agr., Plant Disease Reptr.*, 41:494–495 (1957).

Wells, B. W.: A New Forest Climax: The Salt Spray Climax of Smith Island, N.C. *Bull. Torrey Botan. Club*, 66:629–634 (1939).

Wieler, A.: Die Beteiligung des Bodens an den durch Rauchsäuren hervorgerufenen Vegetationsschaden. *Z. Forst-u. Jagdw.*, 54:534–543 (1922).

Wienhaus, H.: Chemie der Rauchschaden. *Wiss. Z. tech. Hochschule Dresden*, 4(3):19–20 (1954–1955).

Wyman, D.: Salt Water Injury of Woody Plants Resulting from the Hurricane of September 21, 1938. *Arnold Arboretum Bull. Pop. Inform.* IV, 7:45–52 (1939).

Young, H. E.: Fused Needle Disease and Its Relation to the Nutrition of *Pinus*. *Queensland Forest Serv. Bull.*, 13:I-IV, 1–108 (1940).

Zieger, E.: Die heutige Bedeutung der Industrie-Rauchschaden fur den Wald. *Arch. Forstw.*, 4:66–79 (1955).

Zimmerman, P. W.: Chemicals Involved in Air Pollution and Their Effects upon Vegetation. *Boyce Thompson Inst. Plant Research, Professional Paper*, 2(14):124–145 (1955).

Zimmerman, P. W., and A. E. Hitchcock: Susceptibility of Plants to Hydrofluoric Acid and Sulfur Dioxide Gases. *Contribs. Boyce Thompson Inst.*, 18:263–279 (1956).

Zon, R. G.: Effects of Frost upon Forest Vegetation. *Forestry Quart.*, 2:14-21 (1903).

CHAPTER 5

Seedling Diseases

Seedlings are frequently affected by noninfectious and fungous diseases. Young plants are less resistant to disease than older ones, both because of their tender tissues and because they often have difficulty in establishing themselves. Under forest conditions, control measures must be indirect and are now largely impossible of application. In nurseries, direct control measures involving considerable expense are feasible and often necessary. It is doubly important to control disease in nurseries, so that affected trees are not planted out to succumb later or to infect their neighbors. Although little is known about diseases of seedlings growing under natural conditions, seedling diseases in nurseries have been extensively investigated (Davis, Wright, and Hartley 1942), and much is known about suitable soils for nurseries (Wilde 1946). Most of this information concerns coniferous seedlings, since hardwoods have not yet been grown so extensively for forest planting in this country.

Selection of the nursery site is important in controlling disease. Soil is usually so changed, both chemically and physically, by gardens, livestock yards, or human habitation that such sites are unsatisfactory for a nursery for forest conifers. Heavy, wet, poorly drained soils should be avoided. A deep, fresh, light sandy loam meets most requirements, but a soil that is too sandy may induce deficiency diseases and drought. For conifers an acid soil is essential to reduce the incidence of damping-off; consequently a soil just above pH 5.0 is best for most conifers, whereas for eastern red cedar and most broadleaf species the pH range should be from 5.5 to 7.0. An alkaline soil induces damping-off, root rot, and chlorosis in many conifers. Nurseries should be so situated that they are naturally protected from wind, or this should be accomplished by windbreaks. A northerly aspect is usually best because water loss by evaporation is not so rapid and the trees start growth later in the spring, reducing the danger of frost injury. Frost pockets or hollows where early or late frosts are frequent must be avoided. Seed should be collected and nurseries should be located in the same general climatic region in which the species to be grown are to be planted out.

As a protection against attacks by fungi, it is advisable when possible to grow trees in nurseries somewhat removed from natural stands of the same species. If this cannot be done, the natural stand should first be examined, and, if it is infected with a parasite damaging to young trees, measures for the eradication of the parasite in the stand or for its prevention or control in the nursery by spraying or other means should be undertaken. In making an examination of the nursery locality, the alternate hosts of forest-tree rusts must be considered. If the parasite is one that cannot be controlled economically, another location for the nursery should be found. Nursery stock for transplanting and planting out should be grown from seed and not introduced from another region. Stock should not even be interchanged between nurseries in different localities in the same region. To disregard this principle is to risk the introduction of troublesome pathogens into both nurseries and plantations in new regions or localities. Seed can be interchanged with less danger since there are no seed-borne forest-tree diseases of consequence known as yet in this country, and seed, particularly coniferous seed, can usually be disinfected. However, it is inadvisable to sow seed from places with a soil or climate markedly different from that at the nursery.

NONINFECTIOUS DISEASES

Seedlings are particularly subject to many diseases caused by nonparasitic agencies, especially unfavorable weather conditions. Young trees with their relatively large amount of tender tissue and limited root system are much less able to resist heat, cold, and drought than older, larger individuals. Most noninfectious diseases, when their cause is determined, can be controlled in nurseries by cultural methods, but under natural conditions control by silvicultural methods, the only possible means, is difficult or impossible in the present unregulated condition of most of our forests. Consequently these diseases are important limiting factors in the ready establishment of natural reproduction in those forest regions where extremes of temperature occur yearly and the moisture supply is scanty.

HIGH TEMPERATURES

Seedlings and transplants in nursery beds and seedlings under natural conditions are liable to direct injury by excessive heat (Bates and Roeser 1924; Roeser 1932; Daubenmire 1943; Moulopoulos 1947), since their cortical parenchyma cells are killed in 30 minutes by exposure to temperatures between 57 and 59°C (Lorenz 1939). Both conifers and hardwoods suffer. Different species appear to vary somewhat in their resistance to heat, although under controlled laboratory conditions there was little difference in this respect between eastern white, jack, and red pines and white spruce (Shirley 1936). However, eastern white pine and white spruce showed high ability to recover from severe heat injury, whereas jack and red pines were decidedly inferior in recovery power. Heat injury to conifers was induced at a temperature of 120°F, but injury became prevalent and severe at 130°F (Toumey and Neethling 1924:59; Baker 1929:973). Injury occurred even on plants whose roots were in contact with abundant available soil moisture. A surface soil temperature of 175°F has been recorded with a mean of 130°F for an 8½-hour period (Rudolf 1938). Temperatures of 130 to 160°F frequently occur in temperate climates, particularly in loose, sandy, and dark-colored soils. In a soil with a black charcoal surface resulting from fire, there was a 100 per cent loss of Douglas fir seedlings in 3 days, as contrasted to a 16 per cent loss during the same period on a yellow mineral soil (Isaac 1930). Increasing age of seedlings increases their resistance to heat injury. When death is caused by heat alone, affected plants will show some form of girdling injury to the stem at or just above the soil surface, whereas death caused by excessive drying during hot weather will be indicated by general injury to the tissues. Mortality in plantations usually attributed to drought may actually be caused largely by heat (Rudolf 1938).

On the green stems of very young seedlings, short, white, shrunken, watery-looking lesions may appear suddenly, usually just above the soil surface, and originating on the south side of the stems or on the top side of curved or bent stems. This injury has been termed "white spot." Commonly the entire stem is constricted by the lesion, resulting in lopping over and death of the plant. The injury is caused by excessive heat, and entire beds of pine seedlings may be largely destroyed by it. Although the effect on the seedling is similar to that of damping-off, these lesions can be distinguished by their light color, definite boundaries, and limitation to the portion of the stem aboveground according to Hartley (1918:595). Hardwoods as well as conifers may be affected. Fifty per cent shade from the time the seedlings emerge until they are several weeks old will prevent or greatly reduce this injury in nurseries, provided the shade frames are at least 18 inches above the beds and preferably higher so that there will be little interference with air movement. If slat shade is used, the slats should run north and south. During hot quiet days when heat injury is most likely, frequent waterings will serve to keep the surface soil cool.

A disease of older nursery stock or young forest trees, both conifers and hardwoods, termed "basal stem girdle," is ascribed to excessive temperatures of the surface soil, although it has been attributed to *Pestalozzia hartigii* Tub. It is not uncommon to find this fungus associated with stem girdle, probably as a saprophyte, since attempts to prove its parasitism have failed. Two- to four-year-old seedlings and transplants of pines, balsam firs, spruces, Douglas fir, and other conifers are affected. Norway spruce is particularly susceptible.

Fig. 22. Basal stem girdle of a 2-year-old Norway spruce seedling caused by excessive heat at the surface of the soil.

A lesion is formed at the base of the stem, and the girdling and interference with food movement causes a swollen growth of the stem just above the lesion, thus making the lesion itself appear constricted (Fig. 22). The lesion is regularly very short. Affected trees may appear healthy for some time after the lesions develop, but finally they gradually yellow and die, although those not completely girdled may recover. In an English nursery on a light-colored sandy soil, the stock was found free from stem girdle when examined by the writer in September, except in a portion of one bed of 2-year-old Norway spruce where there were heavy losses because charcoal had been added to the surface soil in early summer to prevent damping-off. This blackening of the soil resulted in raising the surface temperature enough to cause injury. Shading will prevent stem girdle. With Norway and Engelmann spruce transplants in a Utah nursery, losses were reduced by inclining the trees slightly to the south when they were transplanted (Korstian and Fetherolf 1921). Transplant rows running north and south, or even better slightly west of south, will reduce stem girdle as compared to rows running east and west.

Injury to black locust seedlings by a tree hopper or membracid, *Stictocephala festina* Say, may be confused with stem girdle caused by heat. The insect in feeding makes a ring of punctures about 1 or 2 inches or more above the ground line around the tender young stems of seedlings (Snyder and Lamb 1935). Many of the younger seedlings wilt and die immediately as a result of the girdling, but, where the girdling is incomplete or on older stems, the trees linger on, and a gall-like swelling or callus is formed above the girdle. Eventually most affected trees succumb. The height of the lesion above the ground distinguishes this at once from heat injury.

LOW TEMPERATURES

Young trees are often injured by low temperature. This injury affects species not adapted to the normal low temperatures of the region to which they have been introduced, as well as species within their natural range when periods of excessively low temperatures occur during the dormant season or unseasonable low temperatures occur during the growing season. The proper maintenance of available potash by soil fertilization is necessary to produce frost-resistant conifers in nurseries (Kopitke 1941).

Frost. Hardwood and coniferous seedlings are subject to frost injury, usually manifest by damage to the unripened tissue above ground. In addition to damaging the tender foliage, there is a possibility that early frosts by freezing the soil may injure roots that are still growing. Hardwoods are more susceptible than conifers to early frosts. In nurseries, damage by early frosts can be largely controlled by avoiding late spring sowing and by cultural methods to discourage growth of the stock in the late summer so that the tender tissues will harden early. Mulching before early frosts occur is effective.

In general late frosts are more serious than early frosts. Damage from late frosts can be avoided by the use of shade to delay too early growth, or, on nights when frost is to be expected, by the use of smudges or heaters, by a temporary mulch of straw or burlap, or by constant sprinkling with water. When frost has occurred unexpectedly, immediate watering of the affected seedlings has apparently reduced the injury in some instances. South exposures should be avoided for species susceptible to frost injury. Above all, nurseries should not be located in frost pockets or hollows where abnormally low temperatures regularly occur in late spring and early fall. Under forest conditions, no direct prevention of frost injury to seedlings is feasible.

Heaving. Young seedlings suffer from heaving caused by repeated freezing and thawing in the upper soil layers with the accompanying expansion and contraction, particularly vertically, owing to the formation of ice crystals and ice lenses (Schramm 1958). The root collar of the affected seedlings may be raised above the soil surface, or the plant may be completely thrown out of the ground. The roots are usually broken off at a depth of a few inches. Naturally reproduced seedlings of ponderosa pine are frequently killed in this way in northern Arizona during an open winter (Haasis 1923). First-year seedlings suffer most. Snow, brush, and ground cover reduce losses from heaving, and in a nursery the trouble can be prevented by mulching (Jones and Peace 1939) or by lowering the water table. A stem girdle of conifers, somewhat similar to that caused by heat, can result from frost heaving of nursery soil (Tryon 1943).

Purple Foliage. Coniferous seedlings in nurseries sometimes develop a bronze to purplish-colored foliage during the winter, which usually disappears during the growing season with little apparent effect on the plants. Seedlings in exposed beds are most likely to discolor. Although the cause of the phenomenon is unknown, it is probably a change in the relative quantities of the different pigments in the foliage brought on by low temperatures. It should not be confused with the lasting purpling and dwarfing that sometimes appears in pine with acute phosphorus deficiency.

WATER DEFICIENCY

Drought takes a heavy toll of seedlings in the relatively arid forest regions in the West, particularly in the ponderosa pine region. In western nurseries without provision for artificial watering, losses may be large during certain seasons. Even where artificial watering is practiced, particularly if water is applied with a hose, losses may result, since the watering may not be sufficiently heavy during a dry season. In one western nursery, losses of 25 per cent of the ponderosa pine and sugar pine seedlings were recorded in August because the water supply furnished by normal sprinkling had been insufficient for the plants during an unusually dry season. Lack of meteorological instruments at the nursery made it impossible to judge conditions correctly and increase the duration of sprinkling. As a rule, seedlings in the tender succulent stage killed by drought will be scattered throughout the seedbeds, whereas those killed by damping-off or other fungous diseases usually occur in definite groups. After the seedlings become old enough to compete for the available water supply, which occurs before the end of the first season in dense stands and where growth is fairly rapid, drought injury most often occurs in strips or patches.

Loss from drought may be cumulative throughout most of a growing season, or it may occur suddenly. This sudden injury in coniferous nurseries, prevalent particularly in the more arid regions, Hartley (1913:2) describes as, "The death of entire seedlings and transplants or parts of them, due to lack of balance between the water absorption by the roots and water loss from the needles during the growing season." Damage may be severe within 48 hours from the first appearance of the injury. When serious, seedlings of all ages are killed outright, but otherwise only part of the plant is killed. The injured needles first become yellowish, next pale-straw color, then gradually turn brown, and finally become nearly red. The fine roots die at the same time or even before the foliage. In hot, dry weather, groups of seedlings in crowded beds, particularly on sandy soil, may be entirely killed very suddenly, and these dead groups may be scattered extensively throughout the seedbeds. In transplant beds the injury does not occur in such distinct patches. Losses can be greatly decreased by partial shade and can be prevented by heavy watering. Crowded beds predispose plants to injury, and sandy soils are most subject to the trouble. An oversupply of nitrogen decreased drought resistance of jack pine and red pine seedlings (Bensend 1943:41; Shirley and Meuli 1939).

Needle curl of pines, from which seedlings usually recover quickly, is an injury induced by a water deficiency just as the needles are emerging from the fascicular sheaths (Jackson 1948).

Winter Drying. This form of drought injury results when the soil and roots are so frozen or cold that water given off by the leaves or twigs cannot be replaced by sufficient absorption from the soil. Essentially it is the same as sudden drought injury, except that it occurs during the dormant season. Although best known on evergreens, deciduous trees may also be affected, water loss from the bare twigs causing death, which becomes apparent when no leaves develop in the spring. Sudden warm weather during the winter or early spring causes this injury, which is not uncommon in western nurseries. Winter drying is most severe in open stands, because in crowded beds the trees protect each other from the drying air and also protect the soil from deep freezing. It is worst during winters with little snow, since the soil is then subject to deep freezing, and the trees are exposed to the air. To prevent winter drying, the nursery beds should be protected by windbreaks and by mulching with straw.

WATER EXCESS

Soil oxygen is necessary for proper development of the roots of most trees. A lack of oxygen will result from excessive soil moisture. The formation of hypertrophied lenticels (Fig. 23), also known as "root warts," on the roots of young conifers is attributed to the effect of unusually wet soil reducing the oxygen supply (Hahn, Hartley, and Rhoads 1920). These excrescences were also found on the basal portions of young trees growing in abnormally wet situations. Tree roots affected by illuminating gas commonly develop the same hypertrophies.

Graves (1915) called attention to heavy losses in nursery-grown seedlings of eastern white pine, red pine, and eastern hemlock from root rot caused by prolonged water saturation of a stiff, clayey soil in the

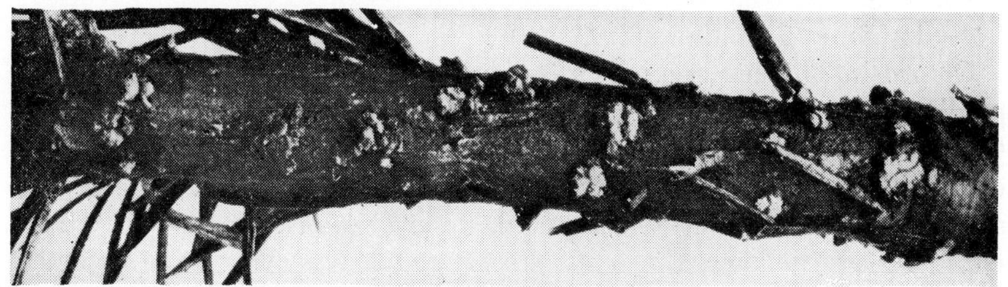

Fig. 23. Hypertrophied lenticels on the basal part of a layering stem of black spruce that had been covered with sphagnum. (*Reproduced from J. Agr. Research* 20, *no.* 4.)

spring, and he charges the death of the trees to a deficiency of oxygen. Although thorough liming of the soil and the addition of more humus was recommended, lime must not be added to a coniferous nursery soil for control because this results in poor growth, chlorosis, damping-off, and root rot. Another nursery site with lighter soil should be chosen, but, if this is not possible, the soil can be lightened by the use of sand or organic matter.

Water excess may predispose some species to infection by semiparasitic fungi.

NUTRITION

A violet color of the foliage of Scotch pine seedlings resulted from a deficiency of phosphoric acid and lime in the soil (Němec 1936). Olson (1930:87) describes a more serious effect of this kind, called "purple top," on western white pine and ponderosa pine seedlings, the buds and uppermost needles of which first turned purple, then dried and became brown. This phenomenon appeared in beds fertilized with dried blood and bone meal, whereas plants in unfertilized beds remained normal.

Zonate chlorosis, or in severe cases reddish-brown to brown discoloration of the foliage and stunting of cedar (*Thuja*), has been attributed to potash deficiency (Meier 1937). Nitrogen deficiency in seedlings of pitch and shortleaf pines was expressed by pallor and severe stunting, in eastern white pine by a tan-colored necrosis, and in red pine by a reddening of the tissues (Hobbs 1944). Phosphorus deficiency resulted in stunting and necrosis of the lower needles, sometimes accompanied by a pinkish-red tinge. Lack of potassium and magnesium also involved changes in the foliage. Needle tip burn of Sitka spruce seedlings, characterized by the tips of the upper needles shriveling and becoming straw-colored, resulted from copper deficiency (Benzian and Warren 1956). Little is known about deficiency symptoms in hardwoods, but the foliage of Siberian elm, ailanthus, and hardy catalpa grown from seed in soil without certain elements varied from pale green to chlorotic (Worley, Lesselbaum, and Mathews 1941).

Chlorosis. This disease, which is characterized by deficiency of chlorophyll in the foliage, occurs on both conifers and hardwoods. Among the hardwoods, black locust seedlings are quite susceptible to it. The disease has appeared in coniferous nurseries in Nebraska, Utah, and southern Idaho, causing significant damage (Korstian et al. 1921). The entire foliage yellows; the roots, stems, and leaves make poor growth: the terminal buds are dwarfed or fail to develop; and chlorotic plants are more susceptible to winter injury. Chlorosis may be due to several causes, some of which are still obscure, but in the Idaho and Nebraska nurseries mentioned it was connected with iron deficiency, probably because the high lime content of the soil and of the water used on the beds lowered the solubility of the iron. In some instances the disease on pines has been controlled by spraying the tops of the plants with a 1 per cent solution of ferrous sulfate in water at 10-day intervals, although in some other instances the disease has not yielded to this treatment. Hardwood foliage, which is easily injured by ferrous sulfate, can be sprayed with ferric citrate or ferric tartrate in ¼ to ¾ per cent solutions. For spruces in dense stands, a chlorosis through the centers of the beds is frequently seen, which responds to nitrogen addition. Ammonium sulfate applied at the rate of 200 to 300 pounds per acre has been beneficial. Ammonium sulfate should be applied during a rain or should be washed off the foliage by immediate sprinkling, otherwise burning may result. If applied late in the growing season, the possibility of winter injury is increased. Its use is also hazardous where damping-off or root rot are known to occur, since it tends to increase the incidence of these diseases. Soils excessively high in calcium carbonate should be avoided for nurseries, especially coniferous nurseries.

TOXIC CHEMICALS

Insecticides, fungicides, herbicides, and fertilizer salts used in excessive concentrations may injure or kill seedlings by acting either directly on the foliage or indirectly through the soil (Voigt 1953; Voigt, Wilde, and Stoeckeler 1958). Even seeds may be affected (Cram and Vaartaja 1955). With the increasing and repeated use of these chemicals more diseases difficult to diagnose can be expected. Particularly severe and recurring injury is caused by arsenicals in the soil.

Arsenic. The larvae of May beetles, known as "white grubs," often cause severe injury in nurseries, especially those just established on grassland or open woodland, by devouring the roots of both conifers and hardwoods. This damage can be prevented by treating the soil with acid lead arsenate (Fleming, Baker, and Koblitsky 1937). However, arsenic is so injurious to both hardwoods and conifers that its use is hazardous. Seedlings and transplants are stunted, the injurious effects of the chemical persisting in the soil for 4 or 5 years or more. Coniferous needles turn red from the tip down for varying distances, the discoloration beginning early in the growing season and developing gradually. Drought injury, with which arsenic injury might be confused, usually occurs rather suddenly.

Although it has been shown by experiments with seedlings grown in nutrient solutions that the toxic effect of arsenic can be inhibited by phosphorus, arsenic injury was not inhibited in clay-loam soil by additions of phosphate, but an inhibiting effect was obtained in sandy loam (Hurd-Karrer 1936). This suggests that applications of phosphates will reduce or prevent arsenic injury in types of soil that permit the phosphates to remain available.

FUNGOUS DISEASES

In this country less is known about fungous diseases of hardwood seedlings than about those of conifers because hardwoods have not been grown so extensively for forest planting, and most hardwoods reproduce satisfactorily under natural conditions so that the seedling diseases that do occur have little or no effect on the development of the stand. Consequently they have not been investigated. Leaf diseases are frequently quite damaging to broadleaf trees in nurseries, but they can usually be controlled by spraying with Bordeaux mixture. Severe losses in ash species resulting from a leaf disease caused by *Marssonia fraxini* Ell. & Davis occur commonly (Davis, Wright, and Hartley 1942:34). *Macrosporium* sp. and *Fusicladium robiniae* Shear are responsible for leaf spots of black locust in the South (Davis and Davidson 1939). Powdery mildew of oaks caused by *Microsphaera alni* (DC.) Wint. might become as severe here as it is in European nurseries if oaks were grown extensively, except that the native oaks have probably a much greater resistance to this indigenous parasite than do the European species. Leaf spots on American elm in nurseries are sometimes severe enough to warrant control by spraying or dusting (Trumbower 1934).

On the other hand there are a number of important fungous diseases of coniferous seedlings, particularly in nurseries, some of which also affect hardwoods. The most serious of these is damping-off.

DAMPING-OFF

Damping-off is a term applied to any disease that results in the rapid decay of young succulent seedlings or other shoots. It is caused by a number of soil-inhabiting fungi that are facultative parasites and not specialized as to host (Hartley 1921).

History and Distribution. Damping-off has been noticed in Europe on tree seedlings since the eighteenth century. It came into prominence in the United States in the early years of the present century when the first attempts were made to grow pine in quantity for forest planting. The disease, primarily found in nurseries and greenhouses, is of worldwide distribution, and among the general publications on it are those of Hansen, Hartley, Pomerleau, Roth, Salisbury, and Ten Houten.

Hosts. A wide range of hosts suffer from damping-off. Many species of conifers are known to be affected, and some hardwoods (Wright 1944). In North America, however, the disease is primarily one of coniferous seedlings. Representatives of all the common genera of conifers have been reported as decidedly susceptible, with the exception of the cedars and their relatives, which have shown a considerable amount of resistance. A few hardwoods, including American elm and black locust, are susceptible. Even with the susceptible trees there will be great variations in susceptibility in different localities, depending on soil and climatic conditions and the fungi encountered.

Damage. Damping-off is an extremely destructive disease, and it causes large losses in coniferous nurseries. Seedlings in the forest are much less affected, unless they are growing under crowded conditions as occurs where they come up from squirrel hoards or artificial seed spots. Damping-off is characterized by rather heavy average losses at many nurseries and by irregular losses. In one season practically all the seedbeds of a species may be destroyed, whereas the following season the damage may be trifling. Furthermore a nursery may remain free from damping-off for many years, and then the disease may suddenly appear.

This suggests the possibility of artificial introduction of these parasites and emphasizes the inadvisability of interchanging stock or soil between nurseries.

Losses are both direct and indirect. There is the actual reduction in the quantity of stock. Then, to maintain production, a much larger area must be sown to seedbeds, with an added cost for seed and sowing, and the expense of caring for the additional seedbeds until transplanting. Indirect losses, though difficult to evaluate, are probably much more serious, since the normal planting program may be temporarily disrupted.

It has been suggested that damping-off may be somewhat valuable as an agent in natural selection by eliminating weak seedlings in the seedbeds and leaving only the best trees for planting, but it is by no means certain that escape from damping-off is correlated with permanently superior vigor.

Symptoms. Damping-off may be defined as fungous invasion leading to the early decay and death of seedlings whose stems are still soft and succulent. Hyphae of the causal fungi spread through the soil, and infection occurs by direct penetration of the tender epidermis covering the succulent tissues of the host. Killing of seedlings when the stems have become stiff enough to continue standing after death, whether from root infection by damping-off fungi or any others in the soil, may be arbitrarily classified as root rot. This killing may even be found in seedbeds during the second season. Since seedlings killed by root rot remain erect, dry up, and turn brown, even shedding their leaves before the stem finally falls over, root rot is easily confused with drought injury.

Two stages of the disease may be recognized: (1) preemergence damping-off, in which the damping-off organisms decay the seed or kill the seedlings before they emerge from the soil (Fisher 1941); and (2) postemergence damping-off, in which the seedlings are affected after they appear above the ground. In postemergence damping-off, if fungous invasion occurs in the roots or lower part of the stem, it can be termed the "soil-infection type," or, if it occurs in the cotyledons or upper part of the stem, it can be termed the "top-infection type," which has also been called "top damping-off." The top-infection type is not common but may be exceedingly destructive in crowded beds following a period of cloudy or rainy weather. Severe losses have occurred to black locust seedlings at several nurseries from the top-infection type (Lambert and Crandall 1936).

In postemergence damping-off of the soil-infection type, the fungi spread rapidly in the tissues, particularly in the roots, and the seedlings either wilt completely or suddenly fall over before wilting. The characteristic lopping over or falling over of the seedlings (Fig. 24) is not the normal type of wilting resulting mainly from a stoppage of the water supply but most commonly occurs when the tissues of the stem just above the soil become decayed while the rest of the stem is still fairly turgid. A bed of seedlings may be completely destroyed within a few days when there are many foci of the disease scattered through it.

Causal Agency. The disease is caused by a number of fungi that live saprophytically in the upper layers of the soil, but which, under favorable conditions, may become virulently pathogenic. Fungi known to be causal

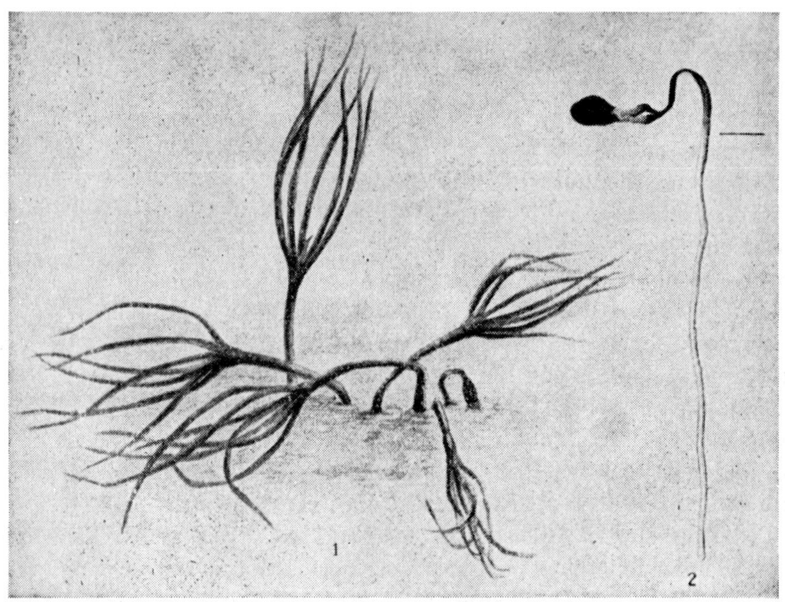

Fig. 24. Symptoms of damping-off on ponderosa pine. (*Reproduced from J. Agr. Research* 15, *no.* 10.)

agents occur in the following major groups: Phycomycetes—*Phytophthora* spp., *Pythium* spp.; Fungi Imperfecti—*Botrytis cinerea* Pers., *Diplodia pinea* (Desm.) Kickx, *Cylindrocladium scoparium* Morg., *Fusarium* spp., *Pestalozzia funerea* Desm., *Rhizoctonia solani* Kühn, and *Sclerotium bataticola* Taub. The most important of these on coniferous seedlings in the United States are *Pythium* and *Rhizoctonia*. Further investigations will add new ones to those now known.

Control. It is impossible to give exact measures for control that will be applicable under all conditions, because damping-off is caused by such a number of widely different fungi that there is necessarily no constant relation between environmental factors and the disease or between various fungicides and their effect on the fungi. Under carefully controlled conditions, *Pythium irregulare* Buis. and *Rhizoctonia solani* attacking red pine were influenced especially, but differently, by soil reaction, temperature, and moisture, so that within limits factors that favored one discouraged the other (Roth and Riker 1943). *Rhizoctonia* was encouraged by more acid soil, dry soil, and warm weather, whereas *Pythium* was favored by more nearly neutral soil, moist soil, and warm weather, but *Pythium* also caused considerable loss at low temperatures. Furthermore, different isolates within species are known to vary in virulence (Hartley 1921:28; Tint 1945:435).

Heavy or excessively wet soil should be avoided and beds kept well ventilated. Nurseries should be located on well-drained sites. It seems unlikely that the damping-off potential of new nursery sites can be predicted by an assay of the microbial population of the soil (Wright 1945).

Broadcast sowing is generally better than drill sowing to prevent damping-off because the tendency in the V-shaped drill is to crowd the seed too closely. However, if flat drills an inch or more broad are used, crowding is less important. Thick sowing should be avoided in any method used. Covering the seed with subsoil sometimes has value. Fall sowing or early spring sowing in northern nurseries where the soil remains practically continuously frozen is less liable to result in loss than late spring sowing.

As soil acidity is generally unfavorable to damping-off (Jackson 1940), coniferous nurseries should have an acid soil. Lime, wood ashes, or any substance that may neutralize the acid should not be used on the seedbeds. Dried blood, poorly rotted manure, or any other materials high in nitrogen are likely to favor the disease. Fertile humus and low light intensity favor damping-off (Vaartaja 1952). If peat is used for composting it should be strongly acid, because weakly acid or neutral peat encourages damping-off (Wilde 1946:182). A cover crop turned under less than 1 month before sowing increases the hazard of damping-off. The choice of a cover crop is important. For broadleaf seedlings it was found that damping-off caused by *Rhizoctonia* and *Pythium* was worse following legumes than following cereals; corn, wheat, and oat soil resulted in the least damping-off (Wright 1941). However, conifer seedlings, especially red pine, following buckwheat suffered severely from damping-off and root rot caused by *Fusarium*, the trouble still persisting 4 years after the buckwheat was cut (Eliason 1938). If a nursery soil is too heavy, sand or organic matter should be mixed with it but not lime (page 79). A nursery soil may be unwittingly made alkaline, thus increasing losses from damping-off and root rots, by covering the seed with alkaline sand or by using alkaline water (Young, Davis, and Latham 1937). There is no practical method known for acidifying, as uniformly as would be needed, soils that have been alkalized by the addition of sand.

Where damping-off is prevalent, i.e., when losses exceed 10 to 15 per cent annually, direct methods of soil disinfection are essential. Heat-treatment with live steam or other medium (Newhall 1955) is rarely practical in forest nurseries. Burning wood on the beds to disinfect them creates wood ashes, which ultimately favors damping-off. Chemical treatments of the soil surface are widely used with reasonable success. One-twelfth to three-sixteenths fluid ounce of sulfuric acid dissolved in a pint of water and then applied to each square foot of seedbed immediately after the seed has been sown and covered has proved valuable for the more susceptible conifers but is injurious to broadleaves. If applied later, severe injury to the seedlings may result. The amount of acid used depends on the acidity and absorptive capacity of the soil. This solution is corrosive and hence inconvenient to handle. Furthermore the acid must be poured into water; water must not be poured into the acid. However, it is so efficient in controlling weeds that reduced weeding charges may more than pay for the cost of treatment. Commercial phosphoric acid has shown promise (Davis, Wright, and Hartley 1942:67). It is safer than sulfuric acid and may be valuable at nurseries where phosphorus is needed for fertilizer. The necessary volume is about 2½ times that of sulfuric acid. Treatment with aluminum or ferrous sulfate is generally more satisfactory than sulfuric acid in coniferous nurseries. The powder, which is not corrosive, is applied just after sowing at the rate of ¼ to 1¼ ounces per square foot, depending on the texture and buffer value of the soil; the heavier the soil and the less total acid it has, the more chemical is required. It is usually dissolved in not less than 1 pint of water per square foot and applied by sprinkling. Until the seedlings are well up, beds treated with aluminum sulfate or sulfuric acid should be kept thoroughly moist because, if the soil surface becomes dry even for a few hours, serious root injury may result. Aluminum sulfate is not so effective as sulfuric acid in controlling weeds, and ferrous sulfate has practically no effect on them. The strongest commercial formaldehyde at the rate of ¼ to ⅜ fluid ounce per pint of water to 1 square foot of seedbed is usually effective and is the safest disinfectant to use in nurseries where no previous experience is available. The covering soil should also be treated. The solution should be applied 4 or 5 days before sowing under usual conditions, but, where the soil is cold, 10 days should elapse between treatment and sowing. With this chemical no special watering is needed during the period of germination and early growth. Acetic acid (Doran 1932), Semesan (Johnson and Linton 1942), basic slag (Gray 1931), zinc compounds, and Bordeaux mixture as soil treatments have all succeeded in some places. Severe damping-off and root rot of pines following green manuring has been controlled by soil fumigation with methyl bromide at the rate of 2 to 4 pounds per 100 square feet of nursery bed (Wycoff 1955). In addition, the seedlings were more vigorous and weeds were effectively suppressed. The use of this poisonous chemical is highly hazardous unless proper safety precautions are taken. Soil treatment with commercial fungicides based on thiram, zineb, and captan are promising, and pelleting seeds with large doses of these compounds has given better control than ordinarily obtained with seed treatments (Vaartaja 1956*a*). To determine the correct treatment for a given nursery, it is essential to test the chemicals in varying amounts on small plots, with nearby untreated plots receiving the same amount of seed for comparison.

Deep acidification to the depth of plowing may be necessary where the soil is so alkaline that the pH value is near 8, in order to bring the value down to about 5.5. Acid peat, humus, sulfuric acid, ferrous sulfate, or aluminum sulfate can be used. To determine the amount to be applied, different amounts will have to be mixed into small test plots and the pH value of the soil determined. After acid or sulfate treatment the soil will reach equilibrium in about 3 days, but a month or more may be needed if acid peat or humus is the acidifying agent. After the soil is acidified, deep treatment should not be needed again, but for effective control it may still be necessary to use surface treatment with a chemical yearly.

To prevent damping-off of the top-infection type, excess moisture, alkaline soil, and the use of unrotted manure should be avoided. In nurseries whose previous history shows top infection to be likely, the seedlings after emergence should be sprayed regularly during the first season with 4–4–50 Bordeaux mixture containing a wetting agent.

As longleaf pine seedlings, because of their persistent rosette habit, have a considerably lengthened period of susceptibility to damping-off, the usual chemical treatments are not satisfactory, but sprinkling infected seedlings with Semesan at the rate of $\frac{1}{10}$ ounce in 1 pint of water per square foot when the loss first begins has been effective (Davis 1941). To avoid injuring the seedlings, the disinfectant should be applied in late afternoon or on cool or cloudy days. In most nurseries, shallow sowing to a depth of about $\frac{1}{4}$ inch, covering the seed with a well-decayed sawdust mulch instead of soil or sand, and removal of sand and dust from the crowns of seedlings will give sufficient control.

For broadleaf trees, soil acidification is not particularly effective. The rows may be ridged after sowing and the ridges removed by hand raking as soon as the seeds begin to germinate (Wright 1944). This provides sufficient moisture for good germination and allows the seedlings to emerge rapidly so that preemergence losses are reduced. The most practical control is shallow sowing as early as possible in the spring. Rotation of susceptible hosts with resistant ones helps to reduce losses. Damping-off of poplar seedlings that could not be controlled by seed treatments, soil additives, or steaming the soil was prevented by sowing the seed in finely divided sphagnum moss (Shea and Kuntz 1956).

It is known that competing organisms in the soil, notable *Trichoderma lignorum* (Tode) Harz, exert a limiting influence on the growth of damping-off fungi (Hartley 1921:82; Allen and Haenseler 1935; Gregory et al. 1952; White 1956). Further investigation may show that biological control of damping-off can be used in some instances in forest nurseries.

TOP KILLING

Hardwood and conifer seedlings are occasionally injured or killed by damping-off fungi that attack the tops after the plants have developed stems stiff enough to continue standing after death. The disease is most severe during moist seasons in dense stands in which the tops of the seedlings are in contact with each other. *Rhizoctonia* caused considerable injury of this type to 1- and 2-year-old red spruce seedlings in a West Virginia nursery, a conspicuous symptom of affected seedlings being angular bending of the stem tips (Latham 1936). The disease was checked when the shade was removed permanently from the 2-year stock and temporarily from the 1-year stock during rainy weather. *Diplodia pinea* (Desm.) Kickx has killed first-year seedlings of Douglas fir and pines, the infection starting in the uppermost or youngest parts of the seedlings (Slagg and Wright 1943*a*).

SORE SHIN

This disease, which appears on both broadleaf and conifer seedlings not less than 1 year old, is characterized by apparently fungous-induced lesions on one side of the stem at or up to 1½ inches above the soil surface or occasionally just below it (Davis, Wright, and Hartley 1942:32, 62). The lesions resemble those caused by heat. Spraying the stems with 4–4–50 Bordeaux mixture is advised for control.

ROOT ROTS

Damping-off fungi may continue their activities by causing root rot in nursery seedlings even into the second or third season (Crandall and Hartley 1938; Davis, Wright, and Hartley 1942:54). The same factors that favor damping-off will favor root rot. A previous crop of buckwheat that resulted in severe damping-off of conifer seedlings caused by *Fusarium* also led to root rot by the same fungus (Eliason 1938). Control of root rots has been effected in some instances by aluminum or ferrous sulfate applied 3 weeks after emergence in the same amounts as, or one-quarter larger than, those used against damping-off, dissolved in at least 1 pint of water per square foot of seedbed. Root rot of longleaf pine seedlings, probably caused by a nematode-fungous complex, was controlled by fall sowing, but control for slash, loblolly, and shortleaf pines required soil treatment with a 20 per cent by volume solution of ethylene dibromide at the rate of 24 gallons per acre applied 2 to 3 weeks before spring seeding (Henry 1953).

Phytophthora cinnamomi Rands causes a dry type of root rot of hardwoods and conifers, accompanied in conifers by resin flow from the bark and resin deposition in the bark and wood (Crandall, Gravatt, and Ryan 1945:173). Infected tissue turns reddish brown except in black walnut, where it becomes black. Diseased tissues beyond the completely infected portion appear as irregular wedge-shaped streaks, and these streaks may form in the stem for a short distance above the ground, although the disease is essentially confined to the roots. The first symptom on conifers is a gradual loss of color by the needles, but on hardwoods it is a sudden wilting of the entire seedling.

Diplodia pinea has caused root rot and death of 3- to 5-year-old red pines in nurseries and of 5- or 6-year-old eastern white pines in a plantation (Crandall 1938). The area attacked extended from the root collar to a point below the soil line. The affected bark was deep red with black streaks continuing into the wood. The fungus can also attack other conifers (Anonymous 1938).

Rhizina inflata (Schaff.) Sacc. (*R. undulata* Fr.; Ascomycetes, Rhizinaceae), causing Rhizina root rot of coniferous seedlings, is widespread but localized. In the Northwest, seedlings up to 5 years old, particularly on burned areas, are killed in isolated, roughly circular patches 2 to 4 feet in diameter. In Europe nurseries have been invaded, and in a nursery in Maryland there was slight damage to 2-0 red pine and white spruce in seedbeds on or near which slash had been burned (Davidson 1935). Infected roots are closely matted together with white mycelium and are more or less resinous. The annual fructifications, developing on the ground, are irregular in shape, up to 2 or 3 inches across, and have an undulating brown upper surface with a narrow white margin. No method of control is known, but the disease is of minor importance.

Fomes root rot (page 108), shoestring root rot (page 104), Phymatotrichum root rot (page 114), and root rot caused by *Polyporus schweinitzii* (page 436) also occur on seedlings.

SNOW BLIGHT OF CONIFERS

Snow blight or Phacidium blight of conifers is caused by *Phacidium infestans* Karst. (Ascomycetes, Phacidiaceae) or forms very similar if not identical with it (Faull 1929, 1930). The fungus is native to the northeastern United States and eastern Canada, where it has been found commonly on spruces and balsam firs. Other conifers, including pines, are attacked. In Europe *P. infestans* is of practical importance to Scotch pine only (Björkman 1948). In nurseries it spreads rapidly and causes great losses. In natural stands the damage is usually not so severe because the young trees are not crowded as in nurseries, and on balsam firs it

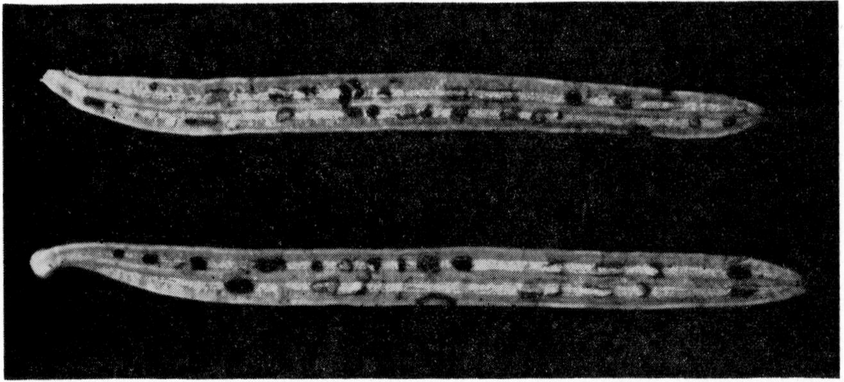

Fig. 25. Apothecia of *Phacidium balsameae* on the underside of balsam fir needles. The light-colored fructifications are still covered by the epidermis of the needle.

kills the previous season's needles or older, not affecting the buds. Furthermore, many of the saplings in natural stands have their leaders above the surface of the snow.

Symptoms. The parasite attacks foliage under the snow, and damage is proportionate to the depth and persistence of the snow cover. Affected foliage assumes a soft and more or less glaucous-brown color. In closely planted nursery beds the disease occurs in subcircular patches that may be 2 feet or more in diameter, or in strips where the plants are in fully separated lines. The line of demarcation between the dead brown trees and the healthy green ones is sharp. Just as the snow melts away, a covering of white mycelium may be observed on the browned foliage.

Causal Agency. In the summer and fall the subepidermal dark-brown to almost black disclike apothecia appear on the underside of needles browned the preceding spring (Fig. 25). The hyaline oval-shaped ascospores are discharged from the clavate asci during periods of mild, damp weather until winter sets in. Tiny black microsclerotia, very resistant to drought or other unfavorable conditions, are also produced abundantly on the surface of affected needles. Presumably they belong to *P. infestans*.

Primary infection occurs in the fall by means of wind-disseminated ascospores that must serve to spread the parasite for considerable distances. It is probable that the microsclerotia also function like spores. In addition an important cause of infection is the mycelium in the browned needles. During the latter part of the winter, for a few weeks before the snow departs, this mycelium grows out under the snow onto adjacent foliage and enters dormant, healthy needles.

Control. Late fall spraying with dormant (strong) lime-sulfur has controlled the disease effectively in nursery beds. Spring spraying was not effective. For control in plantations, only positively healthy stock should be planted, which should first be sprayed or dipped in lime-sulfur. Infected trees in established plantations should be removed and burned. Fall spraying with lime-sulfur should be effective in infected plantations, but the cost may preclude this practice. Dusting, instead of spraying, might be successful, and it can be done at a much lower cost. In Sweden, snow blight is particularly severe on Scotch pines grown from seed of unsuitable origin for the planting site and in dense plantations. Seed from somewhat north of the locality where plantations are to be established is recommended.

A similar disease occurs on Norway spruce in Sweden, but the causal fungus is not yet determined. Red patch of white spruce seedlings, appearing in nurseries in western Canada after the snow melts, has symptoms much like snow blight (Vaartaja 1956*b*). The disease kills needles, twigs, and small seedlings. The cause is unknown, but the disease can be controlled by sprinkling the seedlings with Semesan in September at the rate of 100 pounds per acre.

Other Phacidiums are reported as causing needle blight on conifers (Davis 1922:424; 1924:280; Dearness 1926:237–239; Sydow and Petrak 1924:392). As far as is known, they attack only needles covered by snow so that, even though foliage on lower branches or larger trees may be infected, only seedlings and small saplings are severely injured. *Phacidium infestans* Karst, var. *abietis* Dearn, has been epidemic in northern Idaho and eastern Oregon killing reproduction of lowland white fir, alpine fir, and Douglas fir (Weir 1916). *P. balsameae* Davis is widespread on balsam fir in the Northeastern states, killing or severely suppressing seedlings and small saplings. It has also been reported on white fir in Oregon and on alpine fir in Colorado. *P. abietinellum* occurs on balsam fir, *P. expansum* Davis on black spruce, *P. taxi Fr.* and *P. taxicolum*

Dearn. & House on eastern yew, *P. planum* Davis on eastern white pine and limber pine, and *P. convexum* Dearn. on pitch pine.

MOLDING AND SMOTHERING

The leaves of conifers tightly packed for shipping or covered for prolonged periods by a mulch or by snow may be injured by weakly parasitic fungi. The roots remain healthy for several weeks after the aerial parts of the seedlings are dead (Hartley, Pierce, and Hahn 1919). Large losses from molding and rotting of the roots of broadleaf nursery stock stored over winter have been common, induced by freezing injury to the roots at the time of lifting, in transit, or in inadequate storage (Young 1943).

In coniferous nurseries at high elevations where deep snow covers the ground from fall to late spring, so that the soil does not remain frozen, heavy winter losses from molding may occur in seedlings and transplants. At a Utah nursery, elevation 7450 feet, losses up to 65 per cent of the Douglas fir seedlings and transplants occurred under the snow (Korstian 1923). Other conifers were also affected, but lodgepole pine was practically immune. The disease was checked as soon as the trees were exposed to sunlight in the spring. *Botrytis cinerea* Pers. seemed the most likely causal organism. Satisfactory control was secured by placing a loose framework of boards over the beds so that the weight of the snow could not press the trees to the ground.

Young trees in plantations or naturally reproduced may also suffer, although the losses are much less severe than in nurseries. Spring planting, with large, vigorous stock, is advisable where snow molding may occur.

Although covering nursery beds with a mulch during the winter is necessary in certain regions to prevent injury by heaving or winterkilling, yet severe losses may occur from molding while the mulch is on the beds or immediately after it is removed (Hartley 1913:18). To prevent this disease, avoid fine material that packs tightly, use no more mulch than is needed, and remove it before warm weather begins. Spraying the trees with lime-sulfur before the mulch is put on may be helpful.

Brown felt blight, caused by *Herpotrichia nigra* Hart. and *Neopeckia coulteri* (Pk.) Sacc., which develops under the snow, smothering the foliage of conifers, is frequent in mountainous regions at high elevations (Sturgis 1913). It has not been found damaging coniferous nurseries in this country, although it has caused damage in European nurseries. A somewhat similar disease, but with grayish-brown mycelium, caused by *Rosellinia herpotrichoides* Hepting & Davidson has killed Douglas fir seedlings in a British Columbia nursery (Salisbury and Long 1955).

GRAY MOLD BLIGHT

This disease of the current season's shoots of conifers is caused by *Botrytis cinerea* Pers., with which may be included *B. douglasii* Tub. (Fungi Imperfecti, Moniliaceae). It occurs on seedlings and the lower branches of saplings up to about 5 feet high (Smith 1900; Weir 1912). In the northwestern United States it is found most commonly on Douglas fir, although lowland white fir, alpine fir, western larch, and western hemlock are also affected. The aggregate loss is inconsequential. The fungus has also been associated with a blight of Norway and red spruce seedlings in a North Carolina nursery (Graves 1914:65).

Fig. 26. Gray mold blight of red fir caused by *Botrytis cinerea*. The needles at the tip of the infected twig are curled and withered in contrast to the normal needles on the uninfected shoot.

The most characteristic symptom is the withering, curling, and dying of the current season's shoots (Fig. 26), which may be confused with frost injury. Under moist conditions, a more or less luxuriant aerial mycelium develops, forming a gray mold over the affected twigs and leaves. Wind-disseminated spores are borne abundantly in clumps on the ends of slender upright hyphae. Late in the season, small black somewhat globose bodies about the size of a pinhead, the sclerotia, are formed on the infected twigs and leaves. The parasite overwinters in this form, and in the spring many upright hyphae bearing clumps of spores develop from the sclerotia. Abundant atmospheric moisture favors the fungus, and therefore any measures resulting in better air circulation will help to control it.

BROWN SPOT NEEDLE BLIGHT

Brown spot or brown spot needle blight of pine seedlings (Siggers 1944; Lightle 1960) is caused by *Scirrhia acicola* (Dearn.) Siggers. Brown spot occurs wherever longleaf pine grows in the United States. It is found in all the Coastal states from Virginia to Texas. It probably extends northward as far as the range of shortleaf pine. It has been recorded on other pines, including knobcone pine in Oregon and eastern white pine in the South. The pathogen may attack the foliage of large trees, especially loblolly pine (Boyce 1952). It is by far most serious on longleaf pine seedlings. Where uncontrolled, the disease may kill up to 40 per cent of the seedlings, and the remainder will be so dwarfed that height growth may be delayed ten or more years.

Symptoms, Effects, and Causal Agency. Small, irregularly circular spots of a light gray-green color first appear on the needles. These spots change rapidly to a brown color and then encircle the needle as a narrow brown band, and finally the tips of the needles die as a result of multiple infections. Many brown spots may form on the needles throughout their length, killing them the first season after infection. The elongated curved two- to four-celled brown conidiospores are developed in acervuli throughout the year, and these acervuli appear as very small dark-colored to black elongated spots on the dead portion of the needles. Local dissemination is principally by conidia carried in raindrops splashing from diseased foliage. The air-disseminated, hyaline, oblong-cuneate, unequally bicellular ascospores, with two prominent brown oil drops in each cell, also develop during any month of the year and are probably the means by which the fungus spreads to disease-free stands (Henry 1954). The current season's growth of needles is attacked successively, so that the trees are nearly defoliated yearly, resulting in death or marked retardation of growth. At least three successive annual defoliations are required to kill longleaf pine seedlings. Thrifty seedlings are as readily attacked and killed as less vigorous ones. Dissemination of the brown spot fungus is favored by wet weather, and the disease is more severe on trees in unburned areas because of the large amount of inoculum present.

Control. The judicious use of ground fire is the best method to control brown spot. Fire also creates certain other conditions favorable to the establishment and development of longleaf pine reproduction. Controlled winter burning at three-season intervals, until a sufficient number of seedlings are above 18 inches high (the brown spot danger level), is the best procedure in natural reproduction.

The disease must be controlled in nurseries in order to avoid early infection of plantations. Control can be accomplished by spraying about every 2 weeks, from the last of May into October or November, with 4–4–50 Bordeaux mixture. The spray should be applied at the rate of 4 gallons per 1,000 square feet of bed.

Where the disease is severe, plantations can be protected by burning the planting site a few weeks before planting, which facilitates field work and destroys spores of the brown spot fungus on any diseased seedlings present. Then the planted seedlings may require two semiannual sprays of Bordeaux mixture during each of the first two growing seasons. Although this is expensive, replanting one or more times may be costlier.

OTHER NEEDLE DISEASES

Red spot of Austrian, maritime, and other hard pines, known chiefly in Ohio, is caused by an undescribed fungus (Davis, Wright, and Hartley 1942:62). *Septoria spadicea* Patterson & Charles is associated with a needle disease of eastern white pine in New England somewhat similar to brown spot, resulting in the death of the apical portion of the leaf for one-fourth or one-third of its entire length (Spaulding 1909:4). Trees of all ages from 4 years upward are attacked, and all portions of the crown may be involved. *Ascochyta piniperda* Lind. (*Septoria parasitica* Hart.) has been associated with a blight of the young shoots of red and Norway spruce seedlings in a North Carolina nursery (Graves 1914:64). The fungus fruited on the stems. *Hypoderma lethale* and *Lophodermium pinastri* are somewhat prevalent on loblolly pine and slash pine in the South.

In northern Europe a serious needle cast of pines caused by *Lophodermium pinastri* (Schrad.) Chev. is prevalent in Scotch pine nurseries. The fungus (page 166) is abundant on pines in natural stands throughout the United States, and it has been found killing from 10 to 50 per cent of the seedlings and transplants of red pine in several nurseries in Massachusetts (Spaulding 1935). Eastern white pine was not affected. After heavy frosts in October, the needles turn dull green or purplish red and appear dry. Dead spots form on the diseased needles the following spring, and the needles die back from the tip. Two types of fructifications develop on the dead spots during the spring and summer, and the infected needles fall prematurely, weakening or killing the plants. The late fructifications may appear while the needles are still on the trees or after the needles have fallen. Original infection of nursery stock apparently comes from adjacent pine stands. The disease is favored by mild winters and wet summers.

The disease can be controlled by spraying the nursery stock with double strength (4–4–25) Bordeaux mixture or with lime-sulfur (2 per cent). A sticker should be added to the lime-sulfur. The first application should be made when the new needles are about half grown and the last at about the time growth ceases, with one or occasionally more intermediate applications depending on the weather, frequent rains necessitating spraying at shorter intervals. Spraying should be done in dry weather, and the treatment is effective only so long as the needles remain a bluish color from the spray deposit.

SMOTHERING DISEASE

Thelephora terrestris Ehrh., *T. caryophyllea* (Schaef.) Fr., and *T. fimbriata Schw.* (Basidiomycetes, Thelephoraceae) sometimes smother both coniferous and deciduous seedlings in nurseries and under natural conditions (Anonymous 1906; Weir 1921). Smothering or strangulation is caused by the fruit body of the fungus growing up from the soil around the lower stem of the seedling, and sometimes only the tip of the seedling is left projecting. No living part of the seedling is invaded. The mycelium ramifies in the surface soil and merely penetrates the dead outer tissues of any plant encountered. The disease usually occurs where seedlings are dense enough to shade the ground, but the damage is generally insignificant. Most affected seedlings can be safely transplanted from nursery beds, but it may be necessary to hold such stock in the transplant beds for an extra year until indications of the fungus disappear.

JUNIPER BLIGHT

Juniper blight is caused by *Phomopsis juniperovora* Hahn (Fungi Imperfecti, Sphaerioidaceae), and it is most commonly encountered in nurseries, rarely on wildlings (Hahn 1926, 1943). The disease has been known since 1896 in the Middle West, where it is most serious, and it is distributed east to the Atlantic Coast and south to Alabama. Eastern red cedar, Rocky Mountain red cedar, and their horticultural varieties are particularly susceptible, but other junipers and cedars are also attacked. Closely related strains of the fungus have been found on conifers belonging to several other genera.

The disease works in both seedling and transplant beds. It frequently ruins the value of ornamental plants. Moist seasons favor it. The disease may become epidemic, often destroying entire beds. The appearance of the stock is like that killed by drought. However, on the stems of the diseased plants lesions indicating that the disease is of parasitic origin are found. Lesions on woody stems are often limited, and partial healing may occur, indicating ultimate recovery. The fungus forms definite black pycnidia up to 0.5 mm in diameter, both on the leaves and on the stem lesions.

Since blight once established in a nursery is there to stay, do not bring in nursery stock, especially cedars, from elsewhere. The following control measures are applicable to nurseries in the Great Plains and adjacent areas (Slagg and Wright 1943*b*, 1944). Seedbeds should be well drained. Watering by ditch irrigation is preferable to overhead sprinkling. Since seedlings in thinly sown beds are most liable to blight, about 100 seedlings per square foot should be grown if the stock is to be lifted after 1 year's growth or 50 to 75 per square foot if it is to remain 2 years. All diseased seedlings should be removed and burned early in the season when infection is light. Older trees and hedges at the nursery should be kept covered with a fungicidal spray throughout the growing season, and their blighted parts should be cut out and burned. If possible cedar seedbeds should be located at some distance from older cedar trees and hedges. The location of cedar seedbeds should be changed frequently. Cedar branches and needles must not be used for mulching cedar seedbeds. Control by spraying has proved difficult. Good results have been reported from spraying each week from June 1 to Oct. 1 with Special Semesan, Bordeaux mixture, or Fermate. The first named appears most effective, but regular spraying is expensive, and if either of the first two sprays is used exclusively and continually there is a possibility of poisoning the soil.

Monochaetia sp. has been associated with a similar disease of Port Orford cedar (Hotson and Stuntz 1934). *Phoma sp.* has been associated with a blight of young shoots of red and Norway spruce seedlings (Graves 1914:66).

RUSTS

The foliage or stems of seedlings are frequently attacked by rust fungi (pages 172 and 201). Forest-tree rusts are practically confined to conifers, and all of them, with rare exceptions, are heteroecious, i.e., they require an alternate host to complete their life cycle. Needle rusts rarely kill, but do reduce the growth rate of young trees. Stem rusts, also known as "blister rusts," nearly always kill seedlings. A destructive outbreak of sweetfern blister rust (*Cronartium comptoniae* Arth.) occurred on ponderosa pine in a Michigan nursery (Kauffman and Mains 1915) situated in the midst of a large patch of sweetfern, the alternate host. The western gall rust (*Peridermium harknessii* (Moore) Mein.) was found infecting ponderosa pine transplants in a southern Oregon nursery located in an infected overmature stand of the same species. This blister rust can infect directly from pine to pine. Paintbrush blister rust caused some loss to ponderosa pine in a Montana nursery (Olson 1930:86), but infections were eradicated from ponderosa and lodgepole pines in the forest stand around the nursery, which probably so reduced the amount of infection and subsequent spore production on paintbrush (*Castilleja miniata* Dougl.) that the disease ceased to be of consequence on the nursery stock. Eastern gall rust has been found on nursery stock of Austrian, jack, and shortleaf pines. In southern nurseries up to 20 per cent of 1-0 slash and loblolly pines have been infected by the southern fusiform rust. White pine blister rust (*Cronartium ribicola* Fisch.) has caused severe losses in nurseries.

Spraying the trees with Bordeaux mixture, at the start of the period when telia on the alternate host are discharging spores, should prove effective if telial production is not too heavy in the immediate vicinity of the nursery. The only sure control is to eradicate the alternate host around the nursery, which, however, is frequently impracticable. A radius of ½ mile will usually be sufficient. An exception is white pine blister rust carried by the European black currant (*Ribes nigrum* L.) or, in the West, by certain other very susceptible ribes, when a mile radius will be necessary.

REFERENCES

Allen, M. C., and C. M. Haenseler: Antagonistic Action of *Trichoderma on Rhizoctonia* and Other Soil Fungi. *Phytopathology,* 25:244–252 (1935).

Anonymous: A Tree Strangling Fungus. *J. Board Agr.* (*London*), 12:690–692 (1906).

Anonymous: *Sphaeropsis ellisii* on Conifers. *New Jersey Agr. Expt. Sta., Dept. Plant Pathol., Nursery Disease Notes,* 10:43–46 (1938).

Baker, F. S.: Effect of Excessively High Temperatures on Coniferous Reproduction. *J. Forestry,* 28:949–975 (1929).

Bates, C. G., and J. Roeser, Jr.: Relative Resistance of Tree Seedlings to Excessive Heat. *U.S. Dept. Agr. Dept. Bull.,* 1263:1–16 (1924).

Bensend, D. W.: Effect of Nitrogen on Growth and Drought Resistance of Jack Pine Seedlings. *Minn. Agr. Expt. Sta. Tech. Bull.,* 163:1–63 (1943).

Benzian, B., and R. G. Warren: Copper Deficiency in Sitka Spruce Seedlings. *Nature* (*London*), 178:864–865 (1956).

Björkman, E.: Studier över snöskyttesvampens (*Phacidium infestans* Karst.) biologi samt metoder for snöskyttets bekämpande. [English summary.] *Statens Skogsforskningsinst.* (*Sweden*), *Medd.,* 37:1–136 (1948).

Boyce, J. S., Jr.: A Needle Blight of Loblolly Pine Caused by the Brown-spot Fungus. *J. Forestry,* 50:686–687 (1952).

Cram, W. H., and O. Vaartaja: Toxicity of Eight Pesticides to Spruce and Caragana Seed. *Forestry Chronicle,* 31:247–249 (1955).

Crandall, B. S.: A Root and Collar Disease of Pine Seedlings Caused by *Sphaeropsis ellisii. Phytopathology,* 28:227–229 (1938).

Crandall, B. S., G. F. Gravatt, and Margaret M. Ryan: Root Disease of Castanea Species and Some Coniferous and Broadleaf Nursery Stocks, Caused by *Phytophthora cinnamomi. Phytopathology,* 35:162–180 (1945).

Crandall, B. S., and C. Hartley: *Phytophthora cactorum* Associated with Seedling Diseases in Forest Nurseries. *Phytopathology,* 28:358–360 (1938).

Daubenmire, R. F.: Soil Temperature versus Drought as a Factor Determining Lower Altitudinal Limits of Trees in the Rocky Mountains. *Botan. Gaz.*, 105:1–13 (1943).

Davidson, R. W.: Forest Pathology Notes. 3. *Rhizina inflata* on Red Pine and White Spruce Seedlings. *U.S. Dept. Agr., Plant Disease Reptr.*, 19:96 (1935).

Davis, J. J.: Notes on Parasitic Fungi in Wisconsin—VIII. *Trans. Wisconsin Acad. Sci.*, 20:413–431 (1922).

Davis, J. J.: Notes on Parasitic Fungi in Wisconsin—X. *Trans. Wisconsin Acad. Sci.*, 21:271–286 (1924).

Davis, W. C.: Damping-off of Longleaf Pine. *Phytopathology*, 31:1011–1016 (1941).

Davis, W. C., and R. W. Davidson: *Fusicladium robiniae* and *Macrosporium* sp. in Forest Tree Nurseries. *U.S. Dept. Agr., Plant Disease Reptr.*, 23:63–65 (1939).

Davis, W. C., E. Wright, and C. Hartley: Diseases of Forest-tree Nursery Stock. *Civ. Conserv. Corps Forestry Publ.*, 9:I–IV, 1–79 (1942).

Dearness, J.: New and Noteworthy Fungi—IV. *Mycologia*, 18:236–255 (1926).

Doran, W. L.: Acetic Acid and Pyroligneous Acid in Comparison with Formaldehyde as Soil Disinfectants. *J. Agr. Research*, 44:571–578 (1932).

Eliason, E. J.: Buckwheat as a Factor in the Root Rot of Conifers. *Phytopathology*, 28:7 (1938).

Faull, J. H.: A Fungus Disease of Conifers Related to the Snow Cover. *J. Arnold Arboretum (Harvard Univ.)*, 10:3–8 (1929). *Forestry Chronicle*, 5(2):29–34 (1929).

Faull, J. H.: The Spread and the Control of Phacidium Blight in Spruce Plantations. *J. Arnold Arboretum (Harvard Univ.)*, 11:136–147 (1930).

Fisher, P. L.: Germination Reduction and Radicle Decay of Conifers Caused by Certain Fungi. *J. Agr. Research*, 62:87–95 (1941).

Fleming, W. E., F. E. Baker, and L. Koblitsky: Effect of Applying Acid Lead Arsenate for Control of Japanese Beetle Larvae on the Germination and Development of Evergreen Seedlings. *J. Forestry*, 35:679–688 (1937).

Graves, A. H.: Notes on Diseases of Trees in the Southern Appalachians. *Phytopathology*, 4:63–72 (1914).

Graves, A. H.: Root Rot of Coniferous Seedlings. *Phytopathology*, 5:213–217 (1915).

Gray, W. G.: An Instance of "Damping-off" Retarded by the Use of Basic Slag. *Forestry*, 5:132–135 (1931).

Gregory, K. F., O. N. Allen, A. J. Riker, and W. H. Peterson: Antibiotics as Agents for the Control of Certain Damping-off Fungi. *Am. J. Botany*, 39:405–415 (1952).

Haasis, F. W.: Frost Heaving of Western Yellow Pine Seedlings. *Ecology*, 4:378–390 (1923).

Hahn, G. G.: *Phomopsis juniperovora* and Closely Related Strains on Conifers. *Phytopathology*, 16:899–914 (1926).

Hahn, G. G.: Taxonomy, Distribution, and Pathology of *Phomopsis occulta* and *P. juniperovora*. *Mycologia*, 35:112–129 (1943).

Hahn, G. G., C. Hartley, and A. S. Rhoads: Hypertrophied Lenticels on the Roots of Conifers and Their Relation to Moisture and Aeration. *J. Agr. Research*, 20:253–265 (1920).

Hansen, T. S., et al.: A Study of the Damping-off Disease of Coniferous Seedlings. *Minn. Univ. Agr. Expt. Sta. Tech. Bull.*, 15:1–35 (1923).

Hartley, C.: The Blights of Coniferous Nursery Stock. *U.S. Dept. Agr. Bull.*, 44:1–21 (1913).

Hartley, C.: Stem Lesions Caused by Excessive Heat. *J. Agr. Research,* 14:595–604 (1918).

Hartley, C.: Damping-off in Forest Nurseries. *U.S. Dept. Agr. Bull.,* 934:1–99 (1921).

Hartley, C., and R. G. Pierce: The Control of Damping-off of Coniferous Seedlings. *U.S. Dept. Agr. Bull.,* 453:1–32 (1927).

Hartley, C., R. G. Pierce, and G. G. Hahn: Moulding of Snow-smothered Nursery Stock. *Phytopathology,* 9:521–531 (1919).

Henry, B. W.: A Root Rot of Southern Pine Nursery Seedlings and Its Control by Soil Fumigation. *Phytopathology,* 43:81–88 (1953).

Henry, B. W.: Sporulation by the Brown Spot Fungus on Longleaf Pine Needles. *Phytopathology,* 44:385–386 (1954).

Hobbs, C. H.: Studies on Mineral Deficiency in Pine. *Plant Physiol.,* 19:590–602 (1944).

Hotson, J. W., and D. E. Stuntz: Canker on *Chamaecyparis lawsoniana.* [Abstract]. *Phytopathology,* 24:1145–1146 (1934).

Hurd-Karrer, Annie M.: Inhibition of Arsenic Injury to Plants by Phosphorus. *J. Wash. Acad. Sci.,* 26:180–181 (1936).

Isaac, L. A.: Seedling Survival on Burned and Unburned Surfaces. *J. Forestry,* 28:569–571 (1930).

Jackson, L. W. R.: Effects of H-ion and Al-ion Concentrations on Damping-off of Conifers and Certain Causative Fungi. *Phytopathology,* 30:563–579 (1940).

Jackson, L. W. R.: "Needle Curl" of Shortleaf Pine Seedlings. *Phytopathology,* 38:1028–1029 (1948).

Johnson, L. P. V., and G. M. Linton: Experiments on Chemical Control of Damping-off in *Pinus resinosa* Ait. *Can. J. Research C. Botan. Sci.,* 20:559–564 (1942).

Jones, F. H., and T. R. Peace: Experiments with Frost Heaving. *Quart. J. Forestry,* 33:79–89 (1939).

Kauffman, C. H., and E. B. Mains: An Epidemic of *Cronartium comptoniae* at the Roscommon State Nurseries. *Mich. Acad. Sci. Ann. Rept.,* 17:188–189 (1915).

Kopitke, J. C.: The Effect of Potash Salts upon the Hardening of Coniferous Seedlings. *J. Forestry,* 39:555–558 (1941).

Korstian, C. F.: Control of Snow Molding in Coniferous Nursery Stock. *J. Agr. Research,* 29:741–747 (1923).

Korstian, C. F., and N. J. Fetherolf: Control of Stem Girdle of Spruce Transplants Caused by Excessive Heat. *Phytopathology,* 11:485–490 (1921).

Korstian, C. F., et al.: A Chlorosis of Conifers Corrected by Spraying with Ferrous Sulphate. *J. Agr. Research,* 21:153–171 (1921).

Lambert, E. B., and B. S. Crandall: A Seedling Wilt of Black Locust Caused by *Phytophthora parasitica. J. Agr. Research,* 53:467–476 (1936).

Latham, D. H.: Rhizoctonia Tip Wilt of Red Spruce. *U.S. Dept. Agr., Plant Disease Reptr. Suppl.,* 96:248–249 (1936).

Lightle, P. C.: Brown-spot Needle Blight of Longleaf Pine. *U.S. Dept. Agr., Forest Serv., Forest Pest Leaflet,* 44:1–7 (1960).

Lorenz, R. W.: High Temperature Tolerance of Forest Trees. *Minn. Agr. Expt. Sta. Bull.,* 141:1–25 (1939).

Meier, K.: Über eine durch Kalimangel bedingte "Gelbsucht" an Thujapflanzen. *Landwirtsch. Jahrb. Schweiz,* 51:297–304 (1937).

Moulopoulos, C.: High Summer Temperatures and Reforestation Technique in Hot and Dry Countries. *J. Forestry,* 45:884–893 (1947).

Němec. A.: Beitrag zur Kentnnis der Mangel- ("Karenz"-) Erscheinungen bei Kiefernsämlingen. *Forstwiss. Centr.,* 58:798–808 (1936).

Newhall, A. G.: Disinfestation of Soil by Heat, Flooding and Fumigation. *Botan. Rev.*, 21:189–250 (1955).

Olson, D. S.: Growing Trees for Forest Planting in Montana and Idaho. *U.S. Dept. Agr. Circ.*, 120:1–91 (1930).

Pomerleau, R.: Études sur la fonte des semis de conifères. *Rev. trimestr. can.*, 28:127–153 (1942).

Roeser, J., Jr.: Transpiration Capacity of Coniferous Seedlings and the Problem of Heat Injury. *J. Forestry*, 30:381–395 (1932).

Roth, C.: Untersuchungen über den Wurzelbrand der Fichte (*Picea excelsa Link*). *Phytopathol. Z.*, 8:1–110 (1935).

Roth, L. F., and A. J. Riker: Seasonal Development in the Nursery of Damping-off of Red Pine Seedlings Caused by *Pythium* and *Rhizoctonia*. *J. Agr. Research*, 67:417–431 (1943).

Rudolf, P. O.: Diagnosing Plantation Mortality. *Papers Mich. Acad. Sci.*, 23 (1937): 333–338 (1938).

Salisbury, P. J.: A Review of Damping-off of Douglas Fir Seedlings in British Columbia. *Forestry Chronicle* 30:407–410 (1954).

Salisbury, P. J., and J. R. Long: A New Needle Blight of Douglas Fir Seedlings Caused by *Rosellinia herpotrichoides* Hepting and Davidson. [Abstract.] *Can. Phytopathol. Soc. Proc.*, 23:19 (1955).

Schramm, J. R.: The Mechanism of Frost Heaving of Tree Seedlings. *Proc. Am. Phil. Soc.*, 102:333–350 (1958).

Shea, K. R., and J. E. Kuntz: Prevention of Damping-off of Poplar Seedlings. *Forest Sci.*, 2:54–57 (1956).

Shirley, H. L.: Lethal High Temperatures for Conifers, and the Cooling Effect of Transpiration. *J. Agr. Research*, 53:239–258 (1936).

Shirley, H. L., and L. J. Meuli: The Influence of Soil Nutrients on Drought Resistance of Two-year-old Red Pine. *Am. J. Botany*, 26:355–360 (1939).

Siggers, P. V.: The Brown Spot Needle Blight of Pine Seedlings. *U.S. Dept. Agr. Tech. Bull.*, 870:1–36 (1944).

Slagg, C. M., and E. Wright: Diplodia Blight in Coniferous Seedbeds. *Phytopathology*, 33:390–393 (1943*a*).

Slagg, C. M., and E. Wright: Control of Cedar Blight (*Phomopsis juniperovora*) in Seedbeds. *Am. Nurseryman*, 78(7):22–25 (1943*b*).

Slagg, C. M., and E. Wright: The Control of Phomopsis Blight in Red Cedar Seedbeds. *Kansas State Hort. Soc. Bien. Rept.*, 48:77–82 (1944).

Smith, R. E.: *Botrytis* and *Sclerotinia:* Their Relation to Certain Plant Diseases and to Each Other. *Botan. Gaz.*, 29:369–407 (1900).

Snyder, T. E., and H. N. Lamb: Membracid Girdling of Young Black-locust Seedlings in Southern Louisiana and Mississippi. *Louisiana Conserv. Rev.*, 4(6):9–10, 47 (1935).

Spaulding, P.: The Present Status of the White Pine Blights. *U.S. Dept. Agr., Bur. Plant Ind. Circ.*, 35:1–12 (1909).

Spaulding, P.: *Lophodermium pinastri* Causing Leafcast of Norway Pine in Nurseries. *Northeast. Forest Expt. Sta. Tech. Note*, 18:1–2 (1935).

Strong, F. C.: Damping-off in the Forest Tree Nursery and Its Control. *Mich. Agr. Expt. Sta. Quart. Bull.*, 34:285–296 (1952).

Sturgis, W. C.: *Herpotrichia* and *Neopeckia* on Conifers. *Phytopathology*, 3:152–158 (1913).

Sydow, H., and F. Petrak: Zweiter Beitrag zur Kenntnis der Pilzfloras Nordamerikas, insbesondere der nordwestlichen Staaten. *Ann. Mycol.*, 22:387–409 (1924).

Ten Houten, J. G.: "Kiemplantenziekten van coniferen." [English summary.] Proefschrift, University of Utrecht. Utrecht: J. von Boekhoven (1939).

Tint, H.: Studies in the *Fusarium* Damping-off of Conifers. I. The Comparative Virulence of Certain *Fusaria*. II. Relation of Age of Host, pH and Some Nutritional Factors to the Pathogenicity of *Fusarium*. III. Relation of Temperature and Sunlight to the Pathogenicity of *Fusarium*. *Phytopathology,* 35:421–439, 440–457, 498–510 (1945).

Toumey, J. W., and E. J. Neethling: Insolation a Factor in the Natural Regeneration of Certain Conifers. *Yale Univ. School Forestry Bull.,* 11:1–63 (1924).

Trumbower, J. A.: Control of Elm Leaf Spots in Nurseries. *Phytopathology,* 24:62–73 (1934).

Tryon, E. H.: Stem Girdling of Coniferous Nursery Stock by Frost-heaved Soil. *J. Forestry,* 41:768–769 (1943).

Vaartaja, O.: Forest Humus Quality and Light Conditions as Factors Influencing Damping-off. *Phytopathology,* 42:501–506 (1952).

Vaartaja, O.: Principles and Present Status of Chemical Control of Seedling Diseases. *Forestry Chronicle,* 32:45–48 (1956*a*).

Vaartaja, O.: Notes on Low Temperature Fungi. *Can. Dept. Agr., Sci. Serv., Forest Biol. Div., Bi-monthly Progr. Rept.* 12(4):3 (1956*b*).

Voight, G. K.: The Effects of Fungicides, Insecticides, Herbicides, and Fertilizer Salts on the Respiration of Root Tips of Tree Seedlings. *Soil Sci. Soc. Am. Proc.,* 17:150–152 (1953).

Voight, G. K., S. A. Wilde, and J. H. Stoeckeler: Response of Coniferous Seedlings to Soil Applications of Calcium and Magnesium Fertilizers. *Soil Sci. Soc. Am. Proc.,* 22:343–345 (1958).

Weir, J. R.: A *Botrytis* on Conifers in the Northwest. *Phytopathology,* 2:215 (1912).

Weir, J. R.: *Phacidium infestans* on Western Conifers. *Phytopathology,* 6:413–414 (1916).

Weir, J. R.: *Thelephora terrestris, T. fimbriata,* and *T. caryophyllea* on Forest Tree Seedlings. *Phytopathology,* 11:141–144 (1921).

White, L. T.: Biological Control of Seedling Diseases. *Forestry Chronicle,* 32:49–52 (1956).

Wilde, S. A.: "Forest Soils and Forest Growth." Waltham, Mass.: Chronica Botanica Company (1946).

Worley, C. L., H. R. Lesselbaum, and T. M. Mathews: Deficiency Symptoms for the Major Elements in Seedlings of Three Broad-leaved Trees. *J. Tenn. Acad. Sci.,* 16:239–247 (1941).

Wright, E.: Control of Damping-off of Broadleaf Seedlings. *Phytopathology,* 31:857-858 (1941).

Wright, E.: Damping-off in Broadleaf Nurseries of the Great Plains Region. *J. Agr. Research,* 69:77–94 (1944).

Wright, E.: Relation of Macrofungi and Micro-organisms to Damping-off of Broadleaf Seedlings. *J. Agr. Research,* 70:133–141 (1945).

Wycoff, H. B.: Methyl Bromide Fumigation of an Illinois Nursery Soil. *J. Forestry,* 53:811–813 (1955).

Young, G. Y.: Root Rots in Storage of Deciduous Nursery Stock and Their Control. *Phytopathology,* 33:656–665 (1943).

Young, G. Y., W. C. Davis, and D. H. Latham: Control of Damping-off of Conifers. [Abstract.] *Phytopathology,* 27:144 (1937).

CHAPTER 6

Root Diseases

Accurate diagnosis of root diseases is not easy because the symptoms on the aerial portion of the tree are often similar to the symptoms of other diseases, particularly wilts and diebacks, and because by the time visible symptoms appear most or all of the root system may be involved or destroyed, with secondary agencies obscuring the primary one. Furthermore, investigating the development of soil fungi *in situ* is difficult so that much less is known in general about the means of dissemination and method of infection of fungi attacking roots than of fungi attacking aerial parts of trees (Garrett 1955, 1956*a*). Unfavorable soil conditions, such as impenetrable hardpan or an abnormally high or low moisture content, can cause death of roots, thus in some cases reducing vigor of the tree and certainly affording an entrance for fungi that could not penetrate living roots directly (Day 1934). Wood-destroying fungi entering through dead roots can progress into the trunk, causing decay in the heartwood of the butt without killing the host, so that the resulting loss in merchantable timber above ground is directly chargeable to the diseased condition of the roots, which in itself may be of no practical significance. Extensive decay of the roots does predispose trees to wind throw.

Root diseases, particularly root rots, are more prevalent in planted stands, in stands on sites to which they are not adapted, and in stands of unnatural composition, so that in the future these diseases are likely to increase in this country. Root diseases are especially abundant in tropical regions.

A few seed plants are parasitic on roots of trees, but the effect of these parasites is unknown, although observations indicate that it is insignificant. Certain root infections resulting in abnormalities such as mycorrhiza and root nodules, although not disease according to strict interpretation, nevertheless can be most logically treated in this chapter. Root diseases restricted to seedlings are discussed under seedling diseases.

SHOESTRING ROOT ROT

Armillaria mellea (Vahl.) Quél. (Basidiomycetes, Agaricaceae), known as the "honey mushroom," causes this disease, which is of long standing and world-wide distribution. The disease has also been called "shoestring fungus rot," "Armillaria root rot," "mushroom root rot," "crown rot," "rhizomorphic root rot," and "toadstool disease," and in addition on conifers the names "resin flow" and "resin glut" have been applied. The causal fungus has also been referred to as "honey agaric," "oak fungus," and "shoestring fungus." The German names are *Honigschwamm, Honigpilz* and *Hallimasch;* the French, *maladie du pourridie* and *maladie des racines.* Comprehensive papers are those by Reitsma (1932) and Thomas (1934).

Hosts and Damage. The fungus causes rotting of the bark and wood of the roots and root collar followed by death of the tree. It does not cause disease of thrifty trees, and, since there are strains and races within the species, any susceptibility rating is uncertain, but oaks, chestnuts, spruces, and pines seem to be quite susceptible. Orchard, shade and ornamental trees, and forest trees of reduced vigor are liable to attack. Trees infested by defoliating or bark-boring insects, infected by other fungi, affected by drought, injured by lightning, or weakened in any other way are commonly attacked by *A. mellea.* Day (1929) thinks that *A. mellea* is parasitic only when the vigor of the host has been reduced by unfavorable environmental conditions, particularly poor soil, acting alone or in conjunction with other injurious agents. The disease is most prevalent in plantations because artificially reproduced stands in general are growing under unnatural conditions. It is less frequent in natural forests than in those in which the stand composition has been changed. Trees outside their native range are quite liable to attack. Eastern white pine planted in northern Montana was found more susceptible than western white pine planted on the same area. Drought favors the disease; consequently in recently cutover ponderosa pine forests in eastern Oregon reproduction is commonly killed by *A. mellea* during unusually dry periods. Thomas (1934:216) concludes that "Structural or morphological differences of the host probably exert little influence on resistance. Resistance to *Armillaria mellea* appears to be of the nature of an antagonistic influence exerted upon the fungus by the host only when the latter is in an active, healthy state." Since susceptibility is influenced by so many apparent or obscure factors external to both host and parasite, any stable rating of the relative susceptibility of various species is impossible, and contradictory evidence is to be expected. Owing to its widespread distribution in the forests of this country, if the fungus were a virulent parasite great damage should occur, but there is apparently little relation between the occurrence of the fungus and the prevalence of the disease. Furthermore infections are found commonly in the roots of apparently healthy trees (Christensen 1938; Ehrlich 1939).

Since the fungus is so widespread as a saprophyte, there is a tendency to exaggerate the damage it causes, overlooking determination of the primary factors weakening the trees. Of course, without secondary attack by *A. mellea*, weakened trees in many cases would recover, so that it is difficult to evaluate the destructiveness of this parasite. European literature abounds with references to damage by shoestring root rot, and probably the explanation for the greater incidence of *A. mellea* as a parasite on forest trees in Europe than in the United States is the fact that in Europe a large proportion of the stands are either artificially reproduced or of unnatural composition, often both. Such stands in general are unlikely to be as thrifty as the natural forests of this country. Usually, the presence of shoestring root rot should be interpreted as indicating a debilitated condition of the tree from some prior cause.

In addition to actual killing of trees subnormal in vigor, loss also results from this fungus causing butt rot in trees of merchantable size, e.g., aspen in Minnesota (Schmitz and Jackson 1927:7).

Symptoms. Decline in vigor of the tree as a whole or of certain branches is usually the first external symptom. Leaves may turn yellow and fall prematurely or remain undersized and scanty. Branches may die back. All the foliage may die simultaneously on small trees. Of course, these symptoms are not diagnostic for shoestring root rot only. Vigorous infected Douglas firs often produce heavy crops of small cones on the upper branches coupled with the prolific development of resin blisters on the branches and twigs (Buckland 1953). In conifers, particularly pines, there is commonly an abnormal resin flow from the root collar, sometimes so great that the litter and soil is compacted into a hard crust. The root collar and roots of trees with these symptoms should be examined for areas of decayed bark or wood, for the fan-shaped, veined, white mycelial felts between the bark and wood or in the bark, and for the dark-brown or black rhizomorphs between the bark and wood, on the surface of the roots, or free in the soil. Clusters of honey-colored toadstools may finally appear when the tree is dead or nearly so, but in dry situations the fructifications may develop in only occasional years, and in general they occur only in the autumn. Diseased trees may appear in spots or groups which increase in size as more trees are attacked, but on the other hand diseased individuals may be scattered at random through a stand.

Both sapwood and heartwood are decayed. In the incipient stage rotted wood first shows a faint water soak, followed by a more apparent light-brownish color. In the advanced stage the wood becomes light yellow or white in color, soft and spongy, often stringy in conifers, and marked by numerous black zone lines (Campbell 1934).

Causal Agency. Usually the fungus lives as a saprophyte on stumps and roots of dead trees, but it often becomes parasitic. It seems likely

Fig. 27. Fructifications of the honey mushroom, *Armillaria mellea*. (*Photograph by W. E. Flowers and J. R. Weir.*)

that there are strains within the species varying in parasitic habit (Benton and Ehrlich 1941; Childs and Zeller 1929; Reitsma 1932:513; Ramsbottom 1935:45). The fructifications (Fig. 27) develop in clusters on stumps, at the base or on the butt of dead and dying trees, or on the ground coming up from infected roots. These toadstools have a solid central stalk from 3 to 10 inches long, honey-yellow or brown in color, with a broad cap, the upper surface of which is also honey-yellow and dotted with dark-brown scales; the whitish gills on the lower surface are attached to the stalk as well as the cap. The stalk may or may not have an annulus or ring around it a short distance below the gills. The basidiospores from the gills are wind-borne to infect other stumps or dead trees but do not infect living trees, except possibly rarely through open wounds just at the base of the tree or on exposed roots (Hiley 1919:160). New infection centers have also been established by floodwater washing infected debris from hillsides to lower levels along stream banks (Hewitt 1934). After the fungus has developed for some time in a stump or dead tree, rhizomorphs are formed. These are of two kinds—subcortical and subterranean or free—both of which grow in length apically. The subcortical rhizomorphs are flattened black anastomosing structures formed beneath the bark and wood (Fig. 28), whereas the subterranean rhizomorphs are black cordlike strands $1/25$ to $1/12$ inch in diameter growing in

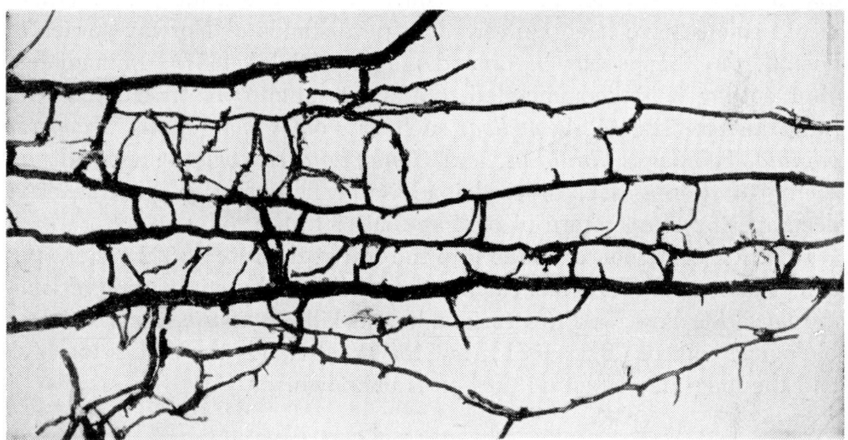

Fig. 28. Subcortical rhizomorphs of the honey mushroom, *Armillaria mellea,* from between the bark and wood of a dead white oak.

the soil. Rhizomorphic development has been discussed by Campbell (1934:17). Rhizomorphs consist of a dark-brown outer layer of closely compacted fungous tissue enclosing a central core of hyaline hyphae arranged in longitudinal rows.

Subterranean rhizomorphs grow through the soil for a limited distance, usually close to the surface but at least not penetrating the subsoil, and are the usual means by which living trees are infected. The likelihood of infection by a rhizomorph decreases as the distance from its food base increases (Garrett 1956*b*). Coming in contact with a living root, the tip of the rhizomorph penetrates the unbroken bark (Woeste 1956:251). Penetration seems to be by a combination of mechanical and chemical means, since several enzymes are secreted by the rhizomorphs (Lanphere 1934). After penetration, the hyphae from the tip of the rhizomorph ramify through the inner bark and wood, destroying them if the host is not in an active, healthy condition. Infection also occurs through direct contact between infected and healthy roots. The hyphae are luminous (Reitsma 1932:503).

Control. Since general experience indicates that trees not growing under optimum conditions are more subject to attack, the most effective control is to maintain vigorous stands. Since *A. mellea* as a saprophyte lives particularly in roots and stumps with a high carbohydrate content, in Nyasaland it has been proposed to ring-bark the trees sometime prior to felling before clearing forest land for planting (Leach 1939). When the trees are finally felled the roots and stumps, by then low in carbohydrates, are quickly invaded by fungi antagonistic to *A. mellea,* so that its inoculum potential in the soil is reduced to a minimum. Where mixed stands have been converted to pure stands of Norway spruce in Poland, the fungus has become damaging (Orłoś 1957). Immediately after cutting it is recommended that spruce stumps be inoculated with fungi antagonistic to *A. mellea,* such as *Fomes marginatus, Trametes odorata, Polyporus fibrillosus,* and *Coniophora cerebella.* Chemical poisoning of stumps is considered inadvisable because this greatly delays decomposition and return of organic matter to the soil.

Injection of carbon disulfide into the soil at the rate of 302 gallons per acre has controlled *A. mellea* in citrus orchards by stimulating *Trichoderma viride* Pers. ex Fries to invade and kill *Armillaria,* thus initiating biological control (Bliss 1951), but for forest stands the cost is too high and the long-time effect on the soil is unknown.

FOMES ROOT AND BUTT ROT

Fomes annosus (Fr.) Cke. (Basidiomycetes, Polyporaceae), often called *Trametes radiciperda* Hartig in continental Europe, causes this disease, which is also known as "spongy sap rot" and "brown root and butt rot."

Distribution, Hosts, and Damage. The fungus is of widespread distribution in the North Temperate Zone at least, usually occurring on conifers but occasionally on hardwoods (Martinez 1943; Rennerfelt 1946, 1953; Rishbeth 1951; Populer 1957). Evidence as to the susceptibility of various conifers is conflicting, but red pine is highly susceptible in North America as is Norway spruce in Europe. The fungus is damaging to planted stands but is of minor consequence in those naturally regenerated. In Europe it is most severe in planted Norway spruce on rich soils, particularly reafforested arable soils, because such soils contain much nitrogenous material that favors the fungus (Hoppfgarten 1933). Drought also seems to favor the fungus. It rarely attacks perfectly healthy trees. Races of *F. annosus* are unknown (Roll-Hansen 1940; Etheridge 1955; Perrson 1957).

The fungus kills young trees by decaying the bark and wood of the roots and root collar. Resistance to killing increases with tree age and is strong from about 25 years of age on. In pine plantations on many acid soils killing practically ceases by the age of 15 years, whereas on alkaline soils killing becomes less frequent by the age of 15 years and commonly stops after 25 years (Rishbeth 1957:80). Thinning plantations increases injury by *F. annosus* to the remaining trees. Damage to older trees is mostly butt rot of the heartwood that can cause heavy losses (Rattsjö and Rennerfelt 1955). The fungus tends to attack the fast-growing trees of Norway spruce, increment is reduced at an early stage in the attack, and stem form is greatly impaired (Arvidson 1954). Infection may be followed by bark beetle attack (Francke-Grosmann 1948; Jorgensen and Petersen 1951). The fungus is widespread as a saprophyte on dead trees, stumps, and cull logs of conifers. In western Oregon and Washington it causes the most important heart rot of overmature western hemlock, usually confined to the butt log but sometimes extending to a height of 40 feet or more with an average of 16.7 feet (Englerth 1942:14). Infection in this species usually occurs through wounds (Rhoads and Wright 1946). Damage by this fungus will increase with the increase in plantations and of stands that deviate from the natural as regards representation of species or ages.

Symptoms and Causal Agency. When young trees are killed, thin, tissue-paper-like mycelium felts, not noticeably veined, form between the bark and wood; small dirty-white outgrowths occur on the underside of the roots. No rhizomorphs develop. Presumably the mycelium grows little, if at all, in the soil. In the incipient stage, decayed wood is pinkish to dull-violet color, depending on the species of wood, but still hard and firm. In the advanced stage, the wood shows small elongated pockets (Fig. 29) sometimes with black spots or flecks in the center, the pockets being separated by areas of firm, brownish wood. The white pockets finally run together, forming a mass of spongy or stringy white fibers flecked with black. Resin flow may occur from the butt. The perennial conks (Fig. 30), which vary from bracket-shaped to flat layers, depending on their position, have a biscuit-colored undersurface with small pores and usually with a lighter-colored sterile margin. The zoned upper surface is grayish-brown to dark brown. The conks usually occur on the crotches of the root collar partially hidden by the litter or even on roots abutting on cavities in the soil. Spores from the conks are wind borne to infect recently cut stumps. The decay progresses from the stump into the roots. The roots of adjacent living trees are infected by contact with a decayed root and through root grafts. The fungus can penetrate unwounded bark (Day 1948; Woeste 1956:241). Furthermore, the fungus can transfer from the root system of one tree species to that of a different species. Experimentally, basidiospores have passed through layers of soil and sand to germinate on wood underneath (Molin 1957).

Control. Recommendations for control are conflicting, but vigorous stands on suitable sites suffer little. A susceptible species in mixture is less affected than when pure, particularly if hardwoods are in the mixture (Rennerfelt 1957). Possible control measures are wider spacing to reduce root contacts, delayed thinning because conifers become more resistant with age, and stump extraction, which is difficult and costly

Fig. 29. Advanced rot caused by *Fomes annosus* in lowland white fir. Notice the black spots in the white areas. (*Reproduced from U.S. Dept. Agr. Bull.* 658.)

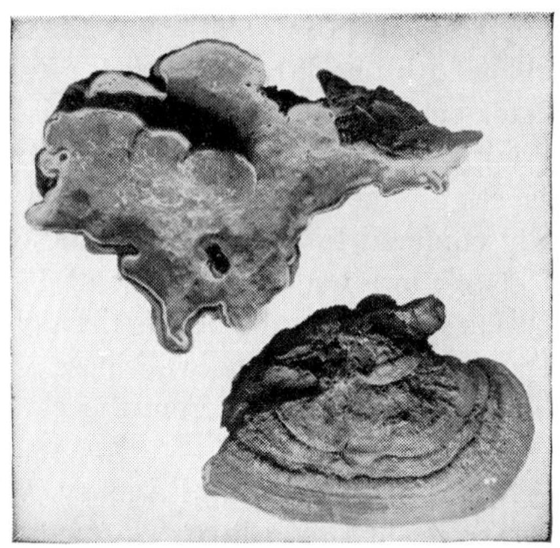

Fig. 30. Typical conks of *Fomes annosus* showing the upper and undersurfaces. (*Reproduced from U.S. Dept. Agr. Bull.* 658.)

(Rishbeth 1952). Brushing stumps with a tar and creosote mixture immediately after felling is considered the most promising method (Rishbeth 1957). This should largely prevent the infection of stumps by wind-borne spores, so that new infection centers should be reduced in numbers. In addition, there is a chance that some fungus antagonistic to *F. annosus* will quickly colonize stumps (Rennerfelt and Paris 1952–1953). The disease has been controlled in a pine arboretum by injecting the soil with carbon disulfide (Anonymous 1955).

PHYTOPHTHORA ROOT ROTS

In general Phytophthora root rots are worst in moist, water-retentive soils (Day 1938:114).

A root rot caused by *Phytophthora lateralis* Tucker & Milbrath (Phycomycetes, Peronosporaceae) has become epidemic in young-growth Port Orford cedar in the heart of the natural range of this tree (Hunt 1959). The disease, known for many years on ornamental trees outside the natural range of the cedar, was found within its natural range in 1952. After the root system is generally infected, the first aboveground symptom is a gradual fading of the foliage until it becomes tan or light brown, crisp, and dry. A cinnamon-brown discoloration of the infected cambium and inner bark extends upward from the root collar a short distance. The fungus spreads locally by the movement of surface water and of soil. Infected trees occur in irregular patches from a few feet to many feet in diameter. Trees up to about 50 years old are attacked and killed. No control is known.

A root rot of American chestnut caused by *Phytophthora cinnamomi* Rands apparently resulted in a steady recession of this tree in the Atlantic and Gulf states for a long time before chestnut blight appeared (Crandall, Gravatt, and Ryan 1945; Crandall 1950). Chinquapins are also susceptible, but Asiatic chestnuts are resistant. In nurseries it occurs on conifers and hardwoods. The symptoms are quite similar to those of the "ink disease" of European chestnut in southern Europe.

Littleleaf. This disease of shortleaf pine, and to a lesser extent loblolly pine, occurs in the South and Southeast, resulting in the decline and death of trees 20 years old and older (Zak 1957). Infected trees have yellowish foliage; needle length and shoot growth decrease; and finally diseased trees are conspicuous because of their short, scant, and chlorotic foliage confined to the ends of the branches. Infected trees commonly produce heavy crops of small cones that persist on the branches for some time. After symptoms appear, the average life of a shortleaf pine is 6 years. Affected trees do not recover naturally. Littleleaf symptoms develop because of a nitrogen deficiency in the tree associated with the dying of new root tips and fine roots brought about largely by *Phytophthora cinnamomi,* a soil fungus widely distributed in the South, where it has become pathogenic in forest soils of low fertility or poor drainage throughout much of the range of shortleaf pine.

For years to come, minimizing losses by modification of cutting practices will be the only control in areas where littleleaf is abundant. Where only an occasional shortleaf pine shows symptoms, cut lightly at 10-year intervals. Where between 10 and 25 per cent of the trees are affected, cut every 6 years, removing all diseased and suspected trees. Where more than 25 per cent of the trees show littleleaf, cut all shortleaf pine when merchantable. On severe littleleaf areas shortleaf pine, and even loblolly pine, should be grown on a short rotation for pulp.

Some shortleaf pines are resistant to littleleaf, and tests indicate that this trait is heritable (Zak 1957). In time resistant strains may be developed for reforestation. However, ultimate control of the disease will depend on rehabilitation of the deteriorated soils by management practices favoring soil-building hardwoods in the stands. Littleleaf took a long time to develop; it will take longer to correct.

A basically similar disease of Monterey pine and other conifers occurs in New Zealand (Newhook 1959).

OTHER ROOT DISEASES

Clitocybe Mushroom Root Rot. Clitocybe mushroom root rot, prevalent in the South particularly on oak, is caused by *Clitocybe tabescens* (Scop.) Bres. (Basidiomycetes, Agaricaceae). It is similar to shoestring root rot, but there are no rhizomorphs accompanying Clitocybe root rot; the younger mycelial sheets are perforate, and they have a less fan-shaped type of development at the advancing margins (Rhoads 1956).

Sparassis Root Rot. Species of *Sparassis* (Basidiomycetes, Clavariaceae) cause root rot of forest trees (Hennig 1952). *S. radicata* Weir causes a yellowish-brown carbonizing decay of old conifers in the West that is confined to the roots (Weir 1917*a*). Yellowish-white mycelial fans develop abundantly under the bark. Resin flows freely from the diseased roots. The annual large white fleshy fructifications with thin flat branches arise from perennial stalks that are connected with underlying diseased roots.

White Root Rot. The weeping conk, *Polyporus dryadeus* Fr. (Basidiomycetes, Polyporaceae), causes a root rot of oaks (Long 1913), occasionally of other hardwoods, and in the West of conifers (Weir 1921). Usually only old or badly suppressed trees are damaged, but the fungus has killed trees ranging from small saplings to large trees in a stand of white fir (Long 1930). In the final stage decayed wood of the roots is pithlike, and white, cream, or straw-colored. The infrequent conks usually appear on the root collar of the diseased tree just at ground level. The annual conks are large, irregularly shaped masses of corky or woody texture. When young, the upper surface is light brown with large drops of water exuding from it, but with age the color changes to darker brown or almost black and the surface becomes rough.

Xylaria Root Rot. Xylaria root rot, also known as "black root rot," is caused by *Xylaria* spp. (Ascomycetes, Xylariaceae). In the East, apple trees in orchards are mostly affected, and occasional trees are killed (Fromme 1928). *X. digitata* (P.) Grev. has been found attacking black cottonwood and hawthorn in Idaho (Weir 1917*b*), and the disease has been reported in the East on various hardwoods, particularly maples, but it is of minor economic importance on forest trees. All tissues of affected roots are attacked. The decayed wood becomes soft, spongy, and dirty white in color, with narrow but conspicuous black zones at times appearing somewhat concentric and forming fantastic patterns (Fig. 140). The decay may extend into the butt, causing some cull. The decay will continue if infected wood after cutting is stored under conditions favoring development of the fungus. The affected roots are at first covered with a thin white compact mycelium that later changes to a black incrustation. The fructifications, arising from old stumps, exposed roots, or diseased roots beneath the ground, are dark brown or black, erect, simple or branched, club-shaped structures from 1 to several inches high.

Mottled Bark Disease. Mottled bark disease is caused by *Stereum sanguinolentum* Alb. & Schw. (Basidiomycetes, Thelephoraceae), which is widespread as a saprophyte on coniferous slash and causes heart rot in living trees (page 424). The fungus has also been associated with the death of planted conifer saplings in Idaho (Hubert 1935). The foliage of diseased trees suddenly turned brown during the growing season, usually preceded by resin flow from the root collar and butt. The inner surface of the bark on roots, root collar, and butt had a white mottled appearance. The underlying sapwood varied from light to dark grayish-brown in color, finally becoming soft and spongy. The small fructifications (page 425) were found on the butts, root crotches, and exposed roots of dead trees. Infection probably occurred through basal wounds.

Laminated Root Rot. This root rot of Douglas fir, known from the Pacific Coast, is caused by *Poria weirii* (page 426) or a specialized form of it (Buckland, Molnar, and Wallis 1954; Childs 1955). It also occurs on other conifers. Douglas fir from 6 years old on is infected and killed, but the greatest damage occurs in stands from 40 to 150 years old. Trees of all degrees of vigor are attacked on good as well as on poor sites. The trees are killed in groups, resulting in stand openings that gradually increase in size. The fungus spreads from tree to tree through root fusions, particularly abundant in Douglas fir, and through root contacts. The first indication of the disease is a reduction in growth, usually accompanied by a crop of smaller than normal cones, then there is a gradual thinning of the foliage for a year or two, and finally all the needles brown and the tree dies. The incipient stage of decay appears on the face of the stump as a brownish stain, crescent-shaped to irregular in outline, and mostly in the heartwood. The advanced decay is soft and flaky, yellowish to brownish, and honeycombed with small pockets, at first filled with whitish fibers, later empty. Brown mycelial felts are present. Decayed wood separates readily along the growth rings. Sporophores develop sparingly on the underside of logs, uprooted stumps, and on the trunks of dead standing trees.

Resinosis Disease. Resinosis disease of eastern white pine and red pine in New York has resulted in the death within 3 to 5 years of planted 15- to 25-year-old trees (Anonymous 1937; York, Wean, and Childs 1936). Scotch pine and Norway spruce were not affected. In some plantations from 50 to 65 per cent of the trees were killed before the attack lessened somewhat. The principal symptoms were resin flow from the butt and white mottling of the inner bark of diseased trees. The resinosis was caused by an unidentified fungus, although associated with it and seemingly responsible for most of the damage in the plantations was *Polyporus schweinitzii* causing root and butt rot. The affected plantations were on heavy, poorly drained, and somewhat alkaline soil, apparently unsuitable for eastern white and red pines.

Ganoderma Root Rots. Root rots caused by *Ganoderma* spp. (Basidiomycetes, Polyporaceae), so destructive in the tropics, as yet have been of minor consequence in the United States, but may become damaging if extensive unnatural stands are created. Several Ganodermas have been associated with root rot and death of trees in this country (Edgerton 1954; Nickell 1952; Norton and Behrens 1956).

Phymatotrichum Root Rot. Phymatotrichum root rot, also known as "Ozonium root rot," "cotton root rot," and "Texas root rot," is caused by *Phymatotrichum omnivorum* (Shear) Duggar (Fungi Imperfecti, Moniliaceae). This soil-inhabiting fungus, limited to the Southwest (Brinkerhoff and Streets 1946), is important to forestry practice because of extensive shelterbelt planting on the western edge of the prairies (Wright and Wells 1948). It is so destructive that planting must be avoided in the root rot area except with the more resistant species, namely, hackberry, western soapberry, desert willow, eastern red cedar, and Rocky Mountain juniper. Seedlings are more susceptible than older trees. Infected trees are stunted, and the leaves yellow, then become bronze and drop off or dry out, become blackish, and remain clinging to the twigs. On the diseased roots, which are usually badly rotted with a shriveled epidermis, are buff to brown fuzzy mycelial strands. Smooth dark-brown strands or rhizomorphs are also produced, by means of which the fungus spreads through the soil during the growing season. During rains, ephemeral cottony-white mycelial mats, on which the spores are borne, form on the soil surface. The spores seem to be functionless; they have not been germinated.

Phloem Necrosis. Phloem necrosis of American elm (Swingle 1942; Tucker 1945) is a virus disease of unknown origin, although probably of many years standing, which is now epidemic in the region of the central and lower Ohio River watershed and extends to northern Mississippi, eastern Oklahoma, Kansas, and Nebraska. Other species of elm are resistant. Both forest and shade trees are affected, and trees of all ages and of all degrees of vigor are susceptible. Trees may be infected for 6 months to 1 or more years before symptoms appear. Then the majority usually die in from 1 to 1½ years, although some may die in 3 to 4 weeks, and others may survive for 2 or more years. All infected trees die eventually.

There is a gradual decline of the entire crown, but in large trees the symptoms may first appear in a single branch or in a part of the top. Leaves droop and curl slightly, the foliage appears sparse, and the leaves yellow and then fall. Death of the tree follows. The roots are killed, and a considerable portion of the root system is affected before there is evidence of disease in other parts of the tree. Before they die, the inner bark or phloem of large roots first becomes yellow, then assumes a typical "butterscotch" color often with small, scattered brown or black flecks, and finally becomes dark brown. The discoloration is usually found only in the large roots and lower trunk, but it may occur in the upper part of the trunk and some of the branches. Moderately discolored inner bark has a faint odor of wintergreen. This discoloration and odor are specific for the disease. The virus is spread by a leafhopper (Baker 1949) and probably also by root grafting. There is no effective control.

Crown Gall. Crown gall, also known as "crown rot," "root knot," "root tumor," "cane gall," "black knot," and "plant cancer," is caused by a motile rod-shaped bacterium, *Pseudomonas tumefaciens* (S. & T.) Duggar (De Ropp 1951). To foresters the name "crown gall" is misleading, since it actually refers to the presence of galls on the root crown or root collar and not on the upper part (or crown) of the tree. The disease, known since 1853, is world wide on a great variety of plants including trees. Hardwoods are commonly infected, but the disease on conifers (Smith 1937) rarely occurs naturally, although it has been produced artificially on several species (Smith 1942). The disease is of no consequence in natural hardwood forests, but it may attack nursery stock. Even seedlings and small saplings are rarely killed, the damage being largely deformation of the host and some retardation of growth. The symptoms are tumorlike swellings (Fig. 31) on the roots, root collar, or even on the main stem and branches. The presence of the crown gall organism is difficult to demonstrate by microscopic examination of infected woody tissue, so that similar hypertrophies caused by other agencies have been commonly classed as crown gall (Riker and Keitt 1926; Lek 1929), and in fact the microscopic aspects of crown gall tissue and callus around graft wounds are quite similar (Sylvester and Countryman 1933). In nurseries, wounding of roots or stems should be avoided, since most infections result from wounds contaminated with infected soil. The bacteria will persist in the soil for not over 2 years in the absence of host plants, but they may be carried for some miles by irrigation water to

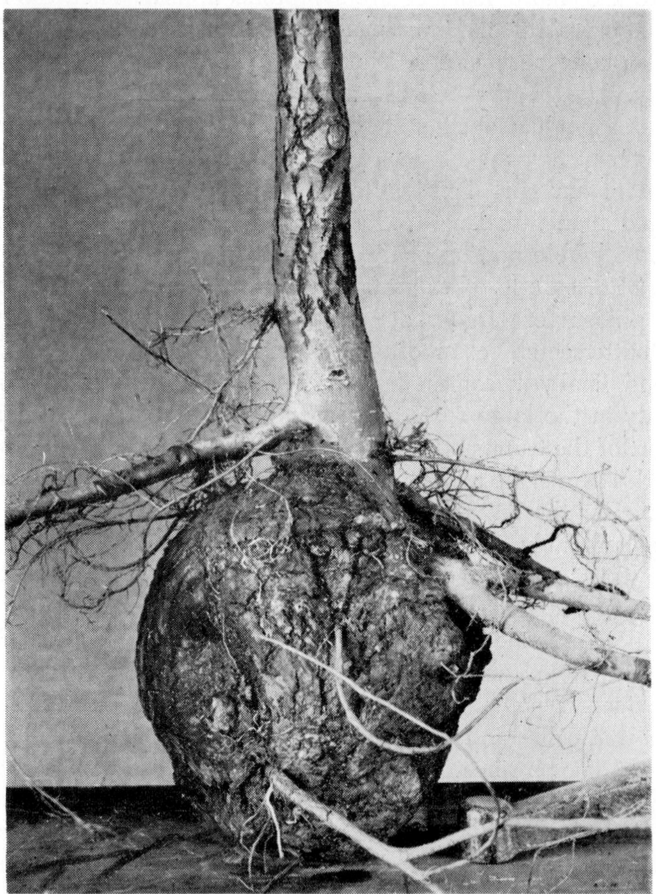

Fig. 31. Crown gall on a willow sapling. (*Photograph by G. G. Hedgcock.*)

contaminate healthy soil (Smith and Cochran 1944). Diseased stock should be destroyed, and, before shipping, the remaining stock should be immersed for 1 hour to a distance of several inches above the root collar in a solution of 7 to 16 ounces of copper sulphate to 26 gallons of water. Galls in the early stages have been eradicated from valuable almond trees by painting the gall with a mixture of 20 volumes of sodium dinitrocresol (Elgetol) and 80 volumes of methanol (Ark 1941).

Hairy Root. Hairy root, also termed "woolly root" or "woolly knot," was long considered to be caused by the same organism causing crown gall, but Riker et al. (1930) demonstrated that on apple it is caused by another motile rod-shaped bacterium, *Pseudomonas rhizogenes* R. B. W. K. & S. Both diseases often occur on the same plant, and the two bacteria cannot be separated except by isolation of single cells (Wright, Hendrickson, and Riker 1930). Hairy root occurs on hardwoods but has not been recorded on conifers, and it is questionable as to whether or not any significant injury results. In the very early stage of the disease

Fig. 32. Hairy root on a grafted apple tree. (*Reproduced from U.S. Dept. Agr. Bull.* 90, *part* II.)

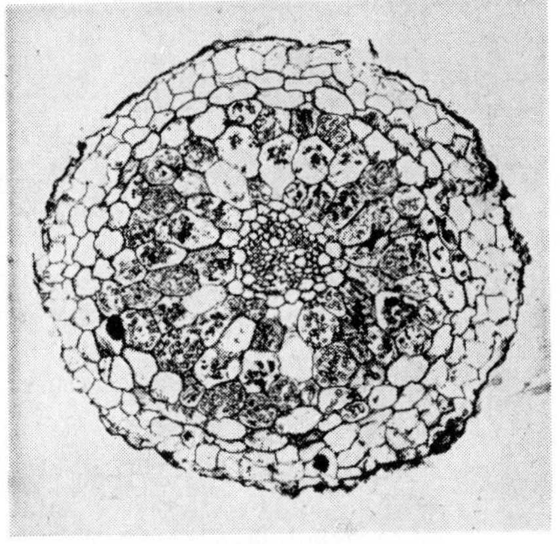

Fig. 33. Photomicrograph of a cross section of an endotrophic mycorrhiza on yellowpoplar showing the intracellular hyphae in the cortical cells and the absence of a fungal mantle. (*Photograph by K. D. Doak.*)

there appears a tumorlike swelling that cannot be distinguished from crown gall, then roots grow out from this swelling, and finally root development is so profuse as to suggest a witches'-broom (Fig. 32). The same control measures are applicable as for crown gall.

MYCORRHIZA

This name is applied to a compound organ consisting of a plant root and a fungous mycelium, the two elements being constantly arranged in an orderly manner in reference to each other. Among the large number of papers on mycorrhiza (Kelley 1950), some deal with mycorrhiza on trees (Bergemann 1956; Björkman 1956; Harley 1956, 1959; Levisohn 1958; Wilde 1954). The methods for investigating these organs have been discussed (Melin 1936). The fungal components of mycorrhiza belong to diverse groups of fungi, the majority belonging to the Hymeno-

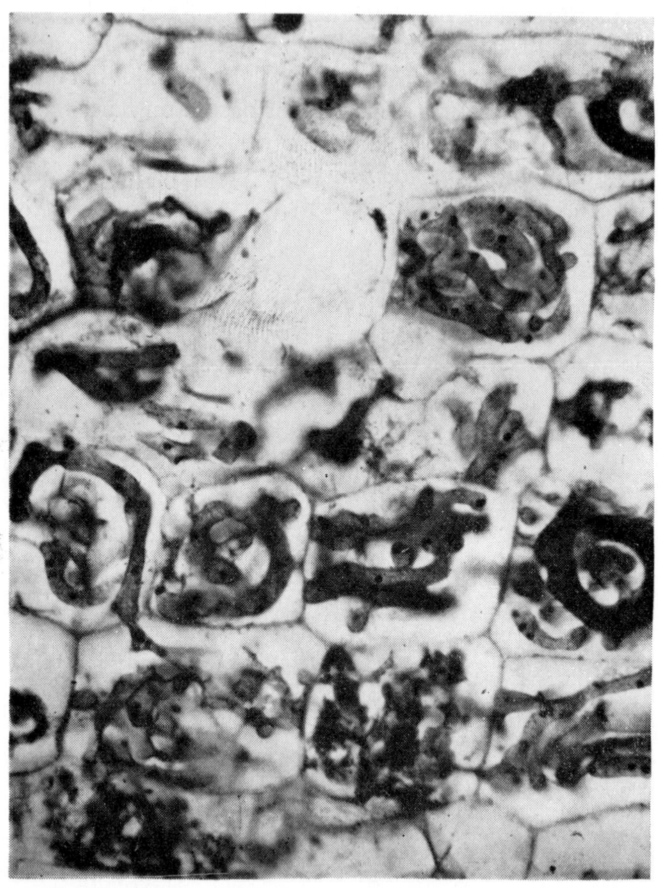

Fig. 34. Photomicrograph of a portion of a longitudinal section of an endotrophic mycorrhiza on yellowpoplar showing details of the intracellular hyphae. (*Photograph by K. D. Doak.*)

mycetes (chiefly members of the families Agaricaceae and Boletaceae), lower Phycomycetes, and the genus *Rhizoctonia* in the Fungi Imperfecti. Mycorrhiza are widespread in vascular plants, occurring normally in all sections of the plant kingdom, with the exception of parasites, water plants, and perhaps a dozen plant families whose members are characteristically found in fertile or water-saturated soils. The term "mycorrhiza" is also extended, with more or less confusion, to plants that lack true roots or where invasion is also in parts other than roots (liverworts, mosses, orchids, and ericaceous plants) but in which the "mycorrhizalike" association seems to exist.

Types. Almost all mycorrhiza may be grouped into three types: endotrophic (endophytic), ectotrophic (ectophytic), and ectendotrophic.

Endotrophic mycorrhiza, found in approximately 75 per cent of all plants, are produced on those plants in which there is apparently little differentiation into root types or in which the primary roots differ chiefly in their fineness or coarseness. Endotrophic mycorrhiza are characterized by the hyphae being intracellular (Figs. 33, 34), producing, within the

Fig. 35. Ectotrophic mycorrhiza on a root of eastern white pine. (*Photograph by A. B. Hatch.*)

cells of the primary root cortex, coils and various swellings that eventually disappear either partially or wholly as a result of digestion by the invaded cells. No fungal mantle is present. There is no noticeable swelling of the roots although heavily infected cells become hypertrophied, and root hairs often develop normally.

Ectotrophic mycorrhiza are usually produced in those groups of plants whose roots are differentiated into so-called "long roots" and "short roots." These include all members of the Fagaceae, Betulaceae, Salicaceae, the hickories, a few other genera among broadleaf trees, and the Abietineae among conifers. In these practically all the short roots in the upper layers of soil are swollen and branched in a corallike manner (Fig. 35), sometimes being brightly colored. Branching is dichotomous in the pines but racemose in other species. In such short roots the fungus is largely confined to the primary root cortex, forming an intercellular net of hyphae between the cortical cells (Hartig's net), and the epidermis and root hairs are replaced by a mantle of mycelium that completely covers the root and effectively isolates it from the soil (Figs. 36, 37). "Knollen" mycorrhiza, sometimes found on pines, are compound ectotrophic mycorrhiza with a superficial resemblance to root nodules on leguminous plants.

Ectendotrophic mycorrhiza combine the characteristics of both forms. Either the same fungus produces both the intercellular and intracellular infections, or two distinct fungi are responsible.

Pseudomycorrhiza usually result from infections by casual soil fungi that invade both the cortical and vascular tissues of the root and appear to retard root development. Except in unusual cases pseudomycorrhizal infections can be distinguished from mycorrhizal infections because of the complete lack of orderly arrangement of the invading hyphae in the former.

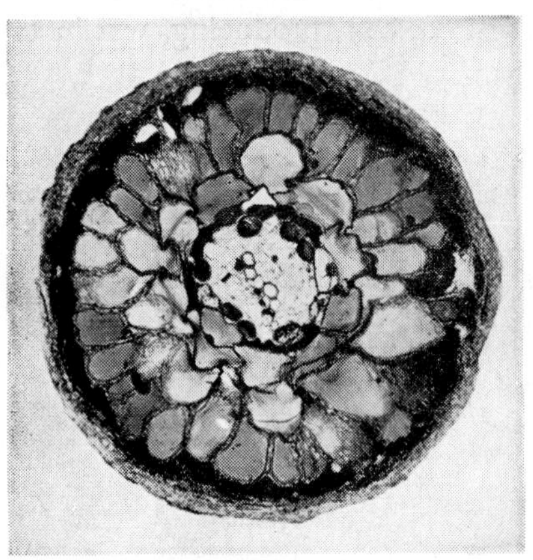

Fig 36. Photomicrograph of a cross section of an ectotrophic mycorrhiza on loblolly pine showing the fungal mantle and the intercellular network of hyphae (Hartig's net) between the cortical cells. (*Photograph by K. D. Doak.*)

Mycorrhiza are produced during all periods of root elongation except when the ground is frozen or when it is too dry for root formation, their activities coinciding with that of other types of roots (long roots). They grow for one season only, secondary growth having been reported only in rare cases. In the spring practically all mycorrhiza that were formed the previous year are brown and shriveled. Warm weather and abundant rainfall are necessary for the development of mycorrhiza. The use of soil biocides delays the development of mycorrhiza (Palmer and Hacskaylo 1958; Wright 1957).

Function. Free atmospheric nitrogen is not fixed by mycorrhiza. The effect of mycorrhizal formation on the higher plant is controversial, the fungal component being considered a weak pathogen by some authorities (Burges 1936; Renner 1935), and it has been suggested that a fungus may penetrate roots by mycorrhizal association and later cause disease (Gosselin 1944; Feliciani and Montefiori 1954). The majority have concluded that the higher plant is benefited (Melin 1953), or if not, that the fungus at least is not pathogenic. The benefits attributed to mycorrhiza are that the absorbing area of the roots is directly increased; the fungal component makes available to the tree organic nitrogen, carbohydrates, mineral nutrients, auxins and enzymes; the fungal component stimulates the metabolic processes of the roots; the fungal component creates favorable soil conditions by releasing nutrients and other substances from organic matter and by destroying toxic soil compounds that may result from organic decomposition; mycorrhizal roots are pro-

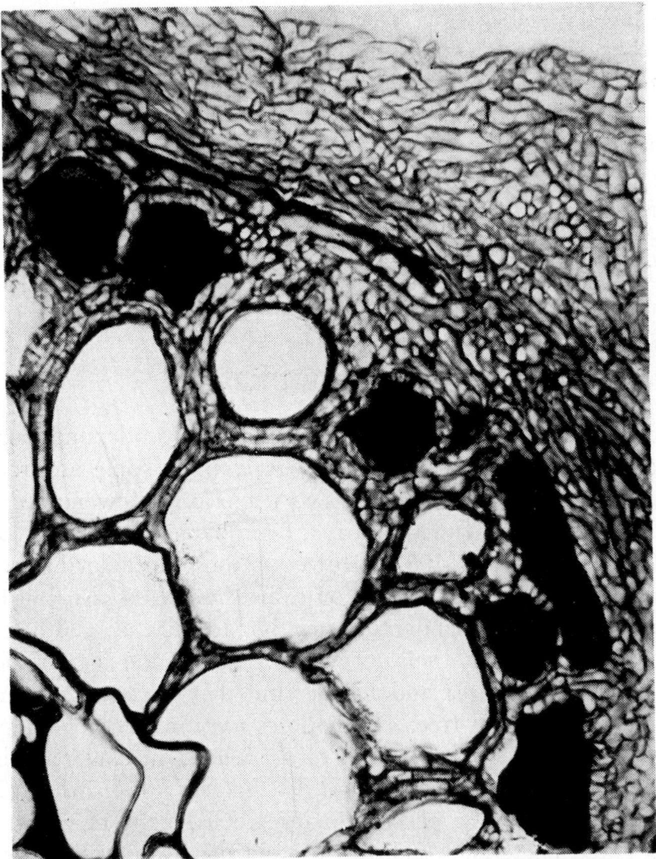

Fig. 37. Photomicrograph of a portion of a cross section of an ectotrophic mycorrhiza on loblolly pine showing details of the fungal mantle and the intercellular network of hyphae (Hartig's net) between the cortical cells. (*Photograph by K. D. Doak.*)

tected from pathogenic fungi; and trees with mycorrhiza are more drought resistant (Cromer 1935). Mycorrhiza are most valuable in poor soils.

Whatever the exact function of mycorrhiza may be, trees in poor soils may fail to grow satisfactorily without them. The absence of mycorrhiza on sickly trees in poor soil may indicate either a lack of essential fungi or inhibition of their activity by some character of the soil. Then the proper fungi must be introduced on trees or soil from another locality, or the factor that prevents the activity of the fungi must be corrected. However, in nursery practice introducing plants or soil from another region or even from another nursery near by is quite likely to result in the introduction of undesirable pathogens that will either jeopardize the existence of the nursery or considerably increase the cost of growing stock because of added expenditures for disease control. A better, although not entirely safe, procedure is to introduce soil from a vigorous natural stand or a thrifty, reasonably old plantation of the same or closely related species to those growing in the nursery and within a radius of not over 100 miles from it. In afforestation projects where nurseries are established in localities remote from natural or planted stands, this is not possible, so that in certain instances the necessity for introducing mycorrhizal fungi might conceivably prove finally to outweigh the danger of bringing in pathogenic organisms.

ROOT NODULES

Root nodules or root tubercles, almost universal on leguminous plants, result from the association of bacteria (*Rhizobium* spp.) and roots, but, although nodules of various types also occur on a few nonleguminous plants, their cause and function are still problematical (Fred, Baldwin, and McCoy 1932; Wilson 1937). However, those on alder fix free nitrogen (Bond 1956). Nodules are not an expression of disease, but they are abnormal structures dependent upon the activity of another agency than the plant itself for their formation. Among the leguminous trees they occur on black locust and acacias, but they do not occur on honey locust, redbud, or coffee-tree. The nodules vary in shape from spherical to elongated and in size from that of a pinhead in clover to that of a pea in trailing wild bean *Strophostyles helvola* (L.) Britt. They are composed chiefly of large parenchymatous cells, many of which contain bacteria. These bacteria, which are present somewhat generally in soil, including forest soil, enter the tissues of the host through root hairs. The nodules generally live but one season, this being so even in perennial herbaceous plants, and consequently there is a new bacterial infection each year, but the nodules of leguminous trees and shrubs may live over from year to year. On black locust they live for 2 or 3 years (Lehotsky 1932:39).

The nodule-forming bacteria are extremely valuable, because, when growing in roots, they fix free atmospheric nitrogen, the resulting nitrogenous compounds becoming available to the plants, which otherwise could not persist in soils free from or poor in nitrogenous material. That is why black locust and pea-tree are such good soil builders. In a nursery, growing of leguminous annuals that are plowed under on a portion of the area on a 2- or 3-year cycle is a most effective method of fertilizing. This is known as "green manuring." However, if done just before sowing it tends to increase damping-off and root rots (page 87).

ROOT GRAFTING

Root grafting between living trees of the same species is common among conifers and hardwoods, but grafting between trees of different species is probably rare (La Rue 1934). Whether trees are weakened by such unions is problematical. After a coniferous stand has been cut, it is common to find callus tissue developing on the tops of stumps as a result of root grafts between the root system of the stump and adjacent living trees (Page 1927), the unions occurring, of course, while the trees were still living. No substantiation has been found for the theory that such callusing is dependent on food stored in the roots before the tree was cut. In Douglas fir, overgrowth of stumps is common, and growth is known to continue for 60 years with the entire cut surface of the stump finally covered by a thick callus (Newins 1916). Cursory examination of Douglas fir stands with many living stumps fails to show any apparent ill effects to the living trees, but no exact investigations have been made. Root grafts between diseased and healthy trees may serve to spread infection. The Dutch elm disease, oak wilt, phloem necrosis, and root rot caused by *Fomes annosus* are sometimes transmitted in this way. Red ring rot caused by *Fomes pini* has passed from decayed trees of Himalayan pine to sound ones by root grafts (Hafiz Khan 1910). It is also possible that diseased trees may obtain considerable sustenance from thrifty trees through root grafts, thus impairing the general vigor of their neighbors.

REFERENCES

Anonymous: Resinosis Disease. *N.Y. State Conserv. Dept. Ann. Rept.,* 26:128–129 (1937).

Anonymous: Root Disease Controlled in Genetics Arboretum. *Calif. Forest and Range Expt. Sta. Ann. Rept.,* 1954:37–38 (1955).

Ark, P. A.: Chemical Eradication of Crown Gall on Almond Trees. *Phytopathology,* 31:956–957 (1941).

Arvidson, B.: En studie av granrotrötans (*Polyporus annosus* Fr.) ekonomiska Konsekvenser. (A Study of the Economic Effects of Root Rot (*Polyporus annosus* Fr.) in the Norway Spruce.) [English summary.] *Svenska Skogsvardsför. Tidsskr.,* 52:381–412 (1954).

Baker, W. L.: Studies on the Transmission of a Virus Causing Phloem Necrosis of American Elm, with Notes on the Biology of Its Insect Vector. *J. Econ. Entomol.,* 42:729–732 (1949).

Bakshi, B. K.: Diseases and Decays of Conifers in the Himalayas. *Indian Forester,* 81:779–797 (1955).

Benton, V. L., and J. Ehrlich: Variation in Culture of Several Isolates of *Armillaria mellea* from Western White Pine. *Phytopathology,* 31:803–811 (1941).

Bergemann, Jutta: Das Mykorrhiza-problem in der Forstwirtschaft. *Allgem. Forstz.,* 11:297–304 (1956).

Björkman, E.: Über die Natur der Mykorrhizabildung unter besonderer Berücksichtigung der Waldbäume und die Anwendung in der forstliche Praxis. *Forstwiss. Centr.,* 75:265–286 (1956).

Bliss, D. E.: The Destruction of *Armillaria mellea* in Citrus Soils. *Phytopathology,* 41:665–683 (1951).

Bond, G.: Evidence for Fixation of Nitrogen by Root Nodules of Alder (*Alnus*) under Field Conditions. *New Phytologist,* 55:147–153 (1956).

Brinkerhoff, L. A., and R. B. Streets: Pathogenicity and Pathological Histology of *Phymatotrichum omnivorum* (the Fungus Causing Cotton or Texas Root Rot) in a Woody Perennial—the Pecan. *Ariz. Univ. Agr. Expt. Sta. Tech. Bull.,* 111:103–126 (1946).

Buckland, D. C.: Observation on *Armillaria mellea* in Immature Douglas Fir. *Forestry Chronicle,* 29:344–346 (1953).

Buckland, D. C., A. C. Molnar, and G. W. Wallis: Yellow Laminated Root Rot of Douglas Fir. *Can. J. Botany,* 32:69–81 (1954).

Burges, A.: On the Significance of Mycorrhiza. *New Phytologist,* 35:117–131 (1936).

Campbell, A. H.: Zone Lines in Plant Tissues II. The Black Lines Formed by *Armillaria mellea* (Vahl.) Quél. *Ann. Appl. Biol.,* 21:1–22 (1934).

Childs, L., and S. M. Zeller: Observations on Armillaria Root Rot of Orchard Trees. *Phytopathology,* 19:869–873 (1929).

Childs, T. W.: Synopsis of Present Information Concerning *Poria weirii* Root Rot in Douglas-fir. *Pacific Northwest Forest and Range Expt. Sta. Research Note,* 116:1–2 (1955).

Christensen, C. M.: Root Rot of Pines Caused by *Armillaria mellea.* [Abstract.] *Phytopathology,* 28:5 (1938).

Crandall, B. S.: The Distribution and Significance of the Chestnut Root Rot Phytophthoras, *P. cinnamomi* and *P. cambivora. U.S. Dept. Agr., Plant Disease Reptr.,* 34:194–196 (1950).

Crandall, B. S., G. F. Gravatt, and Margaret M. Ryan: Root Disease of Castanea Species and Some Coniferous and Broadleaf Nursery Stocks, Caused by *Phytophthora cinnamomi. Phytopathology,* 35:162–180 (1945).

Cromer, D. A. N.: The Significance of the Mycorrhiza of *Pinus radiata. Australia, Commonwealth Forestry Bur. Bull.,* 16:1–17 (1935).

Day, W. R.: Parasitism of *Armillaria mellea* in Relation to Conifers. *Quart. J. Forestry,* 21:9–21 (1927).

Day, W. R.: Environment and Disease. A Discussion on the Parasitism of *Armillaria mellea* (Vahl.) Fr. *Forestry,* 3:94–103 (1929).

Day, W. R.: Development of Disease in Living Trees. *J. Brit. Wood Preserving Assoc.,* 4:25–44 (1934).

Day, W. R.: Root-rot of Sweet Chestnut and Beech Caused by Species of *Phytophthora.* I. Cause and Symptoms of Disease: Its Relation to Soil Conditions. *Forestry,* 12:101–116 (1938).

Day, W. R.: The Penetration of Conifer Roots by *Fomes annosus. Quart. J. Forestry,* 42:99–101 (1948).

De Ropp, R. S.: The Crown-gall Problem. *Botan. Rev.*, 17:629–670 (1951).

Edgerton, C. W.: A *Ganoderma* and Its Association with Root Rot of Trees in Louisiana. [Abstract.] *J. Tenn. Acad. Sci.*, 29:178 (1954).

Ehrlich, J.: A Preliminary Study of Root Diseases in Western White Pine. *North. Rocky Mt. Forest and Range Expt. Sta. Paper*, 1:1–10 (1939).

Englerth, G. H.: Decay of Western Hemlock in Western Oregon and Washington. *Yale Univ., School Forestry Bull.*, 50:1–53 (1942).

Etheridge, D. E.: Comparative Studies of North American and European Cultures of the Root Rot Fungus, *Fomes annosus* (Fr.) Cke. *Can. J. Botany*, 33:416–428 (1955).

Feliciani, A., and R. Montefiori: Allevamento in vivaio di *Pseudotsuga douglasi* e colonizzazione micorrizica delle piante. [English summary.] *Monti e Boschi*, 5:215–217 (1954).

Francke-Grosmann, H.: Rotfäule und Riesenbastkäfer, eine Gefahr für die Sitkafichte auf Öd- und Ackerlandaufforstungen Schleswig-Holsteins. *Forst u. Holz*, 3:232–235 (1948).

Fred, E. B., I. L. Baldwin, and Elizabeth McCoy: Root Nodule Bacteria and Leguminous Plants. *Wisconsin Univ. Studies Sci.*, 5:I–XXII, 1–343 (1932).

Fromme, F. D.: The Black Rootrot Disease of Apple. *Virginia Agr. Expt. Sta. Tech. Bull.*, 34:1–52 (1928).

Garrett, S. D.: A Century of Root-disease Investigation. *Ann. Appl. Biol.*, 42:211–219 (1955).

Garrett, S. D.: "Biology of Root-infecting Fungi." New York: Cambridge University Press (1956*a*).

Garrett, S. D.: Rhizomorph Behaviour in *Armillaria mellea* (Vahl.) Quél. II. Logistics of Infection. *Ann. Botany*, 20:194–209 (1956*b*).

Gosselin, R.: Studies on *Polystictus circinatus* and Its Relation to Butt-rot of Spruce. *Farlowia*, 1:525–568 (1944).

Hafiz Khan, A.: Root Infection of *Trametes pini* (Brot.) Fr. *Indian Forester*, 36:559–562 (1910).

Harley, J. L.: The Mycorrhiza of Forest Trees. *Endeavour*, 15:43–48 (1956).

Harley, J. L.: "The Biology of Mycorrhiza." London: Leonard Hill, Ltd. (1959).

Hennig, R.: Über die bei uns vorkommenden Sparassis-Arten (krause Glucke und Eichen-Glucke) und ihr Parasitismus an Waldbäumen. *Forstwiss. Centr.*, 71:108–116 (1952).

Hewitt, J. L.: A Survey Concerning a Native Pathogen, *Armillaria mellea*. [Abstract.] *Phytopathology*, 24:1142 (1934).

Hiley, W. E.: "The Fungal Diseases of the Common Larch." Oxford: Oxford University Press (1919).

Hopffgarten, E. H. von: Beiträge zur Kenntnis der Stockfäule (*Trametes radiciperda*) *Phytopathol. Z.*, 6:1–48 (1933).

Hubert, E. E.: A Disease of Conifers Caused by *Stereum sanguinolentum*. *J. Forestry*, 33:485–489 (1935).

Hunt, J.: *Phytophthora lateralis* on Port-Orford cedar. *U.S. Dept. Agr., Forest Serv., Pacific Northwest Forest Expt. Sta. Research Note*, 172:1–6 (1959).

Jorgensen, E., and B. B. Petersen: Angreb of *Fomes annosus* (Fr.) Cke. og *Hylesinus piniperda* L. på *Pinus silvestris* i Durslands plantager. [English summary.] *Dansk Skovfor. Tidsskr.*, 36:453–479 (1951).

Kelley, A. P.: "Mycotrophy in Plants." Waltham, Mass.: Chronica Botanica Company (1950).

Lanphere, W. M.: Enzymes of the Rhizomorphs of *Armillaria mellea. Phytopathology,* 24:1244–1249 (1934).

La Rue, C. D.: Root Grafting in Trees. *Am. J. Botany,* 21:121–126 (1934).

Leach, R.: Biological Control and Ecology of *Armillaria mellea* (Vahl.) Fr. *Brit. Mycol. Soc. Trans.,* 23:320–329 (1939).

Lehotsky, K.: Výsledky niektorých pozorovaní Koreňov a nadárov agatu. (*Robinia pseudoacacia* L.). [French resumé.] *Lesnická Práce,* 11:26–39 (1932).

Lek, H. A. A. van der: Over apárasitaire Vitwassen aan hautige Planten. [English summary.] *Tijdschr. Plantenziekten,* 35:25–59 (1929).

Levisohn, Ida: Effects of Mycorrhiza on Tree Growth. *Soils and Fertilizers,* 21:73–82 (1958).

Long, W. H.: *Polyporus dryadeus,* a Root Parasite on the Oak. *J. Agr. Research,* 1:239–250 (1913).

Long, W. H.: *Polyporus dryadeus,* a Root Parasite on White Fir. *Phytopathology,* 20:758–759 (1930).

Marsh, R. W.: Field Observations on the Spread of *Armillaria mellea* in Apple Orchards and in a Black Currant Plantation. *Brit. Mycol. Soc. Trans.,* 35:201–207 (1952).

Martinez, J. B.: El *Fomes annosus* Fr. (*Trametes radiciperda* Hart.) en España. *Madrid Jard. Botan. Ann.,* 3:23–49 (1943).

Melin, E.: Methoden der experimentellen Untersuchung mykotropher Pflanzen. *Handbuch Biol. Arbeitsmethoden,* 11:1015–1108 (1936).

Melin, E.: Physiology of Mycorrhizal Relations in Plants. *Ann. Rev. Plant Physiol.,* 4:325–346 (1953).

Molin, N.: Om *Fomes annosus* spridningsbiologi. [English summary.] *Statens Skogsforskningsinst. (Sweden), Medd.,* 47(3):1–36 (1957).

Newhook, F. J.: The Association of *Phytophthora* spp. with Mortality of *Pinus radiata* and Other Conifers I. Symptoms and Epidemiology in Shelterbelts. *New Zealand J. Agr. Research,* 2:808–843 (1959).

Newins, H. S.: The Natural Root Grafting of Conifers. *Soc. Am. Foresters Proc.,* 11:394–404 (1916).

Nickell, L. G.: A Species of *Ganoderma* Probably Pathogenic to Sassafras in New York. *U. S. Dept. Agr., Plant Disease Reptr.,* 36:28–29 (1952).

Norton, D. C., and R. Behrens: *Ganoderma zonatum* Associated with Dying Mesquite. *U.S. Dept. Agr., Plant Disease Reptr.,* 40:253–254 (1956).

Orłoś, H.: Badania nad zwalczaniem opienki miodowej (*Armillaria mellea* Vahl.) metoda biologiczna. (Investigations upon Combating of *Armillaria mellea* Vahl. with Biological Control.) [English summary.] *Roczniki Nauk Leśnych,* 15:195–235 (1957).

Page, F. S.: Living Stumps. *J. Forestry,* 25:687–690 (1927).

Palmer, J. G., and E. Hacskaylo: Additional Findings as to the Effects of Several Biocides on Growth of Seedling Pines and Incidence of Mycorrhizae in Field Plots. *U.S. Dept. Agr., Plant Disease Reptr.,* 42:536–537 (1958).

Perrson, A.: Über den Stoffwechsel und eine antibiotisch wirksame Substanz von *Polyporus annosus* Fr. *Phytopathol. Z.,* 30:45–86 (1957).

Populer, C.: La pourriture rouge du coeur des resineux (*Fomes annosus* Fr. Cooke). *Soc. roy. forest. de Belg. bull.,* 63:297–329 (1956).

Ramsbottom, J.: Fungi and Forestry. *Scot. Forestry J.,* 49:34–51 (1935).

Rattsjö, H., and E. Rennerfelt. Värdeförlusten på virkesutbytet till följd av rotröta. (Economic Losses in Sawtimber and Pulpwood Yield Caused by Root-rot Fungi.) [English summary.] *Norrlands Skogsvårdsförb. Tidsskr.,* 1955:279–298 (1955).

Reitsma, J.: Studien über *Armillaria mellea* (Vahl.) Quél. *Phytopathol. Z.*, 4:461–522 (1932).

Renner, Sophie: Beitrag zur Kenntnis einiger Wurzelpilze. *Phytopathol. Z.*, 8:457–487 (1935).

Rennerfelt, E.: Om rotrotan (*Polyporus annosus* Fr.) i Sverige. Dess utbredning och sätt att uppträda. [German summary.] *Statens Skogsforskningsinst.* (*Sweden*), *Medd.*, 35(8):1–88 (1946).

Rennerfelt, E.: Om angrepp av rotröta på tall. (On Root Rot Attack on Scots Pine.) (English summary.) *Statens Skogsforskningsinst.* (*Sweden*), *Medd.*, 41(9):1–40 (1953).

Rennerfelt, E.: Untersuchungen über die Wurzelfäule auf Fichte und Kiefer in Schweden. [English summary.] *Phytopathol. Z.*, 28:259–274 (1957).

Rennerfelt, E., and Sheila K. Paris: Some Physiological and Ecological Experiments with *Polyporus annosus* Fr. *Oikos*, 4:58–76 (1952–1953).

Rhoads, A. S.: The Occurrence and Destructiveness of Clitocybe Root Rot of Woody Plants in Florida. *Lloydia*, 19:193–240 (1956).

Rhoads, A. S., and E. Wright: *Fomes annosus* Commonly a Wound Pathogen Rather Than a Root Parasite of Western Hemlock in Western Oregon and Washington. *J. Forestry*, 44:1091–1092 (1946).

Riker, A. J., and G. W. Keitt: Studies of Crowngall and Wound Overgrowth on Apple Nursery Stock. *Phytopathology*, 16:765–808 (1926).

Riker, A. J., et al.: Studies on Infectious Hairy-root of Nursery Apple Trees. *J. Agr. Research*, 41:507–540 (1930).

Rishbeth, J.: Observations on the Biology of *Fommes annosus*, with Particular Reference to East Anglian Pine Plantations. *Ann. Botany* (*London*), 14:365–383 (1950); 15:1–21, 221–246 (1951).

Rishbeth, J.: Control of *Fomes annosus* Fr. *Forestry*, 25:41–50 (1952).

Rishbeth, J.: Some Further Observations on *Fomes annosus* Fr. *Forestry*, 30:69–89 (1957).

Roll-Hansen, F.: Undersøkelser over *Polyporus annosus* Fr., saerlig med henblickk på dens forekomst i det sønnafjelske Norge. [English summary.] *Norske Skogsforsøksv. Medd.*, 7:1–100 (1940).

Roth, L. F., E. J. Trione, and W. H. Ruhmann: Phytophthora Induced Root Rot of Native Port-Orford-cedar. *J. Forestry*, 55:294–298 (1957).

Schmitz, H., and L. W. Jackson: Heartrot of Aspen with Special Reference to Forest Management in Minnesota. *Minn. Agr. Expt. Sta. Tech. Bull.*, 50:1–43 (1927).

Smith, C. O.: Crown Gall on Incense Cedar, *Libocedrus decurrens. Phytopathology*, 27:844–849 (1937).

Smith, C. O.: Crown Gall on Species of Taxaceae, Taxodiaceae, and Pinaceae as Determined by Artificial Inoculations. *Phytopathology*, 32:1005–1009 (1942).

Smith, C. O., and L. C. Cochran: Crown Gall and Irrigation Water. *U.S. Dept. Agr., Plant Disease Reptr.*, 28:160–162 (1944).

Swingle, R. U.: Phloem Necrosis A Virus Disease of the American Elm. *U.S. Dept. Agr. Circ.*, 640:1–8 (1942).

Sylvester, E. P., and Mary C. Countryman: A Comparative Histological Study of Crowngall and Wound Callus on Apple. *Am. J. Botany*, 20:328–340 (1933).

Thomas, H. E.: Studies on *Armillaria mellea* (Vahl.) Quél., Infection, Parasitism, and Host Resistance. *J. Agr. Research*, 48:187–218 (1934).

Tucker, C. M.: Phloem Necrosis, a Destructive Disease of the American Elm. *Missouri Agr. Expt. Sta. Circ.*, 305:1–15 (1945).

Weir, J. R.: *Sparassis radicata*, an Undescribed Fungus on the Roots of Conifers. *Phytopathology*, 7:166–177 (1917*a*).

Weir, J. R.: Note on *Xylaria polymorpha* and *X. digitata*. *Phytopathology*, 7:223–224 (1917*b*).

Weir, J. R.: *Polyporus dryadeus* (Pers.) Fr. on Conifers in the Northwest. *Phytopathology*, 11:99 (1921).

Wilde, S. A.: Mycorrhizal Fungi: Their Distribution and Effect on Tree Growth. *Soil Sci.*, 78:23–31 (1954).

Wilson, P. W.: Symbiotic Nitrogen-fixation by the Leguminosae. *Botan. Rev.*, 3:365–399 (1937).

Woeste, U.: Anatomische Untersuchungen über die Infektionswege einiger Wurzelpilze. *Phytopathol. Z.*, 26:225–272 (1956).

Wright, E.: Importance of Mycorrhizae to Ponderosa Pine Seedlings. *Forest Sci.*, 3:275–280 (1957).

Wright, E., and H. R. Wells: Tests on the Adaptability of Trees and Shrubs to Shelterbelt Planning on Certain Phymatotrichum Root Rot Infested Soils of Oklahoma and Texas. *J. Forestry*, 46:256–262 (1948).

Wright, W. H., A. A. Hendrickson, and A. J. Riker: Studies on the Progeny of Single-cell Isolations from the Hairy-root and Crown-gall Organisms. *J. Agr. Research*, 41:541–547 (1930).

York, H. H., R. E. Wean, and T. W. Childs: Some Results of Investigations on *Polyporus schweinitzii* Fr. *Science*, 84:160–161 (1936).

Zak, B.: Littleleaf of Pine. *U.S. Dept. Agr., Forest Serv., Forest Pest Leaflet*, 20:1–4 (1957).

CHAPTER 7

Foliage Diseases of Hardwoods

There are many diseases of the foliage of broad-leaved trees or hardwoods caused by parasitic fungi, which may also extend their activities to the flowers, fruits, and young twigs. The fact that hardwoods in general have good recuperative power and are not usually seriously affected by foliage diseases is the principal reason why these diseases have not been more extensively investigated. Some hardwoods are attacked by foliage diseases year after year without suffering apparent injury. If an infection is severe and continuous over a period of years, it must be reflected in reduced growth rate of the trees and in a general unhealthy condition that subjects them to attacks of other fungi and insects, or it lowers their resistance to unfavorable weather conditions. Leaf diseases are mainly important when defoliation results or when an attack is so heavy that practically all the leaves are involved and cannot function normally. A single complete defoliation rarely kills hardwoods. Defoliation for several successive years is usually required to bring about their death.

The determination of loss caused by defoliation is not easy unless trees are killed outright. It has been concluded that the growth of hardwood stands attacked by the gypsy moth (*Porthetria dispar* L.) is reduced in exact proportion to the degree of defoliation occurring during June and July, in that a stand with 50 per cent defoliation would make only half-normal growth in volume (Summers and Burgess 1933). Furthermore, diameter growth declines the same year defoliation occurs (Baker 1941). Defoliation in August is much less serious. However, artificial defoliation of several broad-leaved evergreens in autumn greatly increased their susceptibility to cold injury, so that most of the defoliated plants died during the winter (Kramer and Wetmore 1943). The effect on growth of diseases that kill only part of each individual leaf and do not cause defoliation is probably slight.

Leaf Spots. Leaf spots on trees may be caused by insects, toxic gases, bacteria, and fungi (Kerling 1928). Necrotic spots and mottling of leaves are symptoms of virus diseases of American elm (Swingle and Bretz 1950). There are many leaf spots on hardwoods caused by fungi that have not been investigated, so that little is known about the life history and pathogenicity of the fungi causing them. Some of these fungi also attack young twigs and small branches as well as leaves. Most leaf spots are caused by Ascomycetes or Fungi Imperfecti.

Some species among the Fungi Imperfecti are merely known stages of Ascomycetes, but for most of them the perfect stage has not been determined. Sometimes several species of Fungi Imperfecti may be only different stages of the same fungus on the same host or on different hosts to which different names have been given, thus resulting in confusion in classification. In the Fungi Imperfecti, if the fructification is a pycnidium, the fungus is classified and named differently than if the fructification is an acervulus, yet one species may have both kinds of fructifications. Then, if the species also has an ascus stage, three different names may be applied to the one fungus before its life history is fully known.

The damage caused by leaf spots is usually not important, unless defoliation occurs and particularly if defoliation is repeated for several successive seasons.

The characteristic symptom of leaf spots is the formation of dead areas in the leaf. The areas vary in size and shape from small to large, and from circular to angular or irregular. As the name implies, the tissue of an entire leaf is rarely completely killed, except by a number of spots coalescing, but an infected leaf usually has few to many dead spots surrounded by green tissue normal in appearance. Although in some cases the advance of the hyphae in the leaf is checked by the formation of an impenetrable wound tissue around the dead area, in the great majority of cases no such barrier is laid down, and the reason for the pathogenic effect of the parasite being limited to a definite spot is unknown (Cunningham 1928). The killed tissues vary in color from yellowish to all shades of brown and almost black, and different colors may occur as zones in a single spot. Frequently the dead tissue falls out leaving holes. When small circular spots result and the killed tissue falls out completely the effect is known as "shot hole."

Fructifications of the fungi develop on the dead tissue. To the naked eye or through a hand magnifier, they appear as tiny brown or black dots, or larger black spots, scattered over the dead tissue or clustered in the center of the spot. Spores are developed in these fructifications.

The life history of leaf spot fungi is generally similar. First infection in the spring comes either from ascospores that have developed on dead fallen leaves, or from spores produced by fructifications on the bark, if the fungus is one that also attacks the bark tissue of twigs. Then during the remainder of the spring and summer the rapid spread of these fungi is by means of spores produced on the killed tissue of the leaves. Wet seasons favor epidemics of leaf spots.

In forest stands the amount of damage resulting has not warranted intensive investigation of control, and no practical measures are known. For shade and ornamental trees, fallen diseased leaves should be collected and burned in the fall, dead and cankered twigs should be pruned off before the buds open in the spring, and, if adjacent infected trees are not cared for, spraying or dusting should be resorted to. Bordeaux mixture, zineb, captan, or fixed copper are satisfactory sprays. Three applications are necessary; the first when the buds are swelling, the second when the leaves are about 1/4 inch out, and the third from 7 to 10 days later. For those leaf-disease fungi which also live in the twigs, the first application should be made when the first green color appears at the beginning of growth.

American elm particularly is subject to leaf spots (Miles 1921). The most prevalent spot, also occurring occasionally on other elms, is that caused by *Gnomonia ulmea* (Schw.) Thüm. (Pomerleau 1937–1938). The disease appears early in the spring, and, when severe, considerable defoliation results. On seedlings, twig and petiole infection also occurs, sometimes killing all the new growth. The small, irregularly circular spots (Fig. 38), confined largely to the upper surface of the leaf, are grayish in color with the fructifications appearing as black pustules, somewhat concentrically arranged in the center of, or scattered over, the dead area. Spots caused by *Gloeosporium ulmicolum* Miles are elongated along the midrib, veins, and leaf margins, being visible on both leaf surfaces (Trumbower 1934:63). *G. inconspicuum* Cav. causes subcircular brown spots with darker margins and centers, visible on both the upper and lower leaf surfaces.

Large leaf spot of American chestnut and oaks, distributed from Indiana to New Jersey and south to Arkansas and Florida but particularly common in the southern Appalachian region, is caused by *Monochaetia desmazierii* Sacc. (Hedgcock 1929). Red maple, several hickories, witch-hazel, and winged elm are also attacked. Individual trees may suffer a loss of green leaf tissue up to 40 per cent, but the injury usually occurs in the late summer when growth is largely over, thus lessening the injurious effect on the trees. The spots are large, frequently reaching 1 inch in diameter, and more or less circular and have concentric zones of varying gray, yellow, and brown. Another conspicuous leaf spot of oaks, prevalent in the southern Appalachian region, is caused by *Morenoella quercina* (Ellis & Martin) Theiss. (Luttrell 1940). The spots, which may be 1 cm or more in diameter by September, appear as purplish-black roughly circular areas on the upper surface of the leaves, and as irregular brownish discolored areas on the undersurface. *Septoria querceti* Thuem. causes 2- to 3-mm circular spots with straw-colored centers on leaves of the red oak group (Gilman and Wadley 1952). Numerous minute, blackish-brown spots result from infections

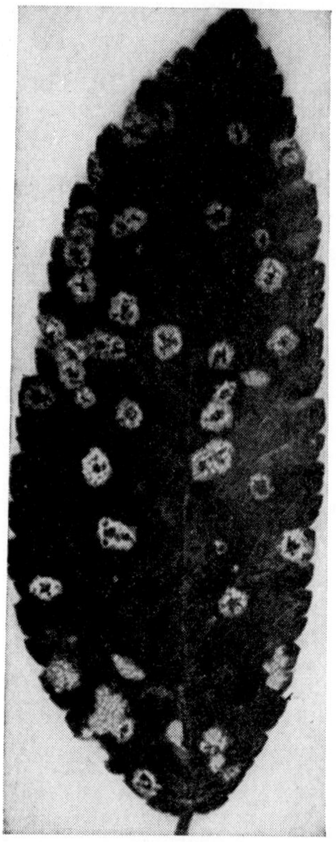

Fig. 38. Leaf spot of elm caused by *Gnomonia ulmea*. (*After F. D. Heald.*)

Fig. 39. Small leaf spot of chestnut caused by *Marssonina ochroleuca*. (*After F. D. Heald.*)

of southern red oak leaves by *Elsinoë quercus-falcatae* Miller (1957). Small leaf spot of American chestnut is caused by *Marssonina ochroleuca* B. & C. (Fig. 39).

Leaf blotch of horse chestnut, which also affects Ohio buckeye, is caused by *Guignardia aesculi* (Pk.) Stew. (Stewart 1916). Both the leaflets and occasionally the petioles are attacked. The irregularly shaped dead areas vary in size from small to so large that they may cause a curling of the leaflet. The central portion of a fully developed spot is from dark red to brown in color with a yellowish margin blending into the green of the healthy tissue. The fructifications of the fungus develop as small black specks on the dead tissue. Anthracnose of horse chestnut, caused by *Glomerella cingulata* (St.) S. & v. S., affects leaflets and petioles (Pierce and Hartley 1916). Lesions on the leaf blades are closely limited to the veins, distinguishing this disease from leaf blotch.

Pecan leaf blotch, caused by *Mycosphaerella dendroides* (Cke.) Dem. & Cole, is prevalent in the South on pecan, and also occurs on hickories (Demaree and Cole 1930). The disease appears in June, and by late summer considerable defoliation may result. The spots are dark brown in color, from 1 to 8 mm in diameter, are usually numerous, and frequently coalesce forming large blotches.

Shoot blight of aspen and largetooth aspen, caused by *Napicladium tremulae* (Frank) Sacc. (*Fusicladium radiosum*), is periodically severe. Young shoots bend double, darken in color, and dry out (McCallum 1920). The leaves discolor and shrivel. *Marssonina populi* (Lib.) Magn. causes a leaf spot and shoot blight of poplars, particularly aspen, that in case of heavy infection results in premature defoliation (Mielke 1957). A leaf blight of balsam poplar, caused by *Linospora tetraspora* Thompson (1939), is widely distributed in Canada. *Septoria musiva* Pk. and *S. populicola* Pk. commonly cause leaf spots on various poplars (Thompson 1941). *S. musiva* also causes cankers (page 264). *Septotinia populiperda* Waterman and Cash (1950) causes a conspicuous leaf blotch on hybrid poplars that may result in premature defoliation. Hybrid poplar leaves are also attacked by *Ciborinia whetzelii* Seaver and *Plagiostoma populi* Cash and Waterman (1957).

A common, small, somewhat angular leaf spot of various maples is caused by *Phleospora aceris* (Lib.) Sacc., an angular to orbicular dark reddish-brown spot of red maple is caused by *Venturia acerina* Plakidas (1942), and a grayish wilt of individual leaves of sugar maple and silver maple is caused by *Cristulariella depraedens* (Cke.) v. Höhn. (Bowen 1930). Red and sugar maples are subject to leaf spots caused by *Phyllosticta minima* (Berk. & Curt.) Ell. & Ev. (Fergus 1954). Species of *Elsinoë, Botrytis, Cercospora, Septoria,* and *Ascochyta* cause leaf spots on flowering dogwood (Jenkins et al. 1953). Large dark-brown spots on older southern magnolia leaves are caused by *Glomerella cingulata* (Ston.) Spauld. & Schrenk (Fowler 1947), whereas small white to gray spots, often coalescing, result from infection by *Elsinoë magnoliae* Miller & Jenkins (1955). Leaf spots and defoliation of species of ash are commonly caused by *Mycosphaerella fraxinicola* (Schw.) House (Wolf 1939) and *M. effigurata* (Schw.) House (Wolf and Davidson 1941). A circular to angular rusty-brown to dark-brown leaf spot of redbud is caused by *M. cercidicola* (Ellis & Kellerm.) Wolf (1940*a*). Large irregular purplish to dark-brown blotches on the leaves of black tupelo and swamp tupelo are caused by *M. nyssaecola* (Cooke) Wolf (1940*b*). *M. mori* (Fkl.) Wolf (1935) causes a small more or less circular leaf spot of red mulberry, and *Linospora gleditsiae* Miller & Wolf (1936) causes a leaf spot of honey locust characterized by numerous flat black fructifications of the fungus on the undersurface of the leaves. Both leaf spots are widely distributed in the South.

Leaf spots and premature defoliation of American sycamore in the Southeastern states is sometimes caused by *Mycosphaerella polymorpha* Smith & Smith (1941), and of California sycamore by *Stigmella platani-racemosae* Dearn. & Barth. Both diseases can be confused with the more common leaf and twig blight. Black angular leaf spots on California laurel are caused by the bacterium *Pseudomonas lauracearum* Harvey (1952).

Leaf and Twig Blight of Sycamores and Oaks. This disease, also known as "anthracnose," is caused by *Gnomonia veneta* (Sacc. & Speg.) Kleb. (Ascomycetes, Gnomoniaceae). It is widespread in the eastern and central United States and in California. Sycamores particularly and to some extent oaks are attacked (Schuldt 1955; Waterman 1951*a*). On oaks the disease is usually on leaves, although twigs may be affected, but on sycamores both leaves and twigs are attacked. In the East, American sycamore may be defoliated for several successive years and the twigs killed back by the latter part of June, but a second crop of leaves is produced each year and few trees die. The damage would be more seriously regarded if sycamores were important timber trees. The premature leaf fall greatly lessens the value of sycamores for shade and ornamental purposes. Individuals within a species vary in susceptibility. Oriental plane is relatively resistant.

Anthracnose appears in the spring. Wet weather when buds are opening and leaves are expanding favors anthracnose development. The sycamore leaves and the growing tips of the twigs may brown and die as they emerge from the bud. In this stage the disease may be confused with late frost injury. Usually the disease develops later, first as brown spots along the main leaf veins, which rapidly enlarge into large dead areas. Two or more of these areas coalesce, finally killing the entire leaf. Small cream-colored dots, about the size of a pinhead, or short tendrils appear on the underside of infected leaves along the veins during moist weather. These are masses of conidiospores extruded from the fructifications (acervuli) in the leaf tissue. In this stage the fungus is known as *Gloesporium nervisequum* (Fckl.) Sacc. Killed leaves soon drop, and severely infected trees remain bare until a second crop of leaves is produced later in the summer.

Small twigs are killed and cankers are formed on branches up to 1 inch in diameter. Below the cankers, water sprouts develop, resulting in an irregular bunchy growth. These cankers may girdle twigs and branches. On stems the acervuli break through the bark the following spring as minute black pimples. The fruiting stages on bark were named as distinct species before it was known that they were merely stages of one fungus.

Control measures for shade trees are to destroy the fallen leaves, prune off and destroy all infected twigs and branches, and spray as directed on page 131. No practical control is known for forest trees.

Anthracnose of river birch caused by *G. betularum* Ell. & Mart. was epidemic in Indiana during 1932 and 1934 (Van Hook and Busteed 1935).

Tar Spots. Large blotches, appearing like spots of tar, are common on leaves of certain hardwoods. When severe, tar spots may cause premature defoliation. In general, injury by tar spots is not serious enough on forest trees to warrant control measures. In nurseries the disease can be controlled by spraying the developing leaves several times with Bordeaux mixture. Destruction of infected leaves after they have fallen to the ground in the autumn is helpful.

Rhytisma acerinum (Pers.) Fr. (Ascomycetes, Phacidiaceae) causes tar spot of maples (Jones 1925). Red and silver maples are commonly affected in the East, whereas Norway and sycamore maples are rarely attacked, although the last two species are quite susceptible in Europe. The first symptoms of tar spot on maple leaves are light yellow-green areas. Later in the summer these areas become shiny black, thickened, and somewhat raised above the upper surface of the leaf. These black blotches, from ⅛ to ½ inch across, are sclerotia (Fig. 40). In the summer, tiny spores, termed "conidia," are developed in the sclerotia, but it is not known how important they are in spreading the disease. The conidial stage is named *Melasmia acerina* Lev. After the infected leaves fall, ascospores are developed in the sclerotia, mature in the spring, and are released during May and June to infect the new foliage. The eight long, threadlike spores per ascus are forcibly ejected into the air to a height of about 1 mm and are then carried by air currents.

R. punctatum (Pers.) Fr. causes specked tar spot of maples. It is rare in the East, occasionally occurring on mountain, silver, and striped maples. It is abundant on bigleaf maple in the Pacific Coast states, but tar spot caused by *R. acerinum* is rare in that region. During the summer, black raised specks about the size of a pinhead are developed in groups on the upper surface of the leaf, occupying light yellow-green areas about ½ inch in diameter. When the remainder of the leaf has faded in the autumn, the yellow-green areas still retain their color.

R. arbuti Phil. causes specked tar spot of madroño on the Pacific Coast. *R. salicinum* (Pers.) Fr. causes unusually thick tar spots on leaves of willows. Several other species of *Rhytisma* cause tar spots on leaves of forest shrubs.

Fig. 40. Tar spot of maple caused by *Rhytisma acerinum.*

Ink Spots. These spots, which are common on aspen and rare on largetooth aspen, are caused by *Sclerotinia whetzelii* Seaver (Ascomycetes, Helotiaceae) in the East and by *S. bifrons* Seaver & Shope (Seaver 1945) in the West. When the disease is severe, small trees may be killed, but those 8 feet or more tall rarely die (Pomerleau 1940). Severely infected leaves are completely killed by midsummer or earlier but persist until autumn. The sclerotia or ink spots, which are black and circular to ellipsoid in shape, varying from 2 to 8 mm in diameter, mostly drop from the leaves during the summer, leaving holes that exactly fit their shape. Apothecia develop in the spring after the sclerotia have lain on the ground through the winter.

Powdery Mildews. The extensively distributed powdery mildews, which occur on a wide variety of plants, are caused by several genera of fungi belonging to the family Erysiphaceae of the Ascomycetes (Heald 1933:565–574). They are obligate parasites, and the existence of strains within a species is not uncommon. Many of them occur on hardwoods, attacking leaves or occasionally fruits and young twigs. They are un-

Fig. 41. Powdery mildew of willow caused by *Uncinula salicis* showing the upper surfaces of a lightly and heavily infected leaf. The perithecia show as tiny black specks.

known on conifers. Usually the damage caused is slight except when young trees, particularly sprouts, are attacked.

On trees the powdery mildews are mostly confined to the upper or lower surface of leaves, forming a white, superficial, cobweblike growth of hyphae visible to the naked eye (Fig. 41). In a few species the mycelium becomes yellowish or brownish with age. By means of special sucking organs (haustoria), the hyphae penetrate the cells of the host to withdraw food. In most genera, only the epidermal or subepidermal cells are invaded, but in the genus *Phyllactinia* the hyphae pass through the stomata into the mesophyll of the leaf, and haustoria are not found in the epidermal cells. Soon after the mycelium develops it appears powdery (hence the common name) through the development of enormous numbers of conidia on the ends of short, erect branches, the conidiophores.

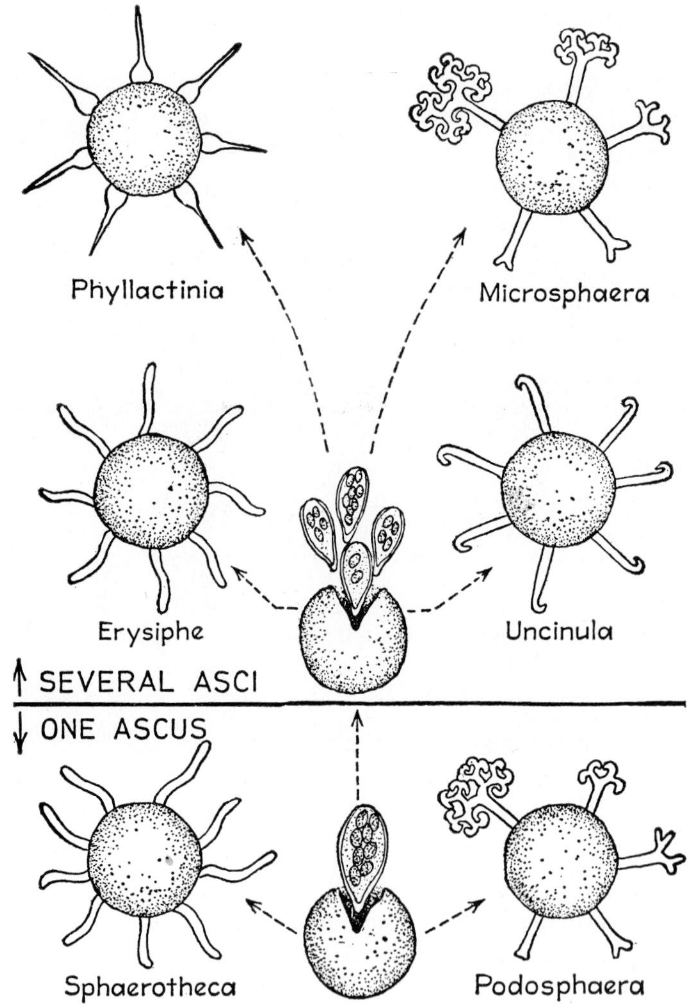

Fig. 42. Semidiagrammatic drawings of the perithecia of the principal genera of powdery mildews. (*After F. D. Heald.*)

The conidia, readily carried by wind, spread the fungus widely during the summer. This stage is referred to the form genus *Oïdium,* and a mildew is so named when the perithecia are unknown. In the late summer or fall the powdery white mycelium becomes less visible, and small globular or slightly flattened perithecia develop, just visible to the unaided eye as minute black dots. At first the perithecia are yellowish in color but later become dark brown to black. The perithecia possess characteristic appendages, which, together with the number of asci they contain, serve to separate the genera of powdery mildews (Fig. 42). The ascospores are developed by a sexual process. The perithecia overwinter on the fallen leaves and then rupture in the spring, discharging the ascospores, which are carried by air currents to infect the new foliage. Some species can overwinter without perithecia, the hyphae either hibernating in the buds or special highly resistant cells being formed.

When control is necessary, dusting with powdered sulfur or spraying with lime-sulfur has proved effective. Sulfur dust should be applied in the early morning while the foliage is still damp. Burning the fallen leaves is also helpful. These measures, of course, can be applied only in nurseries or in small, highly valuable plantations. The powdery mildews are also parasitized by another fungus, *Cicinnobolus cesatii* de By. (Griffiths 1899), which sometimes aids in checking them.

Among the powdery mildew fungi found on hardwoods are the following:

Erysiphe aggregata (Pk.) Farl. on female catkins of alders.
Erysiphe taurica Lév. on leaves of mesquite (Stevenson 1945).
Microsphaera alni (Wall.) Salm. on leaves of various trees, particularly oaks.
Phyllactinia corylea (Pers.) Karst. on leaves of various trees.
Sphaerotheca lanestris Hark. on leaves and shoots of oaks. It causes witches'-brooms on coast live oaks in California (Miller 1942).
Uncinula circinata Cke. & Pk. on leaves of maples.
Uncinula clintonii Pk. on leaves of basswood.
Uncinula flexuosa Pk. on leaves of buckeye.
Uncinula macrospora Pk. on leaves of American elm.
Uncinula salicis (Fr.) Wint. on leaves of poplars and willows.

Oak mildew, caused by *Microsphaera alni* var. *extensa* (Cke. & Pk.) Salm., also known as *M. quercina* (Schw.) Burr., *M. alni* (DC.) Wint., and *M. alphitoides* Griff. & Maubl., is very severe on oak coppice and nursery stock in Europe, and it is also considered to be an important cause of the extensive dying back of oak high forest (Day 1927; Neger 1915; Raymond 1927; Woodward, Waldie, and Steven 1929). Some authorities think that this parasite was originally introduced from North America. In this country it is not of serious consequence.

Sooty Molds. A sooty growth or crust may appear on leaves in isolated patches or covering the entire leaf surface. The black mass is made up entirely of fungous hyphae that develop following the attack of certain aphids and scale insects. The fungi causing sooty molds largely belong to the family Perisporiaceae of the Ascomycetes. Although the hyphae do not penetrate the leaf tissue, living only on the so-called "honeydew" excreted by the insects or on sugary solutions occasionally exuded by the leaves, nevertheless sooty leaves are unsightly, and some interference with their physiological functions must result. Sooty molds are found most frequently in regions with a mild climate such as the coast of California.

Rusts. Leaves of a few broadleaf trees are attacked by heteroecious rusts (Basidiomycetes, Uredinales), the alternate host in most cases being a conifer. In general the damage caused is slight, although occasionally when one of these rusts becomes temporarily epidemic, young trees may be heavily defoliated.

Poplars are hosts for several conspicuous rusts (*Melampsora* spp.), which are similar in appearance and can be separated only by an expert. Susceptible young trees can be severely damaged (Schreiner 1959). In the summer, golden-yellow or orange powdery pustules appear on the undersurface of the leaves, either scattered or so closely crowded that the entire surface seems powdery. These pustules are the uredia. In the late summer and autumn, small slightly raised areas or crusts appear. At first orange yellow in color, they change to dark brown or black (Fig. 43). These are the telia, the stage in which the rust overwinters on the fallen leaves, infecting the alternate host the next spring. *Melampsora abietis-canadensis* has its alternate stage on eastern hemlock. *M. medusae,* common in the East, rare in the West, has larch needles for the alternate host. Douglas fir needles carry the alternate stage of *M. albertensis,* common on aspen, and of *M. occidentalis,* common on black cottonwood.

The leaf rusts of willows are similar in appearance to those of poplars. *M. abieti-capraearum* has the needles of balsam firs for its alternate host; *M. paradoxa,* the needles of larches; *M. ribesii-purpureae* Kleb., the leaves of currants and gooseberries; and *M. arctica* Rostr., species of the Saxifrage family. The rusts of both willows and poplars appear on the leaves of their deciduous hosts year after year at long distances from their coniferous hosts, showing that these rusts can overwinter in the uredial stage. In *M. paradoxa* the mycelium overwinters in catkins, terminal buds, and young stems (Weir and Hubert 1918:55).

Melampsoridium betulinum (Pers.) Kleb. causes a leaf rust of various birches. In the summer the uredia appear on the undersides of the leaves as small round reddish-yellow powdery pustules. These are followed later in the summer or autumn by the telia, which at first are waxy yellow, finally turning dark brown or almost black. The aecia occur on larch needles, but this stage is unknown in North America. *M. alni* (Thüm.) Diet., which causes a similar rust on alder leaves, although common in

Europe, is known in this country only from California where it occurs rarely on red and white alder.

During the summer, small inconspicuous yellow or orange powdery pustules frequently appear on the underside of oak leaves. These are the uredia of *Cronartium cerebrum* or *C. fusiforme*, which cannot be distinguished from each other on oak. Later the telia develop from the same

Fig. 43. Telia of *Melampsora occidentalis* on the undersurface of a leaf of black cottonwood.

pustules as small brown bristles. The alternate stage of these rusts causes the formation of conspicuous swellings on the stems of pine.

The leaves of serviceberries, hawthorns, mountain ashes, and wild apples are attacked by *Gymnosporangium* spp. More or less circular honey-colored slightly thickened spots develop during the summer. On the upper surface of these spots, pycnia first appear as small black dots barely visible to the unaided eye. Later they are followed by the aecia on the underside of the spots, which appear either as cluster cups with orange powdery contents (the aeciospores), or as long whitish or brownish tubes, known as "roestelia," which split to release the spores.

Fruits and small twigs may be affected as well as leaves. Junipers and cedars are the alternate hosts.

Ash rust is caused by *Puccinia sparganoides* Ell. & Barth. It is characterized by swellings on the twigs and petioles (Fig. 44) and distortion of the leaves of various ashes. After the swellings form, they are covered with cluster cups filled with a yellow powder, the aeciospores. The alternate hosts for this rust are marsh and cord grasses (*Spartina* spp.). Partial control has been obtained by spraying with Bordeaux mixture or ziram (Partridge and Rich 1957).

Fig. 44. Distortion of the petioles and twigs of ash caused by *Puccinia sparganoides*. This is the aecial or cluster-cup stage. (*After F. D. Heald.*)

Blisters. Blistering, wrinkling, and curling of hardwood leaves is caused by species of *Taphrina* (Ascomycetes, Exoascaceae) with which the genus *Exoascus* is now included (Mix 1949). The intercellular mycelium in the leaves stimulates the cells to abnormal activity resulting in pronounced hypertrophy. The blistered area may be white, yellow, brown, pink, or red in color. The asci, with four to eight spores, are developed in a palisadelike layer over the affected tissue (Fig. 45). The ascospores bud readily, either within the ascus or after they are expelled, such blastospores serving to overwinter many species.

Although these fungi are widely distributed, the damage caused by them is usually slight. One of the most conspicuous hypertrophies is yellow leaf blister on poplars (Fig. 46) caused by *Taphrina aurea* (Pers.) Fr. The blisters, varying from small to 1 or more inches in diameter, are brilliant yellow on their concave side when the asci are fully developed. Later the color changes to brown. Oak leaf blister (Fig. 47), caused by *T. caerulescens* (Desm.) Tul., is common on oaks in the Eastern states. Cool wet weather favors the development of the fungus. The disease on valuable individual trees can be controlled by a single

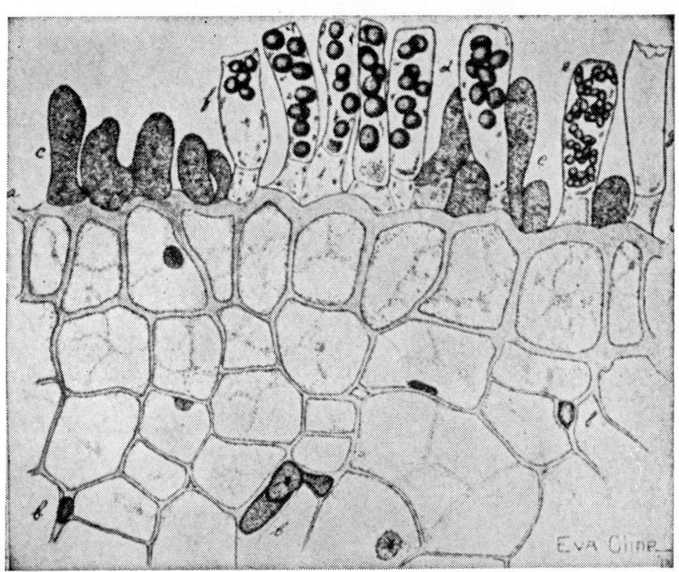

Fig. 45. Cross section of a peach leaf affected with leaf curl caused by *Taphrina deformans* (Fckl.) Tul. Asci containing ascospores are forming on the upper surface: (*a*), cuticle of leaf; (*b*), bits of mycelium of the fungus; (*c*), young asci with spores not yet formed; (*d*), spores just formed; (*e*), spores dividing into smaller ones; (*f*), spores discharging; (*g*), empty ascus. (*After Swingle, Montana Agr. Expt. Sta. Circ.* 37.)

dormant spray in the winter. Many different fungicides, including Bordeaux mixture, are effective (Goode 1953). *T. ulmi* (Fckl.) Johans., epidemic on American elm in a Quebec nursery, was controlled by dusting the trees with sulfur in the spring just before the buds opened (Pomerleau 1934).

Among the many species of *Taphrina* affecting leaves are:

T. aesculi (Patt.) Gies., yellow leaf blister, later turning to a dull red, on California buckeye.

T. castanopsidis Jenkins, leaf blister on golden chinquapin in California.

T. coryli Nishida, leaf curl and yellow leaf blister on hazel.

T. confusa (Atk.) Gies., leaf blister on choke cherry.

T. farlowii (Atk.) Sadeb., leaf curl on black cherry.

T. boycei Mix, yellow leaf blister on water and western paper birches.

T. flava Farl., yellow leaf blister on gray and paper birches.

T. bacteriosperma Johans., yellow to yellowish-red leaf blister and curl on yellow and dwarf birches.

T. japonica Kusano, leaf curl on red alder.

T. virginica Sadeb., leaf blister on hop-hornbeam.

Fig. 46. Yellow leaf blister of black poplar caused by *Taphrina aurea* showing the upper and undersurface of a leaf.

A malformation of the leaves of various shrubs such as blueberry, rhododendron, and other members of the family Ericaceae is caused by *Exobasidium vaccinii* (Fckl.) Wor. (Basidiomycetes, Thelephoraceae). Although leaf blisters sometimes result, the usual hypertrophy is a conspicuous bladderlike enlargement of all or part of the leaf (Burt 1915). Witches'-brooms may also be formed. The fungus does not attack trees.

Flower and Fruit Deformation. Alders throughout their range are attacked by Taphrinas causing deformation of the female catkins (Fig.

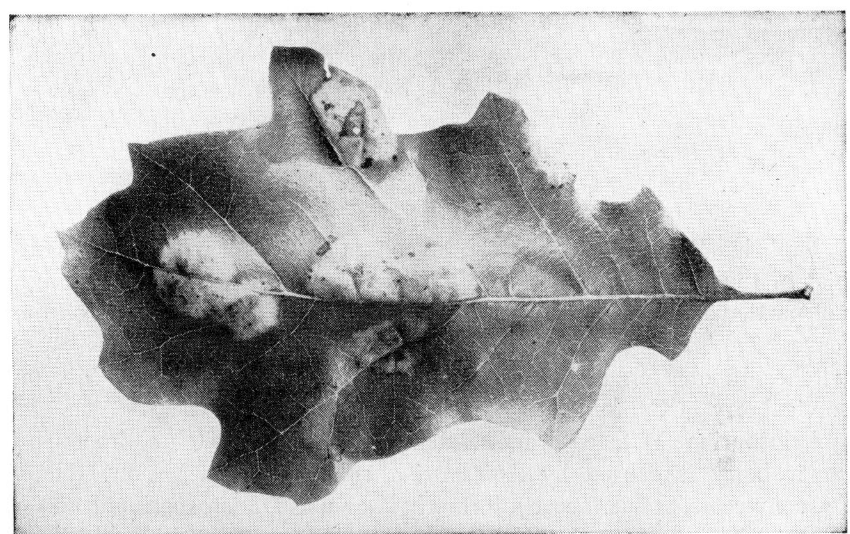

Fig. 47. Upper surface of a leaf of black oak showing blisters caused by *Taphrina caerulescens.* [*Courtesy of Conn.* (*State*) *Agr. Expt. Sta.*]

48). The scales of affected catkins are greatly enlarged and project as curled reddish tongues, which later become covered with a white glistening layer, the asci of the fungus. *Taphrina amentorum* (Sadeb.) Rostr. (*T. alni-incanae, T. alni-torquus*) occurs on red alder in Alaska; *T. occidentalis* Ray on mountain, red, and white alders in the West; and *T. robinsoniana* Gies. on smooth and speckled alders in the East. *T. johan-*

Fig. 48. Deformed catkins of alder caused by *Taphrina amentorum.* (*After E. Rostrup.*)

sonii Sadeb. similarly deforms catkins of poplars, but the enlarged scales are covered with a yellow layer when the asci develop. *T. pruni* (Fckl.) Tul. causes the fruits of plums to become very much swollen and hollow, the hypertrophied fruits being known as plum pockets, bladder plums, or fools. *T. pruni-subcordatae* Zeller causes plum pockets of the Pacific plum in Oregon and California. Several other species also deform fruits of plums.

Taphrina Witches'-brooms. The mycelium of some Taphrinas is perennial in the twigs of the host, causing witches'-brooms, which although not particularly injurious to the tree are unsightly. The fungi fruit on the leaves. *Taphrina cerasi* (Fckl.) Sadeb. and several other species cause witches'-brooms on cherries and plums, and *T. amelanchieri* Mix on serviceberry in California.

Apiosporina Witches'-brooms. This widespread witches'-broom on serviceberry is caused by *Apiosporina collinsii* (Schw.) Von Höhnel (Ascomycetes, Sphaeriaceae), formerly known as *Dimerosporium collinsii* (Schw.) Thüm. (Sartoris and Kauffman 1925; Sprague and Heald 1927). The broom is characterized by numerous stout branches in which the mycelium of the parasite is perennial, and by the development of a sooty growth on the underside of the leaves, which at first is olive brown in color, later changing to black. In the late summer, minute globose perithecia develop in the sooty growth and are so densely aggregated that the undersurface of the leaf appears to be covered with tiny black beads. Injury to the host is slight.

SCAB AND BLACK CANKER OF WILLOW

Scab is caused by *Fusicladium saliciperdum* (All. & Tub.) Lind. (Fungi Imperfecti, Dematiaceae) and black canker by *Physalospora miyabeana Fuk.* (Ascomycetes, Pleosporaceae). The parasitic stage of both is the imperfect or conidial stage, which for *P. miyabeana* is a Gloeosporium. The perfect or ascus stage of *F. saliciperdum* is *Venturia chlorospora* (Ces.) Karst., which so far is known in this country only in culture. The name "willow blight" has been applied collectively to the two diseases. Apparently *P. miyabeana* follows *F. saliciperdum* and increases the damage done by it.

History and Distribution. *P. miyabeana* was first described as causing a willow disease in Japan and later was investigated in England (Nattras 1928). *F. saliciperdum* has been long known in Europe, but it was reported for the first time in North America in 1927 (Waterman 1951*b*). Opinions differ as to the relative pathogenicity of the two. In England Nattras (1928:303) thinks that *F. saliciperdum* is secondary to *P. miya-*

beana; Dennis (1931) states that *F. saliciperdum* is incapable of causing any disease of willow rods, leaves, or tips; but Brooks and Walker (1935) conclude that both fungi can cause serious disease of young willow shoots. In the United States, Clinton and McCormick (1929:452) consider *F. saliciperdum* as the primary parasite, while in Nova Scotia Harrison (1931) finds that *F. saliciperdum* is primary, with *P. miyabeana* attacking leaves and twigs later in the spring.

Apparently both fungi were introduced into North America from Europe at about the same time, spreading from the Maritime Provinces of Canada into New England. *F. saliciperdum* is now present in New England, New York, northern Pennsylvania, and northern New Jersey, while *P. miyabeana* occurs in New England and eastern New York, apparently spreading more slowly. Both fungi have been found in West Virginia and in British Columbia, probably introduced on ornamental willows, and *F. saliciperdum* has been reported in North and South Carolina.

Hosts. So far there have been marked differences in susceptibility. Golden willow, black willow, and heart-leaved willow are susceptible; weeping willow, bay-leaved willow, osier willow, and purple willow are somewhat resistant. Crack willow seems to have both susceptible and resistant types.

Damage. Willow blight is suddenly destructive, particularly when the strong recuperative powers of willows are considered. Trees of any size may die in about 3 years from repeated killing of the leaves and young twigs (Fig. 49). Although willows have been killed extensively in New England and eastern Canada, from the forestry standpoint the disease is not of much consequence in this general region where willows are important only as shade and ornamental trees. Its potentialities are serious to the Middle West where willows are so valuable along the upper reaches of the Mississippi and Missouri Rivers and their tributaries. Growing along the banks, the willows assist in keeping these streams within their channels and also are the most satisfactory material for mattresses in revetments.

Symptoms. Both parasites have the same effect on the host. *F. saliciperdum* attacks the young leaves in the spring, often killing them when their development is just starting. Some leaves escape this initial infection but are killed later. Leaves can be blackened and killed in a few hours (Brooks and Walker 1935:66). From the leaf blades, the fungus proceeds through the petioles into the twigs where it causes cankers. The tips of the young growing shoots are also killed back. *P. miyabeana* acts the same way but attacks the foliage later in the season, often infecting the adventitious growth that develops after the first crop of leaves and shoots have been killed by *F. saliciperdum.* In addition *P. miyabeana* can cause cankers on larger stems. Killed leaves remain on the trees for a while, then they dry up and gradually drop. Severe defoliation for 2 or 3 successive years is usually fatal, because after the first year little adventitious foliage develops. The disease is worst during moist weather.

Fig. 49. Willows along a road in New Hampshire killed by scab and black canker. (*Photograph by R. P. Marshall.*)

Causal Agencies. The two fungi cannot be distinguished from each other on the host until spores appear. *F. saliciperdum* develops its spores commonly on dead leaves and less frequently on the stem cankers and tips of twigs, where during moist weather they appear as small dense olive-brown pustules or velvety masses. The spore masses more or less cover the undersurface of the leaves, being particularly dense along the lower portions of the main veins and midrib. The spores (conidia) are somewhat ellipsoid in shape, rounded at one end and slightly narrowed at the base, usually with two cells of unequal size but occasionally three- or four-celled, and olive to reddish brown in color. They are borne singly on the tips of conidiophores. The fungus overwinters as dormant mycelium in the young twigs infected the previous season. The first spores in the spring develop about the time the willow catkins are opening and are washed down by rain to infect the young leaves in the opening buds.

P. miyabeana develops two spore stages, both of which appear on the stem cankers and on the killed twigs but not on the dead leaves, except that under very moist conditions the imperfect stage may fruit on the leaves. This rarely happens. The spores of the imperfect stage, developed in acervuli, appear to the naked eye as small pink masses. They are most evident in the early summer. The conidia are one-celled, hyaline or faintly greenish in color, ellipsoid to cylindrical in shape, and either straight or slightly curved. They may be rounded at both ends, or rounded at one end only and slightly narrowed at the other. They are borne on the ends of short conidiophores. This stage is quickly followed by the perithecia, which appear as tiny black spots. Under the microscope they are seen to be globose to flask-shaped with short necks and contain numbers of club-shaped asci and slender, thread-shaped paraphyses. The eight spores in each ascus, arranged more or less in two rows, are one-celled, hyaline, oblong to ellipsoid in shape, and sometimes slightly curved. Both conidia and ascospores spread the disease during the summer, and the fungus overwinters in the ascus stage, the first ascospores developing in the spring about 10 days later than the conidia of *F. saliciperdum.*

Control. Since both fungi overwinter on blighted twigs, starting new infections in the spring from this source, the disease can be reduced considerably by removing and destroying all dead branches and as many twigs as possible while the trees are dormant. This should be followed by three or more sprayings with Bordeaux mixture, beginning just before the buds open. The number of spray applications will depend on climatic conditions, more being needed in wet seasons. Such intensive control measures are economically applicable only to valuable shade trees. Control in stands along streams will depend on the establishment of resistant species, if they do not occur naturally.

REFERENCES

Baker, W. L.: Effect of Gypsy Moth Defoliation on Certain Forest Trees. *J. Forestry,* 39:1017–1022 (1941).

Bowen, P. R.: A Maple Leaf Disease Caused by *Cristulariella depraedens. Conn.* (*State*) *Agr. Expt. Sta. Bull.,* 316:625–647 (1930).

Brooks, F. T., and M. M. Walker: Observations on *Fusicladium saliciperdum*. *New Phytologist,* 34:64–67 (1935).

Burt, E. A.: The Thelephoracease of North America. IV. *Exobasidium*. *Ann. Missouri Botan. Garden,* 2:627–658 (1915).

Cash, Edith K., and Alma D. Waterman: A New Species of *Plagiostoma* Associated with a Leaf Disease of Hybrid Aspens. *Mycologia,* 49:756–760 (1957).

Clinton, G. P., and Florence A. McCormick: The Willow Scab Fungus. *Conn. (State) Agr. Expt. Sta. Bull.,* 302:443–469 (1929).

Cunningham, H. S.: A Study of the Histologic Changes Induced in Leaves by Certain Leaf-spotting Fungi. *Phytopathology,* 18:717–751 (1928).

Day, W. R.: The Oak Mildew *Microsphaera quercina* (Schw.) Burrill and *Armillaria mellea* (Vahl.) Quél. in Relation to the Dying Back of Oak. *Forestry,* 1:108–112 (1927).

Demaree, J. B., and J. R. Cole: Pecan Leaf Blotch. *J. Agr. Research,* 40:777–789 (1930).

Dennis, R. G. W.: The Black Canker of Willows. *Brit. Mycol. Soc. Trans.,* 16:76–84 (1931).

Fergus, C. L.: An Epiphytotic of Phyllosticta Leaf Spot of Maple. *U.S. Dept. Agr., Plant Disease Reptr.,* 38:678–679 (1954).

Fowler, M. E.: Glomerella Leaf Spot of Magnolia. *U.S. Dept. Agr., Plant Disease Reptr.,* 31:298 (1947).

Gilman, J. C., and B. N. Wadley: The Ascigerous Stage of *Septoria querceti* Thuem. *Mycologia,* 44:217–220 (1952).

Goode, M. J.: Control of Oak Leaf-blister in Mississippi. [Abstract.] *Phytopathology,* 43:472 (1953).

Griffiths, D.: The Common Parasite of the Powdery Mildews. *Bull. Torrey Botan. Club,* 26:184–188 (1899).

Harrison, K. A.: Willow Disease Investigations—Maritime Provinces. *Can. Dept. Agr., Dominion Botan. Rept.,* 1930:37–38 (1931).

Harvey, J. M.: Bacterial Leaf Spot of *Umbellularia californica*. *Madroño,* 11:195–198 (1952).

Heald, F. D.: "Manual of Plant Diseases," 2d ed. New York: McGraw-Hill Book Company, Inc. (1933).

Hedgcock, G. G.: The Large Leaf Spot of Chestnut and Oak Associated with *Monochaetia desmazierii*. *Mycologia,* 21:324–325 (1929).

Jenkins, Anna E., J. H. Miller, and G. H. Hepting: Spot Anthracnose and Other Leaf and Petal Spots of Flowering Dogwood. *Natl. Hort. Mag.,* 32:57–69 (1953).

Jones, S. G.: Life-history and Cytology of *Rhytisma acerinum* (Pers.) Fries. *Ann. Botany,* 39:41–75 (1925).

Kerling, L. C. P.: De anatomische bouw van bladvleken. *Mededel. Landbouwhoogeschool Wageningen,* 32:1–106 (1928).

Kramer, P. J., and T. H. Wetmore: Effects of Defoliation on Cold Resistance and Diameter Growth of Broad-leaved Evergreens. *Am. J. Botany,* 30:428–431 (1943).

Luttrell, E. S.: *Morenoella quercina,* Cause of Leaf Spot of Oaks. *Mycologia,* 32:652–666 (1940).

McCallum, A. W.: *Napicladium Tremulae*. A New Disease of the Poplar. [Abstract.] *Phytopathology,* 10:318 (1920).

Mielke, J. L.: Aspen Leaf Blight in the Intermountain Region. *Intermountain Forest and Range Expt. Sta. Research Note,* 42:1–5 (1957).

Miles, L. E.: Leaf Spots of Elm. *Botan Gaz.*, 71:161–196 (1921).

Miller, J. H.: *Elsinoë* on Southern Red Oak. *Mycologia,* 49:277–279 (1957).

Miller, J. H., and Anna E. Jenkins: A New Species of *Elsinoë* on Southern Magnolia. *Mycologia,* 47:104–108 (1955).

Miller, J. H., and F. A. Wolf: A Leaf-spot Disease of Honey Locust Caused by a New Species of *Linospora. Mycologia,* 28:171–180 (1936).

Miller, P. A.: Powdery Mildew of the Coast Live Oak. *Western Shade Tree Conf. Proc.*, 9:358–366 (1942).

Mix, A. J.: A Monograph of the Genus *Taphrina. Univ. Kansas Sci. Bull.*, 33 (pt. 1, no. 1): 1–167 (1949).

Nattras, R. M.: The Physalospora Disease of Basket Willow. *Brit. Mycol. Soc. Trans.*, 13:286–304 (1928).

Neger, F. W.: Der Eichenmehltau [*Microsphaera Alni* (Wallr.), var. *quercina*]. *Naturw. Z. Forst- u. Landwirtsch.*, 13:1–30 (1915).

Partridge, A. D., and A. E. Rich: A Study of the Ash Rust Leaf Syndrome in New Hampshire: Suscepts, Incitant, Epidemiology, and Control. [Abstract.] *Phytopathology,* 47:246 (1957).

Pierce, R. G., and C. Hartley: Horse-chestnut Anthracnose. *Phytopathology,* 6:93 (1916).

Plakidas, A. G.: *Venturia acerina,* the Perfect Stage of *Cladosporium humile. Mycologia,* 34:27–37 (1942).

Pomerleau, R.: Notes sur le *Taphrina ulmi. Naturaliste can.*, 61:305–308 (1934).

Pomerleau, R.: Recherches sur le *Gnomonia ulmea* (Schw.) Thüm. *Naturaliste can.*, 64:261–289, 297–318; 65:23–41, 57–70, 89–96, 125–137, 167–178, 178–188, 221–237, 253–279 (1937–1938).

Pomerleau, R.: Studies on the Ink-spot Disease of Poplar. *Can. J. Research. C. Botan. Sci.*, 18:199–214 (1940).

Raymond, J.: Le "blanc" du chêne. *Ann. épiphyt.*, 13:94–129 (1927).

Sartoris, G. B., and C. H. Kaufman: The Development and Taxonomic Position of *Apiosporina collinsii. Papers Mich. Acad. Sci.*, 5:149–162 (1925).

Schreiner, E. J.: Rating Poplars for *Melampsora* Leaf Rust Infection. *U.S. Dept. Agr., Forest Serv., Northeast. Forest Expt. Sta., Forest Research Notes,* 90:1–3 (1959).

Schuldt, P. H.: Comparison of Anthracnose Fungi on Oak, Sycamore, and Other Trees. *Contribs. Boyce Thompson Inst.*, 18(2):85–107 (1955).

Seaver, F. J.: *Sclerotinia bifrons. Mycologia,* 37:641–647 (1945).

Smith, D. J., and C. O. Smith: Species of *Stigmina* and *Stigmella* Occurring on *Platanus. Hilgardia,* 14:205–231 (1941).

Sprague, R., and F. D. Heald: A Witches' Broom of the Service Berry. *Trans. Am. Microscop. Soc.*, 46:219–247 (1927).

Stevenson, J. A.: Powdery Mildew of Mesquite. *U.S. Dept. Agr., Plant Disease Reptr.*, 29:214–215 (1945).

Stewart, V. B.: The Leaf Blotch Disease of Horse-chestnut. *Phytopathology,* 6:5–19 (1916).

Summers, J. N., and A. F. Burgess: A Method of Determining Losses to Forests Caused by Defoliation. *J. Econ. Entomol.*, 26:51–54 (1933).

Swingle, R. U., and T. W. Bretz: Zonate Canker, a Virus Disease of American Elm. *Phytopathology,* 40:1018–1022 (1950).

Thompson, G. E.: A Leaf Blight of *Populus tacamahaca* Mill. Caused by an Undescribed Species of *Linospora. Can. J. Research. C. Botan. Sci.*, 17:232–238 (1939).

Thompson, G. E.: Leaf-spot Diseases of Poplars Caused by *Septoria musiva* and *S. populicola*. *Phytopathology*, 31:241–254 (1941).

Trumbower, J. A.: Control of Elm Leaf Spots in Nurseries. *Phytopathology*, 24:62–73 (1934).

Van Hook, J. M., and R. C. Busteed: Anthracnose of *Betula nigra*. *Proc. Indiana Acad. Sci.*, 44:81 (1935).

Waterman, Alma M.: Sycamore and Oak Anthracnose (Caused by *Gnomonia veneta*). *Soc. Am. Foresters, New England Sect., Tree Pest Leaflets*, 48:[1–3] (1951*a*).

Waterman, Alma M.: Willow Scab (Caused by *Fusicladium saliciperdum*). *Soc. Am. Foresters, New England Sect., Tree Pest Leaflets*, 35 (rev.):[1–3] (1951*b*).

Waterman, Alma M., and Edith K. Cash: A Leaf Blotch of Poplar Caused by a New Species of *Septotinia*. *Mycologia*, 42:374–384 (1950).

Weir, J. R., and E. E. Hubert: Notes on the Overwintering of Forest Tree Rusts. *Phytopathology*, 8:55–59 (1918).

Wolf, F. A.: The Perfect Stage of a Leafspot Fungus on Red Mulberry. *J. Elisha Mitchell Sci. Soc.*, 51:163–166 (1935).

Wolf, F. A.: Leafspot of Ash and *Phyllosticta viridis*. *Mycologia*, 31:258–266 (1939).

Wolf, F. A.: Cercospora Leafspot of Red Bud. *Mycologia*, 32:129–136 (1940*a*).

Wolf, F. A.: A Leafspot Fungus on *Nyssa*. *Mycologia*, 32:331–335 (1940*b*).

Wolf, F. A., and R. W. Davidson: Life Cycle of *Piggotia fraxini*, Causing Leaf Disease of Ash. *Mycologia*, 33:526–539 (1941).

Woodward, R. C., J. S. L. Waldie, and H. M. Steven: Oak Mildew and Its Control in Forest Nurseries. *Forestry*, 3:38–56 (1929).

CHAPTER 8

Foliage Diseases of Conifers

Most pathogenic fungi causing foliage diseases are confined to the leaves, a few attack buds, e.g., *Dematium* (*Pullularia*) *pullulans* de Bary associated with forking of red pine (Jump 1938), and some also invade young twigs. Cone diseases are included in this chapter. Noninfectious diseases affecting foliage are discussed in Chap. 4 and foliage diseases primarily affecting seedlings in Chap. 5. In general foliage diseases are more severe on seedlings, saplings, and small poles than on trees in the larger classes, and new diseases are to be expected on trees established outside their natural range.

Since most conifers depend on 2 to 7 or more years' growth of leaves for their metabolism and leaves once lost are not replaced, complete defoliation is usually fatal. Removal of all needles during the growing season or early in the period of dormancy kills Scotch and jack pines, whereas larch can withstand several years' defoliation. Generally, defoliation immediately reduces increment, the reduction being exactly proportional to the degree of defoliation (Church 1949). Fortunately most pathogenic fungi affect either foliage of the current season or older foliage, rarely both, and it is unusual for all the foliage in either category to be involved. Furthermore the fungi vary in virulence from year to year according to climatic conditions so that a heavy infection extending over a period of years is exceptional. Again, although some trees in a stand are severely attacked, others escape with little or no infection, apparently because of individual resistance, although the reasons for such variation are not known. With age, conifers become more resistant to needle diseases. Consequently foliage diseases caused by fungi rarely cause ultimate understocking of a stand, but they do more or less reduce its rate of growth. Conifers on which all the older needles are killed annually, so that the trees are dependent year after year on needles of the current season only, practically cease growth. Some die soon, but the majority linger for years in a moribund condition. Conifers succumb to defoliation more readily than hardwoods, and their rate of growth is more affected by partial defoliation. Fungous foliage diseases are sometimes very damaging to conifers raised as Christmas trees when they result in unsightly foliage.

CEDAR LEAF BLIGHT

Cedar leaf blight is caused by *Keithia thujina* Durand (Ascomycetes, Phacidiaceae). The genus has been discussed by Durand (1913), and Maire (1927) proposed changing its name to *Didymascella* because *Keithia* has been used previously for a genus of flowering plants. The parasite is a native of North America occurring abundantly in the West throughout the range of western red cedar and is also known from the Lake states and Vermont on northern white cedar. It has been found once on Port Orford cedar in a Washington nursery. The fungus has also been introduced into Great Britain and Ireland, where it has spread rapidly on western red cedar. The time and manner of its introduction are unknown, although field evidence indicates that it may have been introduced on or with seed, since nursery stock of cedar was not imported (Pethybridge 1919). The parasite has also appeared in Denmark and Holland.

Cedar leaf blight is best known on western red cedar, and practically all the information as to its effects, symptoms, and the life history of the causal fungus apply to that host, since it is rarely reported to be of consequence on northern white cedar (Mains 1935).

Damage. Although primarily a disease of seedlings and saplings, yet trees of all age classes are attacked. Seedlings less than 4 years old are killed in a single season when the disease is epidemic. The disease is serious in nurseries in Great Britain (Peace 1955), but in this country has not yet damaged nursery stock. Beyond the seedling stage the effect of the disease is usually to retard growth, although if defoliation is severe enough individual branches or an entire tree may die.

The disease is worst in localities of high atmospheric humidity. Trees in the interior of dense stands or in deep ravines where the air is stagnant and moist suffer most severely. The fact that seedlings are seriously affected indicates that snow may be a factor in promoting the disease, since it is particularly severe where seedlings are covered with snow until late spring.

Symptoms. The lower branches of young trees when growing in dense stands on low ground often appear at a distance as if scorched by fire. The foliage of the upper crowns of mature trees may be very generally infected but never to the same degree as the leaves on lower branches near the ground. In late autumn the young infected leaf twigs drop, leaving the branches somewhat bare. On leaves remaining attached to the older twigs, the presence of the fungus is easily determined because the fructifications drop out leaving deep pits. These leaves turn ashy gray, but the leaves dropping earlier are reddish brown.

Causal Agency. The fructifications (apothecia) of *K. thujina* are usually on the upper surface of the leaves (Figs. 50, 51*a*), although they occasionally occur on the undersurface. They are embedded in the leaf tissue and are exposed by the rupture of the epidermis in a flap- or

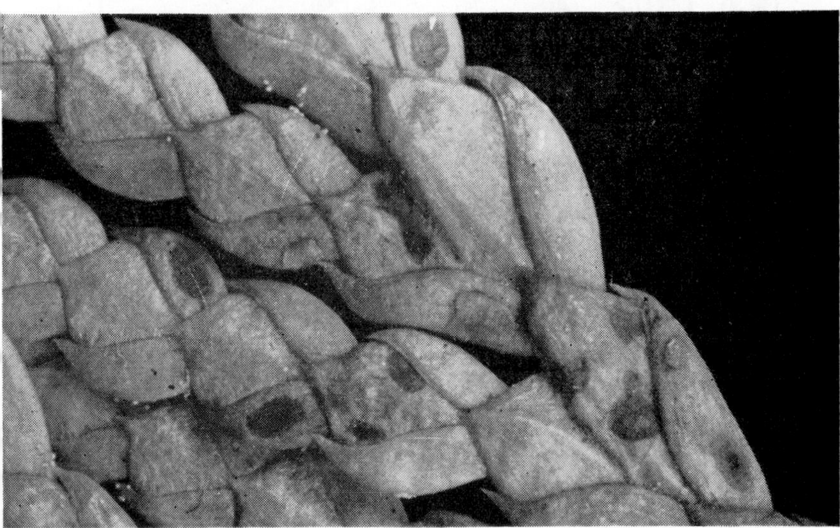

Fig. 50. Dark-colored fructifications or apothecia of *Keithia thujina* on the foliage of western red cedar.

scale-like manner (Fig. 51*b*). The apothecia are cushionlike, depressed when the air is dry and elevated when it is moist, and are circular, elliptical, curved or irregular in outline. From one to three or more may occur on a leaf. The apothecia are olive brown at first but become almost black with age. Each ascus has two hyaline elliptical or pyriform spores (Fig. 51*c*), which when mature become olive brown and nearly circular in shape (Fig. 51*d*). The spores are pitted (Fig. 51*e*) and very unequally two-celled. The paraphyses are club-shaped and single or branched and have cross walls.

Spore discharge begins sometime in June, depending on the locality, and ends about the middle of October while the apothecia are still attached to the leaves. Most of the spores are liberated during late summer and autumn. The substance of the apothecium is hygroscopic, which causes it to swell during moist weather. This aids in spore discharge and also causes a fissure to form around the apothecium, ultimately resulting in complete separation of the apothecium from the leaf. The imperfect or conidial stage of *K. thujina* is unknown.

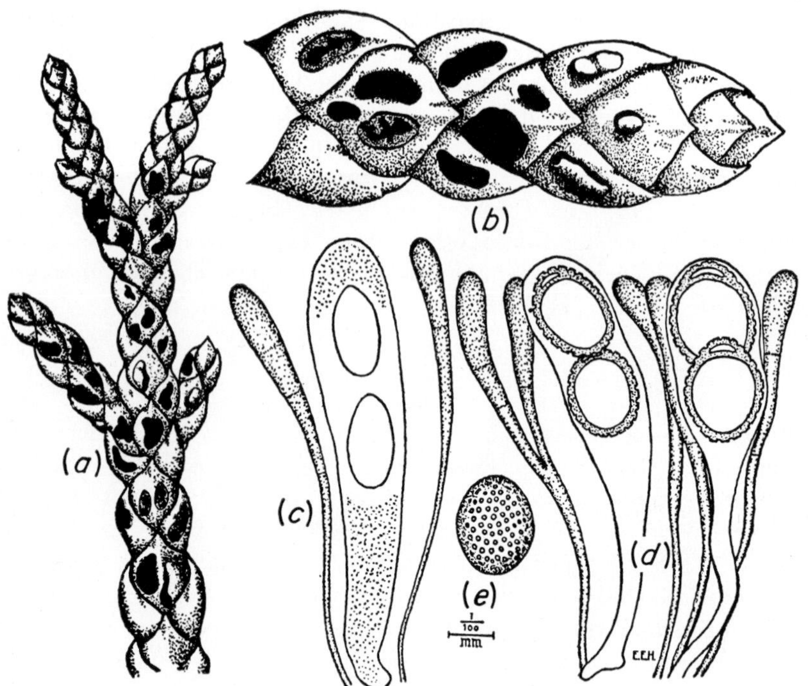

Fig. 51. *Keithia thujina* on western red cedar. (*a*), dark-colored fructifications or apothecia on the foliage; (*b*), a much enlarged portion of the foliage showing apothecia with the epidermal scale or flap falling away; also pits formed in the leaf after the apothecia have dropped out; (*c*), immature asci and spores; (*d*), mature asci and spores showing the small cell at the end of the spore and the club-shaped paraphyses both simple and branched; (*e*), ascospore showing the pitted surface markings. (*Reprinted by permission from "An Outline of Forest Pathology" by Hubert, published by John Wiley & Sons, Inc.*)

Control. For nurseries tentative experiments indicate that several applications of Bordeaux mixture during summer and autumn will check the disease. For forest stands, no control measures have been evolved, since the disease has not prevented the development of merchantable stands. In the future, however, when rate of growth becomes more important, control will have to be found and applied.

A similar, though much less frequent, disease of western red cedar is caused by *Coryneum thujinum* Dearn., but its fructifications are confined to the undersurface of the leaves, appearing as small black eruptions scarcely visible except with a hand magnifier. Both *C. thujinum* and *K. thujina* sometimes occur on the same leaf twigs. Another leaf blight is caused by *Chloroscypha seaveri* (Rehm) Seaver (Ehrlich 1942).

Keithia chamaecyparissi Adams is destructive to the foliage of young trees of southern white cedar (Adams 1918; Korstian and Brush 1931:13). This species has two-spored asci. *K. tsugae* (Farl.) Durand, which has four-spored asci, attacks the needles of eastern hemlock (Spaulding 1914:249), but the injury resulting is usually insignificant. During the summer of 1933 it was locally prevalent in New England. *K. juniperi* Miller, which has eight-spored asci, attacks the needles of eastern red cedar (Miller 1935) and southern red cedar. This is probably not a true *Keithia* but rather it is *Coccodothis sphaeroidea* (Cooke) Theiss. & Syd. (Pantidou and Korf 1954). *Keithia* sp. occurs on Rocky Mountain juniper in Colorado (Shope 1943:33).

NEEDLE CAST OF DOUGLAS FIR

Rhabdocline needle cast or needle blight of Douglas fir (Brandt 1960) is caused by *Rhabdocline pseudotsugae* Syd. (Ascomycetes, Phacidiaceae). It is confined to Douglas fir and is probably distributed throughout the range of its host. The disease is particularly prevalent in Montana, Idaho, Oregon, Washington, and British Columbia, where it becomes epidemic in certain seasons. It has also appeared in plantations in southern New England, but just how it reached there is unknown. Now widely distributed in Great Britain, it first appeared in Scotland, where it was apparently brought directly on trees from western North America prior to 1914 (Wilson and Wilson 1926; Boyce 1927:7). From Great Britain it entered Germany about 1926, presumably on nursery stock, and has spread widely in western Europe where the disease has been extensively investigated (Liese 1932*a*, 1932*b*; Rohde 1932; Tubeuf 1932; Van Vloten 1932). The spread of this parasite is an outstanding example of the danger in the promiscuous shipment of young trees.

Hosts. Although Douglas fir is the only host for *R. pseudotsugae* and in North America all forms of the host are attacked, the habit of this parasite in Europe indicates that there are strains or races within the fungus. Douglas fir in Europe is considered to comprise two species and one variety, namely green or Pacific Coast Douglas fir (*Pseudotsuga douglasii* Carr.), gray or intermountain Douglas fir (*P. douglasii* var. *caesia* Schwer.), and blue or Rocky Mountain Douglas fir (*P. glauca* Mayr). Green Douglas fir is resistant, whereas intermountain and blue Douglas fir may be heavily attacked (Meyer 1954). Liese (1932*b*) attributes the relative resistance of green Douglas fir in Germany to the fact that the buds of this form do not open until after the spores of the fungus have been disseminated, whereas the buds of the other two forms open during the period of maximum spore discharge. However, in the Douglas fir region of western Oregon and Washington, green Douglas fir is subject to severe infection, indicating that the parasite there is adapted to late bud opening. Even so, infection on green Douglas fir within its natural range is not so severe as infection on the intermountain Douglas fir in its natural habitat.

In North America there is great individual variation in susceptibility to the parasite. Many trees in a stand remain completely or nearly free from infection, while others are severely attacked. Of two trees of the same age standing side by side with their branches interlaced, one may be heavily infected and the other remain healthy.

Damage. Needle cast is essentially a disease of young trees from seedlings to about the 30-year age class. Older trees may be attacked, but the injury to them is slight. During seasons of high humidity and considerable rainfall, it is likely to become epidemic over considerable areas, while during dry seasons it may be difficult to find. Dense, pure stands suffer most seriously. Douglas fir in mixed stands is not so severely affected. The fungus attacks needles of the most vigorously growing trees and does not attack any more rapidly the needles of suppressed or otherwise unhealthy trees. When severe, the disease kills all needles except the youngest, causing them to die and drop off. Trees subject to its ravages for consecutive seasons are almost entirely defoliated and either die or merely exist for an indefinite period without making perceptible growth. Fortunately enough trees usually escape severe attack so that a reasonably satisfactory stand develops.

Damage is particularly severe in forest nurseries surrounded by infected stands, and in addition there is the danger of introducing the disease into plantations where the shock of planting combined with the attack of the parasite would cause the death of many trees.

Symptoms. The first symptoms appear in the autumn or more commonly in the early winter as slightly yellow-colored spots on the needles, most frequent near the ends. By spring the spots have changed to a reddish brown, which is sharply delimited from the normal green color, thus giving the leaves a mottled appearance (Fig. 52). A microscopic examination of these spots in their earlier stages reveals abundant coarse intracellular hyphae in the cells of the mesophyll, followed later by fine intercellular hyphae in the mesophyll cells and intracellular hyphae in the chlorenchyma cells (Van Vloten 1932:127). Finally in severe infections the needles assume a more uniform color giving the trees in mass a brown, scorched appearance. During June the fructifications develop, and in late June or July most of the infected needles drop off, although the less severely infected ones may remain on the tree for indefinite periods, dropping at any time during the year. According to Brown (1930) this premature fall seems to be connected with a marked decrease

Fig. 52. Mottled appearance of 1-year-old needles of Douglas fir infected with *Rhabdocline pseudotsugae.* (*Reproduced from J. Agr. Research* 10, *no.* 2.)

in water content of the infected needles, but Van Vloten (1932:138) finds that needle fall is not the result of water loss nor does it depend on the degree of infection. Unless infected trees are examined at the proper time the presence of the disease may be difficult to detect, because in some cases all needles infected the previous year are cast by July, and infection spots have not appeared on needles of the current year, so that the only symptom is sparse foliage.

Causal Agency. The fructifications or apothecia of *R. pseudotsugae* mature during late May or in June depending on the locality, usually on the undersurface of the needle, but occasionally on the upper surface, on either side of the middle nerve. At first they round up as small cushions, but finally the epidermis covering them ruptures with an irregular slit exposing the brownish elongated disc (Fig. 53). The cylindrical to club-shaped asci each contain eight hyaline, one-celled, oblong ascospores, which are rounded at the ends and commonly constricted in the middle (Fig. 54). Paraphyses are present. After the ascospores are discharged from the asci they become two-celled, the cross wall being

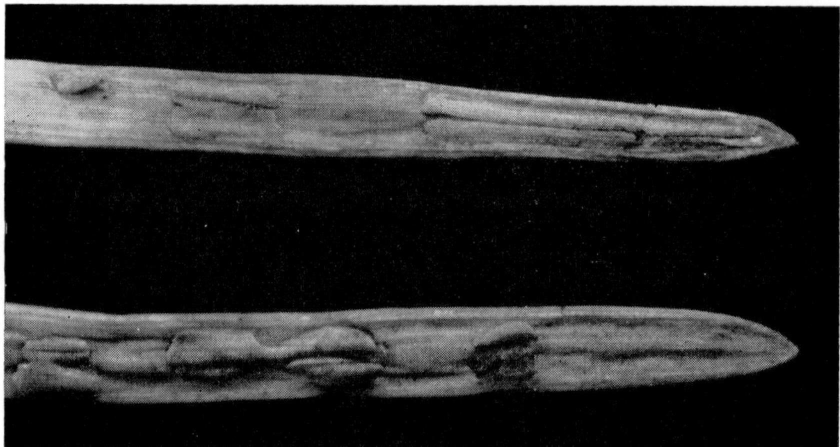

Fig. 53. Fructifications of *Rhabdocline pseudotsugae* on Douglas fir needles showing the manner of their rupture. About 10 ×.

formed at the constriction. Of the two cells, usually only one becomes thick-walled and dark colored, producing a single germ tube on germinating. The other cell remains hyaline in color, thin-walled, and does not germinate (Wilson and Wilson 1926:39).

The ascospores are released and infect the tender young needles of the current year, which are just beginning their development. Apothecia with spores develop on these needles the following spring again to infect the new needles. After the ascospores are discharged the needles drop, so that infected needles usually persist for a little over a year, instead of about 8 years as is the rule for normal needles. Consequently a tree that is attacked year after year is dependent on needles of the current year only for its metabolism, and most of these are diseased. Sometimes, only the needles of every other year are infected, the needles of intervening years remaining free from attack. This would indicate that under certain conditions the mycelium vegetates for nearly 2 years in the leaf tissue before apothecia are developed.

Control. In nurseries it would seem that spraying with Bordeaux mixture or lime-sulfur at the time of ascospore discharge should give adequate protection, although there are no actual results to prove this. Ultimately, control in regulated forests will probably depend on the elimination of susceptible trees early in the life of the stand, on leaving resistant trees to form the final crop, and on the selection of seed for artificial reforestation from resistant mother trees.

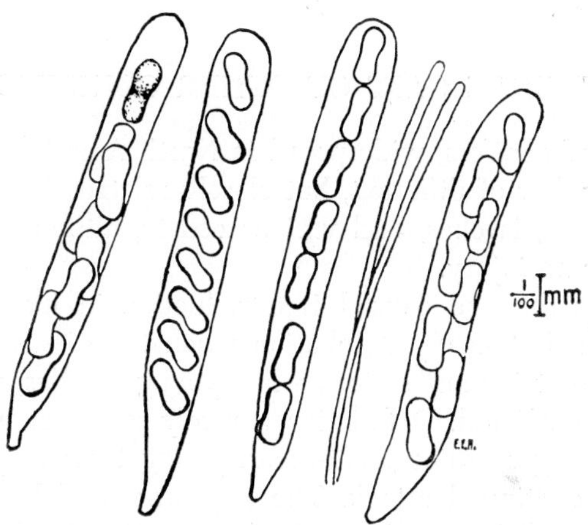

Fig. 54. Asci with ascospores, and paraphyses of *Rhabdocline pseudotsugae.* (*Reprinted by permission from "An Outline of Forest Pathology" by Hubert, published by John Wiley & Sons, Inc.*)

Rhabdogloeum pseudotsugae Sydow, which causes a similar needle cast, has been suggested as the imperfect stage of *Rhabdocline pseudotsugae,* but this is doubtful because *Rhabdogloeum pseudotsugae* is rarely found. *Rhabdogloeum hypophyllum* Ellis & Gill (1945) is commonly associated with *Rhabdocline pseudotsugae* in the Southwest.

Adelopus (*Phaeocryptopus*) *gäumanni* Rohde (1937) causes Adelopus needle cast or Swiss needle cast, which since its discovery in 1925 in Switzerland has become destructive in certain parts of western Europe (Boyce 1940; Merkle 1951). The fungus also occurs on the Pacific Coast, where it is seemingly native, causing no apparent injury to its host. It has also been found somewhat injurious to natural Douglas fir in the Southwest and to plantations in New England (Hahn 1941). Probably introduced to Europe from North America, the fungus there has so far attacked trees up to 55 years old, but in general young plantations have suffered most. Deprived of practically all needles except those of the current season for several consecutive years, severely infected trees become moribund and finally die. Such trees have thin foliage with a general yellowish to brownish color. Individual needles vary from yellow-green or mottled yellow-green to mottled brown or completely brown, depending on the degree and length of time of infection. On the undersurface of diseased needles, numerous tiny black perithecia, issuing from the stomata, appear as sootlike streaks, one on each side of the middle nerve. No method of control is yet known.

Rhizosphaera kalkhoffi Bub. causes a similar needle cast of blue spruce in the Northeast (Waterman 1947).

NEEDLE CAST OF OTHER CONIFERS

This disease of conifers, widespread in North America and Europe, is caused by a number of different, but closely related, fungi. It occurs on pines, spruces, firs, larches, and various cedars, although to the cedars there is little or no injury.

Damage. The damage caused is a reduction in increment brought about by partial but heavy defoliation when the disease becomes epidemic. Needle cast is most serious when young needles, particularly on young trees, are affected. Defoliation is rarely severe enough to kill any trees except young seedlings. However, young trees may be markedly suppressed by continued attack. The appearance of brown foliage on infected trees is often alarming out of proportion to the actual injury resulting when only older needles are attacked, as is often the case. In Europe needle cast is recurrently serious in nurseries and plantations. There is need for study of the disease on various conifers to work out the life histories and virulence of the associated fungi and to determine the actual damage done. Observations indicate that many of the needle cast fungi attack only trees weakened by other unfavorable factors, largely environmental.

Symptoms. The most conspicuous symptom of needle cast is a red or brown discoloration of the foliage, which later may turn to gray. This red or brown discoloration is also a symptom of various diseases caused by nonparasitic agencies, but these usually result in a rapid discoloration of the entire needle and of a definite group of needles, whereas needle cast symptoms appear more irregularly, involving only a portion of the needle in some cases, and leaving healthy needles intermingled with diseased. A single needle in a bundle may be affected while the others remain normal. Furthermore, heavily infected trees may be mixed with others of the same age and apparent vigor which are not attacked. The name "needle cast" is not entirely consistent because in some instances, as for example on larch and on spruce (Campbell and Vines 1938), the diseased needles remain attached to the twigs beyond the normal time.

Causal Agency. Needle cast is largely caused by a characteristic group of fungi (Ascomycetes, Hypodermataceae), which are easy to recognize by their dark, sometimes glossy black, fructifications termed "hysterothecia." The hysterothecia are more or less elongated perithecia extending from a millimeter or so in length to the entire length of the needle (Fig. 55*a*, *c*). At maturity the hysterothecia open during moist weather by a medial slit to discharge the ascospores which are wind borne to infect other needles. Usually there are eight spores per ascus, but a few species have only four. The ascospores are surrounded by a narrow to broad gelatinous sheath. The hysterothecia may be convex, flat, or depressed. Convex or flat pycnidia of various shapes in outline are developed, but the function of the conidia or spermatia produced in them is unknown, although it is probably sexual (Jones 1935:714).

Darker (1932) has monographed the needle cast fungi, describing a number of new species in North America, some of which have appeared in our own literature under the names of well known European species. He recognizes five genera distinguished by differences in the ascospores (Fig. 55*b*). In *Bifusella* the ascospores are one-celled, club-shaped at both ends with the halves joined by a narrow neck, and less than one-half the length of the ascus. In *Hypoderma* the ascospores are one-celled, fusiform (thickest in the middle and tapering toward each end) or rod-shaped, and less than one-fourth the length of the ascus. In *Hypodermella* the ascospores are one-celled and club-shaped at the upper end, taper acutely toward the base, and are one-fourth to three-fourths the length of the ascus. In *Elytroderma* the ascospores are two-celled and broadly fusiform. In *Lophodermium* the ascospores are one-celled, filiform (long, slender, and cylindrical), and fasciculate (arranged in a dense or close bundle or cluster). *Lophodermium* was subsequently monographed by Tehon (1935) and divided into four genera on the basis of differences in the structure and position of the hysterothecium, but the genus *Lophodermium* as defined by Darker is preferred for our purpose.

There are more than forty species of needle cast fungi known in this country, but only those which appear to be, or are known to be, parasitic and those most commonly found will be mentioned.

On the balsam firs there are six needle cast fungi that are quite similar in appearance, in their effects on the host, and as far as is known in their life histories. The dark-brown to black hysterothecia appear on the undersurface of the needle, usually extending the entire length of the middle nerve (Fig. 55*a*) but occasionally interrupted. All these species at some time have been referred to as *Hypodermella nervisequia* (DC.) Lag.,

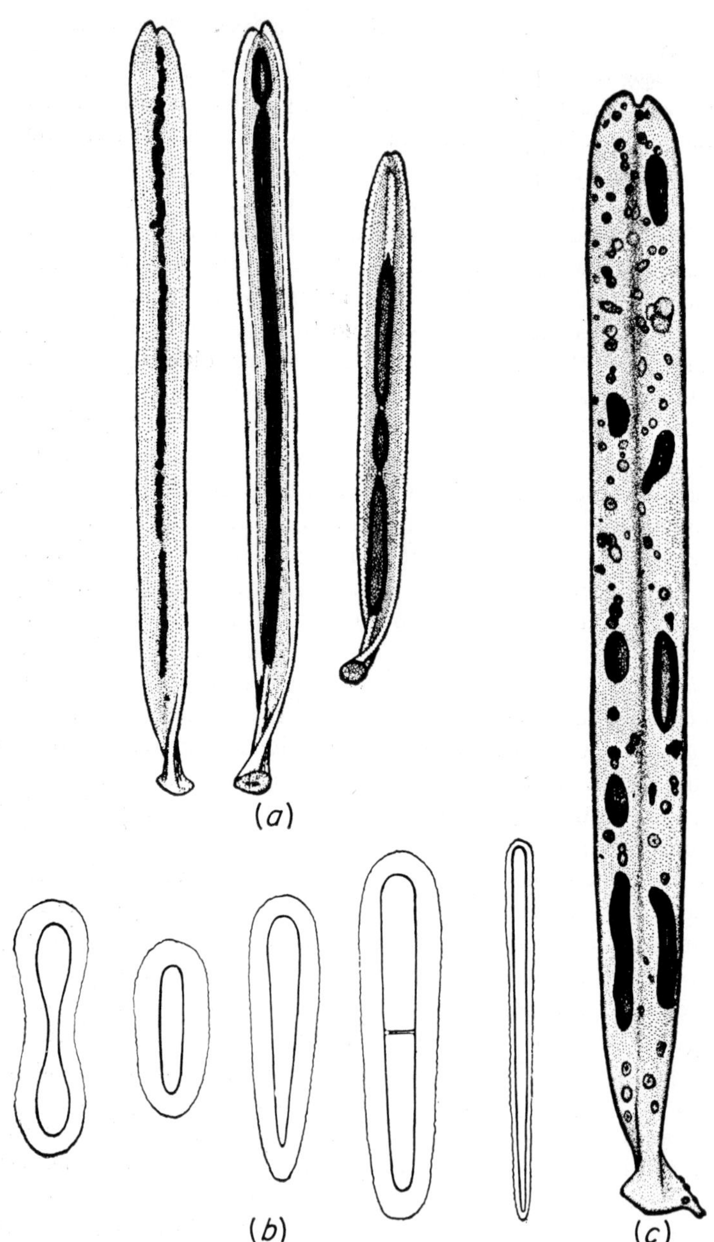

Fig. 55. (*a*), *Hypodermella nervata* on balsam fir. Left to right: pycnidia on the upper surface of a needle, closed hysterothecium of the extremely elongated type, and open hysterothecia. (*After G. D. Darker.*) (*b*), Ascospores of the five genera of the Hypodermataceae. Left to right: *Bifusella, Hypoderma, Hypodermella, Elytroderma,* and *Lophodermium.* (*Drawn by G. D. Darker.*) (*c*), *Lophodermium uncinatum* on Pacific silver fir showing the small, light-colored pycnidia and the larger, black hysterothecia of the shorter type. (*Drawn by G. D. Darker.*)

a species now considered to be confined to Europe where its effect on silver fir has been known for years (Hartig 1874:114). These fungi can be separated without a microscopic examination on the basis of their distribution, hosts, and pycnidial characters.

In the East, three species attack balsam fir, and they may all be found on needles of the same twig. *Bifusella faullii* Darker is the most common and destructive. The ascospores are discharged about July and infect young needles. There is no indication of disease until the following spring, when the infected needles change color. By about the first of July the 1-year-old needles are light brown and later in the season become buff colored. Pycnidia appear on the upper surface of the needles, following the groove. They are concolorous (similar to the needle in color), effused, sinuous, and labyrinthine in form. After the conidia are discharged, the hysterothecia begin to form on the lower surface of the needles. They do not mature and discharge ascospores until about July of the following summer, when the needles are in their third growing season. The hysterothecia are dusky brown in color. Mature pycnidia and perithecia may be found at the same time, but the pycnidia will be on needles in their second growing season and the hysterothecia on needles in their third growing season. *Hypodermella nervata* Darker is next in importance. The pycnidia occupy the groove on the upper surface of the needle in a continuous, or occasionally interrupted, row and are almost black after the conidia have been discharged (Fig. 55*a*). *H. mirabilis* Darker is the least frequent. The concolorous pycnidia form two lines on the wings of the needle, one on each side of the groove.

In the West, *Hypoderma robustum* Tub., which ranges from southern California to southern British Columbia, attacks several firs, but its usual host is white fir. The concolorous pycnidia, which occasionally turn black after the conidia are discharged, form two rows, one on each wing of the needle. *Bifusella abietis* Dearn., ranging from New Mexico to Idaho, occurs on alpine fir. Its pycnidial character is the same as *H. robustum*, but as far as is known the ranges of the two do not overlap. *Hypodermella abietis-concoloris* (Mayr) Dearn., ranging from the Rocky Mountains to the Pacific Coast, is found on several firs. It is easily recognized by its dark-brown pycnidia, either continuous or interrupted, in the groove on the upper surface of the needle.

On spruces *Lophodermium filiforme* Darker is the most serious of these fungi. In stands of red or black spruce severe defoliation of the lower crowns is not unusual. The shining black hysterothecia, usually on the undersurfaces of the needles, may be short or the entire length of the needle in extent. In the East, *Naevia piniperda* Rehm is always associated with it. *L. filiforme* is closely related to the European *L. macro-*

sporum (Hart.) Rehm, which is parasitic on Norway spruce there. *L. piceae* (Fckl.) v. Höhn., long known in this country as *L. abietis* Rostr., which has short shining black hysterothecia developing on all surfaces of the needles, is at worst a weak parasite here, although it has been considered the cause of a serious disease of Norway spruce in Denmark. It also occurs on firs.

Eastern and western larches are diseased by *Hypodermella laricis* (Hubert 1954; Cohen 1957). The parasite not only kills the needles but many of the short secondary shoots that bear them. If the needles on the long terminal shoots were affected, the disease would be more destructive. Severely infected stands have a brown scorched appearance because the killed needles are not cast but remain attached to the spur shoots over winter, in decided contrast to the deciduous habit of normal needles. The very small, oblong to elliptical, dull-black hysterothecia are on the upper surface of the needles. A somewhat similar disease is caused by *Meria laricis* Vuill. (Fungi Imperfecti, Moniliaceae), which in Europe is widespread and injurious on European larch (Peace and Holmes 1933), and which also occurs on western larch in its natural habitat (Ehrlich 1942; Hubert 1954). The first indication of the disease is a discoloration and browning of the needles in the spring or early summer. Infected needles are eventually cast. To detect the extremely inconspicuous fructifications on the undersurface of the leaf, a microscopic examination is necessary. In Great Britain, where the fungus causes damage in nurseries, it has been controlled by spraying with lime-sulfur every 2 to 3 weeks from the time the buds open until the beginning of August (Peace 1936).

Lophodermium juniperinum (Fr.) de Not. occurs abundantly on various cedars, but there is no indication that it is ever parasitic. The short, elliptical, shining black hysterothecia develop on both surfaces of the leaves. *L. thujae* Davis, which is similar in appearance, is known only on northern white cedar. It has four spores in each ascus when mature.

On pines, *L. pinastri* (Schrad.) Chev. is widespread. The pycnidia appear, usually in the spring or early summer, as tiny black spots on the browned needles and they are followed by the dull, or occasionally shining, black elliptical hysterothecia, which are rarely more than 1.5 mm in length and which occur on all surfaces of the needles. The hysterothecia may develop after the dead needles have dropped to the ground. This fungus has been extensively investigated in Europe (Haack 1911; Hagem 1926; Jones 1935; Lagerberg 1913; Schutt 1957; Tubeuf 1901), where some authorities consider that it causes the well-known *Schütte* or needle cast of young Scotch pine, whereas others consider it to be only a saprophyte. Probably the fungus is parasitic to the extent of causing needle cast on trees weakened by unfavorable climatic conditions (Hesselink 1927).

In North America an occasional epidemic has been reported in natural stands (Weir 1913) or in nurseries (page 95), although *L. pinastri* commonly behaves as a saprophyte in the southeastern United States (Boyce 1951).

Jack pine often suffers so severely from needle cast caused by *Hypodermella ampla* (Davis) Dearn. that all needles drop except those of the current season. The parasite is restricted to the one host. The short, elliptical dull-black hysterothecia are scattered over light-buff areas on the needles.

One of the most virulent needle cast fungi is *Hypodermella concolor* (Dearn.) Darker on jack and lodgepole pines. The parasite completes its life cycle in 1 year. Infection occurs in the summer on the young needles, but there are no indications of the disease until late spring or early summer of the following season when the infected needles brown,

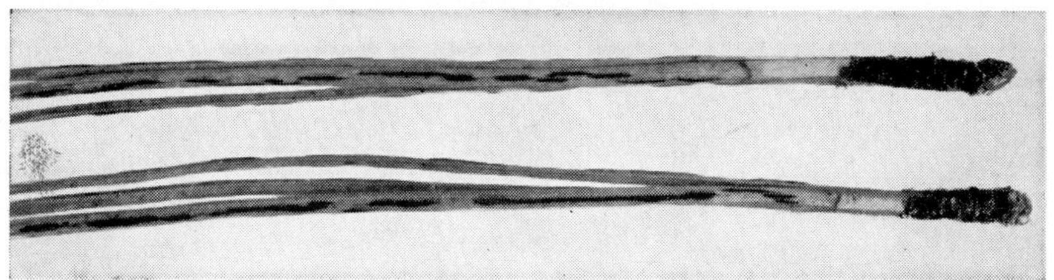

Fig. 56. Hysterothecia of *Elytroderma deformans* on needles of ponderosa pine. (*Reproduced from J. Agr. Research* 6, *no.* 8.)

and by midsummer the short hysterothecia, concolorous with the leaf surface, are mature and appear as shallow depressions. This parasite is epidemic on lodgepole pine over extensive areas in the northern Rocky Mountain region, causing heavy defoliation. Trees infected for successive seasons are reduced to needles of the current season only, instead of retaining their needles from 6 to 8 years. *H. montivaga* (Petrak) Dearn., which has dark-brown elongated hysterothecia, is also parasitic on lodgepole pine in the Rocky Mountains, but little is known as to the amount of injury it causes. This species is closely related to the European *H. sulcigena* (Rostr.) Tub., which causes the serious gray needle disease of Scotch pine in Scandinavia (Lagerberg 1910). *H. montana* Darker, which has shining black hysterothecia, commonly occurs on lodgepole pine in the Sierra Nevada of California.

Elytroderma deformans (Weir) Darker is severely parasitic on ponderosa and Jeffrey pines (Lightle 1954; Childs 1959; Waters 1959). It also attacks lodgepole, piñon, and jack pines and has been recorded once on shortleaf pine. The life cycle is 1 year, and the fungus is easy to recognize by its elongated, dull, dark-colored hysterothecia (Fig. 56), which occur on all surfaces of the needles. The fungus is perennial in the bark of twigs and small branches. Prominent loose witches'-brooms (Fig. 57) are formed on ponderosa and Jeffrey pines. On the other hosts, with the exception of piñon pine, witches'-broom formation is not induced. Young stands of ponderosa or Jeffrey pine may be severely injured. Saplings may have their entire crowns converted into loose, somewhat flat-topped witches'-brooms with stunted, swollen branches. Trees so affected usually

Fig. 57. A ponderosa pine with branches deformed and broomed by *Elytroderma deformans*. (*Photograph by J. R. Weir.*)

die or live for years making little perceptible growth. *Hypodermella medusa* Dearn., which somewhat resembles *E. deformans* but does not cause witches'-brooms, attacks ponderosa, Jeffrey, and lodgepole pines. It causes reduced growth on poor sites (Wagener 1959).

Hypoderma lethale Dearn. causes a needle blight of various hard pines in the eastern United States (Boyce 1958). The hysterothecia are short, narrow, and black. Spraying with ferbam (2 pounds per 100 gallons of water), starting when the needles begin to emerge and continuing until the end of June, will control the disease. Control is usually not necessary in forest stands. *H. hedgcockii* Dearn. also occurs on hard pines from Maryland to Florida. The elliptical shining black hysterothecia occupy discolored areas on green needles. The species is striking in that each ascus contains four normal and four aborted spores.

Naemacyclus niveus (Pers. ex Fr.) Sacc., which has long appeared in literature as *Lophodermium gilvum* Rostr., causes a needle casting of pines (Darker 1932:92). It is found on the Pacific Coast, in Colorado, and in the Northeast. The fructifications are tiny, elliptical, and at first waxy and dark brown in color but later become concolorous with the leaf surface.

On eastern white pine, western white pine, and limber pine, *Bifusella linearis* v. Höhn. is common. The hysterothecia are variable in length, shining black, and usually associated with long sterile black crusts on the same needles. Although the fungus kills needles, its attacks are limited to needles 2 or more years old, so that the resulting loss of foliage is of little consequence to the tree. *Hypoderma desmazierii* Duby (*H. strobicola, H. brachysporum*) occurs on several pines, but it is most frequent on eastern white pine. Infected needles are at first light yellow, then become light reddish brown, and finally a deep brown with occasionally a grayish cast (Graves 1913:133). Usually the tips of the needles are affected first, but finally the entire needle is involved. The shining black hysterothecia are elliptical in outline, short, and usually on the outer surface of the leaf. The fungus, which completes its life cycle in 1 year, appears to be a weak parasite and has not been found affecting more than an occasional tree in a vigorous stand. *Lophodermium nitens* Darker is frequently found on five-needle pines. The shining black hysterothecia are short and elliptical in outline. The fungus seems to be a saprophyte.

Control. There are no control measures known for needle cast under forest conditions in North America. In most cases control measures would not be economically justifiable. Darker (1932:109) suggests the possibility that the study of certain secondary fungi, which follow and sometimes inhibit the development of the needle cast fungi, might be valuable in enabling the introduction of the inhibitors into regions where they do not now occur and where certain needle cast fungi become sporadically epidemic. Probably mixed stands wherever possible instead of pure stands will reduce the incidence of certain needle casts.

Spraying with Bordeaux mixture will control *L. pinastri* in nurseries. It is inadvisable to establish nurseries in the midst of infected natural stands of the same species to be grown in the nursery.

MISCELLANEOUS FOLIAGE DISEASES

Brown Felt Blight. Brown felt blight is caused by *Herpotrichia nigra* Hart. and *Neopeckia coulteri* (Pk.) Sacc. (Ascomycetes, Sphaeriaceae).

These fungi occur only at high elevations in mountainous regions. In North America *N. coulteri* is confined to pines, while *H. nigra* occurs on conifers other than pines (Sturgis 1913; Hedgcock 1914:181); but in Europe *H. nigra* occurs on Swiss mountain pine as well as on juniper, Norway spruce, and silver fir, whereas *N. coulteri* is known only on Swiss mountain pine (Savulescu and Rayss 1928). In this country the damage caused is not economically significant; consequently methods of control

Fig. 58. Brown felt blight of whitebark pine caused by *Neopeckia coulteri.*

have not been developed. Nurseries at high elevations might be injured. In Europe dusting conifers in mountain nurseries with an organic fungicide in the fall has given protection against *H. nigra* (Zobrist 1950).

These fungi develop on foliage under snow. *H. nigra* finds a sufficiently high humidity and a suitable temperature for a protracted period only under snow at high elevations. At lower elevations the winter is too cold and the summer too hot for the fungus (Gäumann, Roth, and Anliker 1934). After the snow melts, lower branches are seen to be covered with a dense, felty or cobwebby growth of brown to almost black mycelium (Fig. 58). The foliage is killed, chiefly by the dense mycelium excluding light and air from it, although the hyphae also penetrate the leaf tissues, so that these fungi are both epiphytic and parasitic. The small black globular fructifications are scattered over the mycelial mass. Macroscopically the two fungi cannot be distinguished from each other, but the ascospores of *H. nigra* are four-celled and olivaceous when mature, whereas those of *N. coulteri* are two-celled and dark brown. Although the two fungi are not known to intermingle on the same plant, their close relationship is suggested by the fact that ascospores intermediate in form have been found (Boyce 1916).

A disease somewhat similar in appearance caused by *Rosellinia herpotrichioides* Hepting & Davidson (1937), but with grayish-brown felty mycelium, occurs on eastern hemlock in North Carolina. It has also damaged Douglas fir seedlings in a nursery in British Columbia (Salisbury 1956).

Sooty Mold. A sooty growth or crust may appear in isolated patches or may cover the entire needle, the black material being made up of fungous hyphae and perithecia. The sooty mold fungi often develop on the honeydew excreted by certain aphids and scale insects. A balsam fir sapling may have its foliage completely blackened as though it had been exposed to dense smoke. The fungi causing sooty mold largely belong to the family Perisporiaceae of the Ascomycetes. Hyphae do not penetrate living leaf tissue. Although sooty leaves are unsightly and some interference with their physiological functions must result, yet there is no apparent injury to affected trees. *Dimerosporium abietis* Dearn. forms black patches on needles of lowland white and Pacific silver firs, whereas *D. tsugae* Dearn. develops similarly on western hemlock. Neither fungus occurs on needles of the current season, and both are usually on the undersurface of the needles.

Tip Blight. This disease of balsam firs is caused by *Rehmiellopsis balsameae* (Waterman 1945). The fungus, which kills the needles and shoots of the current season, shriveling and curling the needles in a characteristic fashion, is found in northern Maine and in Nova Scotia on balsam fir. It also occurs in the Northeast on other firs growing as ornamentals. A similar tip blight in Europe is caused by *R. abietis* (*R. bohemica*), which has been found at one locality in British Columbia causing tip blight of alpine fir. A tip blight and dieback of pines is caused by *Diplodia pinea* (page 316).

Pullularia pullulans (de Bary) Berkh. (Fungi Imperfecti, Moniliales) and a gall midge cause needle blight and late fall browning of planted red pine in Ontario, resulting in considerable defoliation (Haddow 1941). It is associated with a flagging disease of western white pine (Molnar 1954). *Sphaerulina taxi* (Cke.) Mass. (Ascomycetes, Sphaeriaceae) kills needles of the current season on Pacific yew. *Pestalozzia funerea* Desm.

(Fungi Imperfecti, Melanconiaceae) is often associated with leaf blights of conifers, but opinions as to its pathogenicity are conflicting (Guba 1929: 204; Vermillion 1950; Wagener 1935). *Phoma piciena* Pk. (Fungi Imperfecti, Sphaerioidaceae), which occurs on needles of red spruce in the Adirondacks, may defoliate and kill large Norway spruces (Spaulding 1912:151). *Dothistroma pini* Hulb. (Fungi Imperfecti, Phomaceae) causes a needle blight of Austrian, limber, ponderosa, and red pines in the Middle West (Hulbary 1941; Thomas and Lindberg 1954).

There are numerous other fungi associated with needle diseases of conifers (Hedgcock 1932). Their appearance is often sporadic, their pathogenicity has not been proved, and in many cases it probably depends on conditions unfavorable to the host or unusually favorable to the fungus. Certain needle diseases are primarily important on seedlings (page 95). Cones are mostly affected by rust fungi (page 186), but *Sclerotinia kerneri* Wettstein (Ascomycetes, Helotiaceae) causes a thickening of the twigs and an abnormal increase in the number of staminate cones on balsam firs (Cash 1941).

LEAF RUSTS

Many rust fungi attack the needles of conifers (Boyce 1943). Although some idea of their identity can be obtained from the host and certain evident characters, in general, separation of the species is a task for the specialist. Needle rusts are mostly heteroecious with pycnia (0), when present, and aecia (I) on conifer leaves, or rarely on young, tender stems or cones, whereas uredia (II) and telia (III) are on the leaves of ferns or flowering plants and shrubs. A few have uredia or telia on the coniferous host with the other stages lacking or unknown, but in the juniper and cedar leaf rusts the pycnia and aecia occur on rosaceous hosts. In some the pycnial and aecial stages on conifers have not yet been connected with an alternate host, so these are referred to the form genera *Peridermium* or *Caeoma*. When the telial stage is known, a rust takes its scientific name from it. The rust names that follow are given in the check list prepared by Cummins and Stevenson (1956).

The life cycles of most rusts with pycnia and aecia on conifers are similar. The pycnia, appearing in the spring or summer as yellow or orange slightly raised spots on the needles, are very small spherical, hemispherical, or lens-shaped structures that develop beneath the cuticle or epidermis of the leaf and open by means of a narrow or wide pore (Adams 1919:64; Hunter 1936). The minute oval or elongate colorless pycniospores, immersed in a sweetish sticky liquid known as the "pycnial exudate" or "nectar," are exuded through the mouth of the pore. The pycniospores cannot initiate new infections but presumably as in other rusts function in a fertilization process preceding development of the aecia. The aecia appear soon after the pycnia, as white or yellow to orange, flattened or columnar blisters (Figs. 59, 60, 62). Each aecium consists of a white membrane, the peridium, enclosing the white or orange-red aeciospores. This type of aecium allocates a rust in the form genus *Peridermium*, whereas a flattened aecium with the peridium lacking, the epidermis of the host serving as a covering for the aeciospores, characterizes the form genus *Caeoma*. The peridium or the epidermis ruptures and the aeciospores escape, are wind borne, and infect the alternate host. They cannot reinfect the coniferous host. Then the uredia develop as white or orange pustules covered by the epidermis, which ruptures to disclose the white or orange urediospores on the leaves. The wind-borne urediospores can only infect the host on which they are produced, but several generations may develop in a season. Later in the summer or in the autumn, the telia develop on the same host. They vary from reddish to dark-brown waxy cushionlike layers or crusts on the leaves (Fig. 43), or rarely on the stems. The teliospores germinate in place, producing secondary spores termed "sporidia," which are wind borne and infect the conifer needles. The teliospores either germinate in the autumn, immediately infecting conifer needles (the rust thus overwintering as mycelium in the needle tissue), or they remain dormant on the fallen leaves until the following spring before infecting the coniferous host. The sporidia from the teliospores cannot infect the host on which they are produced except in a few species. The life cycle of most needle rusts is thus approximately 1 year from aecium to aecium.

Needle rusts are most important on young trees in the seedling and sapling stages. Larger trees are rarely attacked, infection on them usually being confined to needles on branches near the ground. The effect is to kill individual needles, and in severe cases all needles of one growing season may die, but these rusts vary in intensity from year to year so that a young tree is never defoliated and killed. There must be some reduction in the growth of heavily infected trees.

None of these rusts has proved serious enough to warrant control measures under forest conditions. In nurseries control can be accomplished by eradicating the alternate host from around the nursery site, but keeping an area with a radius of 1,000 or more feet free from light-seeded annual plants is costly and should be avoided by not establishing a nursery in the midst of dangerous alternate hosts. Spraying coniferous stock with a fungicide to prevent infection when the sporidia are being discharged by the rust on the alternate host may be effective, but methods and sprays remain to be determined. If sporidia are developed in the autumn a more powerful fungicide can probably be used than if they are developed in the spring, when the young tender conifer needles would be most susceptible to spray injury. *Darluca filum* (Biv.) Cast. commonly parasitizes needle rusts on conifers (Keener 1934:475), particularly those on balsam firs, but sufficient aecia always escape for the rusts to maintain themselves even when the attack is heavy.

Pine Needle Rusts. The needles of Virginia pine, also known as "scrub pine," are attacked throughout the range of the tree by *Coleosporium pinicola* (Arth.) Arth. (Basidiomycetes, Coleosporiaceae). This rust, which develops vestigial pycnia (Dodge 1925:641) and telia on the pine needles, has no alternate host. The disease is most prevalent on trees 6 to 13 feet high, which in May are characterized by pale-yellowish and sparse foliage (Galloway 1896). At this time the telia appear on the yellowed tips of needles of the previous season, bursting through the epidermis as linear bright orange pustules. The teliospores germinate in place, releasing the sporidia, which immediately infect the young needles just appearing. The 1-year-old infected needles then die gradually and fall within 3 to 6 weeks from the time the telia appear, leaving diseased trees largely dependent on needles of the current year for their metabolism, while healthy trees retain their needles for at least two seasons.

C. crowellii Cummins develops telia, the only spore stage known, on the needles of piñon and limber pines in Colorado, New Mexico, and Arizona.

All the other rusts on pine needles also belong to the genus *Coleosporium* but with their pycnia and aecia on pine needles. The aecia may develop on any face of the needles. The aecia all belong to the form genus *Peridermium,* and even if the alternate stage of a *Peridermium* on pine needles is unknown it will ultimately prove to be a *Coleosporium.*

C. asterum (Diet.) Syd. (*C. solidaginis, P. acicolum*) is widespread on two and three-needle pines, with the alternate stage on various composites, particularly aster (*Aster* spp.) and goldenrod (*Solidago* spp.). The principal pine hosts in the East are jack, red, and pitch pines, and in the West only lodgepole pine. Although originally reported on ponderosa pine in the West, this rust has never since been found there on that host, although it does attack young planted ponderosa pine in the East. This suggests that the western form on lodgepole pine, for which Hedgcock (1933:24) has proposed the name *C. montanum* (Arth. & Kern) Hedgc. (*P. montanum* Arth & Kern), is a distinct species. Older needles of young trees are often severely attacked, particularly red pine in plantations, where a marked retardation in height growth may result even though trees are never killed (Baxter 1931). Older trees with their lower branches infected suffer no apparent injury. The conspicuous white aecia (Fig. 59) appear on needles in the spring or summer. The aeciospores infect the alternate host, such as goldenrod or aster, and uredia soon form on the leaves. The rust may overwinter in the uredial stage on the rosettes of these plants, thus maintaining itself indefinitely without the pine host (Weir and Hubert 1917:334). The urediospores spread infection on the same host, but later in the summer or in the autumn the waxy, reddish-brown crustlike telia appear. When the teliospores germinate, the sporidia infect the pine needles, and the life cycle is complete.

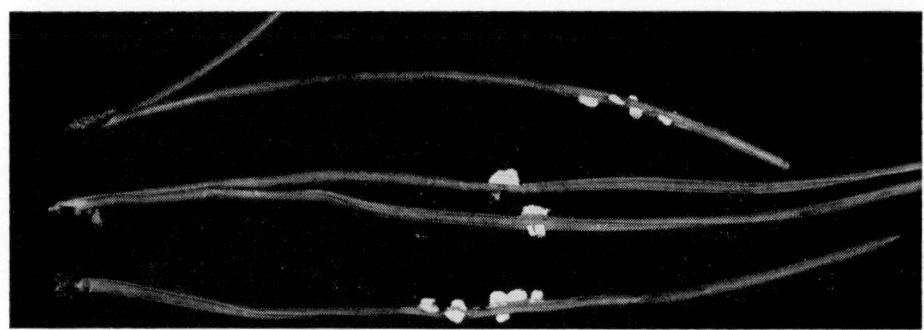

Fig. 59. Aecia of the western form of *Coleosporium asterum* on lodgepole pine. (*Reproduced from Mycologia* 17, *no.* 6.)

There are many other Coleosporiums on pines, and new ones are still being found. The known species are:

C. apocynaceum Cke.: on loblolly, longleaf, and slash pines; on *Amsonia* spp.

C. campanulae (Pers.) Lev.: (*Peridermium rostrupi*) on pitch, red, and probably Scotch pines; on *Campanula* spp., *Lysimachia* sp., and *Specularia* spp.

C. delicatulum (Arth. & Kern) Hedgc. & Long: on two- and three-needle pines; on *Euthamia* spp. Range: East and South.

C. elephantopodis (Schw.) Thüm.: (*P. intermedium*) on southern pines; on *Elephantopus* spp.

C. helianthi (Schw.) Arth.: on pitch and shortleaf pines; on *Helianthus* spp.

C. inconspicuum (Long) Hedgc. & Long: on shortleaf, longleaf, and Virginia pines; on *Coreopsis* spp.

C. ipomoeae (Schw.) Burr.: on southern pines and Chihuahua pine; on five genera of Convolvulaceae, principally *Ipomoea* spp.

C. Jonesii (Peck) Arth.: (*P. ribicola*) on piñon; on *Ribes* spp. and *Grossularia* spp.; particularly prevalent on *R. inebrians* Lindl.

C. lacinariae Arth.: (*P. fragile*) on loblolly, longleaf, and pitch pines; on *Liatris* spp.

C. madiae Cke.: (*P. californicum*) on Monterey, Coulter, and Jeffrey pines; on *Madia* and several other related composites.

C. minutum Hedgc. & Hunt: on loblolly and spruce pines; on *Adelia ligustrina*.

C. sonchi (Strauss) Lév.: (*P. fischeri*) on Scotch and (?) jack pines; on *Sonchus* spp. Found sporadically in the United States and probably introduced from Europe.

C. terebinthinaceae Arth.: on two- and three-needle pines; on *Parthenium* sp., *Polymnia* sp., and *Silphium* spp. Range: East and South.

C. vernoniae B. & C.: (*P. carneum*) on two- and three-needle pines; on *Vernonia* spp. Range: East and South.

Peridermium guatamalense A. & K.: on longleaf and Montezuma pines; alternate host unknown. Range: Ocala, Florida, southern Mexico, and Guatemala.

P. weirii Arth.: on lodgepole pine; alternate host unknown, possibly *Senecio* spp. Known only from Kooskia, Idaho.

Needle Rusts of Balsam Firs. These rusts, which belong to several different genera in one family (Basidiomycetes, Melampsoraceae), have

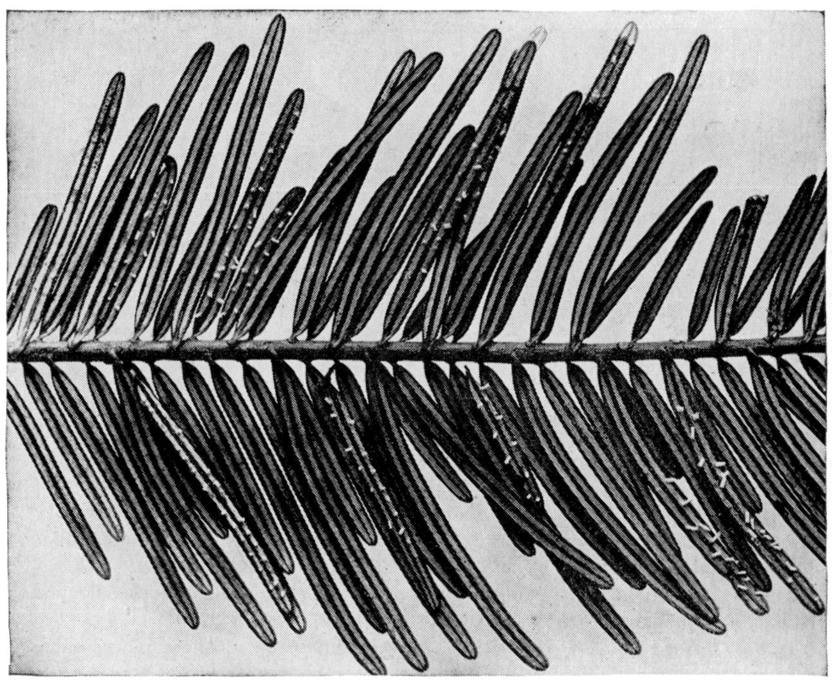

Fig. 60. Aecia of *Peridermium pseudo-balsameum* on 1-year-old needles of lowland white fir. (*Reproduced from Am. J. Botany* 4, *no.* 6.)

their pycnia and aecia on needles of the balsam firs and their uredia and telia on various alternate hosts. They cause some damage to seedlings and small saplings, occasionally killing plants but usually retarding their growth. The pycnia, which appear as small, round, slightly raised orange or yellowish spots, are usually confined to the undersurface of the needle but may occur on both surfaces, depending on the species. On the other hand, the orange-yellow or white aecia, usually columnar in shape (Fig. 60), are confined to the lower surface where they are produced in the spring or summer, preceded by the pycnia. Those species with aecia on needles of the current year overwinter on the telial host, whereas those with aecia on needles 1 year old or older overwinter in the needle tissue and may also overwinter on the telial host. Primary infection of the firs occurs through needles of the current year irrespective of whether the aecia appear the same season or a year or more later.

Although the various species are difficult to distinguish, the character of the mycelia and haustoria (Hunter 1948), the position and size of the pycnia, the color of the aecia, their time of development, and the age of the needles on which they occur are the more obvious diagnostic characters. The alternate hosts comprise a wide variety of annuals and perennials. In some cases uredia and telia on certain alternate hosts have not been connected with the fir host, and in other cases the connection has been established by artificial inoculation, although the aecia have not been found occurring naturally on the fir needles. Furthermore the distribution of certain species is known only partially.

The following needle rusts on balsam firs are now known in the United States:

AECIA ORANGE-YELLOW

Melampsora abieti-capraearum Tub.: (*Caeoma abietis-pectinatae*) on balsam, white, lowland white, and alpine firs; on willows. The aecia, on needles of the current year, are of the caeoma type, flattened and without a visible peridium.

Pucciniastrum abieti-chamaenerii Kleb: on balsam, white, lowland white, alpine, Pacific silver, corkbark, and noble firs; on *Chamaenerion angustifolium*. Aecia on needles of the current year, usually maturing in early summer.

P. epilobii Otth.: on balsam fir; on *Epilobium adenocaulon* and other species. Aecia on needles of the current year. The fir hosts for this and the preceding species are not yet definitely established (Faull 1938*a*).

Calyptospora goeppertiana Kühn (*C. columnaris*): on all the common firs; on *Vaccinium* spp. (Faull 1939). The pycnia are aborted; aecia on needles of the current year usually maturing in late summer. The telial stage causes pronounced swelling of the shoots and witches'-brooms on *Vaccinium*, but no uredia are formed.

Hyalopsora aspidiotus (Magn.) Magn.: (*Peridermium pycnoconspicuum*) on balsam fir; on *Phegopteris dryopteris*. Aecia on 2-year-old needles. Since this rust commonly occurs on its fern host in the West, it will undoubtedly be found there on firs.

Melampsorella caryophyllacearum causes witches'-brooms with deciduous needles (page 199).

Peridermium ornamentale Arth.: on alpine fir and possibly other species; alternate host unknown. The robust-cylindrical to compressed-cylindrical aecia appear in the late summer or early autumn on needles of the current season.

Peridermium holwayi Syd.: on alpine fir; alternate host unknown. The robust compressed-cylindrical aecia appear in the late spring or in the summer on 1-year-old needles.

Peridermium sp.: on Pacific silver fir; alternate host unknown, possibly *Vaccinium* (Boyce 1928). Aecia on 1-year-old needles.

Caeoma faulliana Hunter (1936:118): on alpine fir; alternate host unknown. Aecia on needles of the current year. Distinguished from *M. abieti-capaearum* by differences in the pycnia.

Aecia White

The genera *Uredinopsis* and *Milesia*, to which the species with white aecia belong, have been critically studied by Faull (1932, 1934, 1938*b*, 1938*c*).

Uredinopsis atkinsonii Magn.: (*Peridermium balsameum*) on balsam fir; on *Dryopteris thelypteris* var. *pubescens*.

U. ceratophora Faull: on balsam fir; on *Cystopteris bulbifera*.

U. longimucronata Faull: on balsam fir; on *Athyrium angustum*.

U. longimucronata forma *cyclosora* Faull: on alpine fir; on *Athyrium cyclosorum*.

U. americana Syd.: (*Peridermium balsameum*) on balsam fir; on *Onoclea sensibilis*.

U. osmundae Magn.: (*P. balsameum*) on balsam fir; on *Osmunda* spp.

U. phegopteridis Arth.: (*P. balsameum*) on balsam fir; on *Phegopteris dryopteris*.

U. struthiopteridis Störmer: (*P. balsameum*) on balsam fir; on *Struthiopteris germanica*.

The aecia of the preceding species cannot be separated on the firs. They develop on needles of the current year, usually maturing in midsummer.

U. pteridis D. & H. and *U. hashiokai* Hirats. f.: (*P. pseudo-balsameum*) on Pacific silver, white, lowland white, alpine, noble, and bristlecone firs; on *Pteridium aquilinum* varieties and marginal species (Fig. 61). The pycnia and aecia of the two species are indistinguishable (Ziller 1959*a*). Aecia on 1- to 5-year-old needles (Fig. 60).

Milesia marginalis Faull & Watson: on balsam fir; on *Dryopteris marginalis*. Aecia on needles of the current year, usually maturing by midsummer. The pycnia are borne on both sides of the needles, while in all the other fir needle rusts they are confined to the underside.

M. fructuosa Faull (include *M. intermedia* Faull): (*Peridermium balsameum*) on balsam fir; on *Dryopteris spinulosa*, *D. spinulosa intermedia*, *D. spinulosa americana*, and *D. spinulosa fructuosa*. Aecia on needles of the current year, usually maturing by midsummer.

M. pycnograndis Arth.: (*Peridermium pycnogrande*) on balsam fir; on *Polypodium*

Fig. 61. Uredia of *Uredinopsis* on bracken (brake fern). Notice the coiled spore masses. (*Reproduced from Am. J. Botany* 4, *no.* 6.)

virginianum. Aecia on needles 3 to 9 years old (Faull 1932: 125). The hyphae are perennial in the needles and small stems of balsam fir, causing loose witches'-brooms.

Peridermium rugosum Jacks.: on Pacific silver and lowland white firs. The alternate stage is unknown but is probably *Milesia* sp. Aecia on needles of the current year, usually maturing in late summer or autumn.

M. laeviuscula (Diet. & Holw.) Faull: on lowland white fir; on *Polypodium vulgare* L. var. *occidentalis* Hook. (Ziller 1957). Aecia on needles of the current year, maturing in the fall.

Douglas Fir Needle Rusts. Two needle rusts, *Melampsora occidentalis* Jacks. and *M. albertensis* Arth., are known on Douglas fir (Ziller 1955). The pycnia occur on both surfaces of needles of the current year, whereas the orange-yellow, somewhat elongated aecia of the caeoma type are confined to the undersurface of the same needles. Infected needles die soon after the aecia appear, and they are shed during the same summer. The virulence, symptoms, and gross morphology of the two appear to be the same, but the dimensions of all spores of *M. occidentalis,* except the basidiospores, are appreciably larger than those of *M. albertensis.* The uredia and telia occur on the leaves of various western poplars.

California Torreya Needle Rust. A single needle rust, *Caeoma torreyae* Bonar (1951) occurs on California torreya. Infected needles are thickened and chlorotic to greenish-brown in color. They may be shed prematurely.

Hemlock Needle Rusts. All these rusts occur on needles of the current year and consequently overwinter in the telial stage.

Pucciniastrum vaccinii (Wint.) Jørstad: (*Peridermium peckii*) on eastern and Carolina hemlocks; on various Ericales. This is the most common rust on eastern hemlock, but often only a single leaf on a small twig will be infected (Fig. 62).

Fig. 62. Aecia of *Pucciniastrum vaccinii* on eastern hemlock. [*Courtesy of Conn. (State) Agr. Expt. Sta.*]

Pucciniastrum hydrangeae (B. & C.) Arth.: on eastern and Carolina hemlocks; on *Hydrangea arborescens.* A little-known species.

Melampsora abietis-canadensis (Farl.) Ludw.: (*Caeoma abietis-canadensis*) on eastern hemlock; on various poplars. The aecia, of the caeoma type, occur on both surfaces of the needles, whereas in all the others on hemlock, the aecia are confined to the undersurface. Although infected needles are much aborted, so few of them are attacked that little damage is done. Shoots of the current year may be slightly distorted and cones are also infected (page 188).

Melampsora epitea Thüm. f. sp. *tsugae* Ziller: (*Caeoma dubium*) on western and mountain hemlocks; on willows (Ziller 1959*b*). This species has subepidermal pycnia, which serve to distinguish it from *M. abietis-canadensis* and *U. holwayi,* both of which have subcuticular pycnia.

Uraecium holwayi Arth. (*Uredo holwayi*): on western and mountain hemlocks; alternate stage unknown. This is an unusual rust. The one spore stage known is physiologically an aecium with a peridium, but the spores are borne singly on stalks instead of in chains.

Melampsora farlowii (Arth.) Davis (*Necium farlowii*): on eastern and Carolina hemlocks; alternate stage unknown, probably lacking (Garriss 1959). The reddish, waxy, somewhat raised linear telia occur not only on the undersurface of needles but also on shoots of the current year and on cones (page 188). The young shoots may be twisted and killed. Considerable damage to eastern hemlocks 2 to 15 feet high in nurseries and hedges, and in a lesser degree to Carolina hemlocks, has been caused by this rust in western North Carolina. Control by spraying with lime-sulfur (2 pounds of dry lime-sulfur to 25 gallons of water) or with ferbam 76 per cent (2 pounds to 25 gallons of water). Six or more consecutive applications are necessary, beginning just as soon as the buds burst.

Spruce Needle Rusts. The needles of Engelmann spruce and red spruce are attacked by *Chrysomyxa weirii* Jacks. The waxy orange to orange-brown elongate-elliptical telia occur on 1-year-old needles causing yellowish spots. This is the only spore stage known.

All other rusts on spruces have pycnia and orange-yellow aecia of the familiar columnar type, although often laterally elongated, on needles of the current year, with the uredia and telia on various alternate hosts (Savile 1955). They overwinter in the telial stage. These rusts occasionally become epidemic on spruce causing considerable defoliation (Lindgren 1933).

The following rusts with aecia on spruce needles are now known:

C. ledi var. *cassandrae* (Pk. & Clint.) Savile.: (*Peridermium consimile*) on black, red, and blue spruces; on *Chamadaphne calyculata.*

C. ledi de Bary: (*P. abietinum*) on black, red, and Norway spruces; on the underside of leaves of *Ledum glandulosum, L. groenlandicum,* and *L. decumbens.*

C. ledicola Lagerh.: (*P. decolorans*) on white, black, red, blue, Engelmann, and Sitka spruces; on the upper side of leaves of *Ledum groenlandicum* and *L. decumbens.* This rust frequently discolors infected needles so badly that the trees appear yellowish in color.

C. chiogenis Diet.: on black and white spruces; on the underside of leaves of *Chiogenes hispidula.*

C. empetri Schroet.: on red and white spruces; on *Empetrum* spp. The aecia occur on the upper and lower surfaces of the needles; in all other species of *Chrysomyxa* they are confined to the lower surface.

C. piperiana Sacc. & Trott.: (*P. parksianum*) on Sitka spruce; on the underside of leaves of *Rhododendron californicum.*

C. woroninii Tranz.: on black and white spruces causing shoot rust; on *Ledum groenlandicum* causing witches'-brooms.

Pucciniastrum americanum (Farl.) Arth.: (*P. ingenuum*) on white spruce; on *Rubus leucodermis, R. melanolasius, R. neglectus,* and *R. strigosus.*

P. arcticum Tranz.: (*P. ingenuum*) on white spruce; on *Rubus acaulis, R. chamaemorus, R. pubescens,* and *R. stellatus.*

Peridermium coloradense causes witches'-brooms with deciduous needles (page 200).

Larch Needle Rusts. There are two rusts on the needles of larch in the United States. The aecia are most prevalent on needles of the leading shoots, occurring on the undersurface. The damage to the larches is generally insignificant.

Melampsora paradoxa D. & H.: (*Caemoa bigelowii*) on eastern, European, western, and alpine larches; on many willows. This rust can overwinter in the uredial stage on the stems of willows.

M. medusae Thüm.: (*C. medusae*) on eastern larch; on many poplars.

Cedar and Juniper Leaf Rusts. The leaf rusts on junipers and cedars together with the stem rusts (page 194) are caused by species of the genus *Gymnosporangium* (Basidiomycetes, Pucciniaceae). In this genus the pycnia and aecia, with five exceptions, occur on shrubs and trees of the family Malaceae (Pomaceae). The telia are confined to cedars and junipers, except for one species on cypress.

The life cycle of the Gymnosporangiums is similar. In the early summer, small somewhat swollen light-yellow more or less circular spots appear on the leaves of the pomaceous host. Next, small raised specks appear within this spot on the upper surface. These are the openings of the flask-shaped pycnia, which are embedded in the leaf tissue. The pycnia soon exude a thick orange-colored liquid, the pycnial exudate, containing the pycniospores. After pycniospore discharge, the pycnia appear as small black specks (Fig. 63). Later the aecia push out on the underside of those same spots as prominent dingy-white columns known as "roestelia" (Fig. 63). The outer coating or peridium ruptures by splitting to release the powdery mass of yellowish to brown aeciospores, and sometimes the segments of the peridium resulting from this splitting recurve to give the open aecium a star-shaped appearance. In four species a roestelium is not developed, but the aecium remains largely embedded in the host tissue, barely protruding beyond it, so that, when the peridium ruptures, the open aecium is cup-shaped (a cluster cup or an aecidium), with a powdery mass of orange-yellow aeciospores exposed (Fig. 64). The aecia may also develop on fruits or on tender green stems. The aeciospores are released during the summer and are wind borne to infect cedars and junipers through the leaves or the young green stems.

The mycelium overwinters in the tissues of the coniferous host, where it may remain confined to the leaves or enter the stems. Leaf galls may

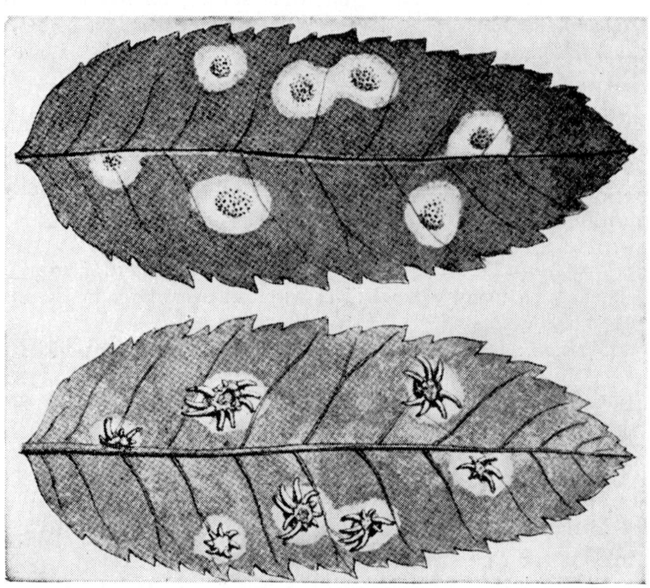

Fig. 63. *Gymnosporangium juniperinum* on mountain ash (*Sorbus aucuparia*). Pycnia on the upper surface of a leaf (top) and aecia (roestelia) on the undersurface. (*After E. Rostrup.*)

result, or various hypertrophies of the stems may be caused when the mycelium persists for many years in the woody tissues (page 194). Usually, by the following spring in the leaf-inhabiting forms or in the spring one or more seasons later in the stem-inhabiting forms, the telia appear in the form of cushions (Fig. 65), ridges, or long hornlike projections (Figs. 66, 71). Uredia are developed in only one species, *G. nootkatense.* At first the telia are dark brown in color, but when moistened they become jellylike and orange-brown. The orange-colored teliospores, more or less embedded in the jellylike matrix formed by the gelatinization of their long colorless stalks, germinate in place, and the secondary spores or sporidia are wind borne to infect the pomaceous host, thus completing the life cycle.

The injury by leaf-inhabiting Gymnosporangiums to susceptible conifers is usually slight, although occasional attacks of *G. juniperi-virginianae* causing "cedar apples" on eastern red cedar are sufficiently severe to retard the normal development of the tree, or even kill it in time. On the other hand, some red cedars are rust resistant (Berg 1940). Damage to the pomaceous host is often serious and becomes of economic consequence when valuable orchard or ornamental trees, such as apples or hawthorns, are attacked. The disease can be controlled on both types of hosts by spraying (Campana and Schneider 1955). Effective control can be accomplished by separating the alternate hosts by a distance of 1 or 2 miles, although light infection of pomaceous hosts may occur

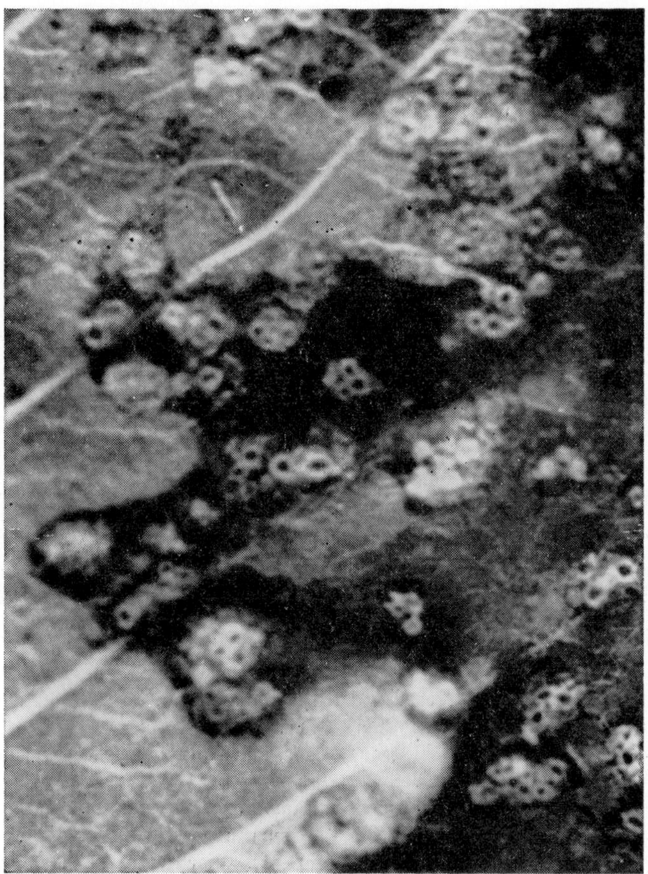

Fig. 64. Aecidia or cluster cups of *Gymnosporangium libocedri* on the undersurface of a serviceberry leaf.

from cedars up to 7 or 8 miles away (MacLachlan 1935*b*). In certain states the destruction of all cedars and junipers within 1 mile of valuable orchards is required by law.

The following are the known species of *Gymnosporangium* in the United States on the leaves of cedars and junipers, with which are included those species attacking tender green stems.

G. davisii Kern: on the leaves of chokeberry; on mountain juniper. The telia are usually on the upper side of the leaves, but occasionally on the small stems at the base of the leaves.

G. cornutum Arth.: the telia occur occasionally on the leaves, but mostly on the stems of mountain juniper (page 198).

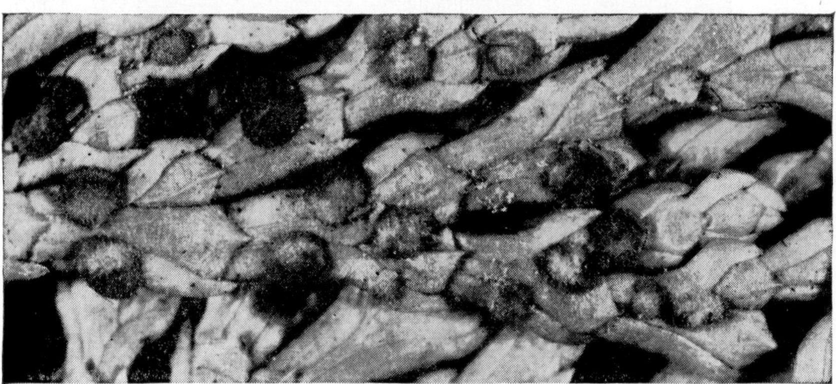

Fig. 65. Telia of *Gymnosporangium libocedri* on the foliage of incense cedar.

G. exiguum Kern: on the leaves and fruits of hawthorn; on alligator and Mexican junipers, on eastern red cedar. Limited to Texas.

G. fraternum Kern: on leaves, young stems, and fruits of chokeberry; on southern white cedar. This species is not common.

G. harknessianum Kern ex Arth.: chiefly on the fruits, occasionally on the stems, of serviceberry; on western juniper. The roestelia are unusually long.

G. inconspicuum Kern: chiefly on the fruits of serviceberry and wild crab; on Utah juniper. This often causes a yellowing of the juniper leaves; more rarely the telia occur on the woody branches.

G. multiporum Kern: aecial host unknown; on western, one-seed, and Utah junipers. The telia arise on the green stems between the scalelike leaves.

G. libocedri (P. Henn.) Kern (*G. blasdaleanum*): mostly on the leaves and fruits of serviceberry and hawthorn, but it also occurs on apple, pear, quince, and mountain ash; on incense cedar. The aecium is a cluster cup (Fig. 64). The incense cedar rust commonly kills small sprays of foliage, particularly on the younger trees, and the telia invariably develop on the leaves (Fig. 65), never on the woody stems, although the fungus causes witches'-brooms and marked swellings of the stems (page 196). In Oregon the fungus damages pear orchards (Jackson 1914).

G. nootkatense Arth.: on the leaves of mountain ash and Oregon crab apple; on

Alaska cedar. The aecium is a cluster cup. This is the only *Gymnosporangium* with urediospores. The uredia are bright orange but soon fade to pale yellow. The teliospores appear later in the uredial pustules.

The following species cause galls of the leaves or tender stems:

G. juniperi-virginianae Schw.: chiefly on the leaves, but occasionally on the fruits of apple and hawthorn; on eastern and southern red cedars, on prostrate and Rocky Mountain junipers. This species causes the well-known and destructive cedar apple rust (Crowell 1934). Cedars are infected during the summer, the young cedar apples or galls appear the following June, but they do not mature and produce the telial horns until the second spring after infection (Fig. 66). At maturity the galls may be 1 or more inches in diameter. After telial production the galls die. Before maturity the galls are greenish brown in color with small pitlike depressions on the surface from which the long cylindrical telial horns appear. Investigations of the origin of the galls has led to contradictory conclusions, but indications are that galls from trees with awn-shaped leaves are of leaf origin while those from trees with scalelike leaves are of stem origin (Miller 1936).

Fig. 66. "Cedar apples" on eastern red cedar caused by *Gymnosporangium juniperi-virginianae.* On the left the telia are dry, on the right moist and jellylike.

G. globosum Farl.: principally on the leaves of hawthorn but also on serviceberry, apple, pear, quince, and mountain ash; on eastern red and southern red cedars, on dwarf, prostrate, and Rocky Mountain junipers (MacLachlan 1935*a*). The galls that are of leaf origin (MacLachlan 1936: 11) are easily confused with those caused by *G. juniperi-virginianae,* but the former are often perennial, producing telia in the spring for several successive years. Furthermore, galls caused by *G. globosum* are mahogany red in color, globose in shape, rarely exceed ¼ inch in diameter, and have small elevated areas on the surface from which the short tongue-shaped telia appear in the spring.

G. bermudianum Earle: on eastern red and southern red cedars. This is a most unusual species in that there is no alternate host, the aecia preceding the telia on small galls of a reddish-brown luster (Thurston 1923). The mycelium is biennial. The fungus is restricted to the Southeast.

G. corniculans Kern: on leaves of serviceberry; on eastern red cedar and prostrate juniper. The galls are irregularly lobed and small.

Fig. 67. Telia of *Cronartium strobilinum* on the undersurface of a live oak leaf.

G. floriforme Thaxter: on leaves of hawthorn; on eastern red cedar. The galls are small.

G. nelsoni Arth.: on leaves and fruits of hawthorn, mountain ash, Oregon crab apple, pear, quince, serviceberry, and wild crab; on one-seed, prostrate, Rocky Mountain, Utah, and western junipers. The galls, which may reach 2 inches in diameter, are firm, woody, and regularly globose.

CONE RUSTS

The cones of a few conifers are attacked by rusts, with a consequent reduction in the amount of seed produced, which must retard somewhat the establishment of reproduction. However, all the cone crop is never destroyed.

Cronartium strobilinum Hedgc. & Hahn (*Caeoma strobilina*) has its pycnia and aecia on cones of longleaf and slash pines, whereas the uredia and telia are on the leaves of several oaks (Fig. 67). On oaks the fungus cannot be distinguished from *C. cerebrum*. Actually the aecium is not a caeoma, but a peridermium, although the peridium is somewhat hidden. Southern cone rust (Maloy and Mathews 1960) may destroy the cone crop over limited areas in the South where it occurs. First-year cones are infected early, and within 2 to 3 months after emergence many of them are larger than second-year uninfected cones on the same tree. The swollen scales have a reddish color. When the aecial cavities rupture, powdery masses of cadmium-yellow aeciospores are exposed entirely covering the cones, which by this time appear as large swollen masses or galls. By autumn the diseased cones are dead, and most of them have dropped, although an occasional mummy cone remains attached to the

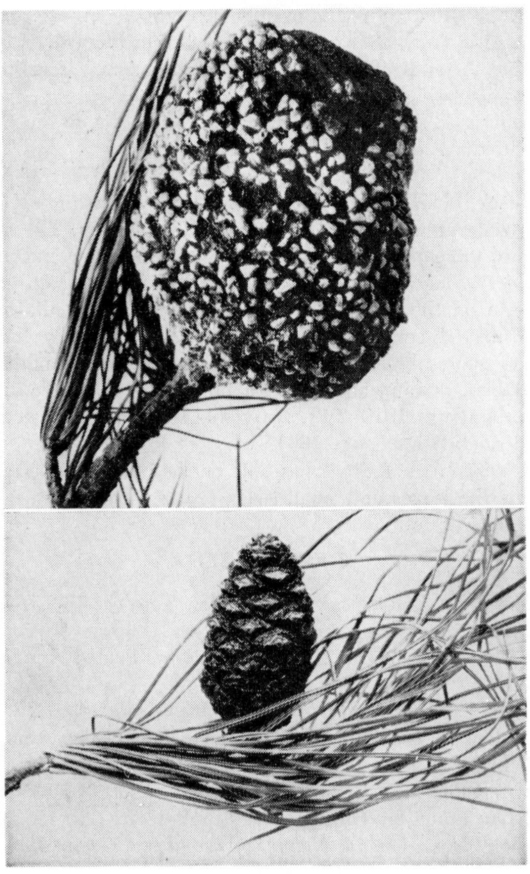

Fig. 68. A normal, immature, 2-year-old cone of Chihuahua pine and a diseased cone bearing mature aecia of *Cronartium conigenum.* (*Reproduced from Phytopathology* 12, *no.* 3.)

tree. In the spring, infected trees can be detected by the presence of nectar-loving insects that feed on the pycnial exudations from the young infected cones. The pycnia are formed the same season in the same tissues as the aecia but mature earlier; in all other Cronartiums the aecia develop a season later than the pycnia.

C. conigenum Hedgc. & Hunt (*Caeoma conigenum*) has its pycnia and aecia on the cones of Chihuahua pine; the uredia and telia are on the leaves of Emory and whiteleaf oaks. On the oaks the fungus cannot be distinguished from *C. cerebrum.* The aecium is a peridermium, with a distinct, erumpent peridium (Fig. 68). According to Hedgcock and Hunt (1922) this rust, in southern Arizona and New Mexico where it occurs, may destroy 50 per cent or more of the cones on groups of trees, and as high as 90 per cent of the cones on an individual tree. In addition to the diminution of seed production, the growth of many limbs is stunted or terminated so that infected trees are likely to present a ragged appearance. Cones are evidently infected during the first year of growth, finally developing into large galls bearing little resemblance to a true cone (Fig. 68). These galls vary greatly in size. The smaller ones develop aecia within 2 years after infection, and the medium-sized ones, at least in part, within 3 years. After fruiting the galls dry up and may remain on the tree as mummies.

Chrysomyxa monesis Ziller (1954) has its pycnia and aecia on the scales of cones of Sitka spruce; the uredia and telia are on *Moneses* spp.

Chrysomyxa pirolata Wint. (*Peridermium conorum-piceae*) has its pycnia and aecia on the cones of black, blue, Engelmann, Norway, red, and white spruces; the uredia and telia are on wintergreen (*Pyrola* spp.) and *Moneses* spp. The aecia are chiefly on the upper side of the cone scales. Infected cones turn yellow and produce no seed.

Melampsora abietis-canadensis Ludw. ex Arth. (*Caeoma abietis-canadensis*) has its pycnia and aecia on the cones of eastern hemlock; and uredia and telia are on various poplars. Infected cones may be recognized in June by the golden-yellow powdery masses of spores breaking out over their surface (Spaulding 1914: 248). They are finally killed, dry up, and blacken, but they still adhere to the twigs as a mummified growth (Adams 1919: 21). No viable seeds are produced. This rust also attacks leaves and small twigs (page 180).

Melampsora farlowii (Arth.) Davis has telia on the cone scales of eastern hemlock. Telia also occur on the leaves and small twigs (page 180).

REFERENCES

Adams, J. F.: *Keithia* on *Chamaecyparis thyoides*. *Torreya*, 18:157–160 (1918).

Adams, J. F.: I. Rusts on Conifers in Pennsylvania. II. Sexual Fusions and Development of the Sexual Organs in the Peridermiums. *Penna. Agr. Expt. Sta. Bull.*, 160:3–77 (1919).

Baxter, D. V.: A Preliminary Study of *Coleosporium solidaginis* (Schw.) Thüm. in Forest Plantations in the Region of the Lake States. *Papers Mich. Acad. Sci.*, 14(1930):245–258 (1931).

Berg, A.: A Rust-resistant Red Cedar. *Phytopathology*, 30:876–878 (1940).

Bonar, L.: Two New Fungi on *Torreya*. *Mycologia*, 43:62–66 (1951).

Boyce, J. S.: Spore Variation in *Neopeckia coulteri*. *Phytopathology*, 6:357–359 (1916).

Boyce, J. S.: Observations on Forest Pathology in Great Britain and Denmark. *Phytopathology*, 17:1–18 (1927).

Boyce, J. S.: A Possible Alternate Stage of *Pucciniastrum myrtilli* (Schum.) Arth. *Phytopathology*, 18:623–625 (1928).

Boyce, J. S.: A Needle Cast of Douglas Fir Associated with *Adelopus gäumanni*. *Phytopathology*, 30:649–659 (1940).

Boyce, J. S.: Host Relationships and Distribution of Conifer Rusts in the United States and Canada. *Trans. Conn. Acad. Arts Sci.*, 35:329–482 (1943).

Boyce, J. S., Jr.: *Lophodermium pinastri* and Needle Browning of Southern Pines. *J. Forestry*, 49:20–24 (1951).

Boyce, J. S., Jr.: Needle Cast of Southern Pines. *U.S. Dept. Agr., Forest Serv., Forest Pest Leaflet,* 28:1–4 (1958).

Brandt, R. W.: The Rhabdocline Needle Cast of Douglas-fir. *State Univ. Coll. Forestry at Syracuse Univ. Tech. Publ.,* 84:1–66 (1960).

Brown, A. B.: Observations on Leaf Fall in the Douglas Fir when Infected with *Rhabdocline Pseudotsugae* Syd. *Ann. Appl. Biol.,* 17:745–754 (1930).

Campana, R. J., and I. R. Schneider: Fungicide Tests for Control of Cedar Apple Rust on Hawthorn. *U.S. Dept. Agr., Plant Disease Reptr.,* 39:985–988 (1955).

Campbell, A. H., and A. E. Vines: The Effect of *Lophodermellina macrospora* (Hartig) Tehon on Leaf-abscission in *Picea excelsa* Link. *New Phytologist,* 37:358–368 (1938).

Cash, Edith K.: An Abnormality of *Abies balsamea. U.S. Dept. Agr., Plant Disease Reptr.,* 25:548 (1941).

Childs, T. W.: Elytroderma Needle Blight of Ponderosa Pine. *U.S. Dept. Agr., Forest Serv., Forest Pest Leaflet,* 42:1–4 (1959).

Church, T. W., Jr.: Effects of Defoliation on Growth of Certain Conifers. A Summary of Research Literature. *Northeast. Forest Expt. Sta., Sta. Paper,* 22:1–12 (1949).

Cohen, L. I.: *Hypodermella laricis,* Its Relation to Spur Shoot Mortality of *Larix occidentalis. Proc. Montana Acad. Sci.,* 17:47–48 (1957).

Crowell, I. H.: The Hosts, Life History and Control of the Cedar-apple Rust Fungus *Gymnosporangium juniperi-virginianae* Schw. *J. Arnold Arboretum (Harvard Univ.),* 15:163–232 (1934).

Cummins, G. B., and J. A. Stevenson: A Check List of North American Rust Fungi (Uredinales). *U.S. Dept. Agr., Plant Disease Reptr. Suppl.* 240:109–193 (1956).

Darker, G. D.: The Hypodermataceae of Conifers. *Arnold Arboretum Contribs.,* 1:1–131 (1932).

Dodge, B. O.: Organization of the Telial Sorus in the Pine Rust, *Gallowaya pinicola* Arth. *J. Agr. Research,* 31:641–651 (1925).

Durand, E. J.: The Genus *Keithia. Mycologia,* 5:6–11 (1913).

Ehrlich, J.: Recently Active Leaf Diseases of Woody Plants in Idaho. *U.S. Dept. Agr., Plant Disease Reptr.,* 26:391–393 (1942).

Ellis, D. E., and L. S. Gill: A New *Rhabdogloeum* Associated with *Rhabdocline pseudotsugae* in the Southwest. *Mycologia,* 37:326–332 (1945).

Faull, J. H.: Taxonomy and Geographical Distribution of the Genus *Milesia. Arnold Arboretum Contribs. (Harvard Univ.),* 2:1–138 (1932).

Faull, J. H.: The Biology of Milesian Rusts. *J. Arnold Arboretum (Harvard Univ.),* 15:50–85 (1934).

Faull, J. H.: *Pucciniastrum* on *Epilobium* and *Abies. J. Arnold Arboretum (Harvard Univ.),* 19:163–173 (1938*a*).

Faull, J. H.: The Biology of the Rusts of the Genus *Uredinopsis. J. Arnold Arboretum (Harvard Univ.),* 19:402–436 (1938*b*).

Faull, J. H.: Taxonomy and Geographical Distribution of the Genus *Uredinopsis. Arnold Arboretum Contribs. (Harvard Univ.),* 11:5–120 (1938*c*).

Faull, J. H.: A Review and Extension of Our Knowledge of *Calyptospora goeppertiana* Kuehn. *J. Arnold Arboretum (Harvard Univ.),* 20:104–113 (1939).

Galloway, B. T.: A Rust and Leaf Casting of Pine Leaves. *Botan. Gaz.,* 22:433–453 (1896).

Garriss, H. R.: Hemlock Twig Rust in North Carolina. *N. Carolina State Coll., Agr. Ext. Serv., Ext. Folder,* 172:[1–7] (1959).

Gäumann, E., C. Roth, and J. Anliker: Ueber die Biologie der *Herpotrichia nigra* Hartig. *Z. Pflanzenkrankh.*, 44:97–116 (1934).

Graves, A. H.: Notes on Tree Diseases in the Southern Appalachians I. *Phytopathology*, 3:129–139 (1913).

Guba, E. F.: Monograph of the Genus *Pestalotia* De Notaris. Phytopathology, 19:191–232 (1929).

Haack, G.: Die Schüttepilz der Kiefer. *Z. Forst u. Jagdw.*, 43:329–357, 402–423, 481–505 (1911).

Haddow, W. R.: Needle Blight and Late Fall Browning of Red Pine. *Trans. Roy. Can. Inst.*, 23:161–189 (1941).

Hagem, O.: Schütteskader paa furuen. *Medd. Vestlandets. Forst. Forsøkssta.*, 3:1–133 (1926).

Hahn, G. G.: New Reports on *Adelopus gäumanni* on Douglas Fir in the United States. *U.S. Dept. Agr., Plant Disease Reptr.*, 25:115–117 (1941).

Hartig, R.: "Wichtige Krankheiten der Waldbäume." Berlin: Springer-Verlag (1874).

Hedgcock, G. G.: Notes on Some Diseases of Trees in Our National Forests. IV. *Phytopathology*, 4:181–188 (1914).

Hedgcock, G. G.: Notes on the Distribution of Some Fungi Associated with Diseases of Conifers. *U.S. Dept. Agr., Plant Disease Reptr.*, 16:28–42 (1932).

Hedgcock, G. G.: Notes on the Distribution of Some Species of *Coleosporium* in the United States and Adjacent Regions. *U.S. Dept. Agr., Plant Disease Reptr.*, 17:20–27 (1933).

Hedgcock, G. G., and G. G. Hahn: Two Important Pine Cone Rusts and Their New Cronartial Stages. Part I. *Cronartium strobilinum* (Arthur) Hedgcock and Hahn, comb. nov. *Phytopathology*, 12:109–116 (1922).

Hedgcock, G. G., and N. R. Hunt: Two Important Pine Cone Rusts and Their New Cronartial Stages. Part II. *Cronartium conigenum* (Pat.) Hedgcock and Hunt, comb. nov. *Phytopathology*, 12:116–122 (1922).

Hepting, G. H., and Davidson, R. W.: A Leaf and Twig Disease of Hemlock Caused by a New Species of *Rosellinia. Phytopathology*, 27:305–310 (1937).

Hesselink, E.: Onder welke omstandigheden doet *Lophodermium pinastri* Chev. te Appelsa, Exloo en Odvorn schade in Dennenbeplantingen? *Tijdschr. Plantenziekten*, 33:105–124 (1927). [*Rev. Appl. Mycol.*, 6:705 (1927).]

Hubert, E. E.: Needle Cast Diseases of Western Larch. *Idaho Agr. Coll. Ext. Bull.*, 215:1–6 (1954).

Hulbary, R. L.: A Needle Blight of Austrian Pine. *Illinois Nat. Hist. Survey, Bull.*, 21:231–236 (1941).

Hunter, Lillian M.: Morphology and Ontogeny of the Spermogonia of the *Melampsoraceae. J. Arnold Arboretum* (*Harvard Univ.*), 17:115–152 (1936).

Hunter, Lillian M.: A Study of the Mycelium and Haustoria of the Rusts of *Abies. Can. J. Research. C. Botan. Sci.*, 26:219–238 (1948).

Jackson, H. S.: A New Pomaceous Rust of Economic Importance, *Gymnosporangium blasdaleanum. Phytopathology*, 4:261–270 (1914).

Jones, S. G.: The Structure of *Lophodermium pinastri* (Schrad.) Chev. *Ann. Botany*, 49:699–728 (1935).

Jump, J. A.: A Study of Forking in Red Pine. *Phytopathology*, 28:798–811 (1938).

Keener, P. D.: Biological Specialization in *Darluca filum. Bull. Torrey Botan. Club*, 61:475–490 (1934).

Korstian, C. F., and W. D. Brush: Southern White Cedar. *U.S. Dept. Agr. Tech. Bull.*, 251:1–75 (1931).

Lagerberg, T.: Om grabarrsjukan hos tallen, dess orsak och verkningar. [German summary.] *Statens Skogsförsöksanst.* (*Sweden*), *Medd.*, 7:127–174 (1910).

Lagerberg, T.: En abnorm barrfällning hos tallen. [German summary.] *Statens Skogsförsöksanst.* (*Sweden*), *Medd.*, 10:139–180 (1913).

Liese, J.: Die Douglasiennadelschütte und die Möglichkeit ihrer Bekämpfung. *Deut. Dendrol. Ges. Mitt.*, 44:294–304 (1932*a*).

Liese, J.: Zur Biologie der Douglasiennadelschütte. *Z. Forst u. Jagdw.* 64:680–693 (1932*b*).

Lightle, P. C.: The Pathology of *Elytroderma deformans* on Ponderosa Pine. *Phytopathology*, 44:557–569 (1954).

Lindgren, R. M.: Field Observations of Needle Rusts of Spruce in Minnesota. *Phytopathology*, 23:613–616 (1933).

MacLachlan, J. D.: The Hosts of *Gymnosporangium globosum* Farl. and Their Relative Susceptibility. *J. Arnold Arboretum* (*Harvard Univ.*), 16:98–142 (1935*a*).

MacLachlan, J. D.: The Dispersal of Viable Basidiospores of the Gymnosporangium Rusts. *J. Arnold Arboretum* (*Harvard Univ.*), 16:411–422 (1935*b*).

MacLachlan, J. D.: Studies on the Biology of *Gymnosporangium globosum* Farl. *J. Arnold Arboretum* (*Harvard Univ.*), 17:1–25 (1936).

Mains, E. B.: Michigan Fungi. I. *Papers Mich. Acad. Sci.*, 20:81–93 (1935).

Maire, K.: Champignons nord-africains nouveaux ou peu connu. *Soc. hist. nat. Afrique du Nord bull.*, 18:117–120 (1927).

Maloy, O. C., and F. R. Mathews: Southern Cone Rust: Distribution and Control. *U.S. Dept. Agr., Plant Disease Reptr.*, 44:36–39 (1960).

Merkle, R.: Über die Douglasien-Vorkommen und die Ausbreitung der Adelopus-Nadelschütte in Württemberg-Hohenzollern. *Allgem. Forst- u. Jagdztg.*, 122:161–192 (1951).

Meyer, H.: Rhabdoclinebefall an Douglasien verschiedener Provenienz. *Forst- u. Holzwirt*, 9A:180–182 (1954).

Miller, J. K.: A New Species of *Keithia* on Red Cedar. *J. Elisha Mitchell Sci. Soc.*, 51:167–171 (1935).

Miller, P. R.: Morphological Aspects of Gymnosporangium Galls. *Phytopathology*, 26:799–801 (1936).

Molnar, A. C.: A Flagging Disease of White Pine (*Pinus monticola*). [Abstract.] *Forestry Chronicle*, 30:231–232 (1954).

Pantidou, Maria, and R. P. Korf: A Revision of the Genus *Keithia*. *Mycologia*, 46:386–388 (1954).

Peace, T. R.: Spraying against *Meria laricis*, the Leaf Cast Disease of Larch. *Forestry*, 10:79–82 (1936).

Peace, T. R.: Observations on *Keithia thujina* and on the Possibility of Avoiding Attack by Growing *Thuja* in Isolated Nurseries. *Gt. Brit. Forestry Comm., Forest Research Rept.*, 1953/54:144–148 (1955).

Peace, T. R., and C. H. Holmes: *Meria laricis*, the Leaf Cast Disease of Larch. *Oxford Forestry Mem.*, 15:1–29 (1933).

Pethybridge, G. H.: A Destructive Disease of Seedling Trees of *Thuja gigantea* Nutt. *Quart. J. Forestry*, 13:93–97 (1919).

Rohde, T.: Die Folgen des Rhabdocline-Befalls in deutschen Douglasienbeständen *Forstarchiv*, 8:317–326 (1932).

Rohde, T.: Ueber die "Schweizer" Douglasienschütte und ihren vermuteten Erreger *Adelopus* spec. *Mitt. Forstw. u. Forstwiss.*, 8:487–514 (1937).

Salisbury, P. J.: *Rosellinia herpotrichioides* Hepting and Davidson on Douglas Fir Seedlings in British Columbia. *Can. Dept. Agr., Sci. Serv., Forest Biol. Div., Forest Biol. Lab., Victoria, B.C., Interim Rept.* (1956). (Mimeographed.)

Savile, D. B. O.: *Chrysomyxa* in North America—Additions and Corrections. *Can. J. Botany,* 33:487–496 (1955).

Savulescu, Tr., and T. Rayss: Un parasite des pins peu connu en Europe *Neopeckia coulteri* (Peck) Sacc. *Ann. épiphyt.,* 14:322–353 (1928).

Schutt, P.: Untersuchungen über Individualunterscheide im Schüttebefall bei *Pinus silvestris. Silvae Genet.,* 6:109–112 (1957).

Shope, P. F.: Some Ascomycetous Foliage Diseases of Colorado Conifers. *Colo. Univ., Studies D,* 2:31–43 (1943).

Spaulding, P.: Notes upon Tree Diseases in the Eastern United States. *Mycologia,* 4:148–151 (1912).

Spaulding, P.: Diseases of the Eastern Hemlock. *Soc. Am. Foresters Proc.,* 9:245–256 (1914).

Sturgis, W. C.: *Herpotrichia* and *Neopeckia* on Conifers. *Phytopathology,* 3:152–158 (1913).

Tehon, L. R.: A Monographic Rearrangement of *Lophodermium. Illinois Univ. Bull.,* 32:1–151 (1935).

Thomas, J. E., and G. D. Lindberg: A Needle Disease of Pines Caused by *Dothistroma pini.* [Abstract.] *Phytopathology,* 44:333 (1954).

Thurston, H. W., Jr.: Intermingling Gametophytic and Sporophytic Mycelium in *Gymnosporangium bermudianum. Botan. Gaz.,* 75:225–248 (1923).

Tubeuf, C. von: Studien über die Schüttekrankheit der Kiefer. *Kaiserl. Gesundh. Biol. Abt. Land- u. Forstw., Arb.,* 2:1–160 (1901).

Tubeuf, C. von: Rhabdocline-Erkrankung an der Douglasie und ihre Bekämpfung. *Z. Pflanzenkrankh.* 42:417–426 (1932).

Van Vloten, H.: "*Rhabdocline pseudotsugae* Sydow, oorzaak eener ziekte van Douglasspar." [German résumé.] Proefschrift, Landbouwhoogeschool te Wageningen (1932).

Vermillion, M. T.: A Needle Blight of Pine. *Lloydia,* 13:196–197 (1950).

Wagener, W. W.: Pestalozzia Blight of Juniper in California. *U.S. Dept. Agr., Plant Disease Reptr.,* 19:97 (1935).

Wagener, W. W.: The Effect of a Western Needle Fungus (*Hypodermella medusae* Dearn.) on Pines and Its Significance in Forest Management. *J. Forestry,* 57:561–564 (1959).

Waterman, Alma M.: Tip Blight of Species of *Abies* Caused by a New Species of *Rehmiellopsis. J. Agr. Research,* 70:315–337 (1945).

Waterman, Alma M.: *Rhizosphaera Kalkhoffi* Associated with a Needle Cast of *Picea pungens. Phytopathology,* 37:507–511 (1947).

Waters, C. W.: Some Studies on Elytroderma-blight of Ponderosa Pine II. *Proc. Montana Acad. Sci.,* 18(1958):7–8 (1959).

Weir, J. R.: An Epidemic of Needle Diseases in Idaho and Western Montana. *Phytopathology,* 3:252–253 (1913).

Weir, J. R.: A Needle Blight of Douglas Fir. *J. Agr. Research,* 10:99–103 (1917).

Weir, J. R., and E. E. Hubert: Observations on Forest Tree Rusts. *Am. J. Botany,* 4:327–335 (1917).

Wilson, M., and Mary J. F.: *Rhabdocline pseudotsugae* Syd.: A New Disease of the Douglas Fir in Scotland. *Trans. Roy. Scot. Arbor. Soc.,* 40:37–40 (1926).

Ziller, W. G.: Studies of Western Tree Rusts I. A New Cone Rust on Sitka Spruce. *Can. J. Botany,* 32:432–439 (1954).

Ziller, W. G.: Studies of Western Tree Rusts II. *Melampsora occidentalis* and *M. albertensis,* Two Needle Rusts of Douglas Fir. *Can. J. Botany,* 33:177–188 (1955).

Ziller, W. G.: Studies of Western Tree Rusts III. *Milesia laeviuscula,* a Needle Rust of Grand Fir. *Can. J. Botany,* 35:885–894 (1957).

Ziller, W. G.: Studies of Western Tree Rusts IV. *Uredinopsis hashiokai* and *U. pteridis* Causing Perennial Needle Rust of Fir. *Can. J. Botany,* 37:93–107 (1959*a*).

Ziller, W. G.: Studies of Western Tree Rusts V. The Rusts of Hemlock and Fir Caused by *Melampsora epitea. Can. J. Botany,* 37:109–119 (1959*b*).

Zobrist, L.: Zehn jahre Versuche zur Bekämpfung des schwarzen Schneeschimmels *Herpotrichia nigra* Hartig. *Schweiz. Z. Forstw.,* 101:632–642 (1950).

CHAPTER 9

Stem Diseases: Rusts of Conifers

Serious stem diseases of conifers are caused by a number of rust fungi (Boyce 1943). The mycelium of these fungi is perennial in the bark or wood of their coniferous hosts, resulting in various malformations such as galls, burls, or swellings, cankers, and witches'-brooms. Trunk infections frequently result in the death of trees, particularly seedlings and saplings. Nearly all these rust fungi are heteroecious with their alternate stages on annual and perennial herbaceous plants, or on broad-leaved trees and shrubs. The most destructive of these diseases are the blister rusts of pines, which in this country were confined to hard pines and piñons, until white pine blister rust was introduced from Europe, threatening the commercial existence of our valuable white (five-needle) pines. Although junipers and cedars are also attacked by a number of caulicolous or stem rusts, these trees do not have the high economic value of pines in our forests.

WITCHES'-BROOMS AND STEM SWELLINGS OF CEDARS AND JUNIPERS

Cedars and junipers are subject to stem diseases caused by *Gymnosporangium* spp. The life cycle of the stem-inhabiting rusts is essentially that of the leaf-inhabiting forms (page 181), except that the mycelium invades the tissues of the stem, where it may be confined to the living bark or it may also occur in the wood (Dodge 1933*a*). Consequently the mycelium is perennial, often attaining a great age (Boyce 1918). The hyphae are always intercellular. The trunk and large limbs are attacked as well as small branches and twigs. The telia in most species push out through the bark as horns, tongues, or flattened pustules and do not usually appear on the leaves, but in *G. libocedri* the telia develop only on the leaves. Two hypertrophies commonly result, witches'-brooms (Fig. 69) and swellings of the stems (Fig. 70). A single species may cause both abnormalities. The resulting damage is sometimes severe. Young trees may be stunted and eventually die; branches on older trees may be killed, and if enough branches are killed the tree will die; or the entire crown may be deformed by witches'-brooms. All of this results in a reduction in growth and in unsightly trees. The only control now known is to eliminate the alternate hosts in the vicinity of susceptible cedars for a distance of at least ½ mile, and under forest conditions this is not yet practicable.

Fig. 69. Witches'-brooms on a branch of southern white cedar caused by *Gymnosporangium ellisii.*

The stem-inhabiting Gymnosporangiums in the United States are grouped here according to their effect on the coniferous hosts.

CAUSING WITCHES'-BROOMS

Gymnosporangium ellisii (Berk.) Ell.: on leaves, fruits, and young stems of *Myrica* spp. and sweetfern; on southern white cedar. The aecia are cluster cups. The cylindrical filiform telia, which are from 3 to 6 mm high, occur only on the stems, except for the first telia developed the spring after a cedar is infected. These appear on the leaf blade or in the axil of the leaf. The mycelium occurs in the inner bark and wood (Dodge 1933*a*:100; Harshberger 1902:487). The hyphae do not invade the wood but are captured by the wood growing around them (Dodge 1934*b*:185).

An infected cedar stem in cross section shows the diseased wood as brown-spotted wedge-shaped areas, which may become more or less confluent involving the entire stem. Witches'-brooms are produced abundantly (Fig. 69). Seedlings 2 to 3 inches high may be infected, their growth retarded, and finally a tree 10 to 15 years old may be only 1 or 2 feet high with the entire crown a witches'-broom (Dodge 1934*a*). Such trees finally die. Larger trees may also be so heavily broomed that death results.

G. kernianum Bethel: on the leaves of serviceberry and pear; on alligator, Utah, and western junipers. The telia arise between the scalelike leaves on the green twigs, but the mycelium is perennial in the stems, causing the formation of dense globose witches'-brooms 6 to 18 inches in diameter, the leaves of which do not take on the juvenile form.

G. vauqueliniae Long & Goodd.: on the leaves of globoid witches'-brooms of *Vauquelinia californica* Sarg.; on one-seed juniper. The telia arise between the scale-like leaves of the green twigs of slight witches'-brooms. This is the only *Gymnosporangium* known to cause witches'-brooms on its aecial host.

CAUSING WITCHES'-BROOMS AND SWELLINGS OF STEMS

G. nidus-avis Thaxter (*G. juvenescens*): on fruits, young stems, and leaves of apple, hawthorn, mountain ash, quince, Japanese quince, and serviceberry; on eastern red and southern red cedars, on prostrate and Rocky Mountain junipers (Prince 1946). The aecia occur on both the upper and undersurfaces of the leaves, instead of being confined to the undersurface. Although telia usually appear on the stems, breaking through the bark, they may also occur in the axils of the leaves (Dodge 1918:294). In addition to witches'-brooms, the leaves of which may take on the awl-shaped juvenile form, this rust also causes long spindle-shaped swellings on the trunks and branches of large trees, which may ultimately result in death of the affected part (Dodge 1931), although these swellings may be caused by *G. effusum* (page 197). Individual witches'-brooms frequently die.

G. libocedri (P. Henn.) Kern (*G. blasdaleanum*): for hosts see page 184. This rust causes the formation of witches'-brooms on incense cedar and slight to pronounced spindle-shaped swellings on the branches or more rarely on the trunks of trees of all sizes. The mycelium invades the wood and may persist for more than two hundred years in a vegetative condition, since telia develop only on the leaves (Boyce 1918). Cross sections through these swellings show the infected wood conspicuously marked with numerous very small dark-brown flecks, often arranged in a fan- or wedge-shaped manner. The same marking is found in the wood of witches'-brooms. Trees with many brooms may finally die. At least their growth is retarded.

CAUSING SWELLINGS OF STEMS

G. biseptatum Ellis: on the leaves of serviceberry; on southern white cedar. The telia appear on the bark or sometimes on the leaves. According to Harshberger (1902:484) the perennial mycelium occurs in the wood as well as the inner bark, but Dodge (1933*a*: 100) states that the mycelium is limited to the inner bark and does not penetrate the cambium. However the cambium is simulated to form an excess of wood, so that a spindle-shaped, but pronounced, burl develops (Fig. 70). If this burl finally encircles the stem, the trunk or branch above the burl dies. A tree may die from the effects of many branch burls, each one the result of a separate infection.

G. hyalinum Kern (Kern and West 1947): on leaves of hawthorn and pear; on southern white cedar. The rust causes slight fusiform swellings on twigs and small branches of the cedar.

G. clavipes Cke. & Pk.: on the fruits and young stems of serviceberry, hawthorn, chokeberry, apple, quince, pear, cotoneaster, and mountain ash; on eastern red cedar and dwarf, mountain, and prostrate junipers. On eastern red cedar this rust causes the formation of slight short fusiform swellings on the twigs and small branches, many of which die from the infection. It is also responsible for less evident swellings on the large branches and main trunk. These swellings are most apparent after a rain as rough blackish rings completely encircling the stem or as blackish patches (Dodge 1933*b*). Although trunk infections rarely kill, trees so affected usually show indications of poor health. The mycelium is confined to the outermost layers of the living inner bark so that an infection can be eliminated by scraping off the bark over and for about 1 inch around the lesion (Crowell 1935:401). *G. clavipes* grows faster across

Fig. 70. Spindle-shaped swelling on a branch of southern white cedar caused by *Gymnosporangium biseptatum*. [*Courtesy of Conn.* (*State*) *Agr. Expt. Sta.*]

the grain than with it, whereas *G. effusum* and *G. nidus-avis* grow much faster with the grain than across it (Hartley 1913).

G. effusum Kern: cultured on chokeberry; on eastern red cedar. On the cedar it causes the formation of long slender fusiform swellings of the smaller branches and sometimes invades the larger trunks. It has been suggested that the fusiform swellings on the branches and trunks of cedar attributed to *G. nidus-avis* are actually caused by *G. effusum* (Long 1945). The rust occurs from southern New York to South Carolina near the coast.

G. exterum Arth. & Kern: on the leaves of *Porteranthus* spp.; on eastern red cedar. The flattened telia usually anastomose over the entire surface of the short fusiform swellings, which are characterized by considerable roughening and exfoliation of the bark. The species is found from Virginia to southeastern Missouri.

G. trachysorum Kern: on leaves of hawthorn; on eastern red cedar. The telia are prominent, varying from 6 to 10 mm high. The swellings, confined to the small branches, are abruptly fusiform or even globoid. The rust is known from Pennsylvania to Louisiana near the coast.

G. clavariiforme (Pers.) DC.: on leaves, fruits, and young stems of serviceberry, chokeberry, pear, and quince; on dwarf and mountain junipers (Fig. 71). Although largely eastern, this species also ranges westward to Saskatchewan and Montana. The slender, cylindrical telia, from 5 to 10 mm high, appear on long fusiform swellings of various sized branches.

G. cornutum Arth.: chiefly on the leaves of mountain ash; on mountain juniper. This species occurs in the East and the West. The telia occasionally appear on the leaves, but usually develop on slight fusiform swellings of the smaller branches.

G. tremelloides Hartig: on the leaves of mountain ash; on mountain juniper. Limited to the West. The telia are flattened, indefinite in outline, and usually of considerable size, often completely covering the swollen portion of the stem. On the larger branches the swellings are hemispherical, breaking forth along the sides, but on the smaller branches they are subglobose galls ranging up to ¾ inch in diameter.

Fig. 71. Telia of *Gymnosporangium clavariaeforme* on the stems of dwarf juniper. [*Courtesy of Conn. (State) Agr. Expt. Sta.*]

G. tubulatum Kern: chiefly on the leaves but also on the fruits of hawthorn; on prostrate and Rocky Mountain junipers. The telia, 3 to 4 mm high, appear on irregular globoid galls, 2 to 10 mm or more in diameter, on the twigs and branches of juniper. Limited to the West. Doubtfully distinct from *G. bethelii.*

G. bethelii Kern: on the leaves and fruits of hawthorn; on Rocky Mountain and western junipers. The roestelia on hawthorn are 3 to 8 mm high. The wedge-shaped telia, 3 to 4 mm high, appear on irregular elongate gall-like knots. Severely infected trees die.

G. speciosum Peck: on leaves of syringa and *Fendlera* spp.; on alligator, one-seed, and Utah junipers. The aecia are cluster cups. The telia appear in longitudinal rows on long fusiform swellings on the juniper branches, which are girdled and soon die (Shope 1940:121). Severely infected trees eventually succumb.

G. inconspicuum Kern: chiefly on the fruits of serviceberry and wild crab; on Utah juniper. The telia are usually on the green twigs (page 184), more rarely on the woody branches.

G. cupressi Long & Goodd.: on the leaves of serviceberry; on Arizona cypress and smooth cypress causing fusiform or globoid swellings on the twigs, branches, and trunks. These swellings range from small up to 3 feet long by 1 foot thick.

YELLOW WITCHES'-BROOM

On balsam firs this disease is caused by the heteroecious rust fungus *Melampsorella caryophyllacearum* Schroet. (*M. cerastii, M. elatina*), which is widely distributed in North America, Europe, and Asia. The telial hosts are chickweed (*Stellaria* spp.) and mouse-ear chickweed (*Cerastium* spp.).

Infected branches develop numerous upright lateral shoots from one point, forming a compact witches'-broom. Infected twigs are dwarfed, the leaves are dwarfed and yellowish, and they shrivel and drop by the following spring leaving the brooms bare until the new shoots develop. The brooms are conspicuous during the growing season since their yellow foliage stands out in striking contrast to the dark-green normal foliage. The fungus is perennial in the stems; consequently in the following season new shoots with the peculiar yellowish leaves develop again. The anatomy of the diseased stems and leaves is abnormal (Anderson 1897). In addition to witches'-brooms, pronounced swellings of the trunk or branches occasionally occur in North America (Fig. 72), but in Europe one to many large open lesions or swellings on the trunks of mature silver firs are common, resulting in considerable cull (Heck 1894). In North America damage to individual trees is generally not severe, because the trunk is not often involved, and usually there are not more than one or two witches'-brooms on the branches, so that infected trees seem to grow as rapidly as sound ones. However, heavily infected trees may be stunted and some may die.

In the spring or early summer the pycnia appear as small raised round orange spots on both surfaces of the dwarfed leaves. By midsummer the aecia develop on the undersurface of the leaves as two rows of orange-yellow blisters. These blisters rupture, discharging the orange-yellow aeciospores, which are wind borne to infect the leaves of chickweed. Small orange-red pustules, the uredia, quickly develop, releasing urediospores to infect other chickweeds. Telia appear later as whitish to pale reddish spots. The teliospores do not germinate and infect firs until the following spring, but the exact method of infection is unknown.

In North America the disease is usually not serious enough to justify special control measures. Based on European experience, removal of trees with main stem infections early in the life of the stand is sufficient for practical purposes (Guinier 1922; Heck 1927), and pruning will reduce the incidence of such infections (Staeger 1928). Witches'-brooms on the branches can be disregarded.

However, in several western forests the fungus is destructive to subalpine fir (Mielke 1957*b*). Trees with 30 to 50 or more witches'-brooms each are common, and in the most severely diseased stands 80 to 90 per cent of the trees are infected. Such stands have no future. Eradication of chickweed from the vicinity would ultimately eliminate the disease, but chickweed is so abundant that this would be costly, although possibly feasible for stands of high value. Chickweed has been killed on cultivated areas with herbicides.

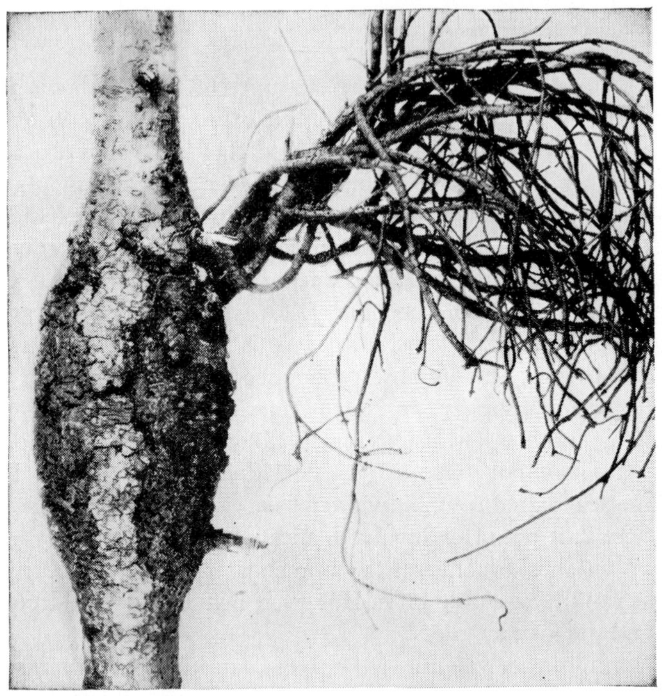

Fig. 72. Swelling of the trunk and a witches'-broom on a balsam fir sapling caused by *Melampsorella cerastii.*

Milesia pycnograndis (page 178) causes loose inconspicuous witches'-brooms on balsam fir. The infected needles are not changed in size or habit, the yellow color is not conspicuous, the pycnia occur only on the undersurface of the leaves, and the aecia are white.

On spruces yellow witches'-broom is caused by *Peridermium coloradense* (Diet.) Arth. & Kern, which is widespread in North America. Although yellow witches'-broom on spruces and on balsam firs has sometimes been considered to be caused by the same fungus, the pycnia, aecia, and distribution of the rust fungi on the two hosts are so different that two distinct species must be involved (Pady 1942). Furthermore, although yellow witches'-broom is abundant on silver fir in Europe, it is unknown there on Norway spruce, but it does occur on Norway spruce in this country. The symptoms are the same as for the disease on balsam firs except that swellings on the trunk and large branches are unknown on spruces. On Engelmann spruce the witches'-brooms usually cause the death of the branches on which they are situated. If brooms are numerous the tree dies.

BLISTER RUSTS OF PINES

Blister rusts of pines are caused by rust fungi belonging to the genus *Cronartium*. Pycnia and aecia are on the stems of pines, whereas uredia and telia occur on a wide variety of annual and perennial plants other than conifers. Invariably a caulicolous or stem rust on pines is a *Cronartium*, and the stage on pines is frequently referred to as a *Peridermium*.

The Cronartiums are widely distributed throughout North America, Europe, and Asia, particularly attacking hard pines. Blister rusts are destructive, especially to seedlings, saplings, and small poles, either killing trees outright, so malforming them that satisfactory timber cannot be produced, or severely reducing their rate of growth. To some extent, blister rusts may function as agents of natural selection in overcrowded stands.

Blister rusts are characterized by the presence of globose, subglobose, or spindle-shaped (fusiform) swellings or galls (Figs. 73, 74, 76), depending on the rust fungus at work. The globose and subglobose swellings increase rapidly in diameter but involve only a small portion of the stem in length, whereas the fusiform swellings grow rapidly in length, spreading for considerable distances up and down the stem before it is girdled and killed. On large stems the spindle-shaped swelling is not evident, for such stems either retain their normal size or become somewhat constricted over the infected area. After aecia are developed, the infected bark cracks into the cambium and the cambium dries up and dies, a larger portion of the stem being involved each year until finally the stem is completely girdled, and the portion from the infection to the tip dies. Small stems are killed rapidly, larger ones more slowly. Infections on the trunk cause the death of the entire tree or the top. Occasionally witches'-brooms may form just above the globose galls. Stems are not readily killed by globose galls and often persist for many years with the galls completely encircling the stem and attaining large size.

On pines the first spores produced are pycniospores, which may appear at any time during the growing season but usually in the summer

(Weir and Hubert 1917*a*). There is no well-defined pycnial structure in blister rusts as there is in needle rusts (Adams 1919). A flat broad layer of interwoven and anastomosed hyphae forms in the bark under the periderm, producing the minute oval or elongated pycniospores and a sweet liquid known as "pycnial nectar" or "exudate." The periderm is forced up in the form of a shallow blister, which finally ruptures, extruding the pycniospores in the sweet liquid, which may collect in orange-yellow to clear drops on the bark (Fig. 78). The pycnial drops quickly disappear, and on stems with rough bark the pycnia are difficult to find. The pycnial nectar is sought by insects, which inadvertently carry the pycniospores to other pycnia. The pycniospores do not cause new infections, but they are concerned with the nuclear development of the fungus (Pierson 1933). Pycnia are developed at least one season before aecia appear (Adams 1921).

In the spring or early summer, aecia are developed over the area occupied by pycnia the previous summer. They push their way through the bark as white or yellowish blisters consisting of a white membrane, the peridium, enclosing an orange-yellow powdery mass of aeciospores (Fig. 73). The peridium soon ruptures irregularly (Fig. 75), releasing the large aeciospores, which have thick, colorless, verrucose walls and orange or red contents (Figs. 77, 82). After the aeciospores are released, the peridium disintegrates, and the bark cracks down to the cambium. The aeciospores are carried by wind, sometimes for great distances, and infect the leaves or tender stems of the alternate host. Aeciospores do not infect other pines except in the species causing western gall rust.

A week or more after infection of the alternate host, the uredia appear on the undersurface of the leaves as tiny blisterlike pustules about the size of a pinhead occupying a yellowish spot on the green leaf. The uredia quickly rupture, disclosing a rounded mass of orange urediospores. The urediospores, about the size of the aeciospores but with spiny walls, are wind borne for short distances to infect the same host, thus spreading and intensifiying the infection during the summer. Urediospores cannot infect pines. In the summer or early autumn the telia are formed both on lesions occupied by uredia and on new lesions. The telia are slender brown bristlelike structures, standing up from the leaf but often slightly curved, and they are composed of vertical rows of elongated angular thin-walled teliospores (Figs. 67, 83, 84). During moist periods the teliospores germinate in place by means of a promycelium bearing four small delicate thin-walled globular sporidia that are carried by wind to infect pines only. The teliospores do not overwinter.

Pine infection in at least one species occurs through the needles (Clinton and McCormick 1919:447) and in certain other species through the tender young shoots and the small side spurs (Meinecke 1929:338; York 1929). However, after infection, it is usually 1 or more years before pycnia develop and the life cycle is completed. The mycelium in pines is largely confined to the inner bark, rarely penetrating the wood. It is sufficiently characteristic to be useful in diagnosis (Jewell 1958).

Most blister rusts can be controlled by eradicating the deciduous[1] hosts from within the infected stand and for some distance around it. Since the width of this safety strip is dependent on the distance to which the delicate sporidia can be blown and still retain sufficient viability to infect pines, which distance is not definitely known for most Cronartiums, the strip should be from 1,000 feet to ½ mile wide. Most of the alternate hosts are abundant and persistent, and some have light seeds that would enable them to quickly reinvade an area from which they have been eradicated. Consequently control by eradicating the alternate host in most cases is not economically feasible, especially since none of the native blister rusts will completely prevent development of a merchantable stand, although the infected stand may be subnormal in rate of growth and understocked. Infected trees should be removed during thinnings or other stand-improvement operations, and this rule should apply particularly to trees with trunk infections.

Where blister rusts occur around nurseries, the alternate hosts should be eradicated for a distance of ½ mile. If this is impossible, the infected pines around the nursery should be cut out for a distance of several hundred yards or more to reduce the amount of infection on the alternate host with the subsequent spread to nursery stock. Spraying susceptible pine nursery stock with a fungicide, such as Bordeaux mixture, at the time sporidia are being released from the alternate host is theoretically promising but has not been demonstrated practically. Infected stock should be destroyed and not shipped out to establish new centers of disease in plantations. Sound practice is not to establish nurseries where blister rusts occur in the immediate vicinity.

Purple mold (*Tuberculina maxima* Rost.) parasitizes the pycnial and aecial stages of blister rusts, inhibiting spore production, but it does not occur in sufficient abundance to be a significant factor in control (Weir and Hubert 1917*a*:139; Hedgcock 1935).

Although the number of blister rusts in North America is small, the life histories of all the fungi causing them are not yet fully understood, so that doubt still exists as to the identity of certain species and their

[1] The term "deciduous," ordinarily applied to woody plants that shed their leaves periodically, is here extended to include herbaceous perennials that die down to the ground and annual plants that disappear completely at the end of the growing season.

host relationships, possibly because there are pathogenic strains within species.

Eastern Gall Rust. This disease is caused by *Cronartium cerebrum* Hedgc. & Long, also called *C. quercuum* (Berk.) Miyabe, a species confined to Asia (Hedgcock and Siggers 1949). The fungus ranges eastward from the Great Plains Region in the United States and Canada, attacking many species of hard pines, particularly Virginia and shortleaf pines. It has been recorded once on red pine, a tree singularly free from blister rusts (Weir and Hubert 1917*b*). It is most abundant in the Southern states. On pines the fungus causes the development of globose or sub-

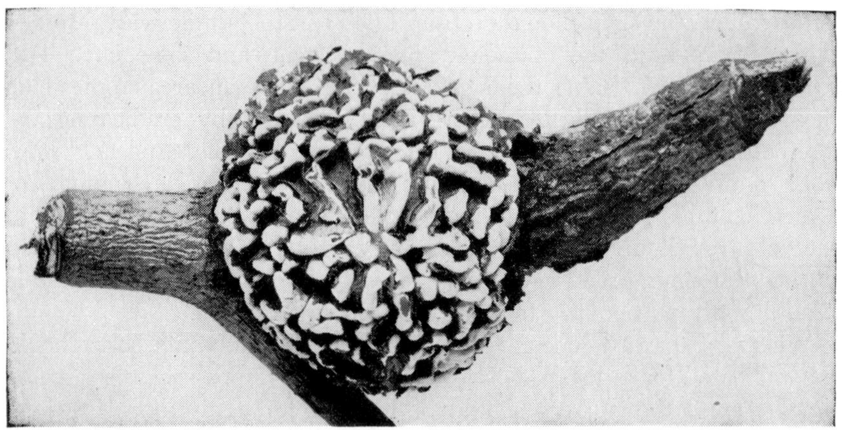

Fig. 73. Eastern gall rust on sand pine caused by *Cronartium cerebrum.* The aecia are mature, but most of them have not yet ruptured.

globose galls on the stems (Fig. 73). Galls on the trunk of Virginia pine, partially grown over, result in lesions that resemble "hip canker" caused by western gall rust. The anatomy of the galls has been studied by Stewart (1916) and their development by Dodge and Adams (1918). Hyphae occur mostly in the rays; abundantly in the bark and sparingly in the wood. In the spring, aecia break through the bark of the galls in a more or less cerebroid arrangement, the season after the development of pycnia. Uredia and telia occur on the underside of the leaves of many oaks, particularly black oaks. From central into southern California similar uredia and telia of an undetermined species are found on evergreen oaks, the mycelium overwintering in the leaves.

Southern Fusiform Rust. This disease is caused by *Cronartium fusiforme* Hedgc. & Hunt (Goggans 1957; Verrall 1958). The fungus is limited to the Southern states, where it attacks a number of hard pines, causing pronounced spindle-shaped (fusiform) swellings of the stems

(Fig. 74), sometimes accompanied by witches'-brooms. The anatomy of the swellings has been investigated by Jackson and Parker (1958). The uredia and telia are found on the underside of leaves of various oaks, especially black oaks, and cannot be distinguished from those of *C.*

Fig. 74. Southern fusiform rust on loblolly pine caused by *Cronartium fusiforme.* Only a few of the mature aecia have ruptured.

cerebrum. Branch infections on pines that never reach the main stem usually have little bad effect, but those that do may kill the tree in time. The smaller the stem at the time of infection the more rapid and severe the damage. In larger trees the stem often is not girdled, but a lesion resembling a canker is formed with a sunken area of rotten wood surrounded by a callus swelling. Such trees are easily broken by wind and ice. This disease has become more prevalent in recent years because slash and loblolly pines, both highly susceptible, have been naturally invading extensive areas formerly occupied by longleaf pine, which is relatively resistant. Moreover, the range of slash pine has been widely extended by planting, frequently on longleaf pine sites where oaks abound. Clear cutting and protection from fire in the longleaf pine region have favored the development of oaks, as well as slash and loblolly pines, and have discouraged longleaf pine. Furthermore, infected seedlings, unrecognized in nurseries, have been planted out, causing serious losses. The mortality among nursery-infected seedlings when planted out is immediately high, and within a few years practically all of them die. Up to 35 per cent of the slash pines in nurseries have been infected and as high as 71 per cent of the trees in plantations have had trunk lesions.

Control measures are necessarily largely preventive (Verrall 1958). When planted where the risk of infection is high, not much can be done to protect a susceptible species from damage. Slash pine should be planted on sites where the chance of infection is low. This can be ascertained by observing the degree of infection on young native slash, loblolly, and longleaf pines within 1 mile of the planting site. Infections on 10 per cent of the longleaf pines would indicate a site of high hazard. Increasing the density of stocking by planting at 4 by 4 feet instead of 6 by 6 feet reduces the amount of trees with stem lesions. Practices such as fertilizing, cultivating, planting on cultivable land before serious soil depletion, and burning that kills pine foliage all increase the amount of fusiform rust on slash pine because the trees break winter dormancy earlier so that susceptibility to the rust is increased. Longleaf and shortleaf pines should be planted on their own sites, rather than be replaced by slash or loblolly pines. Longleaf pine is relatively resistant to the rust, and shortleaf pine is practically immune.

After susceptible species are infected, if intensive control is possible, pruning branch lesions before the swelling reaches the main stem will considerably reduce the damage. This should be done when the stands are 2 to 5 years old and when the trees are less than 10 feet high. It is not necessary to cut off branches with lesions centered more than 15 inches from the main stem. One pruning will usually not be enough. Trees with stem infections should not be pruned. Trees that are stunted or bent because of stem infections should be cut down. However, straight vigorous trees of value in the stand may have stem lesions. If less than 50 per cent of the trunk circumference is killed the tree has an even chance of salvage for 8 years, but if 50 per cent of the circumference is killed the even chance of survival is reduced to 5 years.

Around nurseries, an oak-free zone must be too wide to be practicable.

Spraying twice weekly with ferbam, ziram, or zineb at the rate of 2 pounds in 75 gallons of spray per acre plus a sticker is effective (Foster 1959). Spraying must begin as soon as the seeds germinate and continue to mid-June.

Western Gall Rust. At least two species, possibly containing more than one strain, cause this disease, which is widespread throughout the West. Galls on pines of the mountain region, such as ponderosa, Jeffrey, and lodgepole pines, are caused by *Peridermium harknessii* Moore; those on pines of the coastal region in California, such as Monterey and knobcone pines, are caused by an unnamed fungus tentatively called *P. cerebroides* (?). Digger pine is host to both. This concept, proposed years ago (Meinecke 1920, 1929), has been confirmed by inoculations (Boyce 1957). The globose galls caused by *P. harknessii* have large confluent aecia. The bark sloughs off in large scales, finally exposing the smooth wood with a collar of dead bark standing out around the upper and lower end of the gall. The aecia on the globose to pear-shaped galls caused by *P. cerebroides* (?) are small and irregularly scattered, somewhat suggesting those of the eastern *C. cerebrum*. The bark sloughs away in small scales, not exposing the wood beneath, and the bark collar is absent. Both forms have been doubtfully grouped under *Cronartium coleosporioides* Arth. with the alternate stage on paintbrush and other Scrophulariaceae. However, both fungi can infect directly from pine to pine. *P. harknessii* has also been found on jack pine in Quebec (Pomerleau 1942).

Western gall rust is quite destructive, particularly to seedlings and saplings, killing some trees but stunting and malforming many more. Monterey, lodgepole, and Scotch pines are highly susceptible, whereas ponderosa pine is somewhat less so, although in the Black Hills large ponderosa pines are severely damaged (Peterson 1959). Individuals within a species vary from highly susceptible to almost immune. A single tree may be literally covered with galls. Witches'-brooms often form just above the galls, and infected trees show a reduction in height growth, especially those that have galls on the main stem. In lodgepole pine, an old gall on one side of the trunk that is partially or almost completely grown over forms a cankerlike lesion with a flattened face and bulging sides and rear termed a "hip canker." Galls sometimes occur on branches high up in the crowns of overmature ponderosa pines without noticeable ill effects on the trees. Nursery stock is occasionally infected from diseased trees in the surrounding forest. This disease has invaded the ponderosa pine plantations in the sand hills of Nebraska, with considerable damage resulting. The gall rust prevalent on Scotch pine in the East, and called Woodgate rust (York

1929), is caused by *P. harknessii,* which somehow reached there years ago (Boyce 1957).

Pycnia are rarely produced by either fungus (Gill 1932). Aecia develop abundantly on the galls in spring or early summer. The aeciospores of *P. harknessii* can infect either pines (Meinecke 1920, 1929) or the alternate hosts. That direct infection from pine to pine occurs naturally is shown by the presence of *P. harknessii* on ponderosa pine in localities where the alternate hosts do not occur (Gier 1934).

Because of direct infection from pine to pine, control cannot be accomplished by eliminating the alternate hosts. Infected trees, especially those with galls on the main stem, should be removed from a stand as soon as possible. Around nurseries, trees with galls should be removed for a distance of 300 or more yards to reduce infection in the nursery stock. No infected trees should be shipped from a nursery for planting, because new infection centers can be immediately established without alternate hosts being present.

Other Stem Rusts. There are several stem rusts of pines that, depending on the size of the infected stem, are marked by a slight spindle-shaped swelling or by no enlargement at all. Pronounced hypertrophies, such as characterize southern fusiform rust and the gall rusts, do not develop.

Two or more forms of these rusts, which by present botanical rules must be grouped under *Cronartium coleosporioides* Arth. (Cummins and Stevenson 1956:121), probably represent separate species with differing disease symptoms on the pine hosts. Limb rust, caused by *Peridermium filamentosum* Pk. occurs on the Rocky Mountain variety of ponderosa pine, probably throughout most of the range of the tree (Mielke 1952). The aecia are nonconfluent, long, cylindrical or somewhat flattened, and persistent, with internal filaments extending from top to bottom through the spore mass. Spore production begins in late May or June and continues through the summer, even extending into October. The aecia are mostly confined to twigs and branches not over ¾ inch in diameter and do not develop on the trunk or large branches. Paintbrush is host for the uredia and telia (Hawksworth 1956) in some parts of the range. However, direct infection from pine to pine possibly happens, because the aecial stage is often abundant on pines with no infection on the alternate host. A similar rust occurs on Jeffrey pine (Wagener 1958).

On Rocky Mountain ponderosa pine, *P. filamentosum* mainly attacks trees over 30 years old. Infection occurs through needle-bearing twigs, the fungus grows down the twig to the branch and then to the trunk, entering and killing twigs and side branches along the way. It then grows up and down the trunk, entering the branches and killing twigs and small side branches as it progresses outward. When the end of the main branch is reached, the entire branch dies because of the loss of twigs and smaller branches. In time, the entire tree is killed. Aecia develop only on the twigs and smaller branches, and there is no external sign of the fungus on the larger branches and trunk. Depending upon where infection occurs, trees die slowly from the bottom of the crown upward, from the top down, or both up and down from the middle. Control depends on the removal of infected trees. Trees that have lost 50 per cent or more of their crowns should be cut immediately, whereas

Fig. 75. Blister rust on lodgepole pine caused by *Peridermium stalactiforme*. The aecia have ruptured, discharging the aeciospores.

those that have lost but a small portion of crown can be left for a future cut.

Peridermium stalactiforme Arth. & Kern causes a common filamentous rust of lodgepole pine (Mielke 1956), Jeffrey pine, and probably occasionally of ponderosa pine in the West. It occurs on jack pine in Minnesota (Anderson 1960). The aecia (Fig. 75) are partly confluent or cerebroid, short, and not persistent, with internal filaments projecting part way into the spore mass from the floor and dome of the aecium. On the trunk of lodgepole pine the aecia often form an elongated diamond pattern (Meinecke 1920:282). Aeciospore production begins in May and extends into August. Pycnia develop during April and May of the same season. Uredia and telia are mostly on paintbrush but are also known on other Scrophulariaceae.

The disease is most abundant on seedlings and saplings of lodgepole pine, so that over limited areas up to 80 per cent of the reproduction may be killed, whereas in older stands, although the number of infected trees may be high, fewer succumb because trunk infections girdle the stem less frequently. A slight spindle-shaped swelling characterizes the infection on small branches and stems, which are soon girdled and killed. On larger trees trunk infections result in long and somewhat flattened or sunken areas with resin impregnation of the wood (Mielke 1956). These lesions may be 25 feet or more long. Girdling of small stems and formation of flattened lesions is commonly hastened by the gnawing of rodents that are attracted by the sweet pycnial liquid and the thickened succulent bark of the infected areas. The pycnia appear as small blisterlike swellings on the bark from which exude clear sticky pycnial drops. On drying these drops appear as an orange or brick-red pustular mass. This rust is quite damaging to lodgepole pine. In some stands 30 per cent of the trees may have lesions on the trunk. Losses

Fig. 76. Slight spindle-shaped swelling of a ponderosa pine stem caused by *Cronartium comandrae* showing mature aecia not yet ruptured.

can ultimately be reduced by removing infected trees during stand improvement and cutting operations.

Comandra blister rust is caused by *Cronartium comandrae* Pk. (*C. pyriforme*), which is widely distributed (Hedgcock and Long 1915; Mielke 1957*a*). Ponderosa, Arizona, and lodgepole pines are attacked in the West; and pitch, mountain, jack, loblolly, Austrian, Scotch, and maritime pines in the East. Ponderosa pine suffers most severely in the West, where over limited areas so many seedlings and saplings are killed that understocking results (Meinecke 1928). Large ponderosa pines are also attacked (Boyce 1916*a*). Mature stands of lodgepole pine in the Intermountain Region are seriously damaged. In the East, mountain pine is most commonly attacked. The disease is rare in the southern pine region. Fortunately the destructive effects of the fungus are limited by the distribution of its herbaceous hosts, which are usually restricted to small areas.

The disease is manifest by a spindle-shaped swelling of the infected stem, although the swellings may be so slight as to be obscure (Fig. 76). Occasionally globose swellings occur, suggestive of those caused by *C. cerebrum*. Aecia break through the bark of the swellings, and, after the aecia have disintegrated, numerous openings are left reaching to the cambium, which then dies. In this way the stem is finally killed. Girdling is often hastened by rodents gnawing away the infected bark. On the

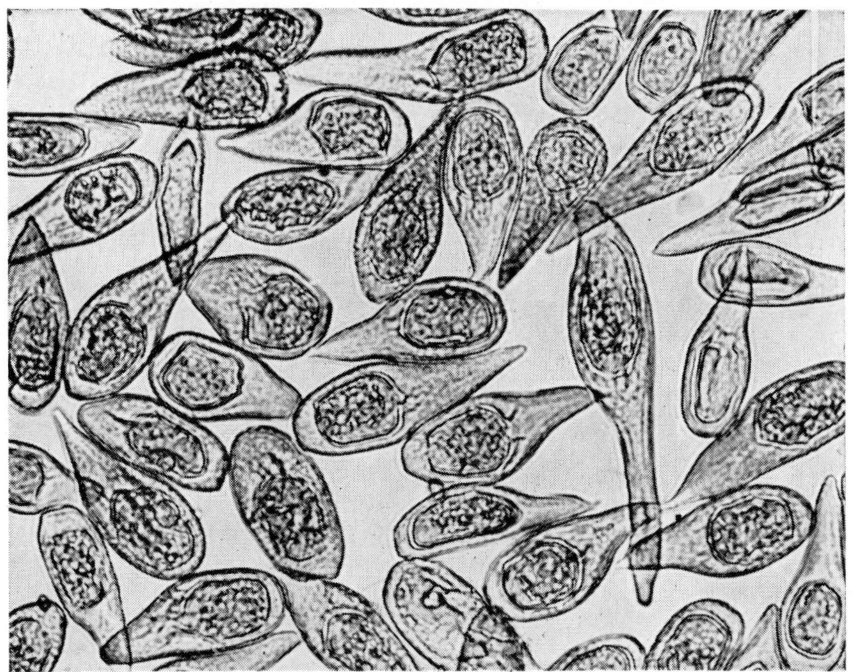

Fig. 77. Photomicrograph of the pyriform or pear-shaped aeciospores of *Cronartium comandrae*. The aeciospores of other Cronartiums are ellipsoid or egg-shaped.

trunks of older lodgepole pines infections are characterized by lesions up to 3½ feet long with roughened bark and abundant resin flow.

This blister rust fungus can be readily separated from all others on pines by its pyriform aeciospores (Fig. 77). The tiny pycniospores have the same general shape (Boyce 1916*b*). Aecia appear the following season over the area occupied by pycnia (Fig. 78). Aeciospores infect the leaves of bastard toad flax (*Comandra* spp.) in late spring or early summer; uredia are produced and followed by telia. Pines are reinfected the same season.

Sweetfern blister rust is caused by *Cronartium comptoniae* Arth. (Spaulding and Hansbrough 1932). The fungus ranges from Nova Scotia west to Saskatchewan and south to North Carolina and Missouri on hard pines. The alternate hosts are sweetfern and sweet gale. In this region the disease is most prevalent on jack pine. The fungus also occurs on the Pacific Coast on lodgepole and ponderosa pines, and on sweet gale. It is destructive to young lodgepole pines (Lachmund 1929). It is not found in the southern pine region.

Fig. 78. The pycnial drops of *Cronartium comandrae* on the stem of a young ponderosa pine. Notice the slight swelling of the bark. Aecia appear the next season over the area occupied by the pycnia. (*Reprinted by permission from "An Outline of Forest Pathology" by Hubert, published by John Wiley & Sons, Inc.*)

The disease damages by its girdling action, and the smaller the tree the more likely it is to be girdled and killed. After a tree has reached a basal diameter of about 3 inches, it is relatively safe. On the whole the disease is not serious in natural stands, since a relatively severe infection may not even result in final understocking because a number of trees always survive infection. However, young seedlings are killed, so that restocking is delayed. Losses in nurseries have ranged from 5 to 20 per cent, but it is in plantations that the worst damage has occurred. Plantations of lodgepole and ponderosa pines in Michigan have been completely ruined, and a loss of 45 per cent in planted ponderosa pine has been recorded in Connecticut. Fortunately sweet gale, the only alternate host in the West, is not associated with ponderosa pine in its natural range there. Plantations of loblolly pine have been threatened in New Jersey.

The rough bark of the hard pines, even at a young stage, makes this disease difficult to detect unless infection is several years old or the blisterlike aecia are present. Affected stems swell slightly. Small trees 2 to 7 years old are usually attacked near the base, and on lodgepole pine only the growth of the first 4 years in the life of the tree is susceptible to infection. On this host long fusiform swellings of infected stems are common, spreading at the rate of 2 to 6 inches a year. On eastern hard pines several inches in diameter, the disease is likely to cause a depressed streak on one side of the stem. Insects attack the affected bark of these cankerous areas, and pitch oozes out. On small trees, killing of the main stem frequently results in multiple-stemmed, shrublike individuals. The mycelium of the fungus in the bark is not sufficiently characteristic to distinguish this from other blister rusts (Hutchinson 1929).

Aecia appear on young seedlings in the third year of the life of the tree, evidently preceded by pycnia in the second year. In British Columbia most aecia are produced in April and May; in the East they are present for 2 or 3 weeks in the latter part of May and early June. Apparently the aeciospores can be wind borne for many miles to infect sweetfern and sweet gale. Bayberry (*M. carolinensis Mill.*) has been artificially inoculated, but California wax-myrtle (*M. californica* Cham. & Sch.) is immune. Uredia are produced on the alternate hosts in the summer, followed by telia in late summer and autumn. Around nurseries no sweetfern or sweet gale should be allowed for several hundred yards, and no large areas of either within 1 mile. These plants cannot be tolerated in or around plantations.

Virginia pine stem rust, confined to branches and main stems of seedlings to small poles in North Carolina, Tennessee, Virginia, and West Virginia, is caused by *Cronartium appalachianum* Hepting (1957). The alternate host is *Buckleya distichophylla* (Nutt.) Torr. Infected areas on pines are not deformed and are only slightly, if at all, enlarged. The aecia are columnar, mostly nonconfluent, and occur singly over the sporulating area. After aecia have erupted over the entire circumference it may take years for a stem to die. Damage to stands is negligible.

Piñon blister rust, a disease of piñon and singleleaf piñon, is caused by *Cronartium occidentale* Hedgcock, Bethel & Hunt (Stillinger 1944). On piñons the disease has been found in Arizona, California, Colorado,

Utah, and Nevada, but on the alternate hosts it is much more widespread in the West. Young trees are attacked on the trunk and branches near the ground. Although numbers are killed, piñons are of little commercial value. The importance of this disease is that the alternate hosts, currants and gooseberries (*Ribes* spp.), for *C. occidentale* are the same as for *C. ribicola* causing white pine blister rust. On ribes the two cannot be separated macroscopically, so that in the West it has been difficult to determine which fungus was present in a new locality on ribes before pines became infected. Both can spread long distances from pines to ribes, and *C. occidentale* may overwinter in the uredial stage on ribes.

In the pycnial and aecial stages the distinctive pine hosts separate the two fungi: piñon pines for *C. occidentale* and white (five-needle) pines for *C. ribicola*. In addition the pycnial exudate of *C. occidentale* is an orange-chrome color, and the inconspicuous pycnial scars are not common, but the pycnial exudate of *C. ribicola* is honey yellow, and the conspicuous pycnial scars are numerous. The aecia of *C. occidentale* vary from distinct sori on singleleaf piñon to broad spore-bearing layers under the bark with thin inconspicuous peridia on piñon, whereas the aecia of *C. ribicola* are usually distinct sori with persistent peridia (Colley, Hartley, and Taylor 1927). There are also differences in the peridia, peridial cells, and aeciospores of the two species.

In the uredial stage the two fungi can be separated by extensive measurements of urediospores (Colley 1925) or by inoculation of differential ribes hosts (Hahn 1930). In the telial stage they can be separated exactly by a microchemical test (Ford and Rawlins 1956) and can be distinguished tentatively by visual examination (Kimmey 1946). Telia of *C. occidentale* may occur so densely as to form furlike mats on the leaf; telia of *C. ribicola* are never so dense. Ungerminated telia of *C. occidentale* are usually a darker shade of brown than those of *C. ribicola*. Telia with germinated teliospores appear faded, but those of *C. occidentale* have a lavender tinge, whereas those of *C. ribicola* do not. *C. occidentale* rarely fails to produce telia on infected leaf areas after producing uredia, whereas such failure is common in *C. ribicola*. Under light infection conditions, i.e., when the initial infection from aeciospores consists of only one or two rust spots per leaf, the pattern of subsequent intensification for *C. occidentale* is larger patches of continuous areas, whereas for *C. ribicola* it is small spots scattered over the leaf surface.

REFERENCES

Adams, J. F.: I. Rusts on Conifers in Pennsylvania. III. Sexual Fusions and Development of the Sexual Organs in the Peridermiums. *Penna. Agr. Expt. Sta. Bull.*, 160:3–77 (1919).

Adams, J. F.: Gametophytic Development of Blister Rusts. *Botan. Gaz.*, 71:131–137 (1921).

Anderson, A. P.: Comparative Anatomy of the Normal and Diseased Organs of *Abies balsamea* Affected with *Aecidium elatinum. Botan. Gaz.*, 24:309–344 (1897).

Anderson, N. A.: Stalactiform Rust on Jack Pine in Minnesota. *Forest Sci.*, 6:40–41 (1960).

Boyce, J. S.: A Note on *Cronartium pyriforme. Phytopathology*, 6:202–203 (1916*a*).

Boyce, J. S.: Pycnia of *Cronartium pyriforme. Phytopathology*, 6:446–447 (1916*b*).

Boyce, J. S.: Perennial Mycelium of *Gymnosporangium blasdaleanum. Phytopathology*, 8:161–162 (1918).

Boyce, J. S.: Host Relationships and Distribution of Conifer Rusts in the United States and Canada. *Trans. Conn. Acad. Arts Sci.*, 35:329–482 (1943).

Boyce, J. S.: The Fungus Causing Western Gall Rust and Woodgate Rust of Pines. *Forest Sci.*, 3:225–234 (1957).

Clinton, G. P., and Florence A. McCormick: Infection Experiments of *Pinus strobus* with *Cronartium ribicola. Conn. Agr. Expt. Sta. Bull.*, 214:428–459 (1919).

Colley, R. H.: A Biometric Comparison of the Urediniospores of *Cronartium ribicola* and *Cronartium occidentale. J. Agr. Research*, 30:283–291 (1925).

Colley, R. H., C. Hartley, and Minnie W. Taylor: A Morphologic and Biometric Comparison of *Cronartium ribicola* and *C. occidentale* in the Aecial Stage. *J. Agr. Research*, 34:511–531 (1927).

Crowell, I. H.: The Hosts, Life History and Control of *Gymnosporangium clavipes* C. and P. *J. Arnold Arboretum (Harvard Univ.)*, 16:367–410 (1935).

Cummins, G. B., and J. A. Stevenson: A Check List of North American Rust Fungi (Uredinales). *U.S. Dept. Agr., Plant Disease Reptr. Suppl.*, 240:109–193 (1956).

Dodge, B. O.: Studies in the Genus *Gymnosporangium*—II. Report on Cultures Made in 1915 and 1916. *Bull. Torrey Botan. Club.*, 45:287–300 (1918).

Dodge, B. O.: A Destructive Red-cedar Disease. *J. N.Y. Botan. Garden*, 32:101–108 (1931).

Dodge, B. O.: The Course of Mycelia of Gymnosporangia in the Trunks of Cedars. *Natl. Shade Tree Conf. Proc.*, 9:94–101 (1933*a*).

Dodge, B. O.: The Orange-rust of Hawthorn and Quince Invades the Trunk of Red Cedar. *J. N.Y. Botan. Garden*, 34:233–237 (1933*b*).

Dodge, B. O.: Witches' Brooms on Southern White Cedars. *J. N.Y. Botan. Garden*, 35:41–45 (1934*a*).

Dodge, B. O.: *Gymnosporangium myricatum* in Relation to Host Parenchyma Strands. *Mycologia*, 26:181–190 (1934*b*).

Dodge, B. O., and J. F. Adams: Some Observations on the Development of *Peridermium cerebrum. Torrey Botan. Club Mem.*, 17:253–261 (1918).

Ford, D. H., and T. E. Rawlins: Improved Cytochemical Methods for Differentiating *Cronartium ribicola* from *Cronartium occidentale* on *Ribes. Phytopathology*, 46:667–668 (1956).

Foster, A. A.: Nursery Diseases of Southern Pines. *U.S. Dept. Agr., Forest Serv., Forest Pest Leaflet*, 32:1–7 (1959).

Gier, L. J.: A Study of Peridermium Gall on Western Yellow Pine. *Trans. Kansas Acad. Sci.*, 37:75–76 (1934).

Gill, L. S.: Notes on the Pycnial Stage of *Peridermium cerebroides. Mycologia*, 24:403–409 (1932).

Goggans, J. F.: Southern Fusiform Rust. *Ala. Agr. Expt. Sta. Bull.*, 304:1–19 (1957).

Guinier, P.: La "dorge" ou "chaudron" du sapin. *Soc. forest. de Franche Comté et Belfort, bull.*, 14:33–347 (1922).

Hahn, G. G.: A Physiological Method of Distinguishing *Cronartium ribicola* and *C. occidentale* in the Uredinial Stages. *J. Agr. Research,* 40:105–120 (1930).

Harshberger, J. W.: Two Fungous Diseases of the White Cedar. *Proc. Acad. Nat. Sci. Phila.,* 54:461–504 (1902).

Hartley, C.: Bark Rusts of *Juniperus virginiana. Phytopathology,* 3:249 (1913).

Hawksworth, F. G.: Note on the Susceptibility of Indian Paintbrush to *Cronartium filamentosum. U.S. Dept. Agr., Plant Disease Reptr.,* 40:581–582 (1956).

Heck, C. R.: "Der Weisstannenkrebs." Berlin: Springer-Verlag (1894).

Heck, R.: Muss man die Hexenbesen der Weisztanne verfolgen? *Forstwiss. Centr.,* 49:132–140 (1927).

Hedgcock, G. G.: Notes on the Occurrence of *Tuberculina maxima* on the Aecia of *Cronartium cerebrum. Phytopathology,* 25:1117–1118 (1935).

Hedgcock, G. G., and W. H. Long: A Disease of Pines Caused by *Cronartium pyriforme. U.S. Dept. Agr. Bull.,* 247:1–20 (1915).

Hedgcock, G. G., and P. V. Siggers: A Comparison of the Pine-Oak Rusts. *U.S. Dept. Agr. Tech. Bull.,* 978:1–30 (1949).

Hepting, G. H.: A Rust on Virginia Pine and *Buckleya. Mycologia,* 49:896–899 (1957).

Hutchinson, W. G.: Studies on the Mycelium of *Cronartium comptoniae* Arthur on *Pinus sylvestris* L. *Phytopathology,* 19:741–744 (1929).

Jackson, L. W. R., and J. R. Parker: Anatomy of Fusiform Rust Galls on Loblolly Pine. *Phytopathology,* 48:637–640 (1958).

Jewell, F. F.: Stain Technique for Diagnosis of Rust in Southern Pines. *Forest Sci.,* 4:42–44 (1958).

Kern, F. D., and E. West: Another Gymnosporangial Connection. *Mycologia,* 39:120–125 (1947).

Kimmey, J. W.: Notes on Visual Differentiation of White Pine Blister Rust from Pinyon Rust in the Telial Stage. *U.S. Dept. Agr., Plant Disease Reptr.,* 30:59–61 (1946).

Lachmund, H. G.: *Cronartium comptoniae* in Western North America. *Phytopathology,* 19:453–466 (1929).

Long, W. H.: Notes on Four Eastern Species of *Gymnosporangium. J. Wash. Acad. Sci.,* 35:182–188 (1945).

Meinecke, E. P.: Facultative Heteroecism in *Peridermium cerebrum* and *Peridermium harknessii. Phytopathology,* 10:279–297 (1920).

Meinecke, E. P.: The Evaluation of Loss from Killing Diseases in the Young Forest. *J. Forestry,* 26:283–298 (1928).

Meinecke, E. P.: Experiments with Repeating Pine Rusts. *Phytopathology,* 19:327–342 (1929).

Mielke, J. L.: The Rust Fungus *Cronartium filamentosum* in Rocky Mountain Ponderosa Pine. *J. Forestry,* 50:365–373 (1952).

Mielke, J. L.: The Rust Fungus (*Cronartium stalactiforme*) in Lodgepole Pine. *J. Forestry,* 54:518–521 (1956).

Mielke, J. L.: The Comandra Blister Rust in Lodgepole Pine. *Intermountain Forest and Range Expt. Sta. Research Note,* 46:1–8 (1957*a*).

Mielke, J. L.: The Yellow Witches' Broom of Subalpine Fir in the Intermountain Region. *Intermountain Forest and Range Expt. Sta. Research Note,* 47:1–5 (1957*b*).

Pady, S. M.: Distribution Patterns in *Melampsorella* in the National Forests and Parks of the Western States. *Mycologia,* 34:606–627 (1942).

Peterson, R. S.: The *Cronartium coleosporioides* Complex in the Black Hills. *U.S. Dept. Agr., Plant Disease Reptr.,* 43:1227–1228 (1959).

Pierson, R. K.: Fusion of Pycniospores with Filamentous Hyphae in the Pycnium of the White Pine Blister Rust. *Nature,* 131:728–729 (1933).

Pomerleau, R.: The Spherical Gall Rust of Jack Pine. *Mycologia,* 34:120–122 (1942).

Prince, A. E.: The Biology of *Gymnosporangium nidus-avis* Thaxter. *Farlowia,* 2:475–525 (1946).

Shope, P. F.: Colorado Rusts of Woody Plants. *Colo. Univ., Studies,* 1:105–127 (1940).

Spaulding, P., and J. R. Hansbrough: *Cronartium comptoniae,* the Sweetfern Blister Rust of Pitch Pines. *U.S. Dept. Agr. Circ.,* 217:1–21 (1932).

Staeger, H.: L'élagage des résineux, traitement complémentaire. *J. forestier suisse,* 79:185–192 (1928).

Stewart, A.: Notes on the Anatomy of Peridermium Galls. I. *Am. J. Botany,* 3:12–22 (1916).

Stillinger, C. R.: Notes on *Cronartium occidentale. Northwest Sci.,* 18:11–16 (1944).

Verrall, A. F.: Fusiform Rust of Southern Pines. *U.S. Dept. Agr., Forest Serv., Forest Pest Leaflet,* 26:1–4 (1958).

Wagener, W. W.: Infection Tests with Two Rusts of Jeffrey Pine. *U.S. Dept. Agr., Plant Disease Reptr.,* 42:888–892 (1958).

Weir, J. R., and E. E. Hubert: Pycnial Stages of Important Forest Tree Rusts. *Phytopathology,* 7:135–139 (1917*a*).

Weir, J. R., and E. E. Hubert: *Cronartium cerebrum* on *Pinus resinosa. Phytopathology,* 7:450–451 (1917*b*).

York, H. H.: The Woodgate Rust. *J. Econ. Entomol.,* 22:482–485 (1929).

CHAPTER 10

Stem Diseases: Rusts of Conifers (Continued)

WHITE PINE BLISTER RUST

This disease is caused by *Cronartium ribicola* Fischer, and no evidence so far shows that there is more than one strain of the fungus (Hahn 1949). Comprehensive accounts of this destructive disease and the causal fungus in North America and western Europe, together with extensive bibliographies, have been given by Spaulding (1911; 1922; 1929), Colley (1918), Moir (1924), Tubeuf (1936), and Mielke (1943). Distinctions between white pine blister rust and the closely related piñon blister rust are given on page 214.

History and Distribution. It is supposed that Swiss stone pine was the original host of the fungus in Asia and that it spread gradually to Europe where it found the introduced eastern white pine to be a very susceptible host. The rust was first certainly found by Dietrich in the Baltic provinces of Russia in 1854 on both hosts, but it was not then known that *Cronartium ribicola* on *Ribes* and *Peridermium strobi* Kleb. on eastern white pine were stages of a single fungus. By 1900 it was widespread over northern and most of western Europe. Although eastern white pine was introduced into Europe in 1705 and grown extensively, the disease was not found on it until 1854. On the basis of experience with this and other introduced plant diseases in North America, this lapse of time between the introduction of the host and discovery of the disease indicates that white pine blister rust was not native to Europe.

In North America the rust was first discovered at Geneva, New York, in 1906 on *Ribes*, but it is known to have been in the Northeastern states by 1898, or perhaps a few years earlier. In 1909 the rust was first found on eastern white pine, and it was discovered that infected 3-year-old white pines from a German nursery had been widely distributed throughout the Northeast. In 1910 diseased trees were found in eastern white pine stock imported from several French nurseries. The rust is now widespread on pines from the New England states west into Minnesota and Iowa and south into North Carolina.

In September 1921 the disease was first found in western North America at Vancouver, British Columbia, where it was introduced in 1910 on a single shipment of 1,000 eastern white pine transplants from a nursery in France. It has spread from this single introduction throughout the range of western white pine in British Columbia, south through Washington and Oregon into the northern half of California, where it occurs on sugar pine, and into northeastern Washington, northern Idaho, and western Montana, where infection is widespread and locally severe on western white pine. It has also appeared on limber pine in Montana east of the Continental Divide, and on western black currant (*Ribes petiolare* Dougl.) in northwestern Wyoming. It is common on whitebark pine.

The widespread introduction of this pathogen from Europe on nursery stock of a native American tree is an anomaly. There was a rapid development of forest planting in the East from 1900 to 1910, with eastern white pine being very popular for this purpose. The cost of eastern white pine stock from our nurseries for forest planting at first was prohibitive from a forestry standpoint. The low-priced stock offered by European nurseries was largely restricted by a high tariff. The tariff was removed and millions of small trees, many of which were infected, were brought in from Europe between 1907 and 1909, principally from Germany. These were established in plantations widely scattered throughout the Northeastern states, the Lake states, and eastern Canada.

Hosts. *C. ribicola* has been found on the majority of the five-needle or white pines, and it will probably attack all species throughout the world (Spaulding 1929:24; Tubeuf 1933). The infection of eastern white, western white, sugar, and whitebark pines under natural conditions has shown the western species to be much more susceptible than eastern white pine, whitebark pine followed by sugar pine especially being extremely susceptible (Childs and Bedwell 1948). This does not necessarily mean that sugar pine within its commercial or botanical range will be more severely damaged than western white pine within its natural limits, because climatic conditions may be of more importance than susceptibility in determining the degree of infection. Observations in Europe and infection experiments have shown limber pine to be about equal to western white pine in susceptibility. Himalayan and Balkan pines have been suggested as possible replacements for our native white pines, but the first, although highly resistant, is not hardy during cold weather, while the second is not resistant (Hirt 1940; Honey, Nelson, and Putnam 1943).

The reasons for the difference in susceptibility of pines are not understood, although it has been suggested that bark thickness, length of time the needles are retained, and number of stomata on the needles are important factors (Spaulding 1925), but later investigation indicates that direct penetration of the epidermis by the hyphae may be the common method of entrance, rather than through stomata (Hirt 1938). Thicker bark and longer retention of needles would explain the greater susceptibility of western white pine as compared with eastern white pine, but occasional vigorous trees of both species either escape or resist infection, although their neighbors are all killed or severely injured (Hirt 1948; Porter 1952). Consequently it would seem that resistance is probably the result of some more fundamental relationship between life processes of fungus and host. In general it is certain that vigor of the host is not a measure of resistance to *C. ribicola,* but rather the contrary, since in natural stands of both eastern and western white pines it is robust dominant trees that are most severely attacked, whereas frail suppressed individuals escape infection or are lightly attacked. This can be explained in part by the fact that the vigorous trees bear many more needles, thus greatly increasing their chances of infection. The number of needles per tree or the target presented can be estimated for western white pines by an alinement chart relating the number of needles to the length and width of the crown (Buchanan 1936).

Since *C. ribicola* is dependent on two groups of hosts for its existence, the susceptibility of the alternate hosts, currants and gooseberries (*Ribes* spp.), termed "ribes" for convenience, is as important in the spread of the disease as the susceptibility of pines, because, disregarding other factors, the degree of infection of ribes determines the amount of damage to pines. The susceptibility of most ribes in North America and Europe, both native and cultivated, has been determined either under natural conditions in test gardens, or in greenhouses (Spaulding 1922:16; Tubeuf 1933; Mielke 1943:46). European black currant (*R. nigrum* L.), which has been extensively cultivated in some parts of the United States, is the most susceptible ribes, spreading severe infection to adjacent pines, although certain hybrids of even this species are immune (Hahn and Miller 1949). Cultivated varieties of red currants on the whole are resistant and cause minimum infection of pines (Hahn 1943; Snell 1950). Two European varieties, Viking and Red Dutch, are practically immune. Male plants of *R. alpinum* L. are immune and female plants susceptible (Hahn 1941).

Native ribes occur in more or less abundance wherever five-needle pines grow naturally. In the Northeast the widely distributed wild gooseberry (*R. cynosbati* L.) and the skunk currant (*R. prostratum* L'Her) are highly susceptible. In the western white pine region of northern Idaho, western black currant (*R. petiolare* Dougl.) is practically as susceptible as European black currant, and it is often associated with the less susceptible white-stemmed gooseberry (*R. inerme* Rydb.) in concentrations along stream banks. The upland species, sticky currant (*R. viscosissimum* Pursh.) and prickly currant [*R. lacústre* (Pers.) Poir.], are also less susceptible. In the Pacific Northwest, stink currant (*R. bracteosum* Dougl.) and red flowering currant (*R. sanguineum* Pursh.), which occur west of the summit of the Cascade Mountains, are highly susceptible. The two principal ribes associates of sugar pine, Sierra gooseberry [*R. roezli* (Reg.) Cov. & Brit.] and Sierra currant (*R. nevadense* Kell.) are high in susceptibility (Kimmey 1945), whereas other ribes in the sugar pine region are less susceptible (Kimmey and Mielke 1944). In fact it can be concluded that any native ribes in North America, given time, will carry sufficient infection to seriously damage adjacent pines.

The greater the susceptibility of a ribes, the more spores will be produced to inoculate pines, but in addition there is the possibility of a difference in the virulence of these spores depending on the ribes from which they originate (York, Snell, and Rathbun-Gravatt 1927:508). The latter has also been suggested as the result of inoculations of eastern white pine with sporidia from the leaves of European black currant and wild gooseberry in Minnesota (Lindgren and Chapman 1933).

Damage. The damage by blister rust is proportional to the number and susceptibility of ribes within and around a stand. Without control of the disease neither eastern white pine, western white pine, nor sugar pine can be perpetuated as commercial species. In general blister rust is not spectacular. On unprotected areas, the disease increases slowly at first and then rapidly. Seedlings and small saplings are infected and killed quickly, while poles and larger trees succumb slowly (Anonymous 1940; Buchanan 1938). Consequently a white pine stand where blister rust is epidemic may appear little affected for many years, judging only by the larger trees, but there will be an almost complete absence of seedlings and small saplings (Snell 1928). The successive crops of seedlings succumb after a few years and nothing is left to perpetuate the stand when the mature trees are cut or finally killed. When infection is heavy, destruction of saplings can be rapid. Sixty-nine per cent of an eastern white pine stand has been killed in 10 years, and practically 100 per cent of a western white pine stand in 16 years (Snell 1931*b*; Lachmund 1934*a*). An adjacent stand of larger western white pine had a 56 per cent mortality during the same 16-year period. Damage to older stands is slower but just as sure, and dying of infected trees may continue for 20 years or more after ribes have been removed (Rusden 1952).

When the main stem of a tree is invaded, death is only a question of time because any stem, no matter how large, will ultimately be girdled. Trees are generally killed by trunk infections, usually entering from diseased branches or occasionally from either adventitious or persistent dwarf shoots. Western white pines and sugar pines sometimes die without trunk infection, the branches being killed by innumerable infections before the disease reaches the main stem. Branch infections can be used as a criterion of damage, since only those branch infections originating within 24 inches from the trunk are likely to reach it (Slipp 1953). The percentage of trees already killed or doomed to die before a stand matures is not a correct estimate of damage, since this is dependent on stocking. A 51 per cent mortality can result in a loss of lumber production of 37 per cent, a 54 per cent mortality in a loss of 80 per cent, and a 26 per cent mortality in no loss (Snell 1931*a*). The true measure of damage is the effect on the productive capacity of a stand. Of course, white pine production will finally cease on any area infested with ribes unless these bushes are eradicated.

Epidemiology. Moist growing seasons greatly favor the intensification and spread of blister rust, and dry seasons retard it. Moist conditions are essential for long-distance spread from pines to highly susceptible ribes and subsequent establishment in a new locality. *C. ribicola* with its five types of spores maturing at different periods during the growing season, is delicately adjusted to climatic conditions, so that any condition unfavorable to it at one period prevents any significant number of new pine infections for that year (Van Arsdel, Riker, and Patton 1956). Heavy rains at the time of aeciospore dispersal are unfavorable, but, when the spores are deposited upon ribes leaves, there must be moist weather for germination of the spores and invasion of the leaf. After infection develops on ribes, there must be some moist periods for further spread of the disease on ribes by urediospores, so that a large volume of teliospores can finally develop to infect pines in the late summer and autumn. Two days (48 hours) with saturated air and with a maximum temperature of not over 68°F are necessary for production of promycelia with their sporidia, casting and germination of the sporidia, and penetration of pine needles. This seems to explain the fact that in the West, blister rust has spread much more rapidly eastward into the western white pine region than it has southward into the sugar pine region, where temperatures are higher and relative humidities lower. Indications are that the disease will spread and increase at a slower rate in the sugar pine region than in the western white pine region. Furthermore in the West, long-distance spread from pines to ribes depends on synchronization of the period of maximum aeciospore production with the period of maximum susceptibility of ribes leaves, but, owing to seasonal variations affecting leaf development more than spore production, the two periods do not always coincide (Lachmund 1934*b*:111). After pines are inoculated it is three seasons or more before aeciospores develop. The disease has a tendency to advance in waves because only occasionally are conditions favorable for the harmonious operation of all these factors over wide areas.

Since in western North America the rust spreads from one point, it has been possible there to determine facts that could not be ascertained in the East where the disease originated at a number of places more or less simultaneously. An outstanding development was the determination of long-distance aecial spread. Aeciospores from pines can be wind borne 300 miles or more to inoculate very susceptible ribes from which in turn near-by pines are inoculated, thus permanently establishing the disease in new and remote localities. The importance of the different species of ribes in the long-distance spread from pines lessens as their relative susceptibility to the disease decreases.

In its invasion of new territory, blister rust goes through the following stages:[1] (1) *Introduction.* Characterized by the initial infection of ribes, often at long distances from infected pines. From these infected ribes, near-by white pines become diseased, and thus new centers of pine infection are formed. No actual damage to pine is apparent in this stage of spread, and usually infected pines are so few that they are difficult to find. (2) *Intensification.* After a few years the pine infection centers established during the introductory stage begin to produce aeciospores. At first the volume of spores is insufficient to result in further long-distance spread of the disease, but all species of ribes in the surrounding locality become thoroughly infected, and each season increases the amount of infection on pines. As the volume of aeciospores from these infected pines increases, the likelihood for long-distance spread to ribes increases. The process also results in the establishment of numerous local centers of severe pine infection. (3) *Damage.* Conditions are now favorable for an epidemic of the disease in this locality. The annual infection of many ribes results in heavy pine infection. As time goes on and the pines begin to die, the effect of this heavy infection in damaging pines becomes more and more striking. Meanwhile the production of aeciospores multiplies in geometrical proportion, until the spread of the disease to ribes in and away from the center becomes so intense that in any

[1] The author is indebted to H. G. Lachmund for the substance of this statement.

favorable season infection may be found on ribes without difficulty for distances up to 100 miles or more. This leads to a general epidemic throughout the surrounding territory.

Symptoms. The disease can be recognized readily during its earlier stages on pines. Although the first indication is a small golden-yellow to reddish-brown spot on the needle at the point of infection, these spots are difficult to distinguish from spots caused by other agencies, except by microscopical examination. During the second season following infection of the needles, or with western white pine to some extent not until the third season, infected bark assumes a yellow to orange color, the orange

Fig. 79. Pycnial scars (the darker patches on the lesion) of white pine blister rust on eastern white pine. (*Reproduced from J. Agr. Research* 15, *no.* 12.)

tinge often being most pronounced around the margin of the discolored area. On seedlings the discoloration often appears the season following infection of the needles. This discoloration is most useful for diagnosing blister rust, particularly when the characteristic spindle-shaped swelling of the stem is slight or irregular. If discoloration and swelling develop in the spring or early summer, pycnia appear the same season during July and August, but if discoloration and swelling are delayed until midsummer or later, pycnial development is postponed until the next season (Lachmund 1933*a*). Pycnia first appear on the bark as small honey-yellow to brownish patches that swell slightly, forming shallow blisters. These blisters rupture, discharging drops of a honey-colored sweet-tasting liquid that has a characteristic starchy odor. The pycnial drops are quickly eaten by insects, dry up, or are washed away by rain. However, the pycnial lesions in the bark gradually assume a dark color like that of clotted blood, being known as "pycnial scars" or "spots," and are visible throughout the year (Figs. 79, 80). On sugar pine the pycnial scars may not always be so prominent. In the spring or early summer of the season following the appearance of pycnia, and over the area previously occupied by pycnia, the aecia push through the bark as white blisters that quickly rupture at the top to disclose a mass of orange-yellow aeciospores (Fig. 81). The aeciospores are soon carried away by wind, the membrane (peridium) originally enclosing them disintegrates, and the bark dries out and cracks, resulting in death of the cambium and underlying wood. This cycle beginning with the infection of needles and ending with the appearance of aecia generally requires about 3 years and 6 months, but it may be shorter or longer, since it is influenced by variable factors such as length of the growing season, time of needle infection, and possibly the distance from the point of infection on the needle to the stem. On seedlings ranging up to 4 or 5 years of age, the cycle is usually 1 year shorter, whereas on western white pines at higher elevations it is frequently 1 year longer.

Production of aecia, preceded by pycnia, occurs yearly until the stem, no matter how large, is killed beyond the lesion. Resin exudation usually becomes copious once the bark has been ruptured by aecia. Since progress of the mycelium in the bark, which is indicated by the discoloration, generally precedes the development of aecia by two seasons or occasionally only one, death of that portion of the stem beyond the infection does not occur for 1 to 3 years or more after the mycelium has girdled the stem. In western white pine, small stems approaching 1 inch in diameter are usually girdled within a year after discoloration and swelling appears, dying 1 or 2 years later when aecia are produced; the number of years required for girdling stems over 3 inches in diameter approximates the number of inches of stem diameter (Lachmund 1934*c*: 493). Because of the conspicuous red-

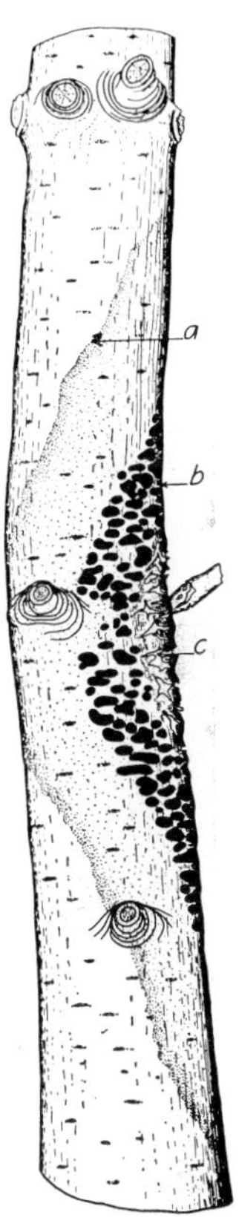

Fig. 80. Drawing of an infected 12-year-old main stem of eastern white pine. The infection entered the main stem along the small branch, the stub of which is shown at the right of the figure. (*a*), the advancing edge of the infection; (*b*), the pycnial area. The black dots are the pycnial scars; (*c*), the aecial area on which the bark is cracked and broken. In another season the aecial area would spread over the pycnial area (*b*), and the pycnial area would be advanced as far as the boundary (*a*) under normal conditions. The boundary (*a*) would be proportionately advanced also. (*Reproduced from J. Agr. Research* 15, *no.* 12.)

brown color of the dead foliage, death of the stem beyond the lesion is known as "flag" formation, and "flags" are the most conspicuous symptoms of blister rust in an invaded area before dead trees appear. Infection progresses longitudinally more rapidly than laterally, the mycelium also growing more rapidly toward the base than toward the tip of a stem. The downward growth enables infection to pass from smaller into larger branches and finally into the trunk, provided infection first appeared at a

Fig. 81. Aecia of *Cronartium ribicola* on eastern white pine. Notice the spindle-shaped swelling of the stem.

point on the branch not too far from the trunk. It is possible for an experienced observer to decide with reasonable accuracy whether or not any branch infection will reach the trunk.

Since it is only the bark that increases in thickness, the wood being little or not at all infected, swelling of large stems is not apparent. On eastern white pine, which has relatively thin bark, no swellings are visible on stems of about 2 inches or more in diameter, but on the thicker-barked western white pine swellings on stems up to 5 inches in diameter are the rule (Rhoads 1920:520; Lachmund 1934*c*:500). In fact constriction of the infected area is common on the larger stems of eastern white pine. This is caused by increased diameter growth of the stem above and below the infected area during the girdling process. Swellings on twigs caused by the entomogenous fungus *Cucurbitaria pityophila* (Fr.) Petrak at a distance somewhat resemble blister rust swellings (Boyce 1952).

Causal Agency. *Cronartium ribicola* infects pines during summer and autumn through needles of all ages. Older needles are more likely to be infected than needles of the current year (Richards 1927; Pierson and Buchanan 1938; Hirt 1944). Only one-fifth as many infections on western white pine occur on needles of the current year as on 1-year-old needles and only one-third as many as on 2-year-old needles, which led Lachmund (1933*b*) to conclude that needles of the current year are resistant to infection, although the nature of this resistance is unknown. The variation in the number of needles of different ages infected during any one growing season enables the year of any wave of infection in a stand to be determined long afterwards (Lachmund 1933*c*; Kimmey 1954). The hyphae grow down the conducting tissue of the needle to the stem, although probably in many instances, particularly in older needles, they never reach the stem before the needles are shed. The rate of growth down the needle is not known. Once in the stem the hyphae develop largely in the inner bark, first invading the phloem parenchyma and rays, but they may also penetrate the ray tissues of the wood to a depth of at least three annual rings. The abundant large intercellular hyphae with prominent irregularly shaped haustoria penetrating into the cell cavities are unlike the hyphae of any other fungus in the bark of white pines, and consequently an accurate microscopic diagnosis of blister rust can be made when macroscopic symptoms are lacking or uncertain (Waterman 1955). Swelling of the bark apparently results almost entirely from the spreading action of numerous hyphae forcing their way between the cells. After the mycelium in the bark has developed for some time, pycnia appear (page 224). Tiny colorless pear-shaped pycniospores mixed with a thin gelatinous fluid are exuded from the pycnia and are distributed by insects. Although the pycniospores are incapable of ini-

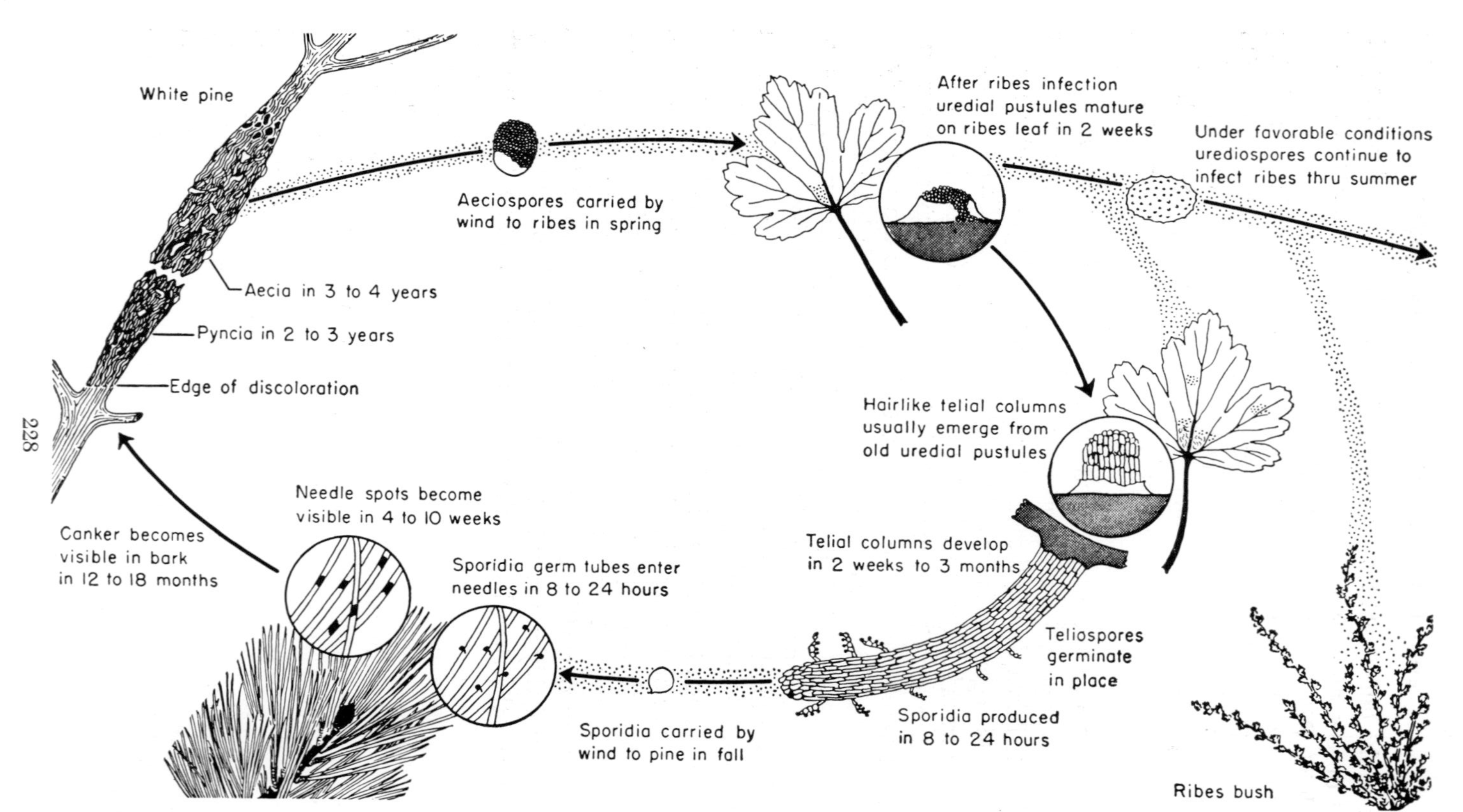

Fig. 82. Life cycle of *Cronartium ribicola.* (*Reproduced from U.S. Dept. Agr., Forest Serv., Forest Pest Leaflet* 36.)

tiating new infections, they probably have a sexual function (Pierson 1933). In the spring and early summer of the season following the production of pycnia, aecia appear over the same area of bark. The large ellipsoid orange aeciospores with thick verrucose colorless walls having a smooth spot at the basal end (Fig. 82) can retain their viability for several months under favorable conditions and for several weeks even when conditions are unfavorable. Produced in enormous numbers, they are wind borne, often for great distances, to infect ribes leaves. They

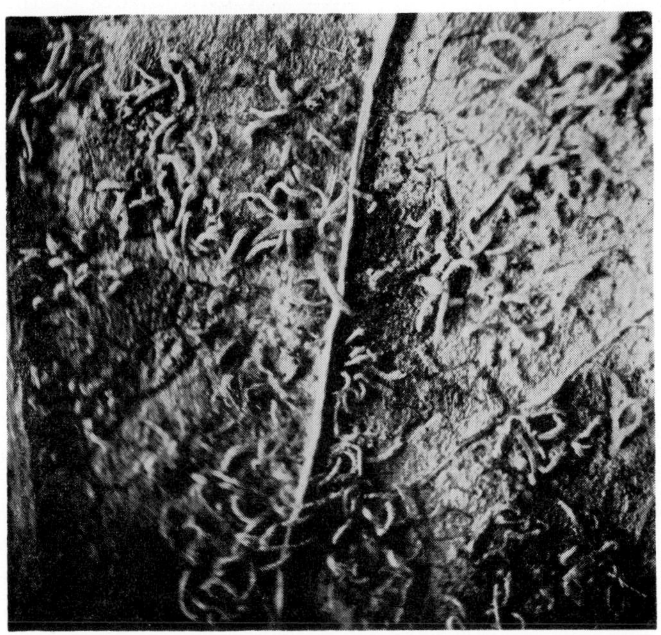

Fig. 83. Telia of *Cronartium ribicola* on the undersurface of a black currant leaf. Much enlarged. [*Courtesy of Conn.* (*State*) *Agr. Expt. Sta.*]

cannot initiate infections on other pines. The germ tubes of the aeciospores enter ribes leaves through the stomata on the undersurface. Within 1 to 3 weeks after infection of ribes, the uredia appear on the undersurface of the leaves as tiny blisters about the size of a pinhead, clustered on small areas of yellowed leaf tissue. The uredia quickly rupture at the top to release the large ellipsoid yellow urediospores, which have thick colorless walls sparsely beset with short sharp spines. The urediospores also remain viable for a long time; in fact, they may occasionally overwinter in the uredia but not in sufficient numbers to be practically significant. The urediospores, which are not produced in quantity, are somewhat moist and sticky, tending to hang together in clumps so that they are carried only short distances by wind, their function being to spread infection to near-by ribes in order that a large volume of telia may be produced. Seven generations or more of uredia may develop during a single summer. The urediospores cannot infect pines. Infection of ribes leaves is by way of the stomata on the underside. Following the uredia, either in the same or new lesions, telia develop usually in the late summer or early autumn. The telia appear as slender brown bristles on the underside of the leaf, often so numerous as to simulate a coarse brown felt, thus giving the common name "felt rust" to this stage (Fig. 83). Each telium is an aggregation of vertical rows of broad, spindle-shaped, somewhat angular teliospores, which germinate *in situ* by means of a five-celled promycelium, each of the four upper cells bearing one thin-walled small globose sporidium (basidiospore) on the tip of a stout sterigma (Fig. 84). The sporidia are then wind borne to infect pine needles. The sporidia cannot infect ribes leaves. Since sporidia are quite delicate, quickly losing their viability when exposed to direct sunlight or dry air, the effective range at which pines can be infected is generally 900 feet from ribes, although aggregations of European black currant, western black currant, and stink currant may spread infection to pines for a distance of about a mile. Although insects, rodents, and birds may carry aeciospores, urediospores, and sporidia, dissemination of the disease by such agents is negligible. Although aeciospores, urediospores, and teliospores with their accompanying sporidia may be roughly classified according to their time of appearance as spring, summer, and late summer to autumn spores, respectively, yet because their development is influenced by climatic conditions there is chronological variation in their appearance from season to season (Spaulding 1922:72; Mielke and Kimmey 1935).

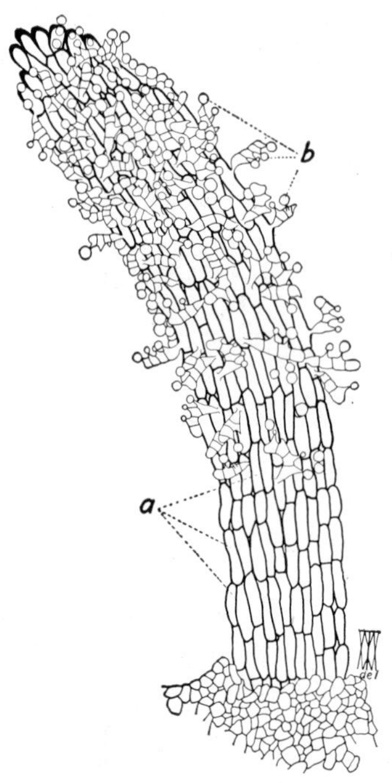

Fig. 84. Drawing of a short mature telial column in which the teliospores (*a*) have germinated, producing promycelia and sporidia (*b*). (*Reproduced from J. Agr. Research* 15, *no.* 12.)

Control. Control of white pine blister rust by the removal of ribes in selected areas of white pine is known as local control. Control is based on the relatively short distance the disease spreads from ribes to pines. Since pines can become infected from European black currants a mile distant or possibly more, this plant should be completely removed from pine-growing regions. This has been accomplished in all important pine-growing regions of the United States. Other ribes in control areas should be reduced to an average of from 1 to 25 feet of live stem per acre, depending on the region and other variable conditions. Stands of reproduction should be given a higher working priority than older growth. Self-incompatibility in ribes, as demonstrated for certain western species, will preclude heavy fruiting in scattered light populations of these plants (Offord, Quick, and Moss 1944). Consequently the chances for new plants to appear should become progressively poorer as the number of ribes bushes per acre is reduced by artificial removal.

The removal of ribes from within and around white pine stands of the East for a distance of 900 feet or less, according to local conditions, is generally successful in preventing further infection or reducing it to so few trees that the stand will not be damaged. In the northern Lake states and most western white pine areas the protection zone must be greater and is determined on the basis of local climate and topography. Concentrations of susceptible ribes favorably located can damage pines up to a mile or more away in all pine-growing regions. One or more reworkings of control areas at periodic intervals following the initial ribes removal job are necessary, because seedling ribes can remain viable in the forest floor for a number of years, germinating soon after any disturbance, such as is caused by ground fires, logging operations, windfalls, and road building. Ribes bushes developing from these sources should be destroyed before they become old enough to produce fruit. This is accomplished by periodically inspecting the control areas to determine ribes growth conditions and to remove any newly developed bushes before seeding takes place. Usually this requires two or, in some cases, more workings. Natural and artificial changes in forest cover are going on continuously and cause wide variations in age, composition, and density of white pine stands, and in the regeneration, size, and abundance of ribes. Under such conditions the elapsed period between workings for any particular area may range from two to many years, and ribes eradication must be practiced continuously in forest areas at properly timed intervals to protect white pine crops from blister rust. Although ribes control is a charge against production of white pines, it is unlikely that any other crop species can be intensively managed without carrying charges for protection against insects and diseases, in addition to fires. This is particularly true for eastern white pine.

Hand pulling and grubbing are the principal methods of ribes removal, taking care to get all the root system capable of sprouting. Sprouting from roots alone is rare, but crown tissue sprouts vigorously. In some places, more commonly in the West, where these methods are not feasible because of associated brush, or abundance, size, or location of the ribes, chemical methods are employed (Offord et al. 1952; Offord, Quick, and Moss 1958). Large or rock-bound bushes can be cut off close to the ground and the stumps treated with an oil-ester solution of 2,4-D or 2,4,5-T composed of 5 pounds of the acid in 100 gallons of diesel oil. Stumps can also be treated with a dry mixture of common salt and borax (1:1). Individual bushes without prior injury can be killed by basal spray treatment with the oil-ester solution, but the strength must be increased up to a maximum of 25 pounds of herbicide to 100 gallons of oil. A sufficient volume of oil and herbicide must be applied as a coarse spray or solid stream to the crown (root collar) and to the lowest 6 to 8 inches of all stems so that the spray runs off the plant parts. Large bushes must be treated from two sides. These chemicals kill the crown and roots, thus preventing sprouting. The treatments are applicable wherever troublesome ribes occur as scattered individuals or small groups.

Western black currant, white-stemmed gooseberry, Sierra gooseberry, skunk currant, prickly currant, and other species often occur in such numbers that they can be destroyed more readily and at less cost with chemicals. Under such conditions the ribes can be dealt with by spraying the foliage with a water solution of 2,4-D, 2,4,5-T, or a combination of the two, depending on the ribes species (Offord, Quick, and Moss 1958). These toxicants are effective in low concentrations and moderate dosages, causing no significant poisoning of the soil. Concentrations of these herbicides vary from 500 to 2,100 parts per million governed by time of year and seasonal plant development. They should be applied as a spray that can be classed between a mist and a drizzle. All aerial parts of the plants, especially the current-season branch tips, must be sprayed just to the point of drip and from opposite sides for large bushes, because an unsprayed branch or branch-tip often results in the survival of a bush. Seedling ribes and young vigorous plants are more readily killed than mature bushes. Foliage spraying should not start in the spring until the primary leaves are fully expanded. Respraying should be done the second season after the first treatment to kill sprouts that have appeared meanwhile.

Ribes removal does not result in the eradication of all ribes no matter how carefully the work is done. Seedlings and some larger plants are missed, particularly on rough and brushy areas. Control does reduce the number of ribes to a point where not enough pine infections can occur to damage a stand. Well-stocked white pine stands, once rid of ribes, will remain so until some major disturbance occurs. Around nurseries growing white pines for forestry purposes, control work should be unusually intensive, the area being reworked at intervals of one or more years to maintain a protective zone of 1,500 feet free from ribes, since it is essential to have rust-free stock for shipment. Where European black currant, western black currant, or other highly susceptible species grow in the environs, removal of these species for a distance of 1 mile or more from the nursery is essential.

As a supplementary method of controlling blister rust in young western white pine, the use of antibiotic fungicides is developing rapidly. Acti-dione BR (cyclohexamide) is now used on a large scale.[1] Trees are individually sprayed at any time of the year with 150 parts per million of Acti-dione BR in No. 1 western fuel oil. A mist-type spray is used, and the solution should be applied to saturation. For trees 12 feet or less in height, spray bark of trunk and branches from ground level up to one-third of the crown height, whereas for trees over 12 feet in height, spray bark of trunk and branches from ground level to a height of 5 feet, spraying the branches out 12 to 18 inches from the trunk. The spray is injurious to terminal shoots. All lesions both on the trunk and on the branches die in a maximum of 18 to 24 months after treatment. This has been effective on trees up to 50 feet high, and evidence indicates that it will be effective in mature trees.

The antibiotic persists 2 years at least in the trunk bark, and it probably lasts longer. Ten to fifteen years will elapse before new infections can cause appreciable damage. In a few more years the necessary frequency of retreatment with this and new antibiotics that are being tried will be known. Furthermore, trials with other white pines should increase the use of this method of control. In addition, experiments on the application of fungicides to foliage by aircraft to eliminate infections and on long-term immunization of pines by treatment with antibiotics in the nursery are under way. Developments are coming fast, so that the next decade should bring revolutionary changes in methods of control.

Blister rust cannot be controlled by silvicultural methods, but control can be materially aided by coordinating silvicultural and disease-control practices. Keeping the land fully stocked with growing timber inhibits

[1] "Protect Your Western White Pine with Acti-dione." U.S. Department of Agriculture, Northern Region, Forest Service, Missoula, Montana. Pamphlet, no page numbers, no date.

the development of ribes. Planting of white pines should be on sites where ribes are absent or have been reduced to protection standards. When stands are harvested, the environmental conditions created for the establishment of white pine reproduction also favor the regeneration of ribes.

Spacing of trees under a partial or selective cutting plan obtains maximum results in the suppression of ribes and devitalization of ribes seed in some stands in Idaho. In most old-growth white pine stands of the West clear-cutting and burning is considered the advisable silvicultural practice (Moss and Wellner 1953). In the open stands of the sugar pine region, all disturbances in the forest, particularly by fire and logging, increase the number of bushes of Sierra gooseberry, so that following a disturbance eradication and reeradication should be timed to prevent any ribes fruits from maturing on the area (Quick 1954). In the eastern white pine region, management operations such as weeding, thinning, and pruning in young stands, which increase either the quantity or more particularly the quality of timber, should be coordinated with disease-control practices. Certain rodents, insects, and other fungi are unfavorable to the blister rust fungus in each stage of its life cycle, which reduces the chances of white pine infection to some extent, but nowhere have these factors been of material significance in control of the rust. In cities and industrial centers where five-needle pines and ribes occur together, blister rust is usually absent, the smoke and fumes in the air apparently being unfavorable to the fungus (Spaulding 1929:21).

To save valuable ornamental trees or to preserve adequate stocking in infected stands, infected branches can be cut off and trunk infections cut out. On eastern white pine the bark on the trunk should be completely removed from a lesion for a distance of 1½ to 2 inches at the sides beyond the colored zone marking the edge of the lesion and for a distance of from 3 to 4 inches at the top and bottom (Martin and Gravatt 1954). For sugar pine these distances for the safety zone should be doubled. The treatment of stem lesions in this way is seldom advisable if the bark must be excised from more than one-half the circumference of the stem. In general, pruned or excised trees should be inspected the first year after treatment and every three years thereafter until no more lesions can be found. Furthermore, no treatment of infected pines should be undertaken unless they are to be protected from further infection by ribes removal. Western white pine can be successfully treated by basal spraying with cyclohexamide as described previously.

The possibility of introducing exotic five-needle pines to replace native species is not encouraging. Trees once suggested were Himalayan and Balkan pines because of their supposed resistance to blister rust, but the first is not hardy during extremely cold weather, and the second is not resistant. Nothing is known about the site requirements, silvicultural characteristics, and properties of the wood of these or any other exotic five-needle pines when grown in this country, and the ever-vexing problem of obtaining seed from high-quality mother trees would have to be solved. Furthermore, any exotic might prove to be susceptible to native diseases such as Atropellis canker on western white pine and Phomopsis canker on eastern white pine. Long-time tests of an exotic should be made before it is grown extensively.

Resistant strains of native five-needle pines may ultimately be obtained (Bingham, Squillace, and Duffield 1953). Rust-free eastern white pines, western white pines, and sugar pines have been found in stands heavily infected with blister rust, and the rust resistance of these trees is heritable. In time, resistant stock in sufficient quantity for extensive planting may be developed from such trees, and when this time comes, control costs can be much reduced. This stock will be particularly valuable in the western white pine type, where ribes removal is so costly and where in some stands the rust hazard is so severe that pine is considerably damaged despite the most thorough ribes removal possible.

REFERENCES

Anonymous: Density of Stocking and Damage from Blister Rust. *N.Y. State Conserv. Dept. Ann. Rept.*, 29:113–114 (1940).

Bingham, R. T., A. E. Squillace, and J. W. Duffield: Breeding Blister-rust-resistant Western White Pine. *J. Forestry*, 51:163–168 (1953).

Boyce, J. S.: *Cucurbitaria pithyophila*, an Entomogenous Fungus. *U.S. Dept. Agr., Plant Disease Reptr.*, 36:62–63 (1952).

Buchanan, T. S.: An Alinement Chart for Estimating Number of Needles on Western White Pine Reproduction. *J. Forestry*, 34:588–593 (1936).

Buchanan, T. S.: Blister Rust Damage to Merchantable Western White Pine. *J. Forestry*, 36:320–328 (1938).

Childs, T. W., and J. L. Bedwell: Susceptibility of Some White Pine Species to *Cronartium ribicola* in the Pacific Northwest. *J. Forestry*, 46:595–599 (1948).

Colley, R. H.: Parasitism, Morphology, and Cytology of *Cronartium ribicola*. *J. Agr. Research*, 15:619–659 (1918).

Hahn, G. G.: Field Tests with a Staminate Clone of Alpine Currant Immune from Blister Rust under Greenhouse Conditions. *U.S. Dept. Agr., Plant Disease Reptr.*, 25:476–478 (1941).

Hahn, G. G.: Blister Rust Relations of Cultivated Species of Red Currants. *Phytopathology*, 33:341–353 (1943).

Hahn, G. G.: Further Evidence that Immune Ribes Do Not Indicate Physiologic Races of *Cronartium ribicola* in North America. *U.S. Dept. Agr., Plant Disease Reptr.*, 33:291–292 (1949).

Hahn, G. G., and A. Miller: Rust Immunity of Canadian Black Currants under Field

Conditions in Connecticut. *U.S. Dept. Agr., Plant Disease Reptr.,* 33:275–276 (1949).

Hirt, R. R.: Relation of Stomata to Infection of *Pinus strobus* by *Cronartium ribicola. Phytopathology,* 28:180–190 (1938).

Hirt, R. R.: Relative Susceptibility to *Cronartium ribicola* of 5-needled Pines Planted in the East. *J. Forestry,* 38:932–937 (1940).

Hirt, R. R.: Distribution of Blister-rust Cankers on Eastern White Pine According to Age of Needle-bearing Wood at Time of Infection. *J. Forestry,* 42:9–14 (1944).

Hirt, R. R.: Evidence of Resistance to Blister Rust by Eastern White Pine Growing in the Northeast. *J. Forestry,* 46:911–913 (1948).

Honey, E. E., L. E. Nelson, and H. N. Putnam: Study of Blister Rust Infection on *Pinus peuce, P. koraiensis, P. strobus* and *P. monticola* at the Cloquet Forest Experiment Station, Cloquet, Minnesota. *U.S. Dept. Agr., Div. Plant Disease Control, Tech. Memo.,* 2:1–17 (1943). (Mim.)

Kimmey, J. W.: The Seasonal Development and the Defoliating Effect of *Cronartium ribicola* on Naturally Infected *Ribes roezli* and *R. nevadense. Phytopathology,* 35:406–416 (1945).

Kimmey, J. W.: Determining the Age of Blister Rust Infection on Sugar Pine. *Calif. Forest and Range Expt. Sta., Forest Research Notes,* 91:1–3 (1954).

Kimmey, J. W., and J. L. Mielke: Susceptibility to White Pine Blister Rust of *Ribes cereum* and Some Other *Ribes* Associated with Sugar Pine in California. *J. Forestry,* 42:752–756 (1944).

Lachmund, H. G.: Mode of Entrance and Periods in the Life Cycle of *Cronartium ribicola* on *Pinus monticola. J. Agr. Research,* 47:791–805 (1933*a*).

Lachmund, H. G.: Resistance of the Current Season's Shoots of *Pinus monticola* to Infection by *Cronartium ribicola. Phytopathology,* 23:917–922 (1933*b*).

Lachmund, H. G.: Method of Determining Age of Blister Rust Infections on Western White Pine. *J. Agr. Research,* 46:675–693 (1933*c*).

Lachmund, H. G.: Damage to *Pinus monticola* by *Cronartium ribicola* at Garibaldi, British Columbia. *J. Agr. Research,* 49:239–249 (1934*a*).

Lachmund, H. G.: Seasonal Development of Ribes in Relation to Spread of *Cronartium ribicola* in the Pacific Northwest. *J. Agr. Research,* 49:93–114 (1934*b*).

Lachmund, H. G.: Growth and Injurious Effects of *Cronartium ribicola* Cankers on *Pinus monticola. J. Agr. Research,* 48:475–503 (1934*c*).

Lindgren, R. M., and A. D. Chapman: Field Inoculations of *Pinus strobus* with Sporidia of *Cronartium ribicola* in Minnesota. *Phytopathology,* 23:105–107 (1933).

Martin, J. F., and G. F. Gravatt: Saving White Pines by Removing Blister Rust Cankers. *U.S. Dept. Agr. Circ.,* 948:1–22 (1954).

Mielke, J. L.: White Pine Blister Rust in Western North America. *Yale Univ., School Forestry Bull.,* 52:1–155 (1943).

Mielke, J. L., and J. W. Kimmey: Dates of Production of the Different Spore Stages of *Cronartium ribicola* in the Pacific Northwest. *Phytopathology,* 25:1104–1108 (1935).

Moir, W. S.: White-pine Blister Rust in Western Europe. *U.S. Dept. Agr., Dept. Bull.,* 1186:1–31 (1924).

Moss, V. D., and C. A. Wellner: Aiding Blister Rust Control by Silvicultural Measures in the Western White Pine Type. *U.S. Dept. Agr. Circ.,* 919:1–32 (1953).

Offord, H. R., C. R. Quick, and V. D. Moss: Self-incompatibility in Several Species of Ribes in the Western States. *J. Agr. Research,* 68:65–71 (1944).

Offord, H. R., et al.: Improvements in the Control of Ribes by Chemical and Mechanical Methods. *U.S. Dept. Agr. Circ.*, 906:1–72 (1952).

Offord, H. R., C. R. Quick, and V. D. Moss: Blister Rust Control Aided by the Use of Chemicals for Killing Ribes. *J. Forestry*, 56: 12–18 (1958).

Pierson, R. K.: Fusion of Pycniospores with Filamentous Hyphae in the Pycnium of White Pine Blister Rust. *Nature (London)*, 131:728–729 (1933).

Pierson, R. K., and T. S. Buchanan: Susceptibility of Needles of Different Ages on *Pinus monticola* Seedlings to *Cronartium ribicola* Infection. *Phytopathology*, 28:833–839 (1938).

Porter, W. A.: Blister Rust Resistant White Pine. *Can. Dept. Agr., Forest Biol. Div., Sci. Serv., Bi-monthly Prog. Rept.*, 8(6):3 (1952).

Quick, C. R.: Ecology of the Sierra Nevada Gooseberry in Relation to Blister Rust Control. *U.S. Dept. Agr. Circ.*, 937:1–30 (1954).

Rhoads, A. S.: Studies on the Rate of Growth and Behavior of the Blister Rust on White Pine in 1918. *Phytopathology*, 10:513–527 (1920).

Richards, J. L.: Susceptibility of Different Aged Pine Needles to Blister Rust and Relation Between the Number of Infections on Pines and the Persistence of Their Needles. *Blister Rust News*, 11:241–247 (1927).

Riker, A. J., et al.: White Pine Selections Tested for Resistance to Blister Rust. *J. Forestry*, 41:753–760 (1943).

Rusden, P. L.: Blister Rust Damage at Waterford, Vt. *J. Forestry*, 50:545–551 (1952).

Slipp, A. W.: Survival Probability and Its Application to Damage Survey in Western White Pine Infected with Blister Rust. *Idaho Univ., Forest, Wildlife and Range Expt. Sta. Research Notes*, 7:1–13 (1953).

Snell, W. H.: Blister Rust in the Adirondacks. *J. Forestry*, 26:472–486 (1928).

Snell, W. H.: Forest Damage and the White Pine Blister Rust. *J. Forestry*, 29:68–78 (1931*a*).

Snell. W. H.: The Kelm Mountain Blister Rust Infestation. *Phytopathology*, 21:919–921 (1931*b*).

Snell, W. H.: Garden Red Currants and White Pine Blister-rust. *J. N.Y. Botan. Garden*, 51:187–190 (1950).

Spaulding, P.: The Blister Rust of White Pine. *U.S. Dept. Agr., Bur. Plant Ind., Bull.*, 206:1–88 (1911).

Spaulding, P.: Investigations of the White-pine Blister Rust. *U.S. Dept. Agr. Bull.*, 957:1–100 (1922).

Spaulding, P.: A Partial Explanation of the Relative Susceptibility of the White Pines to the White Pine Blister Rust (*Cronartium ribicola* Fischer). *Phytopathology*, 15:591–597 (1925).

Spaulding, P.: White-pine Blister Rust: A Comparison of European with North American Conditions. *U.S. Dept. Agr. Tech. Bull.*, 87:1–58 (1929).

Tubeuf, C. von: Studien über Symbiose und Disposition für Parasitenbefall sowie über Vererbung pathologischer Eigenschaften unserer Holzpflanzen. IV. Disposition der fünfnadeligen Pinus-Arten einerseits und der verschiedenen Ribes-Gattungen, Arten, Bastarde und Gartenformen andererseits für den Befall von *Cronartium ribicola*. *Z. Pflanzenkrankh.*, 43:433–471 (1933).

Tubeuf, C. von: Verlauf und Erfolg der Erforschung der Blasenrostkrankheit der Strobe von 1887 bis 1936. *Z. Pflanzenkrankh*, 46:49–103, 113–171 (1936).

Van Arsdel, E. P., A. J. Riker, and R. F. Patton: The Effects of Temperature and Moisture on the Spread of White Pine Blister Rust. *Phytopathology*, 46:307–318 (1956).

Waterman, Alma D.: A Stain Technique for Diagnosing Blister Rust in Cankers on White Pine. *Forest Sci.*, 1:219–221 (1955).

York, H. H., W. H. Snell, and Annie Rathbun-Gravatt: The Results of Inoculating *Pinus strobus* with the Sporidia of *Cronartium ribicola*. *J. Agr. Research*, 34:497–510 (1927).

CHAPTER 11

Stem Diseases: Cankers of Conifers

The term "canker" is broadly used here to mean a disease that causes the death of definite and relatively localized areas of bark on branches or trunks of trees. Strictly speaking, however, repeated callusing is necessary before a lesion can be classed as a canker (Mooi 1948). Although most cankers are of fungus origin, they can be caused by noninfectious agents such as frost (Day and Peace 1934:47), sunscald (page 38), or illuminating gas (May, Walter, and Mook 1941). Death of the bark and cambium is followed by death of the underlying wood, although the causal organism may or may not penetrate the wood. Not only are there some canker diseases that do not ordinarily kill the bark to the cambium, but there are also superficial lesions caused by pathogenic organisms that have been thwarted in their progress by host reactions. Such superficial cankers do little or no damage. Infection may sometimes result only in roughening the bark. For example, rough bark on aspen is abnormal, arising after invasion by fungi, principally *Macrophoma tumefaciens;* mechanical injuries and lichens occasionally appear to be responsible (Kaufert 1937). With those cankers that penetrate to the cambium, sooner or later the bark sloughs off exposing the wood. Cankers may be either annual or perennial, the latter usually being most destructive and conspicuous. With annual cankers (Fig. 92) the agency causing the disease is operative for one season only, the injured tissues then being sloughed off or grown over by a single callus in the same way that a mechanical injury is healed. With bark cankers caused by fungi, Bier (1959) found that the canker grew during the dormant season for the host, but that canker development ceased during the growing season when the relative turgidity of the bark was 80 per cent or higher. With perennial cankers the causal agency functions year after year. If caused by a fungus that is able to penetrate vigorously through living tissues during the growing season of the host, little or no callus is formed, and therefore the stem, no matter how large, is finally girdled. The canker may be either sunken or slightly swollen, but ridges of callus will not be in evidence. Occasionally advance of the mycelium may be checked by the formation of impenetrable cork layers in the bark. Then the stem recovers and the lesion is healed over. If caused by a fungus that grows during the period when the host is dormant but whose development ceases during the growing season while the host forms callous tissue, a perennial canker of long standing is produced that may persist indefinitely on a stem without ever girdling it. Repeatedly the new callous tissue that the host develops around the advancing border of the lesion is killed, until the canker consists of more or less regular concentric ridges around the point of first infection, the so-called "concentric" or "target" canker (Fig. 106). Sometimes these callous ridges are irregular or broken in outline, resulting in a misshapen canker swollen in some places and sunken in others. These cankers of long standing often result in an eccentric or even a crooked bole owing to the killing of the diseased region and the continued growth of the healthy portion of the trunk. Although unusual, a single fungus may cause the formation of relatively rapid-growing cankers without ridges of callus or slower growing cankers with prominent ridges.

Fungi capable of causing long-standing perennial cankers have certain aggregate characteristics that distinguish them from other pathogens (Hiley 1919:33). They kill living tissue either by direct hyphal penetration or by secreting some substance toxic to cells in their vicinity so that the hyphae can penetrate the dead cells. The mycelium grows primarily in the inner bark and cambium, but, if the hyphae penetrate the wood at all, they are unable to spread extensively and reinfect the inner bark; otherwise the hyphae would spread through the stem and girdle it immediately. The mycelium is perennial and spreads slowly from cell to cell, with its rate of circumferential extension approximately equal to the rate of circumferential growth of the stem, for, if the mycelium grew more rapidly, it would soon girdle the stem. Its radial growth is faster than that of the stem, otherwise it would never reach the cambium.

Trees with one to many cankers of the persistent type commonly maintain their position in a stand, showing no apparent reduction in height or diameter growth, even though only a small portion of the trunk is left alive back of the cankers. Such trees frequently break off at a canker, or decay may enter the open wound. The section of a bole occupied by a canker is worthless because of malformation of the stem and discoloration of the wood.

Generally, no matter how old a cankered stem may be, infection occurred when the stem was young. Consequently it is necessary to recognize cankers in their incipiency and remove trees with trunk infections early. If this is done, a mature stand reasonably free from trunk cankers can be expected, even though the causal organism remains abundant in the locality. Of course, this method of control is not effective against a virulent introduced disease such as Endothia canker of chestnut. Branch cankers are not serious unless so many branches are cankered and killed that the crown is much reduced.

Tympanis Canker of Pines. *Tympanis confusa* Nyl. (Ascomycetes, Dermataceae), as defined by Groves (1952), causes this disease (Hansbrough 1936). The canker is sporadic on planted red pine, but it has never been found on either planted or naturally reproduced trees within the optimum natural range of this species, although the fungus is common there as a saprophyte. Cankers occur on both naturally reproduced and plantation-grown trees of eastern white pine within its natural range.

The canker occurs so infrequently on eastern white pine that it causes no damage. On red pine, cankers are confined to the main stems of trees 20 to 30 years old and have occurred in sufficient abundance so as to cause some loss. A few trees are girdled and killed; a few not girdled break off at the point of infection; but the most serious effect is decay of the sapwood and heartwood by secondary fungi. Most cankered trees are infected by wood-destroying fungi entering through the cankers so that even if the trees live to reach merchantable size, the boles will be worthless. In two of the most severely cankered plantations in southern Connecticut, the loss as expressed by the degree of understocking was approximately 5 and 6 per cent.

On red pine the disease has occurred only in plantations south of the optimum natural range of the species. It was generally worse on poorer sites, and more trees in the lower crown classes were infected than in the upper. In southern Connecticut there was a close relation between its outbreak in 1930 and the drought of that year. In New York and Michigan the infected plantations were stagnant, being badly in need of thinning. This disease has appeared only when the vigor of the host has been reduced by some environmental factor or complex of factors. The effects of drought or other unfavorable factors on red pine are probably accentuated south of the optimum range of the tree by the warmer climate. Recurring outbreaks of the disease can be expected, and it is possible that this disease may be a limiting factor in the artificial southward extension of the range of red pine.

The cankers, ranging from 3.5 to 36.0 inches long and from 1 to 8 inches wide, are confined to the main stem on red pine and always center at a branch node or whorl (Fig. 85). Cankers occur at heights of from 3 to 16 feet above ground with the majority ranging between 7 and 10 feet. They are annual, developing during one dormant season of the host, and then enlargement ceases. Consequently stems are rarely girdled.

Owing to the scaly bark, sharp scrutiny is necessary to detect cankers at first. Within a few years the dead bark cracks, sloughs off, and exposes the underlying wood. Then the cankers are easily detected. The underlying wood is resin impregnated, giving it a brownish tinge.

Fig. 85. A canker on the main stem of a red pine caused by *Tympanis confusa*. (*After D. V. Baxter.*)

Pycnidia and apothecia, either single or grouped, develop on the dead bark of cankers (Fig. 86). The dull to shiny black, spherical to cylindrical or conical pycnidia, up to 1 to 2 mm in height and width, develop less frequently. The dull-black apothecia, often with a shiny black hymenial surface or disc, are slightly larger than the pycnidia. Although the canker develops for one season only, the fungus does not die, because fresh fructifications have been collected on cankers originating 7 years previously. Why the mycelium can kill the bark during 1 year only is not known, but since the mycelium retains its viability for a few years at least, there must be some host response that prevents further invasion of

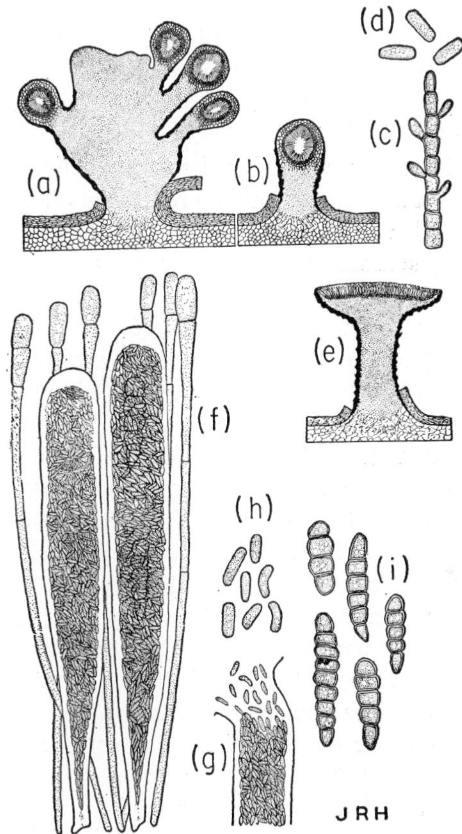

Fig. 86. *Tympanis confusa,* the causal organism of Tympanis canker of red pine. (*a*), medial longitudinal section of imperfect stage showing several pycnidial cavities in one stroma; (*b*), medial longitudinal section of a single pycnidium; (*c*), conidiophore with immature conidia attached; (*d*), mature conidia; (*e*), median longitudinal section of an apothecium of the perfect stage; (*f*), asci and paraphyses. The asci are filled with innumerable small secondary ascospores, while two large primary ascospores are very indistinct in the ascus on the right; (*g*), ruptured ascus with free secondary ascospores; (*h*), secondary ascospores; (*i*), primary ascospores. (*After J. R. Hansbrough.*)

living bark. Infection apparently occurs from dead side branches where the fungus lives as a saprophyte and then grows down into the living main stem.

Since the incidence of Tympanis canker on red pine is higher in pure than in mixed stands, planting red pine with eastern white pine in alternate strips of about three rows each should be valuable. In pure red pine plantations 8-foot rather than 6-foot spacing or less should increase vigor and lessen the likelihood of infection. Because infection occurs through dead branches killed by shading, pruning crop trees should be timed so that most of the branches will still be living when cut off. An 8-foot spacing will increase the probability of most branches being alive when the trees reach pruning size. Thinning is necessary to maintain stand vigor above the level at which crop trees are susceptible to infection, and an 8-foot spacing will delay the time of the first thinning until the trees to be removed are larger and more marketable. Furthermore, the wide spacing coupled with timely thinning may protect the stand against drought, since in dry seasons more soil moisture is present in thinned than in unthinned stands.

Atropellis Cankers of Pines. Several species of *Atropellis* (Ascomycetes, Tryblidiaceae) are associated with cankers of pines (Lohman and Cash 1940) characterized by a grayish-green to blue-black discoloration of the wood beneath each lesion (Fig. 88).

Atropellis pinicola Zeller & Goodding (1930) is constantly associated with a canker of young western white, sugar, and lodgepole pines on the Pacific Coast, particularly in the Pacific Northwest, with the first named being most commonly infected. It has been reported once on planted eastern white pine in Oregon. The canker is prevalent in western Oregon and Washington but occurs only occasionally in the western white pine region. Branches are girdled and killed, being easily mistaken at a distance for "flags" caused by white pine blister rust, but infection is never sufficiently severe to cause the death of trees. The perennial cankers, which rarely if ever occur on the main stem, appear on western white pine as smooth, elongated, flattened depressions covered with bark. The black apothecia on the dead bark of the cankers are from 2 to 4 mm in diameter and either are sessile or have a very short central stalk. The clavate asci, interspersed with longer hairlike paraphyses, each contain eight hyaline, needlelike to slightly club-shaped one-celled ascospores. Infection apparently occurs through uninjured bark or leaf scars.

Atropellis piniphila (Weir) Lohman & Cash (*Cenangium piniphilum*) is associated with cankers of lodgepole and ponderosa pines on the Pacific Coast and in the Southwest. It occurs rarely on whitebark and western white pines. It has been reported once on loblolly pine in Alabama, on Virginia pine in Tennessee, and on jack pine. Young stands of lodgepole and ponderosa pines from 5 to 25 years old are most damaged, not by death of the trees but by deformation of main stems and branches. Infection usually occurs at branch whorls, and a tree may have one or more cankers at every whorl on the trunk. Pruning wounds on ponderosa pine are readily infected. The disease is worst in overcrowded pure stands. The cankers appear as elongated, flattened depressions covered with bark and with a copious exudation of resin (Weir 1921). The apothecia, from 2 to 5 mm in diameter, develop on the cankered bark. They have a short central stalk and are black on the outside with a brownish to black disc. The clavate asci, interspersed with longer filamentous branched paraphyses, each contain eight hyaline oblong to ellipsoid ascospores. The dark-brown mycelium penetrates the bark and sapwood, chiefly following the rays. The affected wood is bluish-black in color.

Atropellis arizonica Lohman & Cash, occurring on cankered ponderosa pine, is known only from the type locality, Safford, Arizona. It is closely allied with *A. piniphila.*

Fig. 87. Canker caused by *Atropellis tingens* on the main stem of a shortleaf pine sapling.

Atropellis tingens Lohman & Cash is constantly associated with a canker of hard pines both native and exotic that is widely distributed from the New England and Lake states to the Gulf states (Diller 1943). Slash pine is the most susceptible species. The disease is primarily one of saplings, quickly girdling smaller branches and causing perennial cankers of the target type on the larger branches and main stems (Figs. 87, 88). Cankers develop slowly. Usually after 10 years canker extension stops, and the lesions heal over. The disease is one of long standing, since a canker 42 years old was found in 1934 on shortleaf pine. The disease was epidemic during 1933 and 1934, but even then only an occasional suppressed seedling in overstocked stands of natural reproduction was killed.

The small black stalkless apothecia that appear on the bark give a characteristic green color reaction when mounted in a 5 per cent aqueous solution of potassium hydroxide, which separates this fungus from associates with similar fructifications. Infection apparently occurs through the needles or through the base of the needle fascicles, since a fascicle of dead needles is commonly present in the center of small incipient cankers while adjacent needles are still green and healthy.

Atropellis apiculata Lohman, Cash, & Davidson (1942) is associated with cankers on the main stems of small Virginia pines in North Carolina and Virginia. It is probably widespread but infrequent. The apothecia give a 5 per cent solution of potassium hydroxide a chocolate-brown color.

Pitch Canker of Pines. This disease of several hard pines is widespread in the South, particularly on Virginia pine. It is caused by *Fusarium*

Fig. 88. Cross section through a canker caused by *Atropellis tingens* on a shortleaf pine stem showing the bluish-black discoloration in the wood.

lateritium Nees em. S. & H. f. *pini* Hept. Branches and leaders of young trees are killed by girdling. Copious resin flow is induced from the constricted lesion on the stem, the underlying wood of which is resin-soaked (Berry and Hepting 1959). No fruiting bodies have been found in nature. Removing diseased trees or pruning diseased branches is suggested for local control.

Larch Canker. This disease is caused, in part at least, by *Dasyscypha willkommii* (Hart.) Rehm (*Ascomycetes, Helotiaceae*). The nomenclature of the fungus has been confusing (Hahn and Ayers 1934:73; Robak 1952; Manners 1953). It causes serious losses to European larch in Europe, where it has been extensively investigated (Gaisberg 1928; Hiley 1919:16–79; Münch 1936; Plassman 1927). It was discovered at two localities in eastern Massachusetts during 1927 on European larch introduced some time previously from Great Britain in a diseased condition. Japanese larch also imported from Great Britain was lightly cankered. Fortunately these infections were not close to natural stands of eastern larch. All infected trees were removed. The fungus was found again in

Fig. 89. Canker on the trunk of a European larch caused by the larch canker fungus.

1935 and in 1952 on a few trees of European larch in the same locality (Fowler and Aldrich 1953).

Japanese larch is relatively resistant, whereas eastern larch and European larch are susceptible, although different races of European larch vary widely in susceptibility (Münch 1933:521). The fungus also attacks western, Dahurian, Siberian, and golden larches. It has not been found on eastern or western larches in North America.

In Europe, young trees up to 5 or 6 years old are frequently killed by girdling, but older trees persist in spite of the canker and show no reduction in growth. Susceptibility to canker decreases with increasing age, and infection usually ceases when trees are between 30 and 40 years old. In severely infected plantations practically every tree may have several cankers scattered along the trunk, making the stand worthless. Cankers also occur on branches, but these would die from shading anyhow.

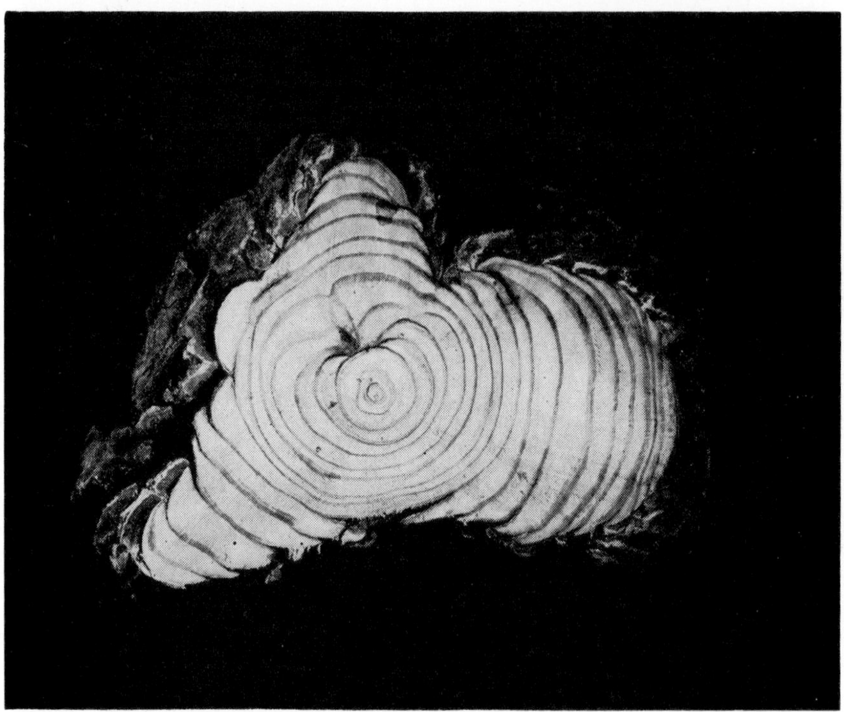

Fig. 90. Cross section through a European larch stem deformed by three cankers caused by the larch canker fungus.

The perennial canker (Figs. 89, 90) appears as a flattened depression on the stem, much swollen on the flanks and the opposite side. The neighboring bark is usually somewhat cracked, dark in color, and with a heavy exudation of resin. In old large cankers the wood is exposed by falling away of the dead bark.

The cup-shaped apothecia, appearing on the bark of the canker usually near the edge, vary from 3 to 6 mm in diameter and have a short stalk not over 1 mm long (Fig. 91). The apothecia are white and hairy on the outside, but the disc or fruiting surface within the cup is apricot orange at first, later changing to a light buff. The asci, containing eight hyaline, unicellular ascospores in a single row, are interspersed with longer thread-like paraphyses. Apothecia may be found throughout the year. Pycnidia are produced in whitish mycelial cushions, preceding the apothecia, but the pycnidiospores do not cause new infections. The manner of infection by the wind-borne ascospores is uncertain, but frost wounds appear to

Fig. 91. Apothecia of the larch canker fungus (*Dasyscypha willkommii*) on the stem of a European larch. The side view of one apothecium shows the short stalk. (*Courtesy of G. G. Hahn.*)

be the principal infection court. Cankers are generally initiated before a stem has passed the age of 10 years. Hyphae occur in both the bark and wood.

According to Day (1958) larch canker is caused by frost, but canker extension can be aided by several fungi, including *D. willkommii.* However, it has been demonstrated that cankers can be initiated by *D. willkommii* in the absence of frost injury (Hahn and Ayers 1943; Manners 1957).

Larch canker in Europe causes negligible damage if suitable races of larches are established on sites that favor vigorous growth. However, European larch is an outstanding example of the difficulties attendant on introducing tree species to new habitats. In the United States the reappearance of the disease in eastern Massachusetts in 1935 and in 1952 after removal of the infected trees found in 1927 indicates that the fungus is tenacious, so that periodic scouting for the disease is necessary in order to find and eliminate new infections. Otherwise the parasite may become established on eastern larch with the possibility of slow northward spread into Canada and finally westward to western larch.

Dasyscypha Canker of Pines. This canker is caused by *Dasyscypha pini* (Brunch.) Hahn & Ayers (1934:479), which has also been referred to as *D. fuscosanguinea* Rehm, a European saprophyte not yet found in North America. The fungus is native to high altitudes and north latitudes in North America and Scandinavia. Its hosts are western white, whitebark, and eastern white pines. Scotch pine is attacked in Norway and Sweden. In the West it is restricted to high elevations, being most prevalent on western white pine at from 4,500 to 6,000 feet, which is near the upper altitudinal limits of this tree. It may also attack trees at lower elevations on unfavorable sites. The fungus can girdle and kill branches and small main stems, but apparently it cannot girdle a stem over 4 inches in diameter. Damage, being mostly restricted to trees at high elevations or on poor sites, is not serious in the aggregate, and more than one canker on a large trunk is rare.

The cankers are similar to larch canker with the same heavy resin flow (Stillinger 1929). A canker may persist on the butt of a tree for 200 years without apparently retarding growth of the tree. The cup-shaped apothecia, appearing on the bark of the cankers, vary from 2 to 4 mm in diameter and have a short stalk. The apothecia are cinnamon brown and hairy on the outside, and the disc is bright orange, later changing to yellow ocher. The asci, containing eight hyaline, unicellular, elongate elliptical or occasionally pyriform ascospores in a single row, are interspersed with longer threadlike paraphyses. Pycnidia are unknown. The presumably wind-borne ascospores infect through living needles or needle fascicles, around the base of which the cankers always begin formation. Cankers are initiated during the first 4 or 5 years in the life of the stem. Trees with trunk cankers should be removed as early as possible in the life of a stand.

Several other *Dasyscyphas* occur on white pines as saprophytes. *D. arida* (Phil.) Sacc., which is widespread in the West, resembles *D. pini* macroscopically. *D. agassizii* (B. & C.) Curt., common on blister rust lesions on eastern white pine (Snell 1929), and *Dasyscypha* sp., common on blister rust lesions on western white pine (Bingham and Ehrlich 1943), both have whitish apothecia.

Dasyscypha Canker of Balsam Fir. *Dasyscypha resinaria* (Cke. & Phil.) Rehm was found causing cankers on the main stem and branches of balsam fir at one locality in northern Minnesota in 1896 (Anderson 1902).

The fungus also attacks Douglas fir in Norway (Jørstad 1931:88), causing similar lesions. According to Anderson, younger stems and branches of balsam fir were girdled and killed. Both the bark and wood became thickened by action of the fungus, resulting in a much swollen, roughened, cankerous lesion centering as a rule around the base of branches killed by shading on the lower part of the main stem. Infection seemingly occurred at the base of these dead branches. The cup-shaped apothecia, which did not appear on the bark of the swollen cankers until after the stem had been killed, were from ½ to 1 mm in diameter and had short stalks. The apothecia were whitish and hairy on the outside, and the disc was slightly orange colored.

Dasyscypha Cankers of Douglas Fir. When larch canker was discovered in eastern Massachusetts, it was feared that Douglas fir was also being attacked, since diseased Douglas firs were found growing in close association with cankered larches, and *Dasyscypha willkommii* had been erroneously reported as parasitic on Douglas fir in Europe (Hahn and Ayers 1938). It was shown (Hahn and Ayers 1934:167) that *D. ellisiana* (Rehm) Sacc., indigenous to the eastern United States, was associated with the diseased Douglas firs, although whether the fungus is entirely responsible for the disease is problematical. *D. ellisiana* is also weakly parasitic on limber, ponderosa, and Swiss stone pines. The lesions on Douglas fir may involve 10 to 15 feet of the basal trunk of trees from 5 to 12 inches dbh. Smaller trees rarely show the disease. In addition to copious resin flow from the diseased bark, under which there may be extensive pitch pockets, the trunk may have numerous swellings giving it a gouty appearance. Lower branches frequently show identical symptoms. The bark is not killed, and therefore an open lesion does not result. Fructifications of *D. ellisiana* commonly appear on the bark swellings and limitedly on normal bark. The short-stalked apothecia are from 0.5 to 2 mm in diameter, white and hairy on the outside, and the disc is light orange-yellow to chrome.

Dasyscypha pseudotsugae Hahn (1940) is associated with open cankers on main stems and branches, up to 1 inch in diameter, of suppressed saplings in California and Oregon. These cankers are from 2 to 3 inches or more in length and much swollen on the flanks. The apothecia are pale orange and hairy on the outside with an orange-colored disc. The same fungus is associated with lesions on the main stem of saplings and poles on poor sites in western British Columbia. The cankers occur on the first 30 feet of the trunk and vary from a few inches to several feet in length. The shorter ones appear like larch cankers, the longer ones are irregular in outline with little or no swelling of the trunk, and sometimes with the bark intact.

Phomopsis Canker of Douglas Fir. This canker, caused by *Phomopsis lokoyae* Hahn (Fungi Imperfecti, Sphaerioidaceae), occurs on Douglas fir in northern California, Oregon, and British Columbia (Boyce 1933; Hahn 1933; Thomas 1950). The canker closely resembles one in Europe caused by *Phomopsis* (*Phacidiopycnis*) *pseudotsugae* Wilson (Hahn 1957). On the Pacific Coast the disease is confined to Douglas fir saplings, usually on mediocre sites, and outbreaks occur following unfavorable conditions for the host, such as drought. Damage is negligible.

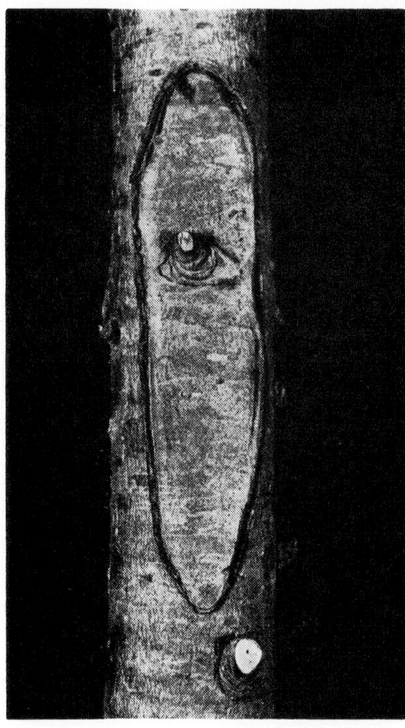

Fig. 92. A canker on the main stem of a young Douglas fir caused by *Phomopsis lokoyae* showing the base of the small branch through which infection probably occurred.

Cankers occur on both main stems and branches with relatively smooth bark. Stem cankers are usually narrowly elliptical in shape (Fig. 92) and 3 to 8 inches long by 1 to 2 inches wide, although they may attain a length of 30 inches or more. These cankers are generally found on the first 10 feet of the main stem. Occasionally the fungus kills back the end of a leader or of a branch, and short shoots on the main stem or branches are commonly killed. The cankers develop during the dormant season of the tree and grow for one dormant season only; consequently, if a stem is not completely or almost completely girdled during that period, the canker is healed over, leaving little trace of its presence (Fig. 93), because only the bark and cambium are killed, the sapwood not being directly attacked.

Tiny black pycnidia break through the bark of cankers, killed-back leaders, and branches during the spring or summer. They may also develop on the bark of saplings dead from other causes. The two kinds of pycnidiospores are presumably wind borne, and infection apparently takes place through young shoots because each canker invariably centers around a dead shoot.

Other Phomopsis Cankers. *Phomopsis pseudotsugae* Wilson, now called *Phacidiopycnis pseudotsugae* (Wilson) Hahn, which is the imperfect stage of *Phacidiella coniferarum* Hahn (1957), is widespread in western Europe. It causes dieback and cankering of introduced Douglas fir and Japanese larch, and occasionally of native European larch. The symptoms are similar to those caused by *Phomopsis lokoyae* on Douglas fir in its

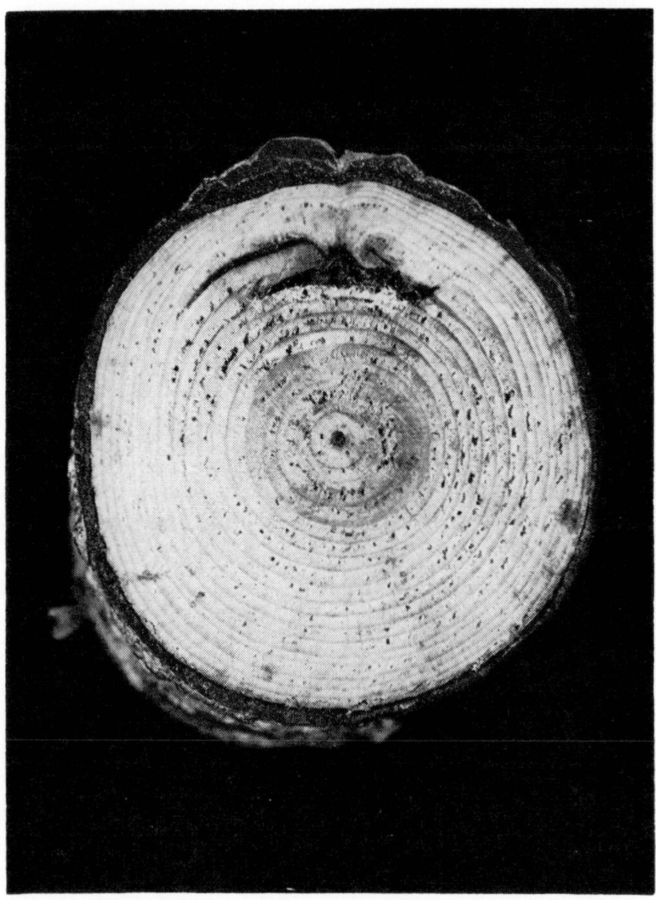

Fig. 93. Cross section through the main stem of a young Douglas fir showing a canker caused by *Phomopsis lokoyae* that has been completely healed over.

natural habitat. *P. pseudotsugae* also sometimes occurs on other conifers in Europe. Occasional small saplings of eastern white pine in the northeastern United States are killed by *P. pseudotsugae*. Bark cankers are formed and the trees are girdled. The fungus also colonizes the bark of the girdling basal constrictions on young eastern white pines caused by mound-building ants (Pierson 1922). Douglas fir is not attacked. The fungus is prevalent in New Zealand, causing dieback and canker of introduced hard pines (Birch 1935).

In northern Idaho, *Phomopsis boycei* Hahn (1930:72) causes cankers on saplings and poles of lowland white fir that girdle and kill many branches during some seasons. Cankers also girdle the main stems of small saplings. The canker is characterized by a constriction of the stem that may be accompanied by a marked swelling at the base of the canker where the dead tissues join the living. The reddish-brown needles of dead branches stand out in striking contrast to the glossy dark-green living foliage. Alpine fir is affected in the same way, but to a lesser degree, by *P. montanensis* Hahn (1930:61). In Europe, *P. abietina* (Hart.) Wilson & Hahn (1928) causes similar injury to silver fir. In the Northeast similar dead branches killed by the balsam sawyer (*Monochamus marmoratus*) are common on balsam fir.

Scleroderris Canker of Balsam Firs. In the West, *Scleroderris abieticola* Zeller & Goodding (Ascomycetes, Tryblidiaceae) is associated with an annual canker on lowland white and Pacific silver firs that occurs on twigs, branches, and trunks of saplings (Zeller and Goodding 1930). Canker development begins in the autumn and stops as soon as the cambium becomes active in the spring, so that, if a stem is not completely girdled in this time, the dead bark falls away during or after the second year and the wound heals over. Twigs and small branches only are girdled, and therefore the damage is insignificant. During the summer the black apothecia, which are from 0.5 to 1.2 mm in diameter and have a short central stalk, develop on the dead bark. Each clavate ascus contains eight colorless, elongated, several-celled hyaline spores. Infection seems to occur through uninjured bark or leaf scars.

Aleurodiscus Canker of Balsam Firs. *Aleurodiscus amorphus* (Pers.) Rab. (Basidiomycetes, Thelephoraceae) causes cankers on the main stem of, and occasionally kills suppressed saplings of, balsam firs (Hansbrough 1934). Branches are not attacked. However, the fungus occurs commonly as a saprophyte on the main stems and branches of dead balsam firs and many other conifers. Although widespread as a saprophyte, it occasionally becomes parasitic. The cankers are narrowly elliptical in shape with a raised border; the light-colored fructifications of the fungus develop abundantly on the dead bark of the cankers, which always center around a dead branch (Fig. 94).

Cephalosporium Cankers. *Cephalosporium* sp. (Fungi Imperfecti, Moniliaceae) causes an inconspicuous canker on the main stems of balsam firs up to 8 inches in diameter in northern Minnesota and Wisconsin (Christensen 1937). The cankers are oval to elliptical in shape, the dead bark is slightly sunken at the border, and resin exudes from broken blisters on the face of the canker. The diseased inner bark is brown, and the brown discoloration may extend into the wood for one or two growth rings. Another *Cephalosporium* causes a similar canker, which appears to be annual, on western hemlock in British Columbia (Denyer 1953). Both cankers occur on suppressed trees, causing slight damage. Fructifications of both fungi are known only from cultures.

Caliciopsis Canker. *Caliciopsis pinea* Peck (Ascomycetes, Coryneliaceae) causes a superficial canker (Ray 1936) that is widespread in the East on eastern white pine, occurring on the main stems of suppressed saplings and on shaded branches. An occasional suppressed sapling seems to succumb. The fungus is also found on mountain, pitch, short-leaf, and Virginia pines, and in Oregon on Douglas fir. On eastern white pine the cankers are sharply marked depressed areas in the bark, with a reddish-brown color and a smoother surface than the surrounding normal bark. The lesions vary from small to several inches in diameter. Occasionally cankers just below branch whorls show a decidedly roughened bark. The small, globose, clustered black pycnidia and the stalked perithecia, which appear as black slender cylindrical bristles or spines about 2 mm high, arise from stroma in the cankered bark (McCormack 1936).

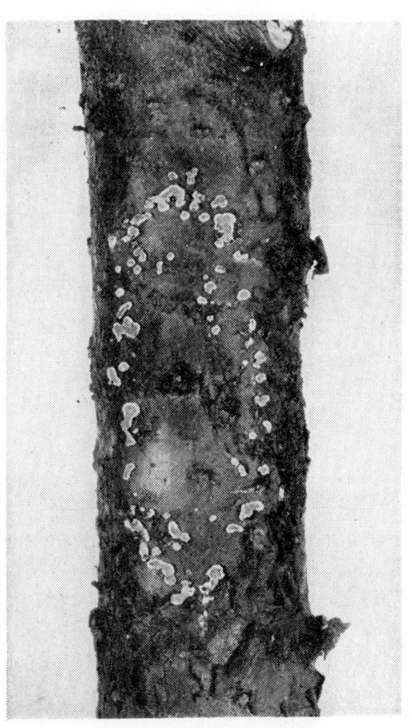

Fig. 94. A canker with fructifications of *Aleurodiscus amorphus* on the main stem of a balsam fir sapling.

Chondropodium Canker. *Chondropodium pseudotsugae* White (Fungi Imperfecti, Sphaerioidaceae) is associated with superficial cankers on the main stems of small saplings of Douglas fir in Oregon and British Columbia (White 1936). The outer layers of bark are killed over circular or elliptical areas about ¼ to ¾ inch in diameter. The pycnidia appear as short, thick, blunt black spines arising from the dead bark. The lesion does not reach to the cambium, and therefore infected trees appear to be as thrifty as uninfected, even though a single stem usually has many cankers.

Coryneum Canker of Cypress. From the region of San Francisco Bay to southern California, Monterey cypress in scattered localities is seriously affected by a canker caused by *Coryneum cardinale* Wagener (Fungi

Imperfecti, Melanconiaceae), the perfect stage of which is *Leptosphaeria sp.* (Hansen 1956). The host, the natural range of which is restricted to the Monterey Peninsula in California, is not important as a forest tree but is widely used for hedges, windbreaks, and ornamental plantings

Fig. 95. Canker on the trunk of a living lowland white fir caused by *Fomes pini.* Conks of the fungus are abundant along the edge of the canker. The tree is approximately 3 feet in diameter, and the canker is about 5 feet long. (*After C. E. Owens.*)

in the coastal region of California. The origin of the disease is unknown, but it has increased and spread rapidly since its discovery in 1927 (Wagener 1939). Branches or portions of the top of infected trees die until the entire tree is killed or becomes so unsightly as to necessitate removal. The dying results from the girdling action of cankers and is accompanied by heavy resin flow from the diseased area. Branch cankers are usually less than 1 foot long, but the trunk cankers may be longer. The tiny blackish pustulelike fructifications of the fungus are irregularly scattered over the surface of the dead discolored bark of the cankers. Cankers should be cut out and destroyed, followed immediately by spraying the surrounding host trees with a fungicide, particularly covering the twigs, branches, and trunks.

Cytospora Cankers. *Cytospora* spp. (Fungi Imperfecti, Sphaerioidaceae) and accompanying perfect forms in the genus *Valsa* are sometimes associated with cankers on conifers first weakened by another unfavorable agent. In eastern Oregon after a severe drought in 1929, many Douglas fir saplings partially died back, and trunk cankers with *Cytospora* sp. on the bark were present on some of the affected trees. *C. kunzei* Sacc. caused pitch-girdle canker of drought-weakened Douglas fir in Colorado (Wright 1957). It also occurs on Engelmann spruce (Hawksworth and Hinds 1960). In the East *C. kunzei* occurs on cankers of spruces and other conifers (Waterman 1955). *C. friesii* Sacc. has been found colonizing fire-injured Douglas firs in western British Columbia (Dearness and Hansbrough 1934:128). In the northern Sierra Nevada of California, *C. abietis* Sacc. causes cankering and death of branches of red and white firs that have been weakened by other factors (Wright 1942).

Cankers and Wood-destroying Fungi. *Fomes pini* (Thore) Lloyd is associated with cankers on lowland white fir in Oregon (Owens 1936:147). Elliptical-shaped, more or less sunken or flattened areas, frequently being several feet in vertical extent and having a swollen margin, are formed around knots or branch stubs (Fig. 95). Over these areas many conks are produced, being most abundant near the margin of the cankers. The fungus works outward from the heartwood, killing the sapwood and inner bark. This fungus also causes large swellings around knots on western larch, and the decayed portion of the trunk often becomes irregular in shape so that it appears somewhat cankered, indicating attack on the sapwood (Boyce 1930:44). *Fomes hartigii* Allesch. usually kills the sapwood and cambium of western and eastern hemlocks so that a trunk canker may result (page 408).

REFERENCES

Anderson, A. P.: *Dasyscypha resinaria* Causing Canker Growth on *Abies balsamea* in Minnesota. *Bull. Torrey Botan. Club,* 29:23–34 (1902).

Berry, C. R., and G. H. Hepting: Pitch Canker of Southern Pine. *U.S. Dept. Agr., Forest Serv., Forest Pest Leaflet,* 35:1–3 (1959).

Bier, J. E.: The Relation of Bark Moisture to the Development of Canker Diseases Caused by Native, Facultative Parasites I, II, III. *Can. J. Botany,* 37:229–238, 781–788, 1140–1142 (1959).

Bingham, R. T., and J. Ehrlich: A Dasyscypha Following *Cronartium ribicola* on *Pinus monticola*. I, II. *Mycologia*, 35:95–111, 294–311 (1943).

Birch, T. T. C.: A Phomopsis Disease of Conifers in New Zealand. *New Zealand State Forest Serv. Bull.*, 7:1–30 (1935).

Boyce, J. S.: Decay in Pacific Northwest Conifers. *Yale Univ., Osborn Botan. Lab. Bull.*, 1:1–51 (1930).

Boyce, J. S.: A Canker of Douglas Fir Associated with *Phomopsis lokoyae*. *J. Forestry*, 31:664–672 (1933).

Christensen, C. M.: Cephalosporium Canker of Balsam Fir. *Phytopathology*, 27:788–791 (1937).

Day, W. R.: The Distribution of Mycelia in European Larch Bark, in Relation to the Development of Canker. *Forestry*, 31:63–86 (1958).

Day, W. R., and T. R. Peace: The Experimental Production and the Diagnosis of Frost Injury on Forest Trees. *Oxford Forestry Mem.*, 16:1–60 (1934).

Dearness, J., and J. R. Hansbrough: *Cytospora* Infection Following Fire Injury in Western British Columbia. *Can. J. Research*, 10:125–128 (1934).

Denyer, W. B. G.: *Cephalosporium* Canker of Western Hemlock. *Can. J. Botany*, 31:361–366 (1953).

Diller, J. D.: A Canker of Eastern Pines Associated with *Atropellis tingens*. *J. Forestry*, 41:41–52 (1943).

Fowler, M. E., and K. F. Aldrich: Re-survey for European Larch Canker in the United States. *U.S. Dept. Agr., Plant Disease Reptr.*, 37:160–161 (1953).

Gaisberg, Elizabeth von: Beiträge zur Biologie des Lärchenkrebspilzes. *Mitt. württemb. forstl. Versuchsanst*, (1928).

Groves, J. W.: The Genus *Tympanis*. *Can. J. Botany*, 30:571–651 (1952).

Hahn, G. G.: Life-history Studies of the Species of *Phomopsis* Occurring on Conifers. Part I. *Brit. Mycol. Soc. Trans.*, 15:32–93 (1930).

Hahn, G. G.: An Undescribed *Phomopsis* from Douglas Fir on the Pacific Coast. *Mycologia*, 25:369–375 (1933).

Hahn, G. G.: Dasyscyphae on Conifers in North America. IV. Two New Species on Douglas Fir from the Pacific Coast. *Mycologia*, 32:137–147 (1940).

Hahn, G. G.: Phacidiopycnis (Phomopsis) Canker and Dieback of Conifers. *U.S. Dept. Agr., Plant Disease Reptr.*, 41:623–633 (1957).

Hahn, G. G., and T. T. Ayers: Dasyscyphae on Conifers in North America. I. The Large-spored, White-excipled species. II. *D. Ellisiana*. III. *Dasyscypha pini*. *Mycologia*, 26:73–101, 167–180, 479–501 (1934).

Hahn, G. G., and T. T. Ayers: Failure of *Dasyscypha willkommii* and Related Large-spore Species to Parasitize Douglas Fir. *Phytopathology*, 28:50–57 (1938).

Hahn, G. G., and T. T. Ayers: Role of *Dasyscypha willkommii* and Related Fungi in the Production of Canker and Die-back of Larches. *J. Forestry*, 41:483–495 (1943).

Hansbrough, J. R.: Occurrence and Parasitism of *Aleurodiscus amorphus* in North America. *J. Forestry*, 32:452–458 (1934).

Hansbrough, J. R.: The Tympanis Canker of Red Pine. *Yale Univ., School of Forestry, Bull.*, 43:1–58 (1936).

Hansen, H. N.: The Perfect Stage of *Coryneum cardinale*. [Abstract.] *Phytopathology*, 46:636–637 (1956).

Hawksworth, F. G., and T. E. Hinds: Cytospora Canker of Engelmann Spruce in Colorado. *U.S. Dept. Agr., Plant Disease Reptr.*, 44:72 (1960).

Hiley, W. E.: "The Fungal Diseases of the Common Larch." New York: Oxford University Press (1919).

Jørstad, I.: Innberetning fra statsmykolog Ivar Jørstad om soppsykdommer på skogtraerne i årene 1926–1930. *Skogdirektoren Innberetning om det Norske Skogvesen,* 1930:78–96 (1931).

Kaufert, F.: Factors Influencing the Formation of Periderm in Aspen. *Am. J. Botany,* 24:24–30 (1937).

Lohman, M. L., and Edith K. Cash: Atropellis Species from Pine Cankers in the United States. *J. Wash. Acad. Sci.,* 30:255–262 (1940).

Lohman, M. L., Edith R. Cash, and R. W. Davidson: An Undescribed *Atropellis* on Cankered *Pinus virginiana. J. Wash. Acad. Sci.,* 32:296–298 (1942).

McCormack, Helene W.: The Morphology and Development of *Caliciopsis pinea. Mycologia,* 28:188–196 (1936).

Manners, J. G.: Studies on Larch Canker I. The Taxonomy and Biology of *Trichoscyphella willkommii* (Hart.) Nannf. and Related Species. *Brit. Mycol. Soc. Trans.,* 36:362–374 (1953).

Manners, J. G.: Studies on Larch Canker II. The Incidence and Anatomy of Cankers Produced Experimentally Either by Inoculation or by Freezing. *Brit. Mycol. Soc. Trans.,* 40:500–508 (1957).

May, C., J. M. Walter, and P. V. Mook: Rosy Canker of London Plane Associated with Illuminating-gas Injury. *Phytopathology,* 31:349–351 (1941).

Mooi, J. C.: "Kanker en takinsterving van de wilg. Veroorzaakt door *Nectria galligena* en *Cryptodiaporthe salicina.*" [English summary.] Proefschrift, Univ. Amsterdam (1948).

Münch, E.: Das Larchenrätsel als Rassenfrage. *Tharandt. forstl. Jahrb.,* 84:437–531 (1933).

Münch, E.: Das Lärchensterben. *Forstwiss. Centr.,* 58:469–494, 537–562, 581–590, 641–671 (1936).

Owens, C. E.: Studies on the Wood-rotting Fungus *Fomes pini.* I. Variations in Morphology and Growth Habit. *Am. J. Botany,* 23:144–149 (1936).

Pierson, H. B.: Mound-building Ants in Forest Plantations. *J. Forestry,* 20:325–336 (1922).

Plassmann, E.: "Untersuchungen über den Lärchenkrebs." Neudamm: J. Neumann (1927).

Ray, W. W.: Pathogenicity and Cultural Experiments with *Caliciopsis pinea. Mycologia,* 28:201–208 (1936).

Robak, H.: Om saprofyttiske og parasittiske raser av lerkekreftsoppen, *Dasyscypha willkommii* (Hart.) Rehm. [On Parasitical and Saprophytical Strains of the Larch Canker Fungus, *Dasyscypha willkommii* (Hart.) Rehm.] [English summary, pp. 188–204.] *Medd. Vestlandets Forst. Forsøkssta.,* 9:117–237 (1952).

Snell, W. H.: *Dasyscypha Agassizii* on *Pinus strobus. Mycologia,* 21:235–242 (1929).

Stillinger, C. R.: *Dasyscypha fuscosanguinea* Rehm on Western White Pine, *Pinus monticola* Dougl. *Phytopathology,* 19:575–584 (1929).

Thomas, G. P.: Two New Outbreaks of *Phomopsis lokoyae* in British Columbia. *Can. J. Research C. Botan. Sci.,* 28:477–481 (1950).

Wagener, W. W.: The Canker of *Cupressus* Induced by *Coryneum cardinale* n. sp. *J. Agr. Research,* 58:1–46 (1939).

Waterman, Alma D.: The Relation of *Valsa Kunzei* to Cankers on Conifers. *Phytopathology,* 45:686–692 (1955).

Weir, J. R.: *Cenangium piniphilum* n. sp., an Undescribed Canker-forming Fungus on *Pinus ponderosa* and *P. contorta*. *Phytopathology*, 11:294–296 (1921).

White, W. L.: A New Species of *Chondropodium* on *Pseudotsuga taxifolia*. *Mycologia*, 28:433–438 (1936).

Wilson, M., and G. G. Hahn: The Identity of *Phoma pitya* Sacc., *Phoma abietina* Hart. and Their Relation to *Phomopsis pseudotsugae* Wilson. *Brit. Mycol. Soc. Trans.*, 13:261–278 (1928).

Wright, E.: *Cytospora abietis*, the Cause of a Canker of True Firs in California and Nevada. *J. Agr. Research*, 65:143–153 (1942).

Wright, E.: Cytospora Canker of Rocky Mountain Douglas-fir. *U.S. Dept. Agr., Plant Disease Reptr.*, 41:811–813 (1957).

Zeller, S. M., and L. N. Goodding: Some Species of *Atropellis* and *Scleroderris* on Conifers in the Pacific Northwest. *Phytopathology*, 20:555–567 (1930).

CHAPTER 12

Stem Diseases: Cankers of Hardwoods

There are more serious canker diseases of hardwoods than of conifers in the United States, but the types of cankers and their manner of formation are essentially the same for both groups of trees (page 238). Besides the cankers that are well known, others, of less importance, with which fungi are associated continue to be found.

Black Knot of Cherry. This disease is caused by *Dibotryon* (*Plowrightia*) *morbosum* (Schw.) T. & S. (Ascomycetes, Dothideaceae). It occurs on cherry and plum trees in orchards, and it also commonly occurs on wild species, only one of which, black cherry, has ever been of any commercial importance in the lumber industry. On twigs and small branches the fungus causes the formation of elongated black swellings commonly confined to one side of the stem and several times the diameter of the normal stem. Small twigs may be killed within a year after infection, but larger branches usually resist attack for several years. On black cherry large cankerous swellings (Fig. 96) 2 or more feet long occur on the trunks of large trees, and where several such lesions are scattered along the bole, the tree is worthless for lumber. Since practically all new infection originates on twigs of the current year, invasion of the trunk or larger branches takes place from a small infected lateral (Koch 1933, 1935:413). Black cherry with cankers on the trunk should be removed from a stand during the course of improvement work.

Didymosphaeria Canker of Alder. *Didymosphaeria oregonensis* Goodding (Ascomycetes, Sphaeriaceae) is constantly associated with this canker of the trunks and branches of mountain, red, and Sitka alders in the Pacific Northwest (Goodding 1931). Young trees only are affected, some being deformed and stunted by numerous cankers, but the aggregate damage is slight. The cankers appear as bands of roughened bark around the stems, sometimes accompanied by a pronounced spindle-shaped swelling. The bands vary from ½ inch to 2 or more feet in length. Cankers develop only on young stems, ceasing to grow after the bark has become hard and thick.

Dothichiza Canker of Poplar. This canker is caused by *Dothichiza populea* Sacc. & Briard (Fungi Imperfecti, Excipulaceae), the perfect stage of which is *Cryptodiaporthe populea* (Sacc.) Butin (1958). The disease, found sporadically in the eastern United States and Canada since

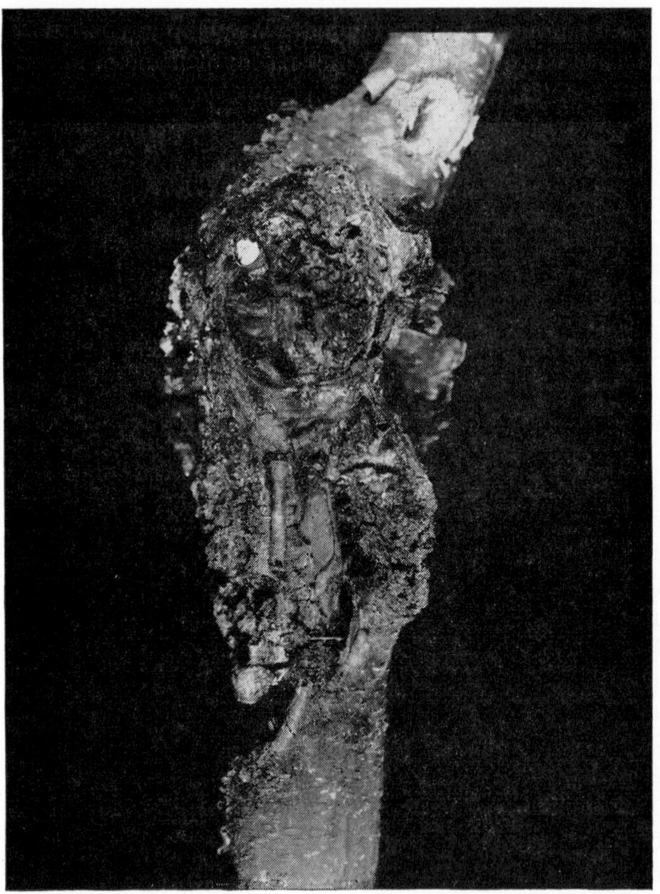

Fig. 96. Cankerous swelling on the trunk of a black cherry tree caused by *Dibotryon morbosum.*

1915 (Waterman 1957), has been known in Europe since 1884, where it is prevalent and injurious (Schmidle 1953*b*; Butin 1957). It was probably introduced into North America from Europe long before its discovery here. The many poplar species, varieties, and hybrids on which it occurs vary from resistant to susceptible. *Populus deltoides* and *P. robusta* are particularly susceptible, whereas *P. canescens* is resistant.

The disease is primarily one of young or newly planted trees and of trees that have been weakened by other factors such as low spring temperatures, transplanting in the autumn, poor drainage, drought, wounding by pruning or by soil cultivation, and infertile soil. It has not yet been

Fig. 97. Cankers on a poplar caused by *Dothichiza populea.*

found in natural forest stands. The virulence of the disease makes it a serious threat to planting poplars.

The fungus attacks the trunk, limbs, and twigs of its host, causing the formation of cankers, which are frequently found around the base of twigs and limbs. The development of cankers varies with the host. In general they first appear as slightly sunken areas with the diseased bark a shade darker in color than healthy bark. The long axis of the elongated cankers parallels the long axis of the stem. The bark is killed to the cambium, the sapwood is invaded, and a brown discoloration results there. Cankers develop during the dormant period for the host. If a stem has been completely encircled it dies, but if not the formation of callous tissue begins immediately, and by July or August the canker may be entirely grown over, in some cases forming a noticeable swelling on the bark. The canker resumes development the following year, finally unhealed cracks appear in the bark (Fig. 97), and the tree dies, although a tree with numerous cankers on the trunk may live for several years. Below the dead limbs and cankered areas water sprouts appear abundantly and grow rapidly, but they in turn are killed, thus destroying the symmetry of the tree, even if it is not finally killed.

In April and May, pycnidia form in the diseased bark as numerous hemispherical pustules about the size of a pinhead, each one with a hole at the top through which the spores extrude in sticky cream- or amber-colored cylindrical tendrils. On drying the tendrils change to a walnut-brown color. The tiny one-celled spores are hyaline, smooth, and subglobose to ovoid, or rarely ellipsoid, in shape. They are probably washed down by rain and carried by insects or birds, and possibly, after the spore tendrils dry, some spores are wind borne. Natural infection probably occurs through wounds and leaf scars.

Infected stock in nurseries and plantations must be destroyed. No stock, even though apparently healthy, should be moved from a nursery known to be infected to a locality where the disease does not occur. Pruning should be avoided if possible, and care exercised to prevent other wounding. Tools used in handling diseased stock should be sterilized each time before using. In Europe the favorable period for artificial inoculation is from September to March, which indicates that spraying with lime-sulfur in the fall and possibly in the spring might be helpful.

Septoria Canker of Poplar. This disease, caused by *Septoria musiva* Pk. (Fungi Imperfecti, Sphaerioidaceae), is widespread (Waterman 1954). The perfect stage is *Mycosphaerella populorum* Thompson. Native poplars within their natural range are not cankered, but certain native hybrids, particularly those with black, balsam, and cottonwood parentage, and certain exotic hybrids, especially those known as "Russian poplar," are highly susceptible. The fungus causes a leaf spot of little consequence on both hybrids and native poplars. Hybridization of poplars to develop rapidly growing trees is widely practiced.

Incipient cankers center around wounds, lenticels, petioles, and stipules. Leading and side shoots are soon girdled. The infection spreads from lateral branches into the main stem, where cankers develop that ultimately girdle and kill the stem. New branches are formed on the stem, and sucker shoots that may develop into trees arise from the stump. After 12 or more years' growth, trees are from 12 to 20 feet high, much branched, and with from one to three stems arising from the original root system. The value of plantings can be completely destroyed. Pycnidia, in which the hyaline, cylindric, straight or slightly curved, septate spores are borne, develop on both cankers and leaves. Control of the disease depends on planting native poplars only or resistant hybrid clones. Dormant cuttings from an infected locality should be surface-sterilized by immersing in a 1 per cent suspension of Semesan (30 per cent hydroxymercurichlorophenol) in water for 20 minutes before shipment into localities where susceptible clones are growing or where the disease does not occur (Ford and Waterman 1954).

Cytospora Canker of Poplar and Willow. This canker is caused by *Valsa sordida* Nit. (Ascomycetes, Valsaceae), although the fungus has usually been discussed under the name of its conidial or imperfect stage, *Cytospora chrysosperma* (Pers.) Fr., a poorly defined species (Schmidle 1953*a*; Schreiner 1931). *V. nivea* (Hoff.) Fr. (*C. nivea*) causes a similar canker, which is found occasionally. Poplars and willows are most commonly affected, but the disease occurs occasionally on mountain ashes, maples, cherries, and elders. In the Southwest, white poplar and eastern cottonwood are highly susceptible. Valley cottonwood is resistant.

Although the fungus is a normal inhabitant of the bark of poplar, willow, and mountain ash trees it may become parasitic if the host is weakened. Under forest conditions it is of no consequence except in stands on an unfavorable site or injured by drought, frost, or fire. Tree vigor is the most important factor in resistance. In the West, poplars and willows are of minor economic importance, but they are valuable shade and ornamental trees, especially in semiarid regions. Heavy losses occur among trees that have been weakened by neglect or by drought and also among trees that have been pollarded. It is to cuttings in nurseries and plantations that the most serious damage is done. In propagating beds, losses up to 75 per cent or more of the cuttings may occur.

The disease appears as lesions or cankers on the trunks and large limbs of affected trees. Small branches and twigs are usually killed without a definite canker being formed. Cankers are developed by gradual killing of the diseased bark in more or less circular areas. On smooth-barked shoots young infections can be recognized by the presence of brownish shrunken patches. The diseased area may be fairly regular or quite irregular in shape, gradually enlarging until the stem is girdled. Water sprouts often develop profusely just below a canker on a large stem; most of them are soon killed. When old trunks and large branches with rough bark are attacked, the typical cankered appearance is seldom developed, and it is consequently difficult to recognize the diseased condition until spore tendrils appear. The diseased inner bark blackens and emits a disagreeable odor. The cambium is killed, the sapwood is also infected, becoming watery and reddish brown in color, and the heartwood is sometimes discolored. Severely attacked trees from 3 to 6 inches dbh die in 2 or 3 years. On white poplar the disease functions more as a dieback, killing trees branch by branch.

The fungus often enters dead tips of twigs or small branches, gradually kills them back to their juncture with a larger branch or with the trunk, and then causes the development of a more or less circular canker on these larger stems. The pycnidia appear on the dead bark as small, somewhat

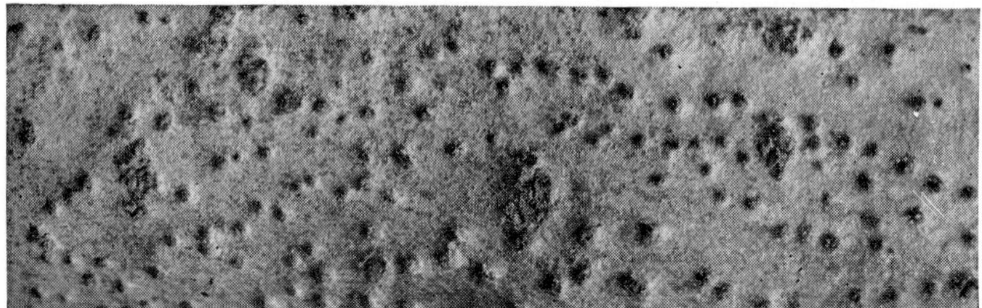

Fig. 98. Pycnidia of *Cytospora chrysosperma* on the bark of a white poplar. (*Reproduced from J. Agr. Research* 13, *no.* 6.)

pimplelike pustules (Fig. 98). The pycnidial stromata develop in the bark, then break through as convex protrusions, exposing a gray to black disc with a single central pore. The minute sausage-shaped hyaline conidia (pycnidiospores) develop in vast numbers on the ends of short hyphae which line the inside walls of the pycnidia. The sticky conidia ooze out through the pores of the pycnidia during moist weather in yellowish, reddish, or reddish-brown long coiled spore masses termed "spore horns," "spore tendrils," or "cirri" (Fig. 111). When there is an excess of moisture, the spore masses are globular or spread out. The perithecia, found infrequently in this country except on aspen, are developed deeper in the bark in groups occupying an indistinct stroma. The flask-shaped perithecia have long necks that push through the surface of the bark, their ostioles usually being arranged more or less circularly around the edge of a blackened or dirty-gray disc. The narrow club-shaped asci contain eight hyaline sausage-shaped ascospores, about twice the size of the pycnidiospores.

Infection of living stems takes place through wounds or dead twigs. Dissemination probably occurs by means of pycnidiospores carried by wind, insects, and birds. Since ascospores occur infrequently in this country it is unlikely that they are important in disseminating the disease.

Because the disease is confined to weakened trees, the most important measure of control is to maintain high vigor of individual trees and of stands. Drought favors the disease, and consequently during dry summers ornamental poplars and willows must be well watered. Wounding should be avoided; consequently pruning, and above all pollarding, are poor practices. In the West, valley cottonwood, since it has shown resistance, should be favored for planting in semiarid regions. More information is needed on the susceptibility of various poplars.

Successful establishment of plantations largely depends on securing a vigorous stand immediately after planting. Trees should be planted with care on a suitable site. Cuttings should not be less than ¼ inch in diameter, and rooted trees with poor root systems should be discarded. Plant immediately after logging and clearing of weed trees, so that competition will not be too severe for the poplars. Sod land is unfavorable for planting.

Cuttings in storage should be kept at a temperature below 35°F. If heeled in, losses are likely to be high or complete. In nursery beds, cuttings should be completely buried so that cut ends will not be exposed to infection. The disease is readily introduced into new localities on infected nursery stock, and consequently stock should not be procured from nurseries where the disease is known to occur.

Other Cytospora Cankers. There are other Cytosporas or related fungi parasitic on hardwoods about which little is known. *C. pulcherrima* Dearn. & Hansbr. has been found attacking fire-injured trees in British Columbia, particularly willow and mountain alder (Dearness and Hansbrough 1934). *C. ambiens* Sacc. is associated with a canker of American elm in Illinois (Harris 1932:52), and *Cytosporina ludibunda* Sacc. causes a dieback and canker of the same tree (Carter 1936). *Cytospora annularis* commonly causes cankers on twigs and branches of ash in Iowa (Reddy 1934).

Neofabraea Canker of Poplar. *Neofabraea populi* Thompson (Ascomycetes, Mollisiaceae) causes this canker of aspen, largetooth aspen, and balsam poplar in Ontario (Thompson 1939). Trees 3 to 6 years old and usually not over 1.5 inches in diameter are affected near the base. A few trees are killed. The cankers first appear as small depressed areas in the bark. Older cankers are elliptical in outline, from 4 to 6 inches in length and with vertical splits in the bark. The small flesh-colored to light-brown disc-shaped apothecia develop on the bark of the canker.

Hypoxylon Canker of Poplar. *Hypoxylon pruinatum* (Klotsche) Cke. (Ascomycetes, Xylariaceae) causes this disease (Anderson 1956). The disease is widely distributed from Alberta east to Nova Scotia, and south to Minnesota and Massachusetts. It also occurs in Colorado (Davidson and Hines 1956). Aspen and largetooth aspen are most commonly attacked; balsam poplar rarely. Infected trees are finally girdled and killed, or the trunk may break off at the canker before girdling is completed.

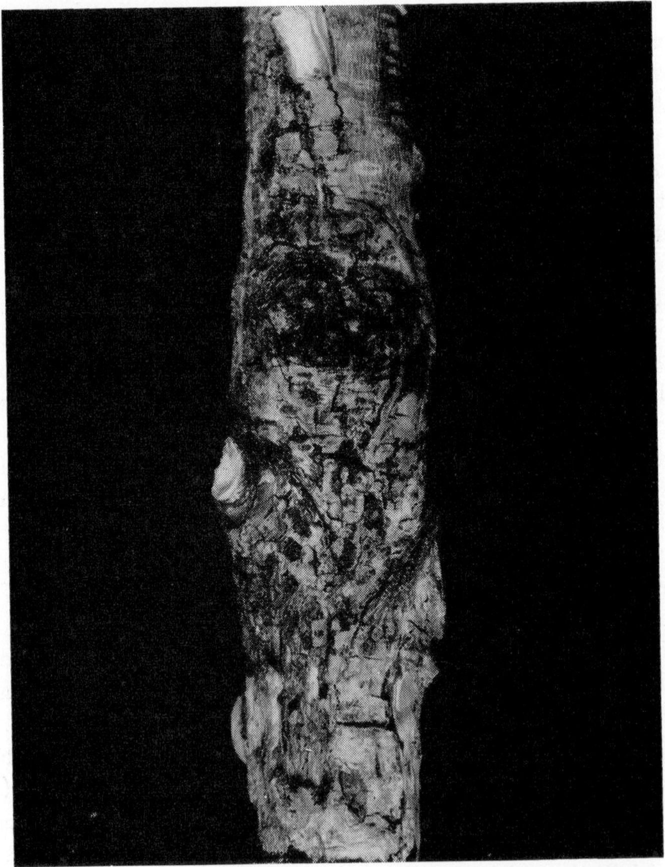

Fig. 99. Canker on the trunk of an aspen caused by *Hypoxylon pruinatum*.

Stands from 15 to 40 years old suffer most. Understocking strongly favors cankering. The disease causes severe losses in aspen. Stands have been reported in which the mortality reached 70 per cent.

The parasite causes conspicuous cankers on the trunk (Fig. 99), which generally attain a length of 3 or more feet before the tree is girdled and killed. The first evidence of the disease is small yellowish-orange slightly sunken areas with irregularly lobed advancing margins, centered around some wound. These areas increase in size until they coalesce to form a canker, delimited by vertical cracks. As the canker develops, the affected bark becomes mottled, taking on a grayish color in which occur black patches owing to the superficial periderm flaking off to expose the blackened cortex. A brownish sap flow may occur at the canker margin. The advancing margin is irregular and wavy and yellow to reddish brown in color.

Conidial fructifications appear as blisters of various sizes at the end of the first or during the second growing season of the cankers. The fungus produces hyphal pegs or pillarlike structures that separate the superficial periderm from the underlying cortical tissue. The periderm ruptures, exposing a grayish layer of conidiophores and conidia solidly covering the outer surface of the cortex and sides of the pillars. The tiny conidia are one-celled, oblong to ovoid in shape, and hyaline, or brownish in mass. The first perithecial stromata appear when the cankers are 3 or more years old as erumpent, flattened, hard black layers covered by a white pruinose coat through which protrude the black ostioles of the sunken perithecia. The cylindrical asci each bear eight one-celled brown, oblong to elliptical ascospores much larger than the conidia. Infection seems to occur through wounds in the living bark.

Infected trees should be removed in thinning. Trunk cankers often occur high enough so that they are difficult to see when the trees are in leaf. Consequently, thinning should be done, or at least the trees marked for removal, after the leaves have fallen. The first thinning should be moderate enough so that the area can be thinned again in about 5 years and a satisfactorily stocked stand maintained. The second thinning is necessary to remove trees with incipient cankers. Cankering occurs least in uniformly well-stocked stands. Clear-cutting of severely infected stands and conversion to other species may sometimes be necessary.

Hypoxylon Canker of Maple. This canker on 10- to 15-year-old red and sugar maples in Ontario is caused by *Hypoxylon blakei* Berk. & Curtis (Bier 1939). The cankers, which are similar to Hypoxylon cankers on poplars, usually attain a length of 2 or more feet before the trees are girdled and killed. The fruiting stages of the fungus are similar to those of *H. pruinatum.*

Black Canker of Aspen. The cause of this canker is unknown, but it is probably infectious. In fact the canker resembles the one on aspen in Michigan figured by Povah (1935:pl. 21), caused by *Nectria galligena* Bres. The disease is found on aspen (Meinecke 1929:4) in the Rocky Mountain region. Affected trees occur in scattered groups, many individuals having such badly cankered trunks that they are unmerchantable, and a few are killed. Larger branches may also be affected. Small trees may be girdled and killed by trunk cankers, but larger trees persist, with the elongated cankers showing a central dead area surrounded by a series of calluses that in turn have been killed, suggesting the parasitic nature of the disease. The old bark stands out raggedly from the calluses, and the canker has a dirty black color. Sometimes growth of the canker stops, and the lesion is healed over. Cankered trees should be removed from a stand.

Botryosphaeria Cankers. *Botryosphaeria ribis* Grossenbacher & Duggar (Ascomycetes, Sphaeriaceae) causes cankers on twigs, larger branches, and trunks of willows that may kill trees within a few years (Wolf and Wolf 1939). Smaller cankers on the trunk are craterlike in form and approximately ¼ to ½ inch in diameter. Up to a hundred may occur in a 1-foot length of stem. Large cankers several inches long with fissured bark may develop by the union of several smaller ones. In spring and early summer there is an exudate from the cankers. This fungus also causes cankers of other hardwoods (Luttrell 1950; Toole 1957; Wester, Davidson, and Fowler 1950).

Sphaeropsis Cankers. Species of *Sphaeropsis, Macrophoma, Diplodia,* and *Botryodiplodia* (Fungi Imperfecti, Sphaeriodaceae), genera so closely related that considerable confusion in naming species has resulted, are frequently associated with canker and dieback of hardwoods weakened by unfavorable environmental conditions (Marshall 1939). So far these cankers have been largely confined to shade and ornamental trees, becoming sporadically abundant, but occasionally damage occurs in forest stands. Oaks, particularly red and chestnut oaks, are subject to a canker and dieback in the East caused by *Sphaeropsis quercina* Cke. & Ell. (*Diplodia longispora* E. & E.), which is the imperfect stage of *Physalospora glandicola* (Schw.) Stevens. The disease has been reported on other oaks. Trees of all ages may be killed. The fungus primarily attacks small twigs and branches, passing from them to the larger branches and trunk, although apparently it may infect any portion of a tree through wounds. The twigs and branches die, the leaves wither and turn brown, and the infected bark becomes sunken and wrinkled, while small black globoid pycnidia break through the bark. Many water sprouts are produced below the dead part of the crown on the trunk and larger branches. On large stems the bark becomes sunken with a ridgelike callous growth bordering the dead areas. Sapwood under the dead bark is dark colored, and the mycelium extends for several inches longitudinally each way from the canker, causing a pronounced black streak. A similar disease caused by *S. malorum* Berk. occurs on various hardwoods. American elm is attacked by *S. ulmicola* E. & E. (Buisman 1931).

Since these fungi overwinter on cankers and dead twigs, producing a new crop of spores in the spring, all infected stems should be removed and burned during the summer when the disease is most easily recognized. Stems should be pruned at least 6 inches below the infected area. Badly infected trees should be completely removed. Fertilization, watering, and other measures to improve tree vigor are valuable in reducing the virulence of attack. These control measures are limited to shade and ornamental trees.

Thyronectria Canker of Honey Locust. *Thyronectria austro-americana* (Speg.) Seeler (Ascomycetes, Hypocreaceae) causes this canker of the smaller branches of honey locust (Seeler 1940; Crandall 1942). The fungus is distributed from Nebraska to Massachusetts and southward to the Gulf states, so the canker is probably also widespread. The slightly depressed bark cankers range from the size of a pinhead to about ½ inch in diameter, eventually enlarging or coalescing to girdle the branch. A gummy exudate occurs on many cankers. The underlying outer wood is streaked with reddish brown for several inches in each direction from a canker. Some trees die, apparently from the multiple branch infections, but there is considerable natural resistance, since many trees remain healthy in close proximity to infected or dying ones.

Canker Stain of Plane. *Ceratocystis fimbriata* Ell. & Halst. f. *platani* Walter (Ascomycetes, Ceratostomataceae) causes this disease of London plane and sycamore or American plane (Walter 1950; Walter, Rex, and Schreiber 1952). It is essentially serious on shade and ornamental trees. It occurs in native stands of sycamore in remote localities, so it is likely that the fungus is a native that has become severely pathogenic on a hybrid and on an indigenous tree growing under unnatural conditions. The first external symptom on smooth bark is a dark-brown or black discoloration in lens-shaped or more elongated areas in line with the grain of the wood. On bark retaining the old plates and scales the first evidence is an elongate depression with the inner bark blackened and dead. During the first year, cankers usually are not more than 2 inches wide, but they may be 30 to 40 inches long. The cankers widen yearly, separate cankers coalesce, and finally the stem is girdled and killed. Once the trunk of a London plane is infected the tree is doomed, although a single infection may take 3 to 5 years to kill a stem 1 foot in diameter. Infected wood is reddish brown or bluish black in color, the discoloration being deepest in the rays. On cross section the stained wood appears in sectors delimited by the rays.

During moist periods from May to October, three kinds of infectious spores are produced in recently killed or cut wood surfaces of infected trees. Infection occurs through wounded bark; even a slight scratch in bark freshly exposed by exfoliation permits entry of the fungus. It is usually disseminated by pruning operations. The disease has not caused sufficient damage in forest stands to necessitate control. For shade and ornamental trees, infected trees or parts thereof must be removed, all unnecessary wounding must be avoided, and pruning tools, climbing shoes, ladders, and climbing ropes must be disinfected before being used on a healthy tree. The fungus can be transmitted in ordinary wound dressings, so a paint of the gilsonite-varnish type containing 0.2 per cent of phenylmercury nitrate should be used. This chemical is poisonous to man.

Phytophthora Cankers. *Phytophthora cactorum* (Leb. & Cohn) Schroet. (Phycomycetes, Peronosporaceae) is responsible for cankers on hardwoods commonly at the base of the trunk and important on shade and ornamental trees (Stuntz and Seliskar 1943; Wagener and Cave 1944). Bleeding canker of hardwoods, characterized by bark fissures at the root collar and upward at intervals on the trunk and branches from which oozes a watery light-brown to thick reddish-brown liquid, is caused by the same or a closely related fungus (Howard 1941; Miller 1941). A deep concentric canker of elms termed "pit canker" is caused by *P. inflata* Caroselli & Tucker (1949).

Hymenochaete Canker. This disease, caused by *Hymenochaete agglutinans* Ellis (Basidiomycetes, Thelephoraceae) is occasional in the East on young hardwoods, and in one instance it has been found on eastern white pine. Seedlings or saplings are attacked. When an infected dead stem comes in contact with a living one the fungus grows rapidly (Fig. 100), the mycelium forming a thin leatherlike fruit body encircling the living stem and binding it firmly to the dead one. The fruit body has a creamy-yellow margin shading into a deep rich-brown central portion. At the point of encirclement, the living stem becomes noticeably constricted, and in 2 or 3 years if the growth of the fungus continues the stem is girdled and dies above the point of infection. Hyphae penetrate the inner bark through lenticels and then invade the rays of the wood (Graves 1914). If the dead stem is separated from the living one before the latter is completely encircled, a sunken canker results on one side, which in time is overgrown by callous tissue and disappears. The disease is of no consequence under natural conditions, but when young hardwood stands are weeded it is advisable not to leave severed stems in contact with living ones of desirable trees.

Smooth Patch of White Oak. Smooth patch, smooth bark, or white patch of white oak (Lair 1946) has been attributed to insects, to *Aleurodiscus oakesii* (B. & C.) Cke. (Basidiomycetes, Thelephoraceae), and to *Corticium maculare* Lair (Basidiomycetes, Thelephoraceae). The irregularly circular, smooth light-gray sunken areas of bark, varying from a few inches to a foot or more in diameter, rarely occur higher than 20 feet on the trunk. They stand out in marked contrast to the rough, ridged, and darker gray normal bark. Fungous mycelium is confined to the dead bark,

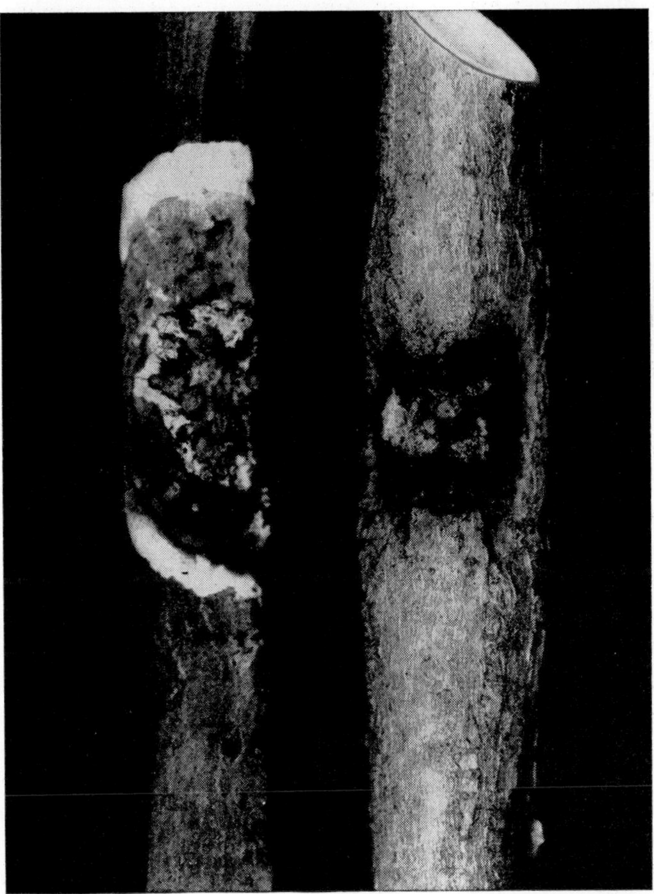

Fig. 100. *Hymenochaete agglutinans.* On the left a stem of smooth alder completely encircled by a fructification showing the light-colored margin. On the right a stem of mountain maple showing a canker with the remains of a fructification in the center.

which disintegrates somewhat with the outer layers flaking off. Infected trees are not injured.

Schizoxylon Canker. *Schizoxylon microsporum* Davidson & Lorenz (Ascomycetes, Stictidaceae) is associated with this trunk canker of red and sugar maples in the Lake states (Davidson and Lorenz 1938). It also occurs on American elm. The canker, which resembles Nectria canker, is of slight importance because it is rare and is usually confined to suppressed trees less than 3 inches dbh.

Eutypella Canker of Maple. *Eutypella parasitica* (Ascomycetes, Sphaeriaceae) is constantly associated with this trunk canker of boxelder, Norway maple, red maple, and sugar maple; most commonly the latter (Davidson and Lorenz 1938). The canker is known from the Lake states, the Northeastern states, and Quebec. Trees in the 1- to 3-inch diameter classes, especially those suppressed, are often killed. Trees over 5 inches dbh are seldom killed. Cankers persist on larger trees for many years, occasionally reaching a length of 5 feet. Such trees are liable to breakage at the canker. Cankers generally occur at 2 to 8 feet above ground, and usually there is one canker per tree. The disease is damaging only to stands containing a high proportion of sugar maple.

The cankers, which resemble Strumella cankers, are characterized by firmly attached bark with broad, slightly raised concentric rings of callous tissue. In the central portion of old cankers, perithecia develop, embedded in the bark, and their long black necks protrude slightly beyond the bark surface. The most characteristic feature is the heavy white to buff mycelial fans under the bark at the margin of the canker. A dead branch stub is usually present in the center of a canker, and initial infection apparently occurs when a tree is young.

Miscellaneous Cankers. *Diaporthe* (imperfect stage, *Phomopsis*) has been associated with a canker of elm in eastern Massachusetts (Richmond 1932) and of yellow birch in eastern Canada (Horner 1955). A similar *Phomopsis* is associated with cankers of American elm in Illinois (Harris 1932:45). *Coniothyrium* sp., *Diplodia ulmi* Dearn., and *Phoma* sp. also occur on elm cankers. *Coniothyrium* also plugs the water-conducting cells of the sapwood. Zonate canker of American elm apparently is a virus disease (Swingle and Bretz 1950). Associated with a prominent sooty-bark trunk canker of aspen in the Rocky Mountains is *Cenangium singulare* (Rehm) Davidson & Cash (1956). *Cucurbitaria elongata* (Fr.) Grev. has been found on small cankers of honey locust in Ohio (Diller and Davidson 1950). *Valsaria megalospora* Auersw. forms blisterlike swollen areas on the bark of red alder in Oregon (Diehl 1955).

STRUMELLA CANKER

This canker is caused by *Strumella coryneoidea* Sacc. & Wint. (Fungi Imperfecti, Tuberculariaceae). The disease is indigenous and has been known since 1899 (Heald and Studhalter 1914; Bidwell and Bramble 1934). It is widely distributed from the southern Appalachians north to Vermont and New Hampshire and westward through New York and

Pennsylvania. It occurs in Minnesota and the fungus has been recorded as a saprophyte in Oregon, in Missouri, and in Ontario, Canada.

Hosts. The canker occurs on many species of oaks, those belonging to the red oak group being most susceptible, and it was not uncommon on American chestnut. It is found occasionally on other hardwoods.

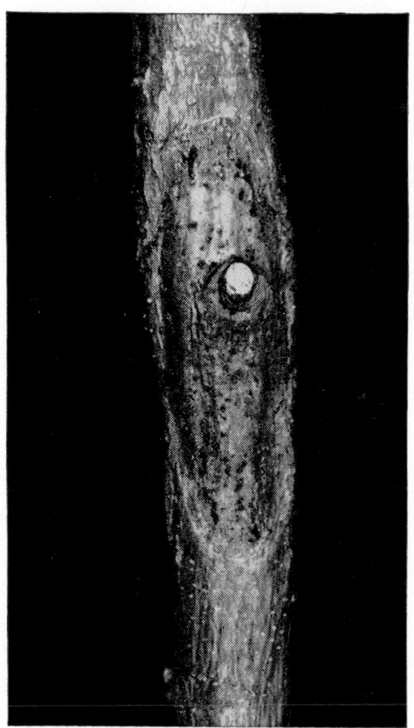

Fig. 101. A young Strumella canker on the main stem of a living red oak sapling. The infection entered through the dead branch stub in the center, and the sterile nodules can be seen on the dead bark of the canker.

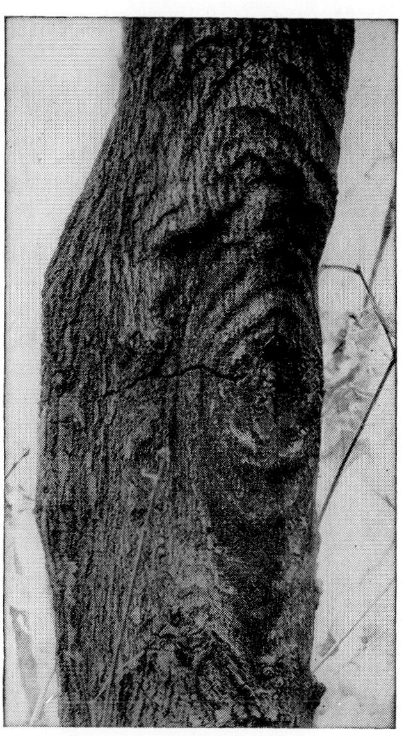

Fig. 102. An old Strumella canker on the trunk of a chestnut oak. The bark is still on the canker and the ridges of callus are plainly visible. (*Photograph by W. C. Bramble and H. E. Stork.*)

Damage. Although it is rare that more than 2 or 3 per cent of the trees in a stand are infected, the damage is of some consequence because infection is practically confined to the valuable oaks. Young trees may be killed, and there are records of severe damage to plantations (Fergus 1951). However, many infected trees survive to maturity, at which time the trunk may break off at the canker, or the most valuable portion of the trunk may be partially or completely worthless for lumber. The great majority of the cankers occur on the first 12 feet of the trunk, thus impairing the butt log.

Infection apparently enters the trunk from a small branch killed by the invading fungus (Sleeth and Lorenz 1945). Trees are usually attacked before attaining the age of 20 to 25 years, since by that time

Fig. 103. An old Strumella canker on red oak showing the distorted trunk. The bark has dropped away from the canker, and the wood is decayed. (*Photograph by C. B. Bidwell.*)

Fig. 104. Sporodochia of *Strumella coryneoidea* on the bark of oak. (*Photograph by C. B. Bidwell.*)

in a forest stand the branches on the lower boles of most oaks are dead and gone. Cankers grow slowly, requiring from 5 to 6 years to girdle stems 2 inches in diameter. Again trees with trunk cankers may live indefinitely, making normal height and diameter growth. Once infected, trees rarely heal out a canker. Trees in the dominant and codominant classes are attacked just as readily as suppressed ones.

Symptoms. The first noticeable symptom is a yellowish or yellowish-brown discoloration of the bark around the point of infection (usually a dead branch), commonly accompanied by a slight depression or raising of the tissue (Fig. 101). On rough bark these symptoms are difficult to detect. Under the outer corky layers of bark, whitish mycelium is present. As infection progresses two types of lesions can result, the target type and the diffuse type, with frequent intergradations. The target type results from the formation of successive ridges of callus in opposition to the slow growth of the fungus. These ridges form concentric circles around the base of the dead branch where the infection centers. Often there is a pronounced distortion or flattening of the cankered portion of the trunk (Figs. 102, 103). A canker may reach a size of 2 feet circumferentially and 5 feet long on a stem 9 inches in diameter. A slow decay of the sapwood takes place beneath the killed bark, and this decay is attributed to *Strumella coryneoidea* (Heald and Studhalter 1914:7). The diffuse type occurs when the fungus grows rapidly enough to girdle the trunk before callus formation takes place, and this type is usually found on small trees up to 3 to 4 inches dbh. A diffuse lesion may be depressed, but it is not zoned.

Water sprouts are often produced below the lesion, either abundantly or sparingly, after a tree has been completely or partially girdled. They in turn are killed by the fungus. When abundant, the sprouts are conspicuous for a considerable distance.

Causal Agency. Previous to the death of a stem, small black carbonaceous nodules made up of a mass of interwoven and anastomosed hyphae are formed on the dead bark of the canker (Fig. 101). These nodules bear no spores. Only occasionally are sporodochia developed on the dead bark of cankers on living trees. But after death of the host, sporodochia are produced abundantly on the dead bark of the lesion and away beyond its limits, frequently scattered for several feet or more along the dead stem. These sporodochia are dark-brown rounded powdery pustles about 1 to 3 mm in diameter (Fig. 104). They bear stout branched conidiophores, from the tips and sides of which irregularly globose to pyriform, brown spiny conidia are produced in abundance. The sporodochia remain on the bark for some years, blackening with age, but the number of years conidia continue to be produced is unknown, although new sporodochia apparently continue to appear year after year as long as the bark remains intact. The conidia are undoubtedly wind borne, but how infection occurs is unknown. Hyphae are found in both the bark and sapwood of lesions. Possibly the perfect stage is *Urnula craterium* (Schw.) Fr. (Davidson 1950).

Control. Control measures cannot be definite and detailed until the life history of the fungus is more fully understood. Meanwhile, since neither thinning nor attempted eradication have controlled the disease, the utilization of cankered trees and the selection of canker-free crop trees should be the objective in cankered stands (Roth and Hepting 1954).

NECTRIA CANKER

On most hardwoods this canker is caused by *Nectria galligena* Bres. (Ascomycetes, Hypocreaceae), but on yellowpoplar and magnolias it is caused by *N. magnoliae* Lohman & Hepting (Lohman and Watson 1943). The Nectria responsible for the elongated bark-covered cankers of sassafras has not been determined.

History and Distribution. Nectria canker probably occurs throughout the temperate zones. It is widespread in Europe and North America. In the United States the name "European canker" has been applied to it, but there is no evidence that the disease was introduced from Europe or elsewhere. The wide distribution of this slowly developing canker in North America indicates that it is native to this continent as well as to Europe. Although known in Europe prior to 1866, the canker was first reported in the United States in 1900 on apple trees and in 1905 on yellow birch in a forest stand. It is prevalent in the Pacific Northwest on pomaceous fruit trees and occurs occasionally there on native hardwoods. It is abundant in the Lake states, in the Northeastern states, and in the southern Appalachians on forest stands.

Hosts and Damage. In the East it can be assumed that Nectria canker will occur on all hardwoods, since it has already been reported on so many. Conifers are not affected. Largetooth aspen, American beech, red maple, and paper, sweet, and yellow birches are particularly susceptible, but the relative abundance of the canker on any hosts varies with the locality. Cross inoculations have shown that *Nectria* causing cankers in the East on one tree species may infect other species of hardwoods (Spaulding, Grant, and Ayers 1936:173). Cankered trees regardless of species, except yellowpoplar, sassafras, and magnolias, must be considered as a possible source of infection for all trees in a mixed hardwood stand.

Nectria canker is the most prevalent and serious canker disease of hardwood stands in the East. From the standpoint of timber production the major loss results from the reduction of merchantable volume. This reduction may be brought about by breakage in the crown or trunk at cankers or by cankers so located that they render the bole partially or completely unmerchantable. This disease does not ordinarily kill many infected trees. Even heavily infected and unmerchantable trees may remain in a stand and occupy space that could be utilized to greater advantage by healthy trees. Branch cankers, unless unusually numerous, do little harm.

In general, heavily cankered stands have been observed to occur frequently at high elevations where the hosts are affected by winter injuries and relatively poor growth conditions. At lower elevations groups of heavily cankered trees often occur in scattered spots. Stands of yellow

Fig. 105. A young Nectria canker on the trunk of a yellow birch.

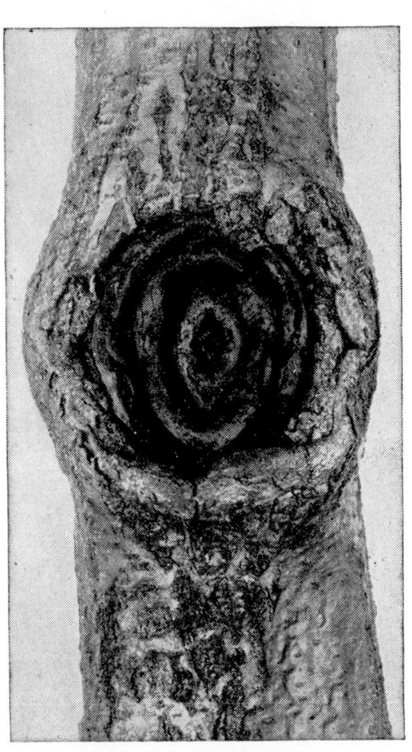

Fig. 106. Nectria canker of the concentric or target type on the main stem of a red oak sapling.

birch unmerchantable because of canker are not uncommon, and sweet birch and yellowpoplar also suffer severely.

Symptoms. The most obvious symptom of this disease is the cankers, which, however, vary considerably in appearance, canker development apparently being influenced not only by the host but also by the environment. Young cankers are easily overlooked, but it is most important that they be recognized so that infected trees can be removed before the cankers become large enough to reduce the value of the bole. The first indication of infection is small depressed or flattened areas of bark in the vicinity of small wounds or around the base of dead side twigs or branches (Fig. 105). These areas may have a darker color and a water-soaked appearance, particularly on yellow birch. Cracking of the bark or the development of callous tissue at the outer edge of the cankered area commonly makes these young cankers more apparent. Such cankers are

Fig. 107. Nectria cankers of the irregular type on sweet birch. The most valuable portion of the bole is unmerchantable. (*Photograph by R. P. Marshall.*)

generally found on young stems up to 10 to 20 years old, although they may be found on larger and older stems of birches.

The older and larger cankers are usually conspicuous. They may be concentric or target-shaped with the bark completely sloughed off, exposing the underlying regular ridges of callous tissue in the wood (Fig. 106). Other open cankers are irregular in shape with indefinite callous ridges

(Fig. 107). Cankers on sugar maple, and yellow and sweet birches are frequently almost circular in outline, those on oaks are often extremely irregular with the horizontal axis sometimes longest, and those on basswood are usually regular, elongate, and pointed at the ends. However, even elongate Nectria cankers do not attain the extreme ratio of length to width that commonly characterizes Strumella cankers. Again a covering of dead bark may remain over the cankered area, and even large cankers covered by bark are not conspicuous. Occasionally the cankered area is partially or completely covered by a roll of callus, indicating that the tree is overcoming the infection. Such healing over is often characterized by pronounced swelling. When a cankered area is completely grown over, it is hard to tell whether the callus is covering a wound or a canker. Decay rarely develops in the wood underlying cankers.

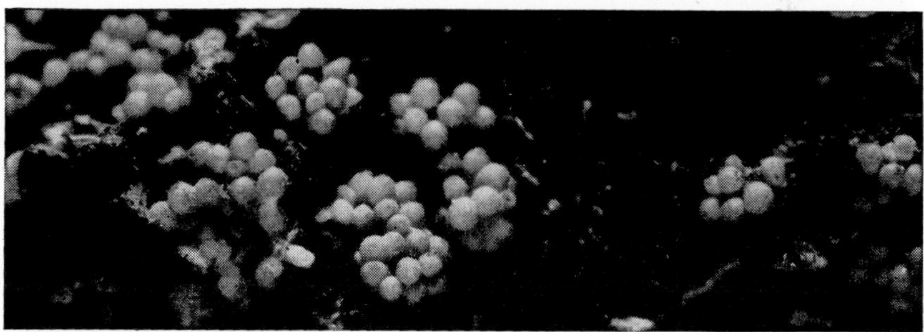

Fig. 108. Perithecia of *Nectria galligena* var. *major* on a black ash canker. About 15×.

On the bark or sometimes on the wood at the margin of the cankers the small red lemon-shaped perithecia may be found either singly or in clusters (Fig. 108). When abundant, they are easily detected because of the reddish color in mass, but, when sparse, they can be found only with the aid of a hand magnifier, since then they are likely to develop in the deeper fissures of the dead bark. However, perithecia are not present at all times on cankers, and on some lesions they develop sparingly or not at all.

Causal Agency. Two types of fructifications are produced. The sporodochia are minute creamy-white pustules on which are developed two types of spores; cylindrical macroconidia and ellipsoid microconidia. Sporodochia are rarely seen and then only on recently killed bark of young cankers protruding through the lenticels. The red perithecia, which darken with age, form during periods of moist weather. The perithecia contain cylindric to club-shaped eight-spored asci, the ascospores being hyaline, smooth, ellipsoid, and unequally two-celled. The ascospores are discharged during moist periods and may be forcibly expelled from the perithecia to a distance of 1 cm from the bark surface. The ascospores are largely disseminated by wind and rain.

Infection cannot occur through living unwounded bark. The fungus enters through unprotected wounds and small injuries, and leaf scars (Grant and Spaulding 1939). It enters through living or dying, but not dead, tissues. Cracks at the branch axils, commonly caused by snow or ice, are frequent points of entry for canker-forming Nectrias. Furthermore, these cracks occur when a tree is in the most susceptible condition, i.e., dormant or near-dormant. The association of branches with cankers emphasizes the fact that this disease usually attacks young trees or the younger parts of older trees.

The extent of the fungus development within a host varies, as is shown by the different symptoms that have been mentioned. There are numerous interrelated environmental factors that may affect the fungus and the host in different ways. In general the fungus enters through a crack in the bark. According to Ashcroft (1934:46), during the dormant period of the host any wound barely penetrating the corky outer layer is sufficient, but, for successful infection during the growing season, wounds must penetrate into the vicinity of the cambium or even to the wood. During the growing season the fungus entering a superficial wound would be cut off by the formation of an impenetrable corky layer before it could reach the wood and eventually would be sloughed off completely. Once established in the host tissues the hyphae are intercellular in the bark and intracellular in the wood, penetrating the latter to a depth of only 2 or 3 mm. The hyphae do not break down the cell walls of the wood but utilize the cell contents of the rays, wood parenchyma, and vessels. In the bark, however, the cell walls are more or less disintegrated. The mycelium is not found in or contiguous to living cells, indicating that the host cells are killed in advance of actual hyphal invasion. During the active growing period of the tree, the fungus on black walnut becomes a parasite in the wood only and is unable to reenter and spread in the living bark tissues until the tree again becomes dormant. The mycelium spreads slowly so that usually not more than ½ inch of tissue is killed annually around the edge of the lesion. Consequently, since callous formation also occurs annually, the chance of any but the smallest stems being girdled is remote, although in yellowpoplar killing is often more rapid. The mechanism of canker formation has been discussed by Ashcroft (1934:25). Canker formation in apple trees results from the effects of indole-β-acetic acid secreted by *N. galligena* on the host tissues, whereas *N. cinnabarina*, which does not cause true cankers, does not secrete this acid (Berducou 1956:102).

Control. The usual method of controlling losses due to this disease is to remove trees with trunk infections from a stand during improvement work (Grant and Childs 1940; Roth and Hepting 1954), except yellow-poplar, because vigorous yellowpoplars will heal out cankers. Therefore, infected dominant or codominant yellowpoplars may be selected for crop trees (Nelson 1940). To remove infected trees intelligently it is essential to recognize the incipient or young cankers and the inconspicuous bark-covered cankers, as well as the open cankers. It is best if the infected trees can be utilized, but if they must be left in the woods they may be felled or girdled. Trees with branch cankers but which are otherwise satisfactory need not be cut. It is economically impossible to completely eradicate Nectria canker from most heavily infected stands, although the removal of as many cankered trees as possible should somewhat reduce spore production by the fungus and subsequent infection of other trees. The fungus cannot be eliminated because it is so prevalent as a saprophyte. Particular care should be given to the treatment of young stands. Cankered trees should be removed, or at least canker-free crop trees should be selected, and, by continually discriminating against infected trees throughout the life of the stand, a final crop practically free from canker should result. Cankers no matter how large almost invariably originated when the stem was young. Since most infections occur at crotches or around the base of dead branches, the main portion of the bole of most hardwoods will be relatively safe from infection by about 30 years of age when the trees have pruned naturally, although an occasional infection will occur higher on the trunk, which may ultimately result in breakage and death of the tree. With yellow and sweet birches the continual development of twigs along the stems, even when old, exposes these species to trunk infections throughout most of their life.

In northern New England, some hardwood forests of pole size and of poor composition and density on inferior sites at high elevations are so severely cankered that conversion to conifers or to a heavy representation of conifers is the only method to assure satisfactory future timber production (Spaulding 1938). This can be done by releasing scattered naturally occurring conifers and by planting red, white, or Norway spruces in groups of 50 or more, with intervals between the groups of at least twice the diameter of the groups.

The present heavily cankered condition of many hardwood stands is probably the result of decades of mistreatment. Repeated cutting has progressively removed the better trees, leaving only the less desirable individuals, whether cankered or not, for the reserve stand and for seed trees. Fire has swept through these stands periodically. The cumulative effect of all this must be a present stocking of less vigorous trees per se and a reduction in site quality that still further reduces vigor of the stand. General observations indicate that for various hardwoods canker incidence is higher on unsatisfactory sites.

BEECH BARK DISEASE

This disease in North America is caused by the woolly beech scale (*Cryptococcus fagi* Baer.) followed by *Nectria coccinea* var. *faginata* Lohman, Watson & Ayers (Ehrlich 1934; Lohman and Watson 1943; Spaulding 1948). This disease, which has been known in northern Europe for many years where usually it is not virulent on European beech, is not caused by the insect acting alone (Thomsen, Buchwald and Hauberg 1949). In North America the disease was first recorded in Nova Scotia about 1920, although the insect, probably introduced from Europe, had been first noticed some 30 years before. The association of the fungus with the disease was not recorded until 1930. The fungus was probably also introduced. The disease, now widespread in the Maritime provinces of Canada and in Maine, has become epidemic on American beech, causing high mortality over extensive areas. It is spreading steadily, making the future of American beech uncertain, although it may be that because of distributional limits of either insect or fungus the disease will not spread throughout the entire range of its host.

Damage. In Canada after the disease had been present in a stand for 14 years, 20 per cent of the beech stems had been killed and 40 per cent of the volume, with no indications that mortality would stop. Large trees succumb most readily.

Symptoms. The insects, mostly in cracks or wounds on the bark, are covered by a woolly white down, easily visible to the naked eye. When an infestation is heavy, trunk and branches appear as though coated by snow. The fungus follows, killing the tissues on which the insect depends for food, so that the insects die, and in time most indications of their presence largely disappear.

Individual lesions in which the bark is killed to the cambium quickly show white sporiferous pustules (sporodochia) of the fungus breaking through the bark. These soon give way to clusters of tiny red, somewhat lemon-shaped, perithecia. When infection of a stem has been heavy, the individual lesions run together, so that large, irregularly shaped areas of dead bark result. Perithecia may then develop so abundantly that the trunk of the tree when wet is red with them. On younger trees and those in open stands where infection is lighter, the lesions remain circular and enlarge slowly if at all. Frequently the lesion becomes surrounded by a callus of healthy wood and bark, ultimately becoming a deeply depressed cavity somewhat cankerous in appearance but with the dead bark tightly appressed. Stems with many of these depressed cavities are pitted and gnarled.

Destruction of the living bark of the trunk results in wilting of the foliage followed by the progressive death of twigs and branches. Finally the entire crown succumbs, next the roots die, and the destruction of the living tree is complete.

Causal Agencies. The minute yellow larvae of *Cryptococcus fagi* establish themselves on the bark surface in the autumn, quickly protecting themselves with a woolly white down. The number of insects increases yearly. Each insect inserts its sucking organ (stylet) into the living bark, and, when a number of insects colonize a small area, groups of parenchyma cells are killed, which shrink on drying, causing minute cracks in the bark. Through these cracks Nectria enters. The mycelium develops periclinally in the bark and cambium, killing the infected tissues. Soon the white pustulelike sporodochia push out through the bark. These are covered with short conidiophores bearing the elongate slightly curved three- to nine-celled macroconidia. Next the perithecia appear in clusters. Each ascus contains eight two-celled, oval to elliptical ascospores that escape through the ostiole of the perithecium. When the ascospores have been discharged, the upper half of the perithecium collapses and sinks into the lower half. Both macroconidia and ascospores are probably largely carried from tree to tree by wind; rain is probably important, with insects playing a minor part, in distributing spores to other portions of the same tree.

Control. Ornamental or shade trees can be protected by spraying with dormant lime-sulfur; either 5 gallons of the liquid form in 95 gallons of water or 12 pounds of the dry form in 100 gallons of water (Crosby and Jones 1950). Thorough saturation by a strong stream of spray solution is important. Infection by *Nectria* does not generally occur when the insects are removed soon enough. In forest stands mortality was greater on steep slopes than on broad ridge tops and was greater among larger trees than among smaller ones, and the severity of infection was found to increase with an increasing proportion of beech in the stand. This suggests that beech should be favored on ridge tops and that selective cutting should aim to reduce the proportion of beech and to remove the larger trees, which succumb most readily.

CHESTNUT BLIGHT

This disease is caused by *Endothia parasitica* (Murr.) A. & A. (Ascomycetes, Sphaeriaceae). Except when small twigs are infected a typical blight is not produced, and consequently the name "Endothia canker" would be more appropriate. However, chestnut blight is the accepted term.

History and Distribution. The disease was first observed in the New York Zoological Park in 1904. The fungus is a native of Asia, probably introduced into the United States on nursery stock. It had probably been present for some years previous to its discovery and in more than one center of infection. The disease spread with deadly rapidity and virulence so that American chestnut throughout its entire natural range has been commercially destroyed. The disease has also been found in British Columbia and the Pacific Coast states. It is established in Europe on European chestnut (Gravatt 1952).

Because of the economic importance of the disease and the rapidity of its attack a large amount of literature quickly accumulated (Beattie 1914). Among the more general papers are those of P. J. Anderson and his coworkers (1913 and 1914), Baxter and Strong (1931), Clinton (1913), Giddings (1912), Gravatt and Gill (1930), Korstian (1915), and Rankin (1914).

Hosts. American chestnut is extremely susceptible, and European chestnut is also susceptible but possibly not to the same degree. Asiatic species such as the Chinese chestnut and the Japanese chestnut although not immune are resistant. Eastern chinquapin is susceptible although not commonly infected. The fungus causes considerable damage to post oak. It grows as a saprophyte on various oaks, red maple, shagbark hickory, and staghorn sumach.

Damage. The loss consists of the effect on the future productive capacity of the land measured by the difference in value between American chestnut and the species replacing it. The reduction in returns from forest lands from which chestnut has been eliminated varies widely according to the composition of the resulting stand, but in general estimates have placed this reduction at from 15 to over 50 per cent from Maryland north. Southward the reduction has been somewhat less.

American chestnut occupied a unique position among hardwoods. There were few, if any, native trees that could compare with it in vigor of sprouting, rapidity of growth, and yield of a great variety of useful products such as high-grade lumber, poles, ties, slack cooperage, paper and fiber board, tannin extract from bark and wood, and nuts. The tree was also highly prized for shade and ornamental purposes. The tannin extract industry, on which a number of communities in the southern Appalachian region are largely dependent, is doomed, although it can continue for a while by utilizing dead trees.

Epidemiology. The fungus spread from diseased trees to nearby healthy ones, and was also carried long distances on shipments of poles or nursery stock, and by birds or wind. Such dissemination resulted in new centers of spread, known as "spot" or "advance" infections, long distances ahead of the main infected area. Spot infections enlarged and ran together, thus forming new continuous infected areas. Spread was retarded by valleys or other areas with little chestnut. The area with 80 per cent of the stand infected moved southwestward across Virginia at the rate of 24 miles per year, but westward across the valleys and ridges it moved more slowly.

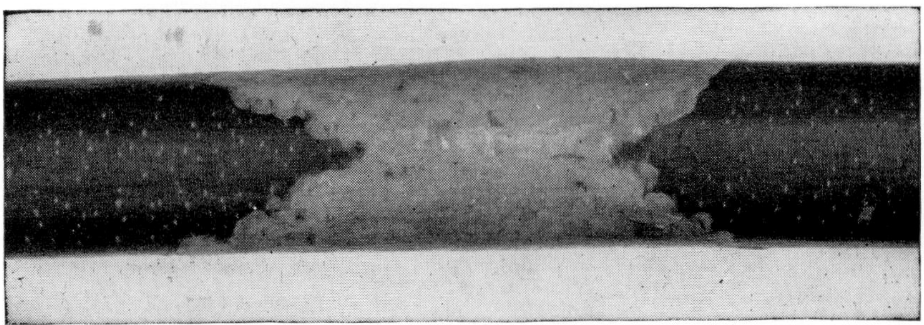

Fig. 109. A young chestnut blight canker on American chestnut. The girdling of the sprout by the meeting of the two edges of the lesion has just been completed. (*Reproduced from U.S. Dept. Agr. Circ.* 370.)

Symptoms. Dead limbs with or without leaves or burrs usually indicate blight. The dead stem will show a canker located on it usually below the lowest killed foliage. Water sprouts frequently develop just below the cankers. Death of stems is caused by a complete stoppage of water conduction (Bramble 1938).

Young cankers on smooth-barked vigorously growing young stems have a yellowish-brown surface color in sharp contrast to the olive-green color of normal bark (Fig. 109). Cankers may be either sunken or swollen (Fig. 110), or an individual canker may show both conditions. Fruiting pustules develop abundantly on the cankered bark some distance back from the advancing edge of the canker. During moist weather long, twisted, bright-yellow to buff spore horns or tendrils may exude from some of the pustules (Fig. 111). Fruiting pustules appear in the cracks or crevices of thick bark (Fig. 112) but not so abundantly as on smooth bark. Tawny-colored, flat, fan-shaped areas of mycelium developed in the inner bark, often as deep as the cambium, are a characteristic symptom.

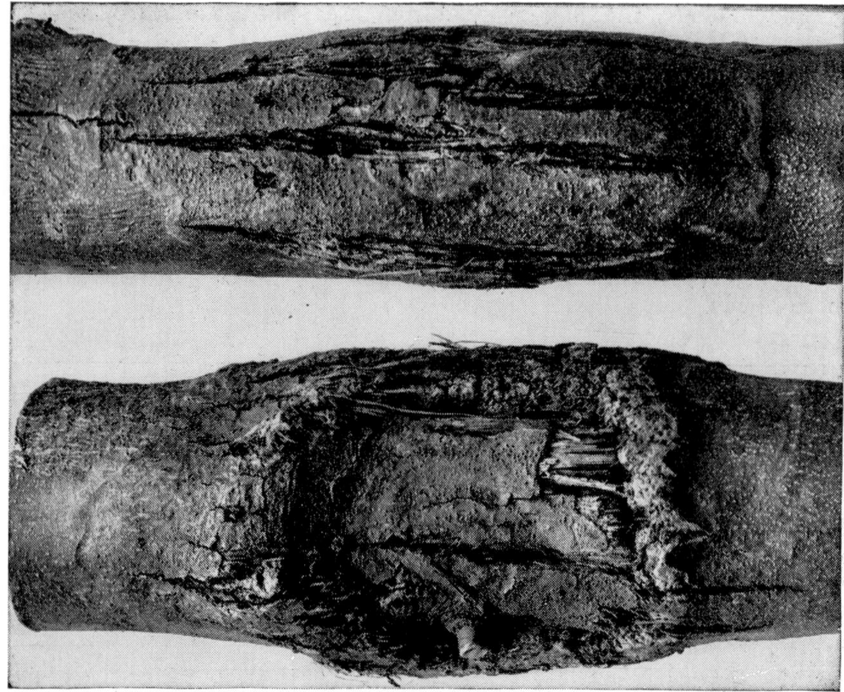

Fig. 110. Older chestnut blight cankers of the swollen and sunken types on American chestnut. (*Reproduced from U.S. Dept. Agr. Bull.* 380.)

Fig. 111. Piece of the infected stem of an American chestnut showing pycnidia and extruded spore horns or spore tendrils of *Endothia parasitica*. (*After F. D. Heald.*)

Fig. 112. Perithecial stromata of *Endothia parasitica* in the crevices of rough bark of American chestnut. (*After F. D. Heald.*)

The effect of the cankers is finally to girdle the stem, no matter how large, thus killing the parts beyond the affected zone.

Causal Agency. Infection occurs on stems of any size by means of pycnidiospores (conidia) and ascospores. Hyphae from germinating spores can enter through any wound in the bark deeper than the outer

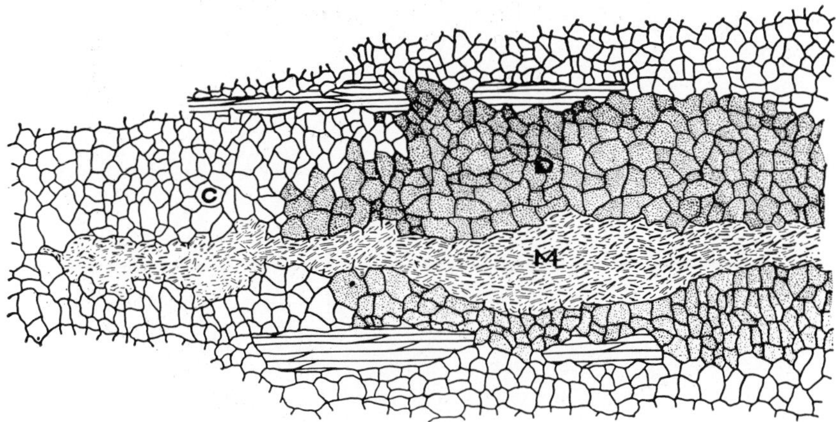

Fig. 113. Mycelial fan (*M*) of *Endothia parasitica* advancing through bark of American chestnut. The tip of the fan on the left is surrounded by cortical tissue (*C*). The contents of the cortical cells back from the tip of the fan are discolored to a yellowish brown, as indicated by the darkened cells (*D*). (*After W. C. Bramble.*)

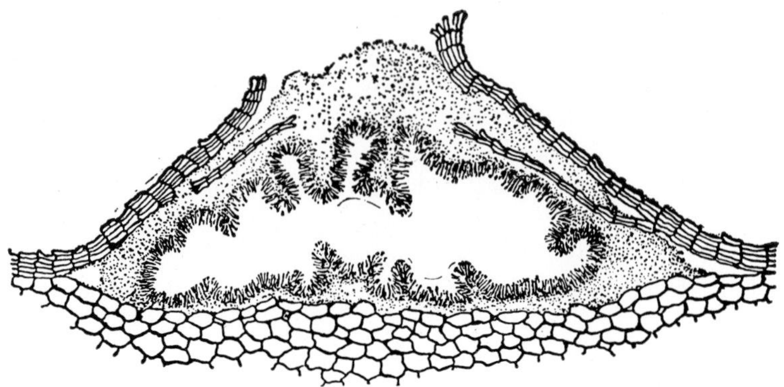

Fig. 114. Section through a pycnidium of *Endothia parasitica.* (*After F. D. Heald.*)

green cortex. Insects are probably responsible for most of the injuries serving as infection courts. The fungus advances through living bark by the mass action of mycelial fans (Fig. 113), whereas bark cells at the sides of the fans are penetrated by individual hyphae (Bramble 1938). The mycelium penetrates the outer rings of the sapwood as individual

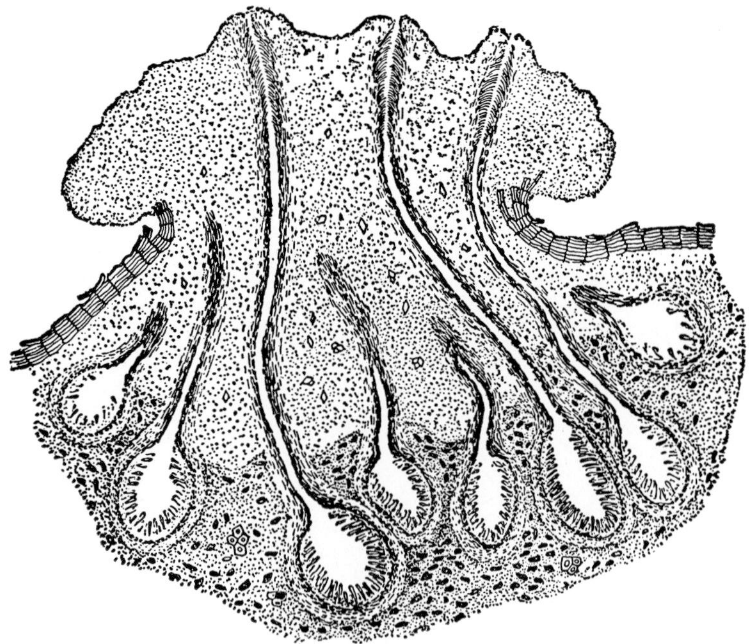

Fig. 115. Section through a perithecial stroma of *Endothia parasitica* showing perithecia opening to the surface by long necks. (*After F. D. Heald.*)

hyphae that advance through the rays and then spread to other tissues. The hyphae have no deleterious effect on the wood for commercial use. Although tannic acid is toxic to the mycelium of some fungi, *Endothia parasitica* produces enzymes that readily break down tannins (Bazzigher 1957).

Pycnidia may break through the bark of cankers at any season of the year as tiny yellowish or orange somewhat globose pustules about the size of a pinhead. Each pycnidium contains one lobulated cavity, rarely more, in which the sticky pycnidiospores develop (Fig. 114). The small

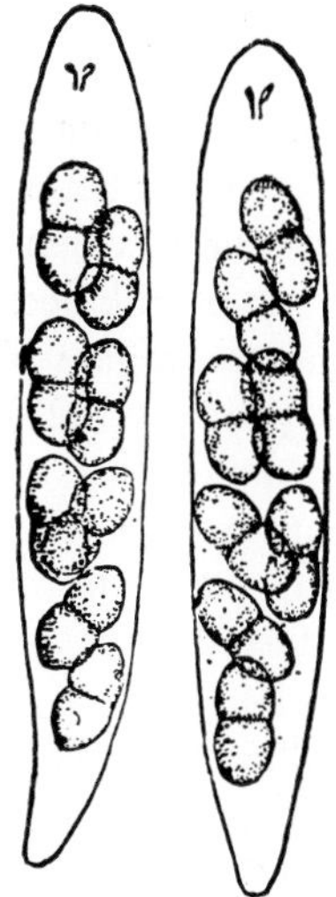

Fig. 116. Asci and ascospores of *Endothia parasitica.* (*After F. D. Heald.*)

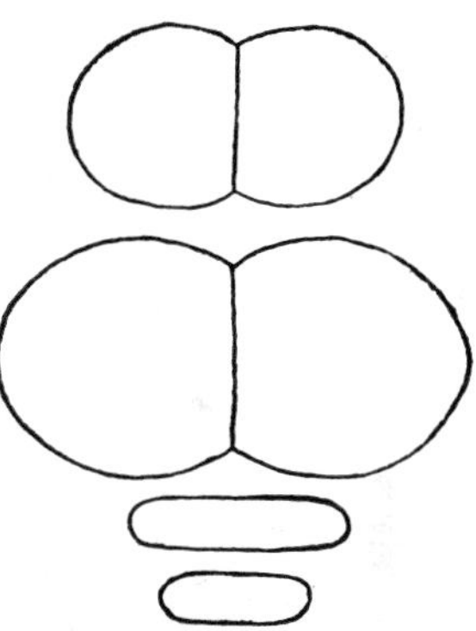

Fig. 117. Shape and relative size of ascospores and pycnidiospores of *Endothia parasitica.* (*After F. D. Heald.*)

one-celled hyaline pycnidiospores ooze out during damp weather as yellowish or orange spore horns or tendrils and are dispersed by rain, insects, and birds.

In time perithecial stromata or pustules may appear, each stroma containing one to many flask-shaped, deeply embedded perithecia with long necks opening to the surface (Fig. 115), the openings or ostioles showing on the yellowish or orange surface of the stroma as minute black raised dots. In the perithecia are developed the two-celled hyaline ascospores (Fig. 116), which are much larger than the pycnidiospores (Fig. 117). The ascospores after discharge from the perithecium are dispersed by wind.

When a tree is killed the fungus can continue to spread throughout the bark, producing pycnidia and perithecia, or these can develop on the bark of logs or down trees. Pycnidia will also develop on the sides of peeled logs, on sapwood at the cut ends, on bark fragments, and on chips of sapwood or lumber.

Control. All efforts to save chestnut timber stands and valuable individual trees failed. When it became apparent that American chestnut was doomed, beginning in 1913 a search was made in Asia, the native home of the fungus, for resistant chestnuts that might be satisfactory substitutes for the American chestnut or for crossbreeding with it, and blight-resistant American chestnut trees and sprouts have also been sought (Anonymous 1954; Beattie and Diller 1954; Clapper 1954; Nienstaedt and Graves 1955). No selections or hybrids satisfactory for forest growing have been obtained yet, but the work is still in progress.

Although American chestnut sprouts vigorously from the stump, the sprouts are soon killed by blight. New sprouts develop, which are killed in turn. The ability of chestnut to sprout after it has been killed by blight is due to the fact that the root tissues, including the region of the root collar, are more resistant than the stem tissues (Graves 1926). Despite reports to the contrary there is no definite evidence that sprouts are developing resistance to blight. Although sprouts may be increasing in size before they are killed, this could well be attributable to reduction in the amount of inoculum by the death of the heavily infected original stand.

The loss of American chestnut is an outstanding example of the destruction that can be wrought by an introduced parasite that finds its new host without any inherent resistance and is a tragic lesson on the danger of introducing living plant material from other countries, particularly when no prior investigation of the diseases that attack the host in its native land has been made.

CANKERS AND WOOD-DESTROYING FUNGI

Although fungi causing decay in living trees are largely saprophytic, invading the dead heartwood only, in addition some are parasitic, attacking the living sapwood and cambium after they have become well established in the heartwood. They are sometimes associated with trunk cankers and other malformations on living trees.

Canker-rots are frequent on oaks (Toole 1959). *Polyporus hispidus* (Bull.) Fr. causes large elongated trunk cankers that are irregular in shape with conspicuous ridges of callus. Species of the black oak group are most commonly cankered. On large trees the cankers resemble somewhat those caused by *Strumella coryneoidea,* but they are more irregular, rarely have a branch stub in the center, and of course have no fructifications of *Strumella* on them. The conks of *P. hispidus* are sometimes found on the ground at the base of trees with cankers, having fallen off the cankered surfaces. *Poria laevigata* (Fr.) Karst. causes a similar canker on bottom-land red oaks. *Poria spiculosa* Campbell & Davidson, which causes a white trunk rot of oaks, hickories, and honey locust, also causes cankers. On hickories the cankers are conspicuous, developing around a branch wound and resembling a nearly healed swollen wound. Sterile brown fungous material is always present in the cankers and often protrudes in the cracks between the callus folds. On oaks the lesions are less conspicuous. *Irpex mollis* Berk. & Curt. causes cankers and a tough spongy white rot of the heartwood of oaks in the Southeast. *Fomes everhartii* has been associated with gnarly swellings on the base of the trunk of pin oak (Graves 1919:120), and it undoubtedly occurs on other oaks in the same way. *F. robustus* Karst. causes trunk cankers on oak in Europe (Lohwag 1937). Although it occurs on oak in the United States it is not recorded as parasitic. *Stereum rugosum* Pers., although known only as a saprophyte in this country, is associated with trunk cankers on oak, including American northern red oak, in Europe (Banerjee 1956; Schönhar 1951).

Fomes igniarius var. *laevigatus* (Fr.) Overh. causes cankers on yellow birch characterized by a sunken area of considerable size to which the bark remains firmly attached and by a slight callus development (True, Tryon, and King 1955). *Poria obliqua* also causes conspicuous cankers on yellow birch (page 415). *Stereum murraii* causes elongated cankers with depressed centers and swollen margins accompanied by a considerable distortion of the trunk on several hardwoods, particularly yellow birch and sugar maple (Davidson, Campbell, and Lorenz 1941). *Daedalea unicolor* is associated with cankers of maples and other hardwoods (Campbell 1939). The canker is outlined by faint to prominent callusing, and conks of the fungus often develop on the canker face. *Polyporus glomeratus* causes cankers on maples and American beech (page 417). *Fomes fomentarius* attacks the sapwood of European beech in Germany causing long grooves and ridges on the trunk (Lohwag 1931). Although it is known to attack sapwood of living hardwoods in this country, no such hypertrophies have been recorded.

REFERENCES

Anderson, P. J.: The Morphology and Life History of the Chestnut Blight Fungus. *Penna. Chestnut Tree Blight Comm. Bull.*, 7:1–44 (1914).

Anderson, P. J., and H. W. Anderson: The Chestnut Blight Fungus and a Related Saprophyte, *Penna. Chestnut Tree Blight Comm. Bull.*, 4:1–26 (1913).

Anderson, P. J., and D. C. Babcock: Field Studies on the Dissemination and Growth of the Chestnut Blight Fungus. *Penna. Chestnut Tree Blight Comm. Bull.,* 3:1–45 (1913).

Anderson, P. J., and W. H. Rankin: Endothia Canker of Chestnut. *N.Y.* (*Cornell*) *Agr. Expt. Sta. Bull.,* 347:530–618 (1914).

Anderson, R. L.: Hypoxylon Canker of Aspen. *U.S. Dept. Agr., Forest Serv., Forest Pest Leaflet,* 6:1–3 (1956).

Anonymous: Chestnut Blight and Resistant Chestnuts. *U.S. Dept. Agr., Farmers' Bull.,* 2068:1–21 (1954).

Ashcroft, J. M.: European Canker of Black Walnut and Other Trees. *West Va. Agr. Expt. Sta. Bull.,* 261:1–52 (1934).

Banerjee, S.: An Oak (*Quercus robur* L.) Canker Caused by *Stereum rugosum* (Pers.) Fr. *Brit. Mycol. Soc. Trans.,* 39:267–277 (1956).

Baxter, D. V., and F. C. Strong: Chestnut Blight in Michigan. *Mich. Agr. Expt. Sta. Circ. Bull.,* 135:1–18 (1931).

Bazzigher, G.: Tannin- und phenolspaltende Fermente dreier parasitischer Pilze. *Phytopathol. Z.,* 29:299–304 (1957).

Beattie, R. K.: Bibliography of the Chestnut Bark Disease. *Penna. Chestnut Tree Blight Comm., Final Rept.,* Jan. 1 to Dec. 15, 1913:95–121 (1914).

Beattie, R. K., and J. D. Diller: Fifty Years of Chestnut Blight in America. *J. Forestry,* 52:323–329 (1954).

Berducou, Jeanne: Mécanisme de la formation des chancres a *Nectria* du pommier; rôle de l'acide indole-B-acétique dans la biologie de Nectria *galligena* Bres. et de *Nectria cinnabarina* Tode. *Ann. École natl. supérieure agron.* (*Toulouse*), 4(2):1–179 (1956).

Bidwell, C. B., and W. C. Bramble: The Strumella Disease in Southern Connecticut. *J. Forestry,* 32:15–23 (1934).

Bier, J. E.: Hypoxylon Canker of Maple. *Forestry Chronicle,* 15:122–123 (1939).

Bramble, W. C.: Effect of *Endothia parasitica* on Conduction. *Am. J. Botany,* 25:61–65 (1938).

Buisman, Christine: Three Species of *Botryodiplodia* (Sacc.) on Elm Trees in the United States. *J. Arnold Arboretum* (Harvard Univ.), 12:289–296 (1931).

Butin, H.: Untersuchungen über Resistenz und Krankheitsfälligkeit der Pappel gegenüber *Dothichiza populea* Sacc. et Br. *Phytopathol. Z.,* 28:253–374 (1957).

Butin, H.: Über die auf *Salix* und *Populus* vorkommenden Arten der Gattung *Cryptodiaporthe* Petr. *Phytopathol. Z.,* 32:399–415 (1958).

Campbell, W. A.: *Daedalea unicolor* Decay and Associated Cankers of Maples and Other Hardwoods. *J. Forestry,* 37:974–977 (1939).

Caroselli, N. E., and C. M. Tucker: Pit Canker of Elm. *Phytopathology,* 39:481–488 (1949).

Carter, J. C.: *Cytosporina ludibunda* on American Elm. *Phytopathology,* 26:805–806 (1936).

Clapper, R. B.: Chestnut Breeding, Techniques and Results. *J. Heredity,* 45:107–114, 201–208 (1954).

Clinton, G. P.: Chestnut Bark Disease. *Conn. Agr. Expt. Sta. Rept.,* 36(1912):359–453 (1913).

Crandall, B. S.: Thyronectria Disease of Honey Locust in the South. *U.S. Dept. Agr., Plant Disease Reptr.,* 26:376 (1942).

Crosby, D., and T. W. Jones: Beech Scale-Nectria. *Soc. Am. Foresters, New England Sect., Tree Pest Leaflets,* 4(rev):[1–4] (1950).

Davidson, R. W.: *Urnula craterium* Is Possibly the Perfect Stage of *Strumella coryneoidea*. *Mycologia*, 42:735–742 (1950).

Davidson, R. W., W. A. Campbell, and R. C. Lorenz: Association of *Stereum murrayi* with Heart Rot and Cankers of Living Hardwoods. *Phytopathology*, 31:82–87 (1941).

Davidson, R. W., and Edith K. Cash: A Cenangium Associated with Sooty-bark Canker of Aspen. *Phytopathology*, 46:34–36 (1956).

Davidson, R. W., and T. E. Hinds: Hypoxylon Canker of Aspen in Colorado. *U.S. Dept. Agr., Plant Disease Reptr.*, 40:157 (1956).

Davidson, R. W., and R. C. Lorenz: Species of *Eutypella* and *Schizoxylon* Associated with Cankers of Maple. *Phytopathology*, 28:733–745 (1938).

Dearness, J., and J. R. Hansbrough: *Cytospora* Infection Following Fire Injury in Western British Columbia. *Can. J. Research*, 10:125–128 (1934).

Diehl, W. W.: *Valsaria megalospora* on Red Alder. *U.S. Dept. Agr., Plant Disease Reptr.*, 39:334 (1955).

Diller, J. D., and R. W. Davidson: *Cucurbitaria elongata* Associated with a Canker of Honeylocust in Ohio. *U.S. Dept. Agr., Plant Disease Reptr.*, 34:224 (1950).

Ehrlich, J.: The Beech Bark Disease a *Nectria* Disease of *Fagus*, Following *Cryptococcus fagi* (Baer.). *Can. J. Research*, 10:593–692 (1934).

Fergus, C. L.: Strumella Canker on Bur Oak in Pennsylvania. *Phytopathology*, 41:101–103 (1951).

Ford, H. F., and Alma M. Waterman: Effect of Surface Sterilization on Survival and Growth of Field-planted Hybrid Poplar Cuttings. *U.S. Dept. Agr., Plant Disease Reptr.*, 38:101–105 (1954).

Giddings, N. J.: The Chestnut Bark Disease. *West Va. Agr. Expt. Sta. Bull.*, 137:209–225 (1912).

Goodding, L. N.: *Didymosphaeria oregonensis*, a New Canker Organism on Alder. *Phytopathology*, 21:913–918 (1931).

Grant, T. J., and T. W. Childs: Nectria Canker of Northeastern Hardwoods in Relation to Stand Improvement. *J. Forestry*, 38:797–802 (1940).

Grant, T. J., and P. Spaulding: Avenues of Entrance for Canker-forming Nectrias of New England Hardwoods. *Phytopathology*, 29:351–358 (1939).

Gravatt, G. F.: Blight on Chestnut and Oaks in Europe in 1951. *U.S. Dept. Agr., Plant Disease Reptr.*, 36:111–115 (1952).

Gravatt, G. F., and L. S. Gill: Chestnut Blight. *U.S. Dept. Agr. Farmers' Bull.*, 1641:1–18 (1930).

Graves, A. H.: Parasitism in *Hymenochaete agglutinans*. *Mycologia*, 6:279–284 (1914).

Graves, A. H.: Some Diseases of Trees in Greater New York. *Mycologia*, 11:111–124 (1919).

Graves, A. H.: The Cause of the Persistent Development of Basal Shoots from Blighted Chestnut Trees. *Phytopathology*, 16:615–621 (1926).

Harris, H. A.: Initial Studies of American Elm Diseases in Illinois. *Illinois Nat. Hist. Survey Bull.*, 20(1):1–70 (1932).

Heald, F. D., and R. A. Studhalter: The Strumella Disease of Oak and Chestnut Trees. *Penna. Dept. Forestry Bull.*, 10:1–15 (1914).

Horner, Ruth M.: A Diaporthe Canker of *Betula lutea*. [Abstract.] *Can. Phytopathol. Soc. Proc.*, 23:16–17 (1955).

Howard, F. L.: The Bleeding Canker Disease of Hardwoods and Possibilities of Control. *Western Shade Tree Conf. Proc.*, 8:46–55 (1941).

Koch, L. W.: Investigations on Black Knot of Plums and Cherries I–IV. *Sci. Agr.*, 13:576–590; 15:80–95, 411–423, 729–744 (1933, 1935).

Korstian, C. F.: Pathogenicity of the Chestnut Bark Disease. *Nebraska Univ. Forest Club Ann.*, 6:45–66 (1915).

Lair, Eugenie D.: Smooth Patch, a Bark Disease of White Oak. *J. Elisha Mitchell Sci. Soc.*, 62:212–220 (1946).

Lohman, M. L., and Alice J. Watson: Identity and Host Relations of Nectria Associated with Diseases of Hardwoods in the Eastern States. *Lloydia*, 6:77–108 (1943).

Lohwag, H.: Zur Rinnigkeit der Buchenstämme. *Z. Pflanzenkrankh.*, 41:371–385 (1931).

Lohwag, K. *Fomes hartigii* (Allesch.) Sacc. et Trav. und *Fomes robustus* Karst. *Ann. Mycol.*, 35:339–349 (1937).

Luttrell, E. S.: Botryosphaeria Stem Canker of Elm. *U.S. Dept. Agr., Plant Disease Reptr.*, 34:138–139 (1950).

Marshall, R. P.: Sphaeropsis Canker and Dieback of Shade Trees. *Natl. Shade Tree Conf. Proc.*, 15:65–69 (1939).

Meinecke, E. P.: Quaking Aspen: A Study in Applied Forest Pathology. *U.S. Dept. Agr. Tech. Bull.*, 155:1–33 (1929).

Miller, P. A.: Bleeding Canker Disease of California Trees. *Western Shade Tree Conf. Proc.*, 8:39–45 (1941).

Münch, E.: Untersuchungen über Eichenkrankheiten. *Naturw. Z. Forst- u. Landwirtsch.*, 13:509–522 (1915).

Nelson, R. M.: Vigorous Young Yellow Poplar Trees Can Recover from Injury by Nectria Cankers. *J. Forestry*, 38:587–588 (1940).

Nienstaedt, H., and A. H. Graves: Blight Resistant Chestnuts, Culture and Care. *Conn. Agr. Expt. Sta. Circ.*, 192:1–18 (1955).

Povah, A. H. W.: The Fungi of Isle Royale, Lake Superior. *Papers Mich. Acad. Sci.*, 20:113–156 (1935).

Rankin, W. H.: Field Studies on the Endothia Canker of Chestnut in New York State. *Phytopathology*, 4:233–260 (1914).

Reddy, C. S., et al.: Tree Diseases of Iowa. *Iowa Agr. Expt. Sta. Rept., Agr. Research* for the year ending June 30, 1934:78–79 (1934).

Richmond, B. G.: A Diaporthe Canker of American Elm. *Science*, 75:110–111 (1932).

Roth, E. R., and G. H. Hepting: Eradication and Thinning Tests for Nectria and Strumella Canker Control in Maryland. *J. Forestry*, 52:253–256 (1954).

Schmidle, A.: Die Cytospora-Krankheit der Pappel und die Bedingungen für ihr Auftreten. *Phytopathol. Z.*, 21:83–96 (1953*a*).

Schmidle, A.: Zur Kenntnis der Biologie und der Pathogenität von *Dothichiza Populea* Sacc. et Briard, dem Erreger eines Rindenbrandes der Pappel. *Phytopathol. Z.*, 21:189–209 (1953*b*).

Schönhar, S.: Eichen- und Roteichenkrebs in Württemberg. *Allgem. Forstz.*, 6:369–370 (1951).

Schreiner, E. J.: Two Species of *Valsa* Causing Disease in *Populus*. *Am. J. Botany*, 18:1–29 (1931).

Seeler, E. V., Jr.: Two Diseases of *Gleditsia* Caused by a Species of *Thyronectria*. *J. Arnold Arboretum* (*Harvard Univ.*), 21:405–427 (1940).

Sleeth, B., and R. C. Lorenz: Strumella Canker of Oak. *Phytopathology*, 35:671–674 (1945).

Spaulding, P.: A Suggested Method of Converting Some Heavily Nectria-cankered Hardwood Stands of Northern New England to Softwoods. *J. Forestry,* 36:72 (1938).

Spaulding, P.: The Role of Nectria in the Beech Bark Disease. *J. Forestry,* 46:449–453 (1948).

Spaulding, P., T. J. Grant, and T. T. Ayers: Investigations of Nectria Diseases in Hardwoods of New England. *J. Forestry,* 34:169–179 (1936).

Stuntz, D. E., and C. E. Seliskar: A Stem Canker of Dogwood and Madrona. *Mycologia,* 35:207–221 (1943).

Swingle, R. U., and T. W. Bretz: Zonate Canker, a Virus Disease of American Elm. *Phytopathology,* 40:1018–1022 (1950).

Thompson, G. E.: A Canker Disease of Poplars Caused by a New Species of *Neofabraea. Mycologia,* 31:455–465 (1939).

Thomsen, M., N. F. Buchwald, and P. A. Hauberg: Angreb af *Cryptococcus fagi, Nectria galligena* og andre Parasiter paa Bøg i Danmark 1939–43. (English summary.) *Forstl. Forsøgsv. i Danmark,* 18:97–326 (1949).

Toole, E. R.: Twig Canker of Sweetgum. *U.S. Dept. Agr., Plant Disease Reptr.,* 41:808–809 (1957).

Toole, E. R.: Canker-rots in Southern Hardwoods. *U.S. Dept. Agr., Forest Serv., Forest Pest Leaflet,* 33:1–4 (1959).

True, R. P., E. H. Tryon, and J. F. King: Cankers and Decays of Birch Associated with Two Poria Species. *J. Forestry,* 53:412–415 (1955).

Wagener, W. W., and M. S. Cave: Phytophthora Canker of Madrone in California. *U.S. Dept. Agr., Plant Disease Reptr.,* 28:328 (1944).

Walter, J. M.: Canker Stain of Planetrees. *U.S. Dept. Agr. Circ.,* 742 (slightly rev.):1–12 (1950).

Walter, J. M., E. G. Rex, and R. Schreiber: The Rate of Progress and Destructiveness of Canker Stain of Planetrees. *Phytopathology,* 42:236–239 (1952).

Waterman, Alma M.: Septoria Canker of Poplars in the United States. *U.S. Dept. Agr. Circ.,* 947:1–24 (1954).

Waterman, Alma M.: Canker and Dieback of Poplars Caused by *Dothichiza populea. Forest Sci.,* 3:175–183 (1957).

Wester, H. V., R. W. Davidson, and M. E. Fowler: Cankers of Linden and Redbud. *U.S. Dept. Agr., Plant Disease Reptr.,* 34:219–223 (1950).

Wolf, F. T., and F. A. Wolf: A Study of *Botryosphaeria ribis* on Willow. *Mycologia,* 31:217–227 (1939).

CHAPTER 13

Stem Diseases: Galls, Witches'-brooms, and Fasciations

Galls and witches'-brooms are common on hardwoods and conifers. These malformations are so striking that they have long been the most commonly noticed pathological phenomena on plants, leading to the study of galls and their causal agents becoming a specialized field known as cecidology.

GALLS

Globose or subglobose swellings (hypertrophies) known as galls or burls are common on trees, although usually they occur on an occasional tree rather than on the majority of the trees in a stand, except that galls caused by rust fungi and mistletoes, which are discussed elsewhere, may be abundant (pages 194, 201, and 322). Insects also cause a great variety of galls (Felt 1940). The anatomy of galls has been discussed by Küster (1930). In general, trees with galls on either the branches or trunk, except those with many galls, seem to grow vigorously. However, trees with trunk galls should be removed from a stand as soon as possible because the wood of the galls has such a distorted grain that it is worthless, except in a few instances where it is valuable for furniture or novelties.

Galls with bark the same as or closely similar to bark of normal stems have so far proved to be caused by noninfectious agents, whereas those with bark markedly changed in character from normal have been caused by pathogens.

NONINFECTIOUS GALLS

Certain large galls or burls on hardwood trunks are apparently the result of some injury. These galls contain the pith of innumerable buds that rarely develop, and the grain of the entire burl is greatly distorted. Black walnut and black cherry burls are highly valued for furniture, the figure being somewhat similar to that of bird's-eye maple. However, the burl figure is due to conical elevations, each with a dark speck, whereas bird's-eye, largely confined to sugar maple, is due to sharp depressions in the growth rings, the same contour being followed by the succeeding growth rings for many years. Bird's-eye is not associated with galls and occurs in trees in which the outward appearance of the trunk gives no indication of the abnormal wood. Bird's-eye in sugar maple

Fig. 118. Galls on the trunk of white spruce, cause unknown. (*Photograph by R. P. Marshall.*)

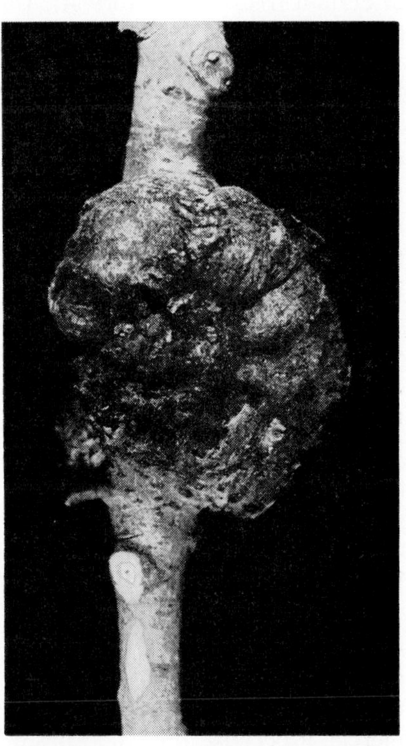

Fig. 119. Gall on the main stem of large-tooth aspen presumably caused by bacteria.

may be caused by fungi (Hale 1932:15), although a similar abnormality of birch in northern Europe is thought to result from external climatic or soil conditions causing an internal gummosis, but in which the cell walls or contents are not dissolved (Hintikka 1926). Galls, cause unknown, consisting of a dense aggregate of proliferating buds growing from a short, stout, stalklike branch occur sparingly on ponderosa pine (Cohen and Waters 1957) and on sweet birch.

The cause of the large galls or burls so common on the trunks of redwood, locally frequent on Sitka, white, and other spruces (Fig. 118), and occasionally on other conifers, is unknown (White and Millington 1954*a, b*). Douglas fir frequently has smaller burls originating at a wound on the trunk, particularly at bruises caused by falling trees. Smaller galls may occur on the trunks and branches of conifers either singly or in groups. It has been suggested that this phenomenon is explained by mutation (Tubeuf 1930), or by a hereditary predisposition of certain trees to gall formation in response to unknown environmental factors (Sprengel 1936).

INFECTIOUS GALLS

Infectious galls are largely caused by bacteria (Riker, Spoerl, and Gutsche 1946) and to some extent by fungi, and there are infectious galls the causal agent of which is still unknown. In this category are the globose to fusiform galls, often bearing one or more cankers, that occur on the trunks of sugar maple (Davidson and Campbell 1944). They are seemingly caused by an infectious agent of low virulence.

Bacterial Galls. On some hardwoods tumorlike enlargements on the stems are found, which are usually referred to as "crown gall" because of their common location on the trunk at or near ground level, i.e., at the root crown or root collar of the tree (page 115). Crown gall is caused by *Pseudomonas tumefaciens* (S. & T.) Duggar, but whether all the various galls with the same general characteristics that are found well above ground level on trunks and on branches are caused by this or any other bacterium is not definitely proved. Largetooth aspen is often subject to a branch and trunk gall (Fig. 119), somewhat globose in shape and with a rough, irregular surface. On a branch or trunk a number of adjacent galls may more or less fuse together in a cankerous fashion, or individual galls may be scattered thickly along a stem. When this occurs on the trunk the section occupied by the galls is worthless commercially because the wood has a badly twisted and disorganized grain, is sometimes discolored, and checks severely while seasoning. Similar galls caused by *P. tumefaciens* occur on branches of other poplars.

The common cankerous gall on ash in Europe (Tubeuf 1936; Riggenbach 1956) caused by *Bacterium savastanoi* var. *fraxini* (Brown) Dowson is unknown in this country, although white ash is susceptible to it.

Globose or subglobose galls caused by *Pseudomonas pini* (Vuill.) Petri and other bacteria are well known in Europe on trunks and branches of conifers (Dufrenoy 1925; Petri 1924), but if they occur in this country they have been overlooked.

Douglas fir saplings from about 3 to 15 years old in the West are occasionally affected by a gall caused by *Bacterium pseudotsugae* (Hansen and Smith 1937). Young trees may be killed, or the leader may die owing to girdling of the main stem by a gall. The galls (Fig. 121), which occur on the main stem and branches of trees of reduced vigor or on suppressed branches are roughly globular in shape with a rough spongy surface. They may persist and increase in size for many years without killing a

Fig. 120. A gall on the trunk of pin oak caused by *Phomopsis* sp. (*Photograph by R. P. Marshall.*)

stem, and they vary from the size of a pinhead to several inches in diameter.

Fungal Galls. *Phomopsis* sp. (Fungi Imperfecti, Sphaerioidaceae) causes galls on oaks, hickories, maples, and American elm that resemble crown gall (Brown 1938, 1941). Nothing is known about the life history of the fungus in nature. On oaks, particularly of the black and red oak groups, an almost globose gall with a rough surface covered with small knobs of tissue is formed (Fig. 120). On branches up to 1 or more inches in diameter, the galls usually encircle the stem and attain a diameter eight or more times that of the branch, while on the trunk of large trees the hypertrophies reach considerable size but may not encircle the stem. That portion of a smaller branch beyond the gall frequently dies. Trees are rarely killed by trunk galls.

Macrophoma tumefaciens Shear (Fungi Imperfecti, Sphaerioidaceae) is associated with a branch gall on poplars in Montana and adjacent states

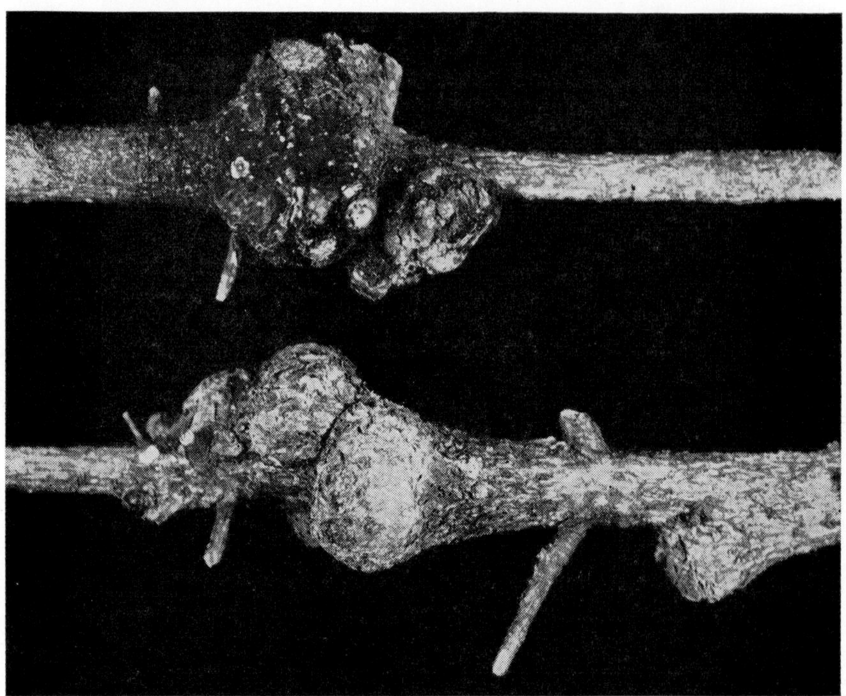

Fig. 121. Galls on the main stem of a Douglas fir sapling caused by *Bacterium pseudotsugae.*

(Hubert 1915) and in the Lake states (Kaufert 1937). On black cottonwood, individual branches may bear a large number of galls almost beadlike in arrangement that ultimately seem to cause their death. The disease, although striking in appearance, in the aggregate is not serious. The galls are almost spherical in shape and variable in size, usually not exceeding 1½ inches in diameter. Although both wood and bark of galls are abnormally thickened, the greatest increase is in the bark (Hyde 1922). The galls generally form on a stem at the base of a twig, which usually dies. Initial infection apparently occurs in the region of the buds during the time of their formation. The black pycnidia of the fungus are embedded in the bark of the gall, commonly along the bark fissures.

WITCHES'-BROOMS

A striking and common malformation of woody plants is the development of witches'-brooms, which may attain great size involving the entire crown of a tree. Witches'-brooms are characterized by dense or loose clusters of branches, usually more or less upright but occasionally with a drooping habit, often originating from an enlarged axis (Fig. 69). A somewhat similar proliferation of roots is known as hairy root (page 117). The leaves may be reduced in size, the buds small and close together, and the internodes short, or all these structures may be larger than normal. Just what physiological reaction leads to the development of witches'-brooms is unknown, but certainly its inception is not limited to any single agent or group, since these malformations, which occur on both hardwoods and conifers, are caused by fungi, mistletoes, viruses, insects, and mites, and in addition the cause of many brooms is still unknown (Liernur 1927; Tubeuf 1933). Witches'-brooms are considered by Tubeuf as parasitizing the normally branched host in the same way as mistletoe plants.

Among the fungi, rusts commonly cause witches'-brooms (page 195), as do certain leaf fungi that become perennial in the stem tissue (page 146). A witches'-broom on shagbark hickory is caused by *Microstroma juglandis* (Bereng.) Sacc., which fruits on the leaves but invades the stem (Stewart 1917). Although no brooms caused by bacteria are known on forest trees in the United States, brooms on certain conifers in Europe have been ascribed to bacteria, but the pathogenicity of the organisms has not actually been proved (Dufrenoy 1925:91). The dwarf mistletoes (page 326) cause brooming on many conifers, a condition that has been fully discussed by Tubeuf (1919). The brooming disease of black locust and honey locust is caused by a virus (Grant, Stout, and Readey 1942). The brooms are short lived, and the branches bearing them die back from the tips. Dense large brooms of unknown cause with short needles and twigs are occasionally encountered on pines, balsam firs, Douglas fir, and other conifers (Rhoads 1945; Buckland and Kuijt 1957). In Europe similar brooms on Norway spruce and Scotch pine are thought to arise as bud mutations and to be heritable, since many of the progeny from the seed in cones of such witches'-brooms have been dwarfed and have developed the broom character (Tubeuf 1933:235).

FASCIATIONS

Fasciation (White 1948) is characterized by a cylindrical stem becoming broad and flattened (Fig. 122). The cause has been ascribed to local overnutrition; a fasciation in sweet peas is known to be caused by bacteria (Tilford 1936); but the cause of most fasciations is unknown. Fasciation

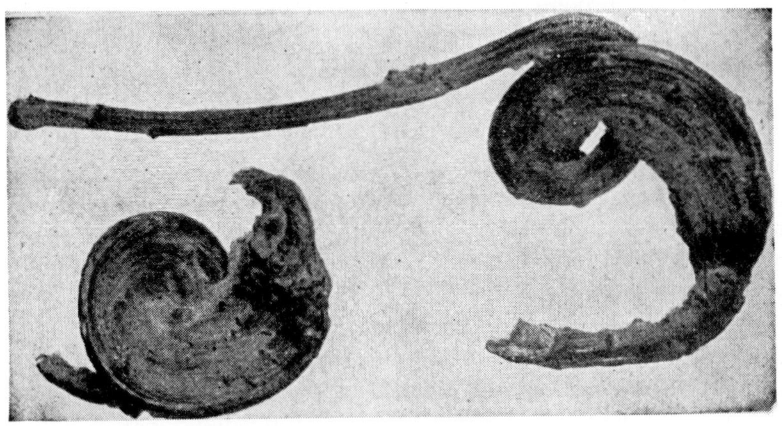

Fig. 122. Fasciation of apple twigs. (*After F. D. Heald.*)

is hereditary in some plants. Fasciation occurs on both conifers and hardwoods, but only an occasional stem is affected, and therefore it is of no practical importance on forest trees. The malformation is usually confined to the growth of one season; perennial fasciation is rare.

REFERENCES

Brown, Nellie A.: The Tumor Disease of Oak and Hickory Trees. *Phytopathology,* 28:401–411 (1938).

Brown, Nellie A.: Tumors on Elm and Maple Trees. *Phytopathology,* 31:541–548 (1941).

Buckland, D. C., and J. Kuijt: Unexplained Brooming of Douglas-fir and Other Conifers in British Columbia and Alberta. *Forest Sci.,* 3:236–242 (1957).

Cohen, L. I., and C. W. Waters: Some Observations of an Undescribed Disease of Ponderosa Pine Twigs. *J. Forestry,* 55:515–517 (1957).

Davidson, R. W., and W. A. Campbell: Observations on a Gall of Sugar Maple. *Phytopathology,* 34:132–135 (1944).

Dufrenoy, J.: Les tumeurs des résineux. *Ann. inst. natl. agron. Paris,* 19:33–201 (1925).

Felt, E. P.: "Plant Galls and Gall Makers." Ithaca, N.Y.: Comstock Publishing Associates, Inc. (1940).

Grant, T. J., D. C. Stout, and J. C. Readey: Systemic Brooming, a Virus Disease of Black Locust. *J. Forestry,* 40:253–260 (1942).

Hale, J. D.: The Identification of Woods Commonly Used in Canada. *Can. Dept. Interior, Forest Serv., Bull.*, 81:1–48 (1932).

Hansen, H. N., and R. E. Smith: A Bacterial Gall Disease of Douglas Fir, *Pseudotsuga taxifolia. Hilgardia,* 10:569–577 (1937).

Hintikka, T. J.: Über den Habitus und die Wachtumsart der Wisabirken. *Deut. Dendrol. Ges. Mitt.*, 36:209–214 (1926).

Hubert, E. E.: A New *Macrophoma* on Galls of *Populus trichocarpa. Phytopathology,* 5:182–185 (1915).

Hyde, K. C.: Anatomy of a Gall on *Populus trichocarpa. Botan. Gaz.*, 74:186–196 (1922).

Kaufert, F.: Factors Influencing the Formation of periderm in Aspen. *Am. J. Botany,* 24:24–30 (1937).

Küster, E.: "Anatomie der Gallen." Handbuch der Pflanzeanatomie . . . hrsg. von K. Linsbauer, Bd. V/1, 1 Abt, 3 Teil. Berlin: Verlagsbuchhandlung Gebrüder Borntraeger (1930).

Liernur, A. G. M.: "Hexenbesen. Ihre Morphologie, Anatomie und Entstehung." Rotterdam: Nijgh & Van Ditmar (1927).

Petri, L.: I tumori batterici del pino d'Aleppo. *Ann. regio ist. super. forest. nazl. Firenze,* 9:1–43 (1924).

Rhoads, A. S.: Witches'-brooms of Pine Trees in Florida. *U.S. Dept. Agr., Plant Disease Reptr.*, 29:26–28 (1945).

Riggenbach, A.: Untersuchung über den Eschenkrebs. *Phytopathol.* Z., 27:1–40 (1956).

Riker, A. J., E. Spoerl, and Alice E. Gutsche: Some Comparisons of Bacterial Plant Galls, and of Their Causal Agents. *Botan. Rev.*, 12:57–82 (1946).

Sprengel, F.: Ueber die Kropfkrankheit an Eiche, Kiefer, und Fichte. *Phytopathol.* Z., 9:583–635 (1936).

Stewart, F. C.: Witches-brooms on Hickory Trees. *Phytopathology,* 7:185–187 (1917).

Tilford, P. E.: Fasciation of Sweet Peas Caused by *Phytomonas fascians* n. sp. *J. Agr. Research,* 53:383–394 (1936).

Tubeuf, C. von: Überblick über die Arten der Gattung *Arceuthobium* (*Razoumowskia*) mit besonderer Berücksichtigung ihrer Biologie und praktischen Bedeutung. I. Zwergmisteln und Hexenbesen der Nadelhölzer. *Naturw. Z. Forst- u. Landwirtsch.*, 17:167–189 (1919).

Tubeuf, C. von: Das Problem der Knollenkiefer. Z. *Pflanzenkrankh.*, 40:225–251 (1930).

Tubeuf, C. von: Studien über Symbiose und Disposition für Parasitenbefall sowie über vererbung pathologischer Eigenschaften unserer Holzpflanze. I. Das Problem der Hexenbesen. Z. *Pflanzenkrankh.*, 43:193–242 (1933).

Tubeuf, C. von: Tuberkulose, Krebs und Rindengrind der Eschen (*Fraxinus*) Arten und die sie veranlassenden Bakterien, Nektriapilze und Borkenkäfer. Z. *Pflanzenkrankh.*, 46:449–483 (1936).

White, O. E.: Fasciation. *Botan. Rev.*, 14:319–358 (1948).

White, P. R., and W. F. Millington: The Distribution and Possible Importance of a Woody Tumor on Trees of White Spruce, *Picea glauca. Cancer Research,* 14:128–134 (1954*a*).

White, P. R., and W. F. Millington: The Structure and Development of a Woody Tumor Affecting *Picea glauca. Am. J. Botany,* 41:353–361 (1954*b*).

CHAPTER 14

Stem Diseases: Diebacks and Wilts

Although it is not always possible to make an exact distinction, dieback is characterized by a progressive dying back of a stem from the tip, whereas wilt is characterized by simultaneous death of part or all of a stem accompanied by wilting of the foliage. Dieback is commonly caused by a fungus invading a stem at or near the tip and then growing downward, killing the tissue as it advances. Wilt, however, usually results from a fungus invading a stem lower down or even first entering the roots and then reducing or inhibiting water conduction. Consequently wilt may be termed a "thrombotic" disease. Many wilt fungi produce toxins that disrupt the normal functioning of the hosts' cells (Gäumann 1954). Although serious on shade and ornamental trees, wilts and diebacks have caused less damage to forest trees in this country.

Dieback is a normal condition in woody plants having indefinite annual growth, i.e., when the twigs do not complete their elongation and develop terminal buds by the end of the growing season but continue to elongate until stopped by cold. The tender terminal internodes are killed by the first frosts so that new growth the next spring comes from lower lateral buds. Honey locust exemplifies this indefinite or indeterminate annual growth. However, dieback is more often caused by fungi. Species of *Coniothyrium*, *Cytosporina*, *Sphaeropsis* and other genera are associated with both dieback and canker.

Introduction of the Dutch elm disease into the United States has given impetus to the study of wilts, so that new wilts and diebacks are being found. A characteristic symptom of several wilts, i.e., a streaky discoloration in the outer layers of sapwood, is not so distinctive for the different wilts that this character can be used to determine the exact fungus responsible. For example, culture technique is necessary definitely to distinguish discolorations symptomatic for the Dutch elm disease from discolorations symptomatic for the Verticillium and Dothiorella wilts of elm.

Cenangium Dieback of Pines. This disease, also called "pruning disease," is caused by *Cenangium ferruginosum* Fr. (Ascomycetes, Dermataceae). It is questionable whether the fungus can initiate infections by itself (Van Vloten and Gremmen 1953). Periodic epidemics on Scotch pine over extensive areas in northern Europe depend on a gall midge (*Cecidomya brachyntera* Schw.) that mines the lower part of needles down to the stem (Liese 1935). Infection by *Cenangium ferruginosum* follows almost immediately. The gall midge is prevalent during periods of drought. In the United States the fungus is associated with extensive flagging injury on ponderosa pine in the Southwest caused by a scale insect *Matsucoccus vexillorum* Morrison although no lesions are found on twigs on which scales are not established (McKenzie, Gill, and Ellis 1948). Apparently *Cenangium ferruginosum* invades the twigs well in advance of their death.

Typically the disease is a dieback of twigs that is most prominent in spring and early summer. Small dusty-brown or black apothecia appear on the dead twigs in late summer or autumn. During wet weather they open and are cup-shaped, exposing the dirty-greenish disc. The asci, intermingled with slightly longer paraphyses, contain eight hyaline one-celled elliptical ascospores. A conidial stage is not known. There is no bluing of infected wood, but there may be decided resin impregnation.

It is unlikely that control measures will be necessary for pines in natural stands. For exotics the only known control is to cut off and destroy infected twigs.

Persimmon Wilt. *Cephalosporium diospyri* Crandall (Fungi Imperfecti, Moniliaceae) causes this lethal disease of the common persimmon (Toole and Lightle 1960). The wilt occurs in scattered localities from North Carolina south into Florida and west into Texas and Oklahoma. Damage may be locally severe, but the wilt has not been found in timber-producing areas. The origin of the fungus is obscure, but its behavior suggests that it has been introduced. The most common symptom of the disease is an abrupt wilting of the foliage which starts in the top of the tree, an infected tree often defoliating and dying within two months. The fungus fruits in the fall beneath the bark, producing masses of salmon-colored spores that are readily wind disseminated. Internally, fine brownish-black streaks are present throughout the wood of the outer growth rings. No control is known.

Mimosa Wilt. This extremely destructive wilt of the mimosa or silk tree, an importation from eastern Asia widely grown in the South as an ornamental, is caused by *Fusarium oxysporum* forma *perniciosum* (Hepting) Toole (Fungi Imperfecti, Tuberculariaceae). The origin of the disease is unknown, but it behaves as though caused by an introduced fungus. The symptoms are wilting and yellowing of the foliage followed by death of the tree, together with a brown ring or partial ring of discoloration in the outer growth ring. The fungus spreads through the soil, entering the host by way of the roots. Once established in a locality, there is no means of control other than planting resistant strains of mimosa that have been selected and propagated for this purpose (Toole and Hepting 1949).

Verticillium Wilt. *Verticillium* spp. (Fungi Imperfecti, Moniliaceae) including *V. albo-atrum* R. & B. cause this disease, which is widespread

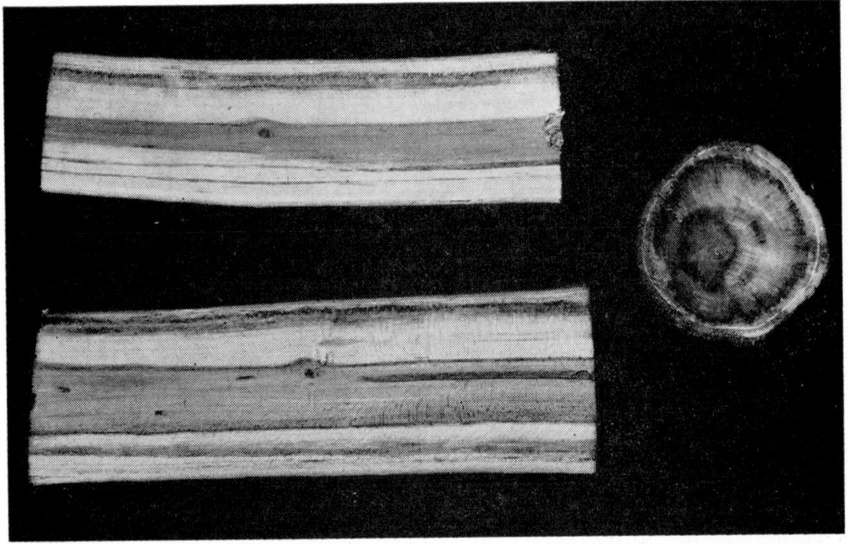

Fig. 123. Cross and longitudinal sections of a small stem of silver maple showing discoloration caused by *Verticillium* sp. (*Photograph by R. P. Marshall.*)

in Europe and North America, where a large number of unrelated annual and perennial plants are attacked (Rudolph 1931). Among the hardwoods, elms and maples are common hosts. Conifers are not attacked. So far the disease has been of no consequence in forest stands.

In the United States silver, Norway, red, and sugar maples are frequently killed or seriously injured (Dochinger 1956). The conspicuous external symptom is sudden wilting followed by yellowing and finally browning of the foliage, which usually involves individual branches or part of the crown. Internally there is extensive streaking in the outer sapwood, commonly extending from the roots into the branches (Fig. 123). In maples the green streaks resulting from infection can be confused with the dark green stain so commonly found around wounds, but this wound stain is localized. In elms the discoloration is similar to that accompanying Dothiorella wilt and the Dutch elm disease so that a positive diagnosis can be made only by means of cultures.

The causal fungus can live in the soil, and it is likely that infection usually occurs through the roots. However, *Verticillium* is capable of producing the disease when it is introduced into the vascular system through wounds made at any point in the stem or root (Caroselli 1957). When or where fruiting takes place on infected trees is not known.

Infected trees, particularly maples, should be removed and burned, or if it is desired to preserve a diseased tree as long as possible, all dead or dying branches should be cut off and burned. Such trees should be inspected periodically so that newly affected parts may be promptly destroyed. All cuts and other wounds should be kept covered with a mixture of equal parts of creosote and coal tar, first applying a coating of shellac to prevent injury to living tissues by the creosote tar. Tools used on infected trees should be sterilized in a solution of 1 part of mercuric chloride (corrosive sublimate) in 1,000 parts of water before they are used again. Maples removed because of wilt should be replaced by other species. Hardwoods planted on land formerly occupied by susceptible hosts like potatoes are particularly liable to infection.

Dothiorella Wilt of Elm. *Dothiorella ulmi* Verrall & May (Fungi Imperfecti, Sphaerioidaceae) causes this disease, which has also been referred to as a dieback and has been termed "Cephalosporium wilt" (Creager 1937; McKenzie and Johnson 1939). The origin of the fungus is unknown, but the disease is probably one of long standing for it is general throughout the central and eastern portions of the United States, commonly affecting American elm, and occasionally slippery elm and Siberian elm. Shade and ornamental trees from seedlings to veterans may be killed or seriously injured by death of part of the crown. After the first symptoms appear, older trees usually die in from 3 to several years, and young nursery trees die in 1 or 2 years. However, a tree may be infected for some years without injury, and diseased trees may recover.

Wilting and yellowing of the foliage and gradual dying back of the crown together with a brownish discoloration in the outer growth rings of the wood are the principal symptoms. Flat cankers with dead bark may develop on small branches. Pycnidia form abundantly on the bark of killed twigs and sometimes on the bark covering cankers. From them the spores are extruded in a sticky mass through the ostiole to the surface of the bark where they are apparently disseminated by wind, rain, and possibly insects. Infection occurs through wounds in the leaves commonly made by insects; localized yellowish areas appear near the infected wounds, and darkened veins extend out from the yellowish areas. Later the yellowish tissues turn brown and die. Mycelium can be traced from the leaves through the vascular tissues of the petioles into the wood, where it is abundant in and confined to the vessels.

Infected trees should be removed and destroyed or infected branches pruned out, making the cut a foot or more below the lowest point of discoloration in the wood. All branches removed should be burned. Pruning for several successive seasons will usually be necessary. A combination of an insecticidal and fungicidal spray may be valuable to reduce the number of leaf infections.

OAK WILT

Ceratocystis fagacearum (Bretz) Hunt (Ascomycetes, Sphaeriaceae), formerly known as *Endoconidiophora fagacearum,* causes this disease. It is of long standing, being distributed over much of the eastern United States (Fowler 1958). Trees of all sizes may be infected—those of the red oak group are particularly susceptible, often dying the year they first show symptoms, whereas those of the white oak group may not die for several years after infection.

Hosts. At least 50 species and varieties of *Quercus,* three species of *Castanea,* including American, Chinese, and European chestnut, one species of *Castanopsis,* and one species of *Lithocarpus* are susceptible. In addition, the wilt fungus has persisted and spread in artificially inoculated apple, ash, beech, dogwood, hop-hornbeam, hickory, sassafras, sourwood, and wild cherry, although wilt symptoms and death occurred only in apple (Bart 1957).

Damage. In parts of the Middle West where oak wilt has been present for many years damage has been considerable. For example, over 4 per cent of the total oak area in four Wisconsin counties is infected, and in this area 0.6 per cent of the total oak volume is killed annually. An important but less obvious source of loss than mortality is the failure of oaks to regenerate in infected areas. In the Wisconsin counties cited above it is estimated that, if the current rate of oak wilt spread and mortality continues for the next 100 years, from 21 to 30 per cent of the oak type will have been invaded (Anderson and Skilling 1955). Probably much of this area will have been removed from oak production.

Symptoms. Oak wilt symptoms are most noticeable during late spring and early summer. In dry years they may be confused with drought symptoms. First, the leaves of infected red oaks crinkle and pale slightly. Next, they become bronze or brown progressively from the edge or tip toward the midrib and base. Mature leaves may fall at any symptom stage from green to brown, premature leaf shedding being an outstanding symptom. A brown to black discoloration is often found in the outer growth ring of the sapwood, but this is not a dependable diagnostic character. Symptoms may progress over the entire tree in a few weeks, the lower branches being affected last. Leaf discoloration and shedding are usually confined to a few branches of white oaks each year so that infected trees become stagheaded before death.

Causal Agency. The fungus invades oaks through wounds, and it becomes established in the outer vessels of the sapwood. Tyloses and gums develop in the vessels a few days before visible foliage symptoms occur. Scattered mycelial masses of the fungus, known as mats, may form beneath the bark within several months after death of the tree. Some mats raise and crack the bark, giving off a fermenting odor. Conidia are produced abundantly on mats, and, under certain conditions, perithecia arise on them, extruding ascospores in a sticky matrix.

Complete knowledge of overland spread of the disease is lacking, but several kinds of insects that feed on mats, including Nitidulid beetles and fruit flies, may carry inoculum from them to wounds in healthy trees (Himelick and Curl 1958). Certain wood-boring insects may carry the wilt fungus upon emergence from diseased trees. Squirrels which eat the mats may be responsible for limited spread. Evidently the main period of infection is in the spring when the new vessel wood is being formed. Oaks wounded at this time, such as by ax blazes, pruning, logging, climbing irons, and lightning, may be liable to infection. The fungus is also transmitted through root grafts between diseased and healthy trees. The prevalence of root grafts varies between localities, and, in some areas, they may be less important in local spread than in others.

Control. Although various control methods are in use, more time is necessary to evaluate their effectiveness. One method of stopping root graft spread is to sever the roots of all living oaks around infected trees by using a ditchdigger or tractor-drawn blade. Another means of isolating the root systems of diseased oaks is to cut and poison the stumps of living oaks within about 50 feet of them. Theoretically, complete destruction of the aboveground parts of wilting oaks by burning is the surest way of destroying inoculum for overland transmission, but this is often not practicable. Some of the less intensive measures in use are summer cutting or deep girdling of infected trees to prevent mat formation by causing them to dry quickly and summer cutting followed by the application of an insecticidal spray to reduce insect infestation of mats, wood, and bark.

DUTCH ELM DISEASE

Dutch elm disease is caused by *Ceratocystis ulmi* (Buism.) C. Moreau, formerly known as *Ceratostomella ulmi*. Literature on the disease is extensive (Clinton and McCormick 1936; Goss and Moses 1937; McKenzie and Becker 1937; Peace 1960; Walter and May 1943; Welch and Matthysse 1955).

History and Distribution. The misleading name "Dutch elm disease" does not describe the country of origin but was given because of the early and excellent investigation of it by pathologists in Holland. The disease was first noticed in northern France, in Belgium, and in Holland in 1918. Since then it has spread north to Sweden; east to Russia, the Balkan countries, and Italy; south to Portugal and Spain; and west to Great Britain. Although the origin of the pathogen is obscure, it was probably brought into Europe from Asia during the First World War. It seems significant that certain Asiatic elms are resistant to the fungus just as Asiatic chestnuts are resistant to the chestnut blight fungus, which is known to be native to Asia.

The disease was discovered in Ohio during the summer of 1930 on four American elms at Cleveland and one at Cincinnati. In 1933 it was found in northern New Jersey, and the major infection now extends into New Jersey, Pennsylvania, New York, and Massachusetts, with outlying infections far more widespread. There is an extensive infection in Quebec, probably a direct introduction from Europe (McCallum 1946). The disease was introduced into the United States from Europe on elm burl logs for veneer manufacture. Infections in this country have centered around ports of entry, railroad distributing yards, or veneer plants. Both the fungus and the large and small European elm bark beetles, *Scolytus scolytus* Fab. and *S. multistriatus* (Marsh.), which carry the fungus, have been introduced on logs, but only the latter insect is established here.

Hosts. All the American and European elms are attacked. *Ulmus hollandica* var. *belgica* (Burgsd.) Rehd., the common elm in the Netherlands, is extremely susceptible. A seedling selection from *U. carpinifolia* Gleditsch, discovered in the Netherlands and named the Christine Buisman, is so highly resistant that it is now being tested in experimental plantings. Among the Asiatic elms *U. parvifolia Jacq.*, *U. pumila L.*, and its variety *arborea* Litv. are resistant. Because of their small size and less luxuriant foliage, these Asiatic elms are poor substitutes for American elm. The fact that elms hybridize freely probably accounts for the conflicting statements on the susceptibilty of some species.

Damage. Trees of all ages can be killed. In the United States elms are primarily important as shade and ornamental trees, occupying a minor position in forest stands, where they form a relatively small proportion in mixture with other hardwoods. A single shade tree is usually more valuable than all the elms in several or more acres of forest. The disease has been worst in the Netherlands, Belgium, and northern France because of the extreme susceptibility of the elms there. In some districts 60 to 70 per cent of the trees have been destroyed. Elsewhere, losses have not been nearly so severe, and therefore a considerable percentage of elms survive. In the worst infected area in Connecticut after 13 years of the disease, less than 25 per cent of the elms were infected or killed. Some infected trees recover. Furthermore, the rate of killing is highest in a locality when the disease first reaches epidemic proportions.

Fig. 124. Cross and longitudinal sections of small branches of elm showing the discoloration caused by *Ceratocystis ulmi*. (*Reproduced from U.S. Dept. Agr. Circ.* 170.)

Symptoms. The effects of the disease vary from slight dwarfing of the leaves, through various stages of yellowing and degrees of defoliation, to death of branches or the entire tree. The attack is more obvious on young, vigorously growing trees than on old, slower growing ones. Internally the vessels (water-conducting tubes) in the wood are discolored brown (Fig. 124), being clogged with tyloses and brown gumlike substances. These symptoms are not distinct from symptoms caused by *Dothiorella ulmi* and *Verticillium*. Consequently a laboratory test is necessary for exact diagnosis.

Causal Agency. The fungus lives in the sapwood, fruiting on the wood and bark of logs and dead or dying trees. The fructifications (coremia) develop only in sheltered places, such as cracks between the wood and loosened bark. They develop abundantly in insect galleries beneath the bark. A coremium, scarcely visible to the unaided eye, consists of a black stalk about 1 mm high with an enlarged head bearing vast numbers of minute egg-shaped to pear-shaped spores embedded in a translucent drop of sticky fluid. This is the Graphium stage. In the vessels, spores increase abundantly by budding in a yeastlike manner and are distributed rapidly for long distances in the wood by movements of the sap (Banfield 1941). The perithecial stage is unknown in the United States.

The fungus produces a soluble toxin that presumably is transported by sap movements to cause discoloration and death of the wood in advance of the mycelium. Since dying stems occur in which there is only a little discoloration in part of the last growth ring, apparently the toxin can be lethal to the host tissue before discoloration appears.

The fungus is transmitted from diseased to healthy trees largely by insects. In the United States, the small European elm bark beetle, established since before 1909, and the native dark elm bark beetle (*Hylurgopinus rufipes* Eichh.) are responsible for spreading the disease. In Quebec only the native beetle has been found. The adult beetles emerge from under the bark of dead or dying stems inadvertently carrying the sticky spores of the fungus on their bodies. After emergence the beetles usually feed on nearby healthy trees in the crotches of young twigs or leaf axils, chewing through the bark into the wood and introducing the spores of the fungus. The beetles then colonize weakened, dying, or dead stems, constructing their brood galleries under the bark, but they do not colonize vigorous stems. However, stems affected with the disease are so weakened that they are suitable for development of beetle larvae. Usually there are two broods of beetles each year. Those of the overwintering brood, emerging in May and June, cause the most damaging inoculations. For crotch feeding the beetles usually do not travel more than a few hundred feet from the place of emergence, but for breeding purposes they may travel more than 2 miles. The disease spread through Connecticut at an approximately constant rate of 5.4 miles per year. It is possible that occasional beetles may be carried involuntarily for much greater distances by high winds.

The only other way the disease is transmitted from elm to elm is by root grafts and the direct connections between trees that have originated as root sprouts. Since root grafts between elms are common, it is likely that infection frequently spreads in this way.

Control. To protect elms at present, sanitation against the bark beetles that spread the fungus is necessary. The emergence of beetles from fallen, storm-damaged, cut, or diseased elms can be prevented by burning or debarking infested wood by May. Bark removed should be destroyed.

Infested wood with the bark on can be made safe by spraying with a mixture of 8 pounds of technical DDT dissolved in 100 gallons of No. 2 fuel oil, which will destroy the larval broods of the beetles. The spray should be applied so as to wet the bark surface completely; 2 ounces of spray will usually treat 1 square foot of bark surface. The same spray on noninfested wood will prevent breeding attacks by the beetles. When applied to the bark of living trees a spray containing DDT in the form of an emulsion is effective for preventing elm bark beetle feeding. Concentration of the spray varies according to whether it is used in a hydraulic sprayer or a mist blower (Whitten and Swingle 1958).

The disease can often be eliminated from trees with one or two infected branches by pruning, provided this is done when the first external symptoms appear. Since the trees are subject to reinfection, pruning does not obviate the necessity for sanitation in a locality to get rid of the breeding places for beetles. Chemotherapy has been tried using several organic chemicals but with indifferent success. One of them, 8-quinolinol benzoate, suppressed symptoms in trees already diseased but did not prevent them from dying (Dimond et al. 1949).

Since within the various species of elms susceptible to the fungus there is considerable variation in the reaction of individual trees to the disease, it may eventually be controlled by finding or developing a resistant tree.

OTHER WILTS AND DIEBACKS

Besides *Cerotocystis ulmi* on elms, other species of *Cerotocystis* and closely related genera are pathogenic on various trees (Buisman 1933:434), the activity of these fungi in the sapwood inhibiting or reducing water conduction, thus causing death of the stem. *Cerotocystis fimbriata* f. *platani* kills London planes (page 271). *Cerotocystis coerulescens* (Münch) Bakshi causes sapstreak of sugar maple and yellow-poplar, resulting in the death of large trees (Roth, Hepting, and Toole 1959). Crowns of affected trees die from the top down, death usually being complete in from 2 to 4 years. At the same time the trunk wood becomes water-soaked in wide radial streaks from the inner into the outer sapwood, and in the middle of the water-soaked streaks are narrow reddish or gray streaks that are all gray when the wood is dry.

In living pines attacked by certain bark beetles, blue stain in the sapwood, caused by *Ceratocystis* spp. and related fungi introduced by the beetles, invariably appears shortly after the trees are infested. Blue stain is not found in uninfested trees. Although the girdling effect of the beetle tunnels would finally kill most of the trees, their death is hastened by the blue stain and other fungi, and some trees that might recover from the beetle attack are likewise killed (Bramble and Holst 1940; Craighead and St. George 1940). *Trichosporium symbioticum* Wright and *Spicaria anomala* (Corda) Harz are associated with engraver beetles killing white firs in California (Wright 1938). These fungi cause a brown stain in the sapwood and inner bark, kill the cambium, and reduce the moisture content of the cambial region ahead of the beetle larvae.

Several fungi of the genera *Coniothyrium, Diplodia, Phoma, Phomopsis,* and *Sphaeropsis* are associated with dieback of American elms in Illinois (Harris 1932; Tehon 1934). In some of these diebacks, cankers appear in the early stages, and in others cankers do not develop until later. *Vermicularia* sp. causes a dieback of branches with accompanying dark-brown discolored areas or blotches arranged circularly throughout the wood in cross section but without any canker formation. *Cytosporina ludibunda* Sacc. causes a dieback of branchlets followed by canker formation when the infection reaches larger stems (Carter 1936). *Fusarium* spp. have been found in diseased elm branches, and since these fungi cause wilts in many plants it is probable they will sometimes attack trees beyond the seedling stage. The Fusaria usually enter the roots, and infection progresses upward through the vascular tissues. *Sphaeropsis* spp. are associated with canker and dieback of hardwoods (page 270). *Diplodia natalensis* Evans causes a dieback of sycamore in which the disease starts in the small branches and finally progresses into the trunk (Thompson 1951). *Diplodia pinea* (Desm.) Kickx. (*Sphaeropsis ellisii* Sacc.) often causes dieback of the current season's growth of hard pines, which if repeated, results in stunting (Capretti 1956; Waterman 1943*a*, 1943*b*). In association with the pine spittle bug (*Aphrophora parallela* Say) the same fungus has caused severe injury to Scotch pine (Haddow and Newman 1942). Burn-blight, caused by *Chilonectria cucurbitula* (Curr.) Sacc. following injury by the Saratoga spittle bug (*A. saratogensis* Fitch), has damaged jack and red pine severely in northeastern Wisconsin, particularly on poor sites (Gruenhagen, Riker, and Richards 1947). It is to be expected that various fungi, usually of little consequence, will be found occasionally causing disease of trees growing outside their natural range or under otherwise unfavorable conditions.

Dying poplars and willows sometimes show a stain in the wood, at first water-soaked, later changing to brown, and commonly with an intermediate red stage (Hartley and Crandall 1935). From the margin of the discoloration, short rod bacteria are readily isolated. The symptoms are somewhat similar to the watermark disease of willow in Europe caused by *Bacterium salicis* Day (Metcalfe 1940–1941).

Nectria cinnabarina (Tode) Fr. is widespread as a hardwood saprophyte and occasionally becomes weakly parasitic, causing either cankers around wounds and the base of dead branches or dieback of twigs and branches, especially of maples (Cook 1917; Uri 1948). The cankers are not the perennial type caused by *N. galligena,* but are characterized by discolored and sunken bark. *Nectria* sp. is associated with a dieback and canker of sapling balsam firs in Ontario (Quirke and Hord 1955).

Dieback of butternut is caused by *Melanconis juglandis* (E. & E.) Graves, which is a weak parasite. Ordinarily the progress of the disease is slow, and in final stages trees are markedly stagheaded (Graves 1923). Under forest conditions infected trees should be removed during improvement cuttings. For shade and ornamental trees diseased branches should be pruned off well below the apparent infected portions of the wood.

Cytophoma pruinosa (Fries) Höhn. is commonly associated with cankers on stems and branches of white ash suffering from dieback (Silverborg and Brandt 1957). Several fungi of unknown importance are associated with yellowpoplar dieback, characterized in early stages by dwarfed and chlorotic leaves and dead twigs in the upper crown (Johnson et al. 1957).

REFERENCES

Anderson, R. L., and D. D. Skilling: Oak Wilt Damage—A Survey in Central Wisconsin. *Lake States Forest Expt. Sta., Sta. Paper,* 33:1–11 (1955).

Banfield, W. M.: Distribution by the Sap Stream of Three Fungi that Induce Vascular Wilt Diseases of Elm. *J. Agr. Research,* 62:637–681 (1941).

Bart, G. J.: Susceptibility of Non-Oak Species to *Endoconidiophora fagacearum.* [Abstract.] *Phytopathology,* 47:3 (1957).

Bramble, W. C., and E. C. Holst: Fungi Associated with *Dendroctonus frontalis* in Killing Shortleaf Pines and Their Effect on Conduction. *Phytopathology,* 30:881–899 (1940).

Buisman, Christine: Über die Biologie und den Parasitismus der Gattung *Ceratostomella* Sacc. *Phytopathol. Z.,* 6:429–439 (1933).

Capretti, C.: *Diplodia pinea* (Desm.) Kickx agente del dissecamento die varie specie del gen. *Pinus* e di altre conifere. [French summary.] *Ann. accad. ital. sci. forestali,* 5:171–202 (1956).

Caroselli, N. E.: Verticillium Wilt of Maples. *Rhode Island Univ. Agr. Expt. Sta. Bull.,* 335:1–84 (1957).

Carter, J. C.: *Cytosporina ludibunda* on American Elm. *Phytopathology,* 26:805–806 (1936).

Clinton, G. P., and Florence A. McCormick: Dutch Elm Disease, *Graphium ulmi. Conn. (State) Agr. Expt. Sta. Bull.,* 389:701–752 (1936).

Cook, M. T.: A Nectria Parasitic on Norway Maple. *Phytopathology,* 7:313–314 (1917).

Craighead, F. C., and R. A. St. George: Field Observations on the Dying of Pines Infected with the Blue-stain Fungus, *Ceratostomella pini* Münch. *Phytopathology,* 30:976–979 (1940).

Creager, D. B.: The Cephalosporium Disease of Elms. *Contribs. Arnold Arboretum,* 10:1–91 (1937).

Dimond, A. E., et al.: An Evaluation of Chemotherapy and Vector Control by Insecticides for Combating Dutch Elm Disease. *Conn. (State) Agr. Expt. Sta. Bull.,* 531:1–69 (1949).

Dochinger, L. S.: New Concepts of the Verticillium Wilt Disease of Maple. [Abstract.] *Phytopathology,* 46:467 (1956).

Fowler, M. E.: Oak Wilt. *U.S. Dept. Agr., Forest Serv., Forest Pest Leaflet,* 29:1–7 (1958).

Gäumann, E. Toxins and Plant Diseases. *Endeavour,* 13:198–204 (1954).

Goss, Marie C., and C. S. Moses: "A Bibliography of the Dutch Elm Disease." (Mim.) U.S. Dept. Agr., Div. Forest Pathol. (1937).

Graves, A. H.: The Melanconis Disease of the Butternut (*Juglans cinerea* L.). *Phytopathology,* 13:411–435 (1923).

Gruenhagen, R. H., A. J. Riker, and C. Audrey Richards: Burn-blight of Jack and Red Pine Following Spittle Insect Attack. *Phytopathology,* 37:757–772 (1947).

Haddow, W. R., and F. S. Newman: A Disease of the Scots Pine Caused by *Diplodia pinea* Associated with the Pine Spittle-bug. *Trans. Roy. Can. Inst.,* 24:1–17 (1942).

Harris, H. A.: Initial Studies of American Elm Diseases in Illinois. *Illinois Nat. Hist. Survey Bull.,* 20(1):1–70 (1932).

Hartley, C., and B. S. Crandall: Vascular Disease in Poplar and Willow. [Abstract.] *Phytopathology,* 25:18–19 (1935).

Hepting, G. H.: Sapstreak, a New Killing Disease of Sugar Maple. *Phytopathology,* 34:1069–1076 (1944).

Himelick, E. B., and E. A. Curl: Transmission of *Ceratocystis fagacearum* by Insects and Mites. *U.S. Dept. Agr., Plant Disease Reptr.,* 42:538–545 (1958).

Johnson, T. W., Jr., et al.: Observations on Yellow-poplar Dieback. *Forest Sci.,* 3:84–89 (1957).

Liese, J.: Zum Triebsterben der Kiefer. *Deut. Forstwirtsch.,* 17:381–383 (1935).

McCallum, A. W.: The Occurrence of the Dutch Elm Disease in Canada. *Forestry Chronicle,* 22:203–209 (1946).

McKenzie, H. L., L. S. Gill, and D. E. Ellis: The Prescott Scale (*Matsucoccus vexillorum*) and Associated Organisms That Cause Flagging Injury to Ponderosa Pine in the Southwest. *J. Agr. Research,* 76:33–51 (1948).

McKenzie, M. A., and W. B. Becker: The Dutch Elm Disease. A New Threat to the Elm. *Mass. Agr. Expt. Sta. Bull.,* 343:1–16 (1937).

McKenzie, M. A., and E. M. Johnson: Cephalosporium Elm Wilt in Massachusetts. *Mass. Agr. Expt. Sta. Bull.,* 368:1–24 (1939).

Metcalfe, G.: The Watermark Disease of Willows. *New Phytologist,* 39:322–332; 40:97–107 (1940–1941).

Peace, T. R.: The Status and Development of Elm Disease in Britain. [*Gt. Brit.*] *Forestry Comm. Bull.,* 33:1–44 (1960).

Quirke, D. A., and H. H. V. Hord: A Canker and Dieback of Balsam Fir in Ontario. *Can. Dept. Agr., Sci. Serv., Forest Biol. Div., Bi-monthly Progr. Rept.*, 11(6):2 (1955).

Roth, E. R., G. H. Hepting, and E. R. Toole: Sapstreak Disease of Sugar Maple and Yellow-poplar in North Carolina. *U.S. Dept. Agr., Forest Serv., Southeastern Forest Expt. Sta. Research Notes*, 134:[1–2] (1959).

Rudolph, B. A.: Verticillium Hadromycosis. *Hilgardia*, 5:197–353 (1931).

Silverborg, S. B., and R. W. Brandt: Association of *Cytophoma pruinosa* with Dying Ash. *Forest Sci.*, 3:75–78 (1957).

Tehon, L. R.: Elm Diseases in Illinois. *Natl. Shade Tree Conf. Proc.*, 10:105–111 (1934).

Thompson, G. E.: Die-back of Sycamore. *U.S. Dept. Agr., Plant Disease Reptr.*, 35:29–30 (1951).

Toole, E. R., and G. H. Hepting: Selection and Propagation of *Albizzia* for Resistance to Fusarium Wilt. *Phytopathology*, 39:63–70 (1949).

Toole, E. R., and P. C. Lightle: Status of Persimmon Wilt, 1950. *U.S. Dept. Agr., Plant Disease Reptr.*, 44:45 (1960).

Uri, Jeanne: Het parasitisme van *Nectria cinnabarina* (Tode) Fr. [The Parasitism of *Nectria cinnabarina* (Tode) Fr.] [English summary.] *Tijdschr. Plantenziekten*, 54:29–73 (1948).

Van Vloten, H., and J. Gremmen: Studies in the Discomycete Genera *Crumenula* de Not. and *Cenangium* Fr. *Acta Botan. Neerl.*, 2:226–241 (1953).

Walter, J. M., and C. May: Dutch Elm Disease and Its Control. *U.S. Dept. Agr. Circ.*, 677:1–12 (1943).

Waterman, Alma M.: *Diplodia pinea* and *Spaeropsis malorum* on Soft Pines. *Phytopathology*, 33:828–831 (1943*a*).

Waterman, Alma M.: *Diplodia pinea*, the Cause of a Disease of Hard Pines. *Phytopathology*, 33:1018–1031 (1943*b*).

Welch, D. S., and J. G. Matthysse: Control of the Dutch Elm Disease in New York State. *Cornell Ext. Bull.*, 932:1–14 (1955).

Whitten, R. R., and R. U. Swingle: The Dutch Elm Disease and Its Control. *U.S. Dept. Agr., Agr. Inform. Bull.*, 193:1–12 (1958).

Wright, E.: Further Investigations of Brown-staining Fungi Associated with Engraver Beetles (*Scolytus*) in White Fir. *J. Agr. Research*, 57:759–773 (1938).

CHAPTER 15

Stem Diseases Caused by Mistletoes, Dwarf Mistletoes, Lichens, and Climbers

Mistletoes and dwarf mistletoes are semiparasites or partial parasites, but certain other seed plants and lichens found on trees are epiphytes, i.e., nutritively independent plants that receive mechanical support from other plants but are not parasitic on them and have no connection with the soil. Climbers ascend by climbing upon trees or any other support but have their own root system in the soil. Dodders (*Cuscuta* spp.) are parasitic on other plants, but they are of no importance on forest trees, except occasionally on broadleaf species in nurseries (Latham 1938).

AMERICAN MISTLETOES

Mistletoes, also termed "true mistletoes," are seed plants belonging to the family Loranthaceae. Of the many genera of mistletoes throughout the world, mostly tropical, two occur in Europe. *Loranthus europaeus* Jacq. on oaks and *Viscum cruciatum* Sieb. on various hardwoods are limited in distribution, whereas *V. album* L. on hardwoods and conifers is widespread (Tubeuf 1923:94). In the United States only *Phoradendron* (tree thief), a genus restricted to the Americas, is represented.

Distribution, Species, and Hosts. *Phoradendron* or American mistletoes (Trelease 1916) range from Oregon, Colorado, the mouth of the Ohio River, and southern New Jersey southward into Mexico, Central America, South America, and the West Indies. There are several species in the United States, mainly on hardwoods but a few on conifers, mostly restricted to the Southwest and mostly on trees of minor economic importance. Whether any hardwood is immune to attack by mistletoe is questionable. The following are the more common species (Fosberg 1941):

Eastern mistletoe [*Phoradendron flavescens* (Pursh.) Nutt. var. *glabriusculum* Engelm.] is the only mistletoe occurring in the eastern United States north of Florida. It has white berries and is found on many hardwoods, although not reported in yellowpoplar. Black locust is highly susceptible. This mistletoe is usually confined to a single host in a given region (Trelease 1916:16), which suggests host races, strains, or forms, such as occur in *Viscum album* (Tubeuf 1923:100). Varieties within a species, age of twigs, and time of year also affect susceptibility to *V. album* (Scholl 1957).

In the West, hairy mistletoe [*Phoradendron flavescens* var. *villosum* (Nutt.) Engelm.], chiefly on the Pacific Coast from Oregon through California to Lower California and sparingly east to Texas, occurs on various hardwoods and especially on oaks, the principal hosts. It causes large hypertrophies on oaks. This variety is distinguished by its straight velvety pubescence. The berries are pinkish white. *P. flavescens* var. *pubescens* Engelm. occurs in Texas, Arizona, and New Mexico south to central Mexico on a wide variety of broadleaf trees. It is characterized by a stellate or tufted, frosty-appearing pubescence. The color varies from yellow-green to whitish and the berries are white. *P. flavescens* var. *macrophyllum* Engelm. ranges from central California through Arizona and New Mexico to western Texas on various broadleaf trees. It is glabrous or glabrate with white berries that may be tinged with pink.

Desert mistletoe (*P. californicum* Nutt.) occurs in southern California, Nevada, Arizona, and northwestern Mexico. It is a leafless species, generally pendent, with long stems and reddish-pink berries, chiefly on Leguminosae such as mesquite, but never on conifers.

Fir mistletoe [*P. bolleanum* (Seem.) Eichler var. *pauciflorum* (Torr.) Fosberg] is found in the Sierra Nevada of California, in lower California, and in Arizona. The plant is glabrous or sparsely hispidulous, has straw-colored berries, an open habit, and attacks white fir, commonly killing the leader and subsequent volunteers. It usually occurs high up in tall trees. It also occurs on *Cupressus*.

Dense mistletoe [*P. bolleanum* var. *densum* (Torr.) Fosberg] ranges from southern Oregon through California into Arizona and Mexico. It is glabrous or sparsely hispidulous, has straw- or wine-colored berries and a compact habit, and occurs on *Juniperus* and *Cupressus*.

P. bolleanum var. *bolleanum* Fosberg occurs from Texas to central Mexico on *Juniperus* and *Arbutus*. It is glabrous or sparsely hispidulous.

P. bolleanum var. *capitellatum* (Torr.) Kearney and Peebles is found from western Texas through New Mexico and Arizona, and south into Mexico. It is stellate tomentulose with linear-spatulate leaves and occurs on *Juniperus*.

Juniper mistletoe (*P. juniperinum* Engelm.) ranges from Colorado and Utah through New Mexico, Arizona, and Texas into Mexico. It is a leafless species with straw- or wine-colored berries attacking *Juniperus*. It has been found once on fernbush (Hawksworth 1952).

Constricted mistletoe [*P. juniperinum* var. *ligatum* (Trel.) Fosberg] ranges from southern Oregon through California and western Nevada into Mexico. It is leafless with the stem scales sharply constricted and grooved at the base, which serves to distinguish it from *P. juniperinum*. It occurs on *Juniperus*.

Incense cedar mistletoe (*P. juniperinum* var. *libocedri* Engelm.) occurs throughout the range of incense cedar, to which it is confined, in Oregon, California, Nevada, and Lower California. The pendent plants are leafless and have straw- to salmon-colored berries. This mistletoe suppresses its host severely when infection is heavy, and it causes noticeable spindle-shaped swellings on limbs and trunks at the point of attack. It may exist in the trunk of an infected tree for over four hundred years in a completely parasitic manner, all aerial portions of the mistletoe disappearing, leaving only living haustoria and sinkers persisting in the inner bark and sapwood (Wagener 1925).

Damage. On the whole the damage caused by mistletoes in forest stands is not serious, with the exception of the incense cedar mistletoe. Stands of incense cedar may be so heavily infected that growth is markedly retarded. In addition some trees have noticeable swellings on the trunks, the wood of which is riddled with old sinkers and consequently worthless, whereas others seem to die from the effect of supporting so many plants of the parasite. In Texas and adjacent Oklahoma where tree growth is not vigorous, mistletoes are at their worst on hardwoods. Even there, York (1909:21) doubts that mistletoes ever kill trees, but Bray (1910:20) states that hackberry (sugarberry) and water oaks are often killed. Blumer (1910:244) reports frequent killing of velvet ash in Arizona. Fir mistletoe, by killing the tops of large trees, reduces the seed crop since fir cones are borne almost entirely in the top of the crown (Fowells and Schubert 1956:29).

Symptoms. The best indication of infection is the presence of the mistletoe plant. A common effect on hardwoods is atrophy and final death of that portion of a branch beyond the point of infection. Frequently at the point of attachment of the mistletoe plant, a branch is stimulated to excessive growth, resulting in swellings of various shapes on different hosts. The hairy mistletoe causes unusually prominent hypertrophies on oaks. Occasionally, as in Osage orange, a branch is stimulated to an excessive formation of twigs resulting in a loose witches'-broom. If infection occurs when a tree is young and persists, the trunk and all the branches may be greatly deformed. Again trees may be infected for years without showing any significant deformation. In any case the vitality of the infected branch is reduced, and, if infection is sufficiently general, i.e., of the whole tree, reduced growth results. Mistletoes do not kill seedlings and saplings.

The Parasite. As is generally characteristic for the Loranthaceae, the "seed" is a naked embryo and endosperm invested with a fibrous coat, the endocarp of the fruit. The seeds are borne in white, straw-colored, pink, or red fruits known as "berries" and are embedded in a sticky, gelatinous pulp enabling them to adhere tightly to the bark of trees. The seeds are largely distributed by sticking to beaks and feet of birds or by being eaten and dropped in the excrement. Some may fall naturally to lower branches or be washed down by rain. Under favorable conditions the seeds germinate almost anywhere but they can penetrate only young, thin bark. The radicle on coming in contact with its host flattens out, forming a circular attachment disc. From the underside of this disc develops a papillalike projection, the primary haustorium, which penetrates the outer bark through lenticels or axillary buds. Whether unbroken bark can be penetrated is not known (Cannon 1904), but *Viscum album* can do so. On reaching the living tissues of the host, branches termed "cortical haustoria" are sent out, which ramify irregularly through the inner bark, longitudinal extension being most rapid. From the cortical haustoria are developed the sinkers that grow radially through the inner

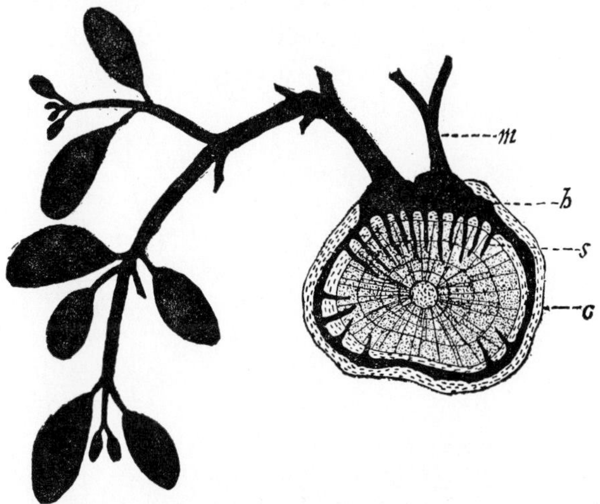

Fig. 125. Cross section of a branch infested with mistletoe: *m*, shoot of mistletoe; *h*, haustorium; *s*, sinker; *c*, bark of host. (*After F. D. Heald.*)

bark to the cambium, later becoming embedded in the wood by the formation of new growth rings (Fig. 125). The sinkers occupy the rays particularly.

The aerial portions of the mistletoes appear as leafy or sometimes leafless, perennial, evergreen, bunched tufts of shoots on the stems of their hosts, being particularly conspicuous on hardwoods after leaf fall (Fig. 126). All species are opposite leaved, although in a few the leaves are reduced to mere scales at the stem joints. The stems and leaves are green, but the plants may have a yellowish-green, golden, brown, or olive cast depending on the species and the season, variations from green being particularly prevalent during the winter. The plants are dioecious with the three-parted or occasionally two-, four-, or five-merous inconspicuous flowers without petals, borne on axillary, or occasionally terminal, several-jointed spikes. The chromosome sex mechanism of several California mistletoes has been investigated (Baldwin and Speese 1957). The first-formed shoots usually arise from buds that develop on the attachment disc, with occasionally more than one shoot arising from a single disc. Buds also arise later on the upper surface of the cortical haustoria, and

Fig. 126. Hairy mistletoe on Oregon white oak. (*Photograph by R. P. Marshall.*)

these buds may produce additional shoots. The stems are round and jointed.

Rate of growth is slow at first, and the primary shoot may not grow more than ¼ or ½ inch in length during the first season. After cortical haustoria are thoroughly established, growth is more rapid. Shoots from dormant buds on oaks have been known to grow 4 to 6 inches in a single season after spring pruning of the previously developed crop of shoots.

On water oaks in wet bottom-land soil in the Southwest, plants have developed in 6 to 8 years with a spread of nearly 3 feet and with a stem more than 1 inch in diameter just above the buttressed point of attachment to the tree. The longevity of the haustoria appears limited only by the life of the host. However, the aerial part of any plant on the average does not survive more than 8 or 10 years, or in exceptional cases as long as 20 years, being killed by cold (Wagener 1957), broken off by storms, or otherwise mechanically destroyed. Destruction of the aerial part of the parasite usually stimulates the development of adventitious buds and multiplies the number of shoots upon a widening area.

The stems and leaves contain normal chlorophyll (Cannon 1901:376), so that mistletoes can elaborate their own food from water and mineral salts obtained from the host; hence the plants are semiparasites or water parasites. According to Harris and his colleagues (1930), in general the tissue fluids of mistletoes are more highly concentrated than those of their hosts. This is considered a necessary condition for permanent survival of the parasite where the supply of water in the conducting tissues of the host is too limited to meet the requirements of both host and parasite at all times, but if water in the conducting tissues is always ample for both, the parasite may exist with a lower osmotic concentration of its tissue fluids than prevails in the host.

American mistletoes require much sunlight for their best development. This fact probably explains why in the eastern hardwood forests and on white fir in California the plants are usually restricted to the upper crown of tall trees and also why the parasites make their best growth in the open stands of the drier Southwest and southern California.

Control. Because of the traditional sentiment surrounding mistletoe, it is not generally recognized as a harmful parasite on trees, and, even if it is, the mistletoe may be desired for its decorative effect. Consequently attempts at control may meet with antagonism.

On young shade and ornamental trees, the outer end of each infected branch should be cut off 1 or more feet closer to the trunk than the area of infection as soon as the mistletoe shoots appear. On trees with infection already heavily established, the smaller branches should be cut off and the plants should be removed from the larger stems by cutting out the underlying bark and wood for 1 or more feet each way from the point of attachment. The cut surface should be treated with a disinfectant. Simply knocking off mistletoe plants merely results in the development of new shoots over a widening area, although if the successive crops of shoots are in turn removed every year or two the injurious effects of the parasite are reduced. Winter spraying with 2,4-D has killed from 50 to 85 per cent of the Phoradendron plants on walnut in California (Graser

1954). Trunk injections with 2,4-D are reported to kill *Loranthus* on *Eucalyptus* in Australia (Greenham et al. 1951).

In managed forests infected trees should be removed as early as possible during intermediate cuttings. In untreated stands infected trees should if possible all be removed in the first cutting. Trees with trunk infections are particularly undesirable. Since mistletoe requires much light, reasonably dense stands will suffer little, but mistletoe is most severe on trees occupying poor sites where it is difficult or impossible to maintain adequate density.

Mistletoes have few enemies to keep them in check (Ellis 1946:46). *Macrophoma phoradendri* Wolf causes partial or complete defoliation of mistletoe in Texas, but new shoots and leaves are formed again. *Protocoronospora phoradendri* Darling and *Phyllosticta phoradendri* Bonar are destructive to long-spiked mistletoe in California. *Uredo phoradendri* Jacks. kills shoots of juniper mistletoe beyond the point of infection (Hawksworth 1953). Juniper mistletoe can be completely killed by smelter smoke (Long 1922).

DWARF MISTLETOES

Dwarf mistletoes are seed plants belonging to the genus *Arceuthobium* (*Razoumofskya*) of the family Loranthaceae. The extensive literature was reviewed by Kuijt (1955). The genus is restricted to conifers and is confined to the northern hemisphere, the greatest variety of forms occurring in western North America. Only four species are known outside of North America. *Arceuthobium* in the United States was monographed by Gill (1935) and the 13 poorly defined species formerly recognized were reduced to 5, with 2 of them embracing several forms based on host relationships.

Distribution, Species, and Hosts. In North America dwarf mistletoes range from Alaska to Guatemala and eastward in Canada and the United States to the eastern slope of the Rocky Mountains. They are unknown in the Black Hills of South Dakota. Only one species occurs in the East.

Key to the Species of the United States
(Gill 1935:150)

Fruit maturing in the autumn immediately following pollination; floral processes on a shoot all in the same stage of development. *A. pusillum*

Fruit maturing in the second autumn after pollination; floral processes on a shoot often in two stages of development, i.e., buds and flowers or flowers and fruit.

- Flowers blooming in the late summer (August–September) . . . . . *A. campylopodum*
- Flowers blooming in the spring (April–June)

Accessory branches arising from collateral buds, forming a whorl . . . . *A. americanum*
Accessory branches arising from superposed buds in fanlike arrangement
Plants greenish, slender, stems never more than 1 mm in diameter at the base or over 35 mm long . . . . *A. douglasii*
Plants yellowish, robust, stems more than 2 mm in diameter at the base and 20–150 mm long . . . . *A. vaginatum*

In general the various species and forms are confined to a definite host or group of hosts, but transfers from one host species or genus to another do occur (Kuijt 1955:603).

Eastern dwarf mistletoe (*A. pusillum Peck*) is the only species occurring in the East. It ranges from Minnesota east to New Jersey and north into Canada, being common on black spruce, infrequent on red and white spruces, rare on tamarack, and

Fig. 127. Shoots of eastern dwarf mistletoe on black spruce.

found once on eastern white pine and on jack pine (Pomerleau 1942). The shoots are extremely inconspicuous, rarely attaining a height of more than ¾ of an inch (Fig. 127).

Lodgepole pine dwarf mistletoe (*A. americanum* Nutt.) is common on the mountain form of lodgepole pine but has not been found on the coast form (Gill 1957). It occurs on jack pine in western Canada but does not follow this host to the East. Occasionally it may occur on other pines when these are associated with infected lodgepole pine.

Douglas fir dwarf mistletoe (*A. douglasii* Engelm.) is confined to and common on Douglas fir throughout the range of the tree except that it does not occur in western Oregon and Washington, where Douglas fir attains its optimum development. The plants are small, rarely attaining a height of 1½ inches (Fig. 128).

Southwestern dwarf mistletoe [*A. vaginatum* (Willd.) Presl.] is prevalent in Colorado, Utah, Arizona, New Mexico, and Mexico on ponderosa pine (Andrews 1957). It has been reported on Apache and Chihuahua pines.

Western dwarf mistletoe (*A. campylopodum* Engelm.) is widespread on the Pacific Coast from Canada south into Mexico, and in Idaho and western Montana (Kimmey and Mielke 1959). The species is divided into a number of forms. *A. campylopodum* forma *typicum* occurs on Coulter, digger, Jeffrey, Monterey, and ponderosa pines but is rare on knobcone and lodgepole pines; f. *divaricatum* occurs on piñons; f. *cyanocarpum* on bristlecone, foxtail, limber, and whitebark pines; f.

blumeri on Mexican white, sugar, and western white pines; f. *abietinum* on alpine, corkbark, lowland white, noble, Pacific silver, red, and white firs; f. *tsugensis* on mountain and western hemlocks; f. *laricis* on alpine and western larches; and f. *microcarpum* on blue, Engelmann, and weeping spruces. Sitka spruce is not attacked.

Damage. Dwarf mistletoes cause extensive damage in western coniferous forests (Korstian and Long 1922; Weir 1916*b*). In the future when the overmature virgin stands have been largely cut over, these parasites, if not controlled, will be far more damaging than fungi. Trees of any age may be deformed or killed, with the greatest mortality among seedlings and saplings; lodgepole and ponderosa pines, and red fir being particularly susceptible to killing. In general dwarf mistletoes reduce the vigor and rate of growth of their hosts, so that infected stands require a longer time to mature and even then often produce a lower quality of timber. In a cutover ponderosa pine stand in Arizona over a 30-year period, mistletoe was the greatest single factor in retarding growth, and it was also responsible for over half the mortality of trees of merchantable size during the same period (Pearson and Wadsworth 1941). Douglas fir, red fir, lodgepole pine, ponderosa pine, western hemlock in Alaska and British Columbia, and western larch are most severely attacked, and stands may be so heavily infected as to be practically worthless.

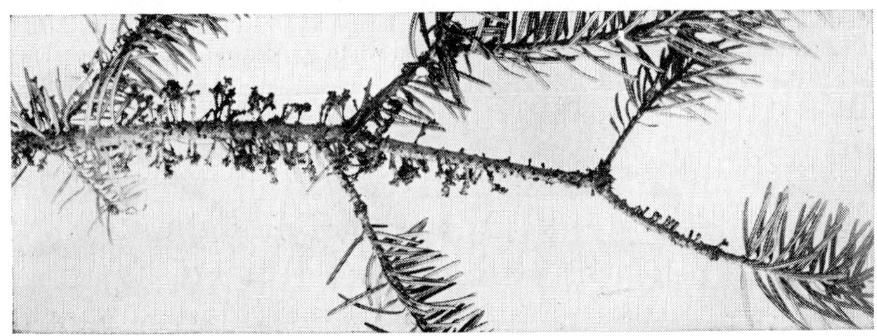

Fig. 128. Shoots of Douglas fir dwarf mistletoe.

Dwarf mistletoes also reduce the quality and quantity of merchantable timber. The formation of witches'-brooms leads to the development of stouter limbs to support the brooms, thus resulting in larger knots in the lumber. Trunk infections increase the number of branches. These increases in size and number of knots may appreciably lower the grade of the lumber. Trunk infections commonly produce large swellings or burls with abnormally grained, spongy wood that is frequently stained, pitchy, or decayed, so that part or all of a log is worthless. Trunk infections are particularly abundant and damaging in western larch (Weir 1916*a*:22) and western hemlock (Wellwood 1956).

In addition the seed crop is affected. For ponderosa pine in Arizona in 1913 the reproductive value per tree was reduced by 60 per cent in the moderately infected trees and by 75 per cent in the heavily infected as compared to the uninfected (Korstian and Long 1922: 26). For Douglas fir, lodgepole pine, and western larch the percentage of germination was reduced by mistletoe infection (Weir 1916*b*:30). For Jeffrey pine, germination was 20 per cent lower, and the resulting seedlings were less vigorous when seed from infected and uninfected trees was compared (Munns 1919). The effect of mistletoe on the progeny is a strong reason for the elimination of infected trees.

Fig. 129. Witches'-brooms on Douglas fir caused by dwarf mistletoe. The crown is practically one large broom, and the top is dead.

Heavy mistletoe infection predisposes ponderosa pine and Douglas fir to attack by bark beetles, and therefore mistletoe areas may form centers for the spread of insect infestations (Weir 1916*b*:29). After infected stands are logged, the heavy witches'-brooms increase the fire hazard in slash. Branches with brooms, because of the abnormal amount of compression wood and resin they contain, decay more slowly than normal branches, thus maintaining the fire hazard for a longer time if the slash is not burned after logging. Furthermore, living or dead standing trees, particularly Douglas firs, with large brooms often promote crown fires.

Symptoms. The surest sign of infection is the presence of the mistletoe plant. But if the shoots have disappeared, the small basal "cups" in which they were inserted remain on the bark, or a cross section of the branch examined under a hand magnifier will usually show the yellowish, wedge-shaped sinkers. The most striking symptom is the formation of witches'-brooms, which are particularly dense in Douglas fir and eastern spruces. In these species most or all of the crown may be transformed into a huge broom (Fig. 129), often characterized in Douglas fir by drooping branches but in the eastern spruces by more or less upright branches. In

Fig. 130. Witches'-broom on ponderosa pine caused by western dwarf mistletoe. Notice the pronounced taper of the branches in the dead broom. (*Photograph by S. L. Frost.*)

Fig. 131. Witches'-brooms on ponderosa pine caused by western dwarf mistletoe. The tree is nearly dead.

pines (Figs. 130, 131) and balsam firs the brooms are not so dense, and the affected branches are markedly increased in diameter with a pronounced taper from base to tip; in hemlocks the brooms are usually horizontally flattened; and in western larch the branches tend to radiate in several directions from an approximate center. Broomed branches of western larch break off readily during storms; consequently severely infected trees may ultimately lose part or all of their crowns, death finally resulting.

Distinct fusiform swellings are regularly produced on infected stems of all species except Douglas fir and eastern spruces, on which they may or may not occur. Swellings resulting from trunk infections often persist throughout the life of the tree and finally become somewhat cankerous through death of part of the infected area. These cankerous swellings of the trunk are common on white fir (Fig. 132), red fir, and western hemlock, affording an entrance for wood-destroying fungi. Not uncommonly the trunk breaks off at the swelling when this is decayed. On small main stems and branches of white fir, bark-inhabiting fungi may enter the killed portion of the swelling, rapidly girdling the stem. Mistletoe-infected branches of red fir in California are commonly invaded by *Cytospora* sp.

Fig. 132. Swelling and canker on the trunk of white fir caused by western dwarf mistletoe. (*After L. S. Gill.*)

Mistletoe also causes spiketop. Infection occurs first and is heaviest in the lower crown, where numerous brooms develop. These brooms appropriate more and more food materials that are needed for the tree as a whole, resulting in partial starvation and sometimes death of the leader and upper crown (Fig. 129).

The foliage of infected trees is reduced. For ponderosa pine Korstian and Long (1922:22) conclude that

> Trees heavily infected with mistletoe not only have considerably shorter leaves and leaf tufts than healthy trees, but the leaves of mistletoe-infected trees are lighter in color than those of healthy trees, being a yellowish-green as compared with the olive-green on the healthy trees. Each heavy infection causes a localization and reduction of the photosynthetic or assimilatory leaf surface of the tree, which in turn results in a marked decrease in the rate of growth.

For Douglas fir Weir (1916*b*:18) found that, although needles on young brooms may be abnormally long, on old brooms the needles were a little less than one-half as long as those on normal branches, and the foliage on the part of the branch beyond the brooms was thin. Needles on very old witches'-brooms of western larch yellowed earlier than those on branches of uninfected trees. Dowding (1929:99) found that infection by male plants on branches of jack pine caused no apparent change in the needles, but infection by female plants resulted in distinctly yellowish needles of about half normal length, and the branch internodes were

Fig. 133. Female plant of western dwarf mistletoe, with mature berries, on ponderosa pine.

markedly shortened. Needles on witches'-brooms of eastern spruces are shorter and yellowish in color. When weaker and more shaded branches are attacked, although the effect on the needles is the same, growth in length of the twigs is increased. This accentuation of length growth is more common to Douglas fir, resulting in the drooping willowy branches so often seen in witches'-brooms on this tree.

The Parasite. The ovoid olive-green to dark-blue fruits or berries, each containing a single seed (rarely two) embedded in a mucilaginous pulp, are borne on short pedicels that become reflexed at maturity (Fig. 133). On ripening the berries develop a considerable inside pressure, which increases until any slight disturbance causes them to explode and eject the seed with some force to a maximum horizontal distance of 53 feet, but usually about 15 feet or less. In dense young stands dispersal is even more limited. Consequently, in one-storied stands the spread to new trees is slow. However, in uneven-aged stands an infected tree in the overstory can scatter seeds into the understory over a radius of 60 feet, with occasional seeds going greater distances (Gill 1954). Local distribution largely occurs in this way, but animals or birds must be responsible for long-distance spread, since the seeds cannot be carried far by wind. The seeds with their sticky coating are scattered at random. Many perish, and others adhere to the bark of young branches. On germination the radicle grows along the surface of the stem until it is obstructed by a bud or a leaf base, in which case an irregular mound of tissue develops, functioning as a holdfast, from the underside of which the primary haustorium develops. The primary haustorium penetrates the bark as far as the cambium, and penetration usually does not occur through bark more than 3 years old, except in lodgepole pine with its thin bark, where the majority of infections may be established through bark over 5 years old and occasionally through bark over 60 years old (Hawksworth 1954). From the primary haustorium the cortical haustoria develop, longitudinal extension being more rapid than growth in other directions. *A. pusillum* appears to be able to grow longitudinally only toward the growing tips of the host branches, whereas the other species grow in both directions. The sinkers originate at those points where the cortical haustoria touch the cambium. The sinkers usually follow the rays, but they do not actually grow into the wood, merely becoming embedded as new growth rings are added. The life of the cortical system is limited only by the life of the host part in which it occurs.

The aerial portions of dwarf mistletoes appear as perennial shoots, either simple or branched, on the host stem. Even in the largest species the shoots rarely attain a maximum length of 8 inches. In *A. pusillum* they average only from ½ to ¾ inch in length and in *A. douglasii* from ¼ to ¾ inch, rarely attaining to 1½ inches (Figs. 127, 128). The shoots are jointed. The segments are four-angled at the base and are usually angular throughout their length, although occasionally they may be round at the upper end. The leaves are reduced to opposite pairs of scales at the top of each segment. In color the stems are extremely variable, this character being influenced by light, temperature, and latitude. Color varies from yellowish to brownish to brownish green and olive-green in many shades. The plants are dioecious. The male flowers are three parted (Fig. 134), rarely two or five parted, arise at the stem nodes, and are sessile except invariably in *A. americanum* and occasionally in *A. doug-*

lasii, where they are borne on small pedicel-like segments. The female flowers are two parted, rarely three parted, sessile, and arise at the stem nodes or terminally. *A. pusillum, A. americanum,* and *A. douglasii* bloom in the spring from April to June, *A. campylopodum* blooms in the late summer from August to September, and *A. vaginatum* reaches its peak of flowering between the other two groups, although it overlaps largely with the spring-flowering group. In *A. pusillum* the fruits mature the autumn following pollination, in *A. vaginatum* the second summer following pollination, and in the other three species the second autumn after pollination. Pollination is largely effected by insects.

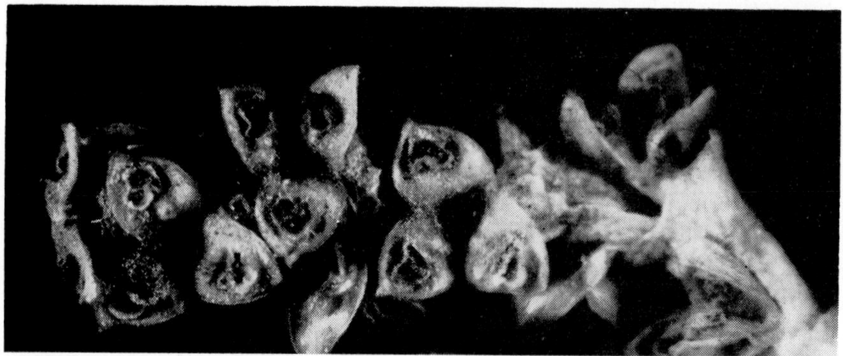

Fig. 134. Portion of a male plant of western dwarf mistletoe on white fir showing the fully mature male (staminate) flowers.

The shoots always arise from buds in the cortical system after the latter is well established. Usually they occur in tufts, but occasionally they are scattered along the young twigs. This variation is probably a reaction to the host. In *A. pusillum* on spruces the shoots are always scattered, but on tamarack the shoots are often clustered. *A. douglasii* has this scattered tendency, whereas the other species show both habits. The development of *A. pusillum* is somewhat different from the other species. According to Gill (1935:153),

> In the case of well established infections on spruce, the shoots appear in the autumn as tiny eruptions in the bark of the host growth two seasons old. The following year they attain practically full size and by autumn bear well-developed flower buds which bloom the following spring. Flowering shoots, therefore, are found on 3-year-old host growth. There are exceptions to this order when length-growth of the host is poor, when it is stunted by witch-broom formation, and in the rare cases when the shoots of the parasite produce two crops of flowers. Under such circumstances the flowering shoots may be found on older wood, but they have never been reported on younger growth.

After flowering the male shoots of *A. pusillum* die, as do the female shoots after the seeds are discharged. The cortical system, however, spreads into each new internode of the host stem, and the cycle is repeated. Shoots of the other species probably persist no longer than two or three fruiting seasons, and consequently the life of their shoots does not exceed 5 or 6 years, but new shoots continually arise from the persistent cortical system.

Dwarf mistletoes are not so exacting in their light requirements as *Phoradendron*, although their development is favored by light. Even though the aerial parts of dwarf mistletoes contain some chlorophyll, it is usually recognized that they take the larger part of the elaborated food required from their hosts, being more parasitic than *Phoradendron*. From cytological evidence, Dufrenoy (1936) infers that, where a haustorium of *A. campylopodum* is in contact with the rays of Jeffrey pine, active translocation of nutrients occurs. Cannon (1901:388) states that *Arceuthobium* is a total parasite.

Dwarf mistletoes occur under widely varying conditions. Western hemlock is severely parasitized in Alaska, whereas in western Oregon and Washington where this tree attains its optimum it is only occasionally infected and little damaged. Injury to Douglas fir is severe in eastern Oregon and Washington, yet Douglas fir dwarf mistletoe is unknown in western Oregon and Washington in the optimum range of Douglas fir, the western limit of the parasite being the summit of the Cascade Mountains. Eastern dwarf mistletoe is severe and abundant on spruces in acid bogs where the trees grow poorly. Himalayan pine in India is killed by *A. minutissimum* Hook. f. on poor sites at high elevations where the tree is valuable only as a cover, but on good sites timber crops are produced in spite of the continued presence of the parasite (Gorrie 1929). Against this, white fir is vigorously attacked in southern Oregon and California where the tree attains its largest size, but it is rarely infected elsewhere (Gill 1935:136). Furthermore the vigor of mistletoe shoots increases with the vigor of the tree or individual part on which they occur.

Control. Reduction of damage caused by dwarf mistletoes in coniferous forests of the West is the most important problem in applied forest pathology in that region. The parasites can be eliminated by cutting all infected trees from seedlings to veterans in a stand. In some stands this will necessitate clear-cutting followed by artificial regeneration. If a stand is freed from dwarf mistletoe, the spread into it again by mistletoe from surrounding infected stands will be slow and can be prevented by periodic cutting out of new infections. From residual stands, *A. americanum* on lodgepole pine can be expected to spread into reproduction on adjacent clear-cut areas less than 30 feet in about 20 years (Hawksworth

1958). It has been suggested that a protective zone cleared of all trees, or cleared of the susceptible tree species only, may be maintained between the diseased stand and the stand to be protected (Kimmey 1957).

Recommendations for control of dwarf mistletoes can be applied to most species of the parasite, although modifications may be necessary to suit different species. Dwarf mistletoes cannot be eliminated in one cutting because of latent or invisible infections established before the operation. For *A. vaginatum* on ponderosa pine the shortest time before the first mistletoe shoots appear after the seed alights is 26 months, and some may not appear for 10 years (Gill 1954:3). A second cutting, 5 years later, although it will not eliminate the mistletoe, should reduce it to a negligible loss factor for 20 years or more. Not until intermediate cuttings become practicable will it be possible to make complete progress against damage by dwarf mistletoes. For a time control will be largely limited to whatever can be accomplished at the time of logging when cutting is limited to trees of merchantable size. However, effort must be made to leave the reserve stand in better condition both immediately and for the future. Since the opening up of a stand stimulates growth of mistletoe on the infected trees remaining, it is necessary to cut as many infected trees as possible without removing all seed trees or completely exposing the soil on the most adverse sites where trees are needed for protective purposes. All heavily infected trees should be cut regardless of the condition of the stand after cutting. Trees with trunk infections, whether heavy or light, should be removed. Moderately and lightly infected trees should be cut, unless there are no other available seed trees. In stands with satisfactory advance reproduction all infected trees should be cut, but, if advance reproduction is inadequate, it may be necessary to leave some lightly or moderately infected trees for seed trees.

Control of dwarf mistletoes for species clear-cut is relatively simple provided that infected trees of all sizes are removed. In western hemlock, the rate of infection and ultimate damage by mistletoe can be reduced by clear-cutting to maintain relatively even-aged stands (Buckland and Marples 1952). In black spruce in the initial stage of infection of a stand, all trees with witches'-brooms should be cut, together with all apparently healthy trees within 40 feet of them (Anderson and Kaufert 1953). During the second stage, when a small opening surrounded by heavily broomed and dying trees develops in the stand, the infested area plus a one-chain strip beyond the broomed trees should be clear-cut. During the later stage, when the areas of infection are large and occupied by a new stand heavily infected, the infected surrounding old stand together with a strip one chain wide outside the broomed trees should be clear-cut to protect the remaining healthy stand.

Dwarf mistletoe can be eradicated by branch pruning of individual trees. Trunk infections cannot be satisfactorily cut out; trees thus affected must be removed. If shoots are on a branch within 1 foot of the trunk, the mistletoe is probably already established in the trunk. Infected branches can be cut off, but one pruning operation will not suffice because of latent infections that cannot be detected. Pruning one whorl above the highest infected limb will eliminate most of the latent infections. A second pruning must follow 5 years later, and probably a third after the same interval. In exceptionally valuable sites with a scarcity of overstory trees, pruning may be limited to the removal of female plants, thus preserving a canopy and still protecting the growing stock. However, control by pruning is impractical in most forest stands.

Operations against mistletoe are most timely just before the berries mature, when the maximum number of young shoots can be detected.

Chemical control of dwarf mistletoe has not yet been developed. Sprays are known that kill the aerial portion of the plants but the absorbing system is not affected, so that new shoots develop soon (Gill 1954).

The fact that even in heavily infected stands individual trees remain free from dwarf mistletoe indicates resistant host strains as found in ponderosa pine by Bates (1927:141), which leads to the hope that continual elimination of infected trees may finally result in relatively resistant stands. At least the indications of resistance so far observed warrant research on developing resistant trees.

Dwarf mistletoe infections should be eradicated from the immediate vicinity of nurseries by felling and pruning diseased trees. Although it is poor practice to plant infected stock even in areas where the parasite is already established, the greatest danger is the introduction of these parasites into regions where they do not exist. The extensive and valuable ponderosa pine stands in the isolated Black Hills region are now mistletoe free, but there is no apparent condition to preclude the establishment of dwarf mistletoe once it is introduced into the region. Aside from spruces and tamarack, eastern conifers are not attacked, but *A. campylopodum* has been successfully inoculated on red, slash, and Virginia pines and on eastern hemlock as well as on several exotic conifers sometimes planted in this country (Hedgcock and Hunt 1917; Weir 1918). Whether dwarf mistletoes from the West could become established in eastern forests is problematical, but reasonable precautions should be observed to prevent finding out by an unexpected and uncontrolled experiment.

Dwarf mistletoes have a few enemies (Gill 1935:219), but their attack is not sufficiently effective to prevent extensive damage to trees. Bark-eating rodents often destroy young infections. Spittle bugs sometimes mine out the stems, killing branches or entire shoots. *Septogloeum gillii*

Ellis kills large numbers of shoots in some areas (Mielke 1959). Another fungus, *Wallrothiella arceuthobii* (Pk.) Sacc., attacks the female flowers and prevents seed maturation (Weir 1915; Dowding 1931). Since *W. arceuthobii* is most prevalent in damp localities there is little likelihood

Fig. 135. A foliose lichen (*Parmelia* sp.) on the bark of a fallen tree.

that it will ever be of more than local significance in checking the development of dwarf mistletoes.

LICHENS

Lichens, which are of world-wide distribution, are so specialized as to structure and physiology that they are generally recognized as a distinct class of plants (Perez-Llano 1944). A lichen is composed of a fungus body with a green or blue-green alga enclosed by the hyphae. The fungus obtains elaborated food from the alga, whereas the latter receives some food materials and protection from the fungus. Some lichens grow on the dryest and barest rocks. In temperate regions the fungi living with algae are Ascomycetes with fructifications usually of the apothecial type.

The three forms of lichens—crustose, foliose, and fruticose—frequently grow on living or dead trees. The crustose species are mostly found on the bark of the trunk or large limbs. They grow as a crust so closely appressed to the bark that they cannot be removed from it. The foliose forms (Fig. 135) are leaflike and prostrate but not so firmly attached to the substratum, and the fruticose forms (Fig. 136) are bushlike and erect or hanging. The effect of lichens on trees is only slightly detrimental. Actually, abundant development of lichens on trees indicates low vigor induced by some other cause (Romell 1922; Münch 1936:557). Lichens are often abundant on conifers occupying poor sites. The living tissues of the host are not usually penetrated by the fungus component. In South Africa one of the beard lichens (*Usnea* sp.) is harmful to *Podocarpus thunbergii* and *P. elongata*, its fungus component being parasitic on the tissues external to and sometimes internal to the cork cambium (Phillips 1929). When lichens develop abundantly on trees, twigs or small branches may be killed, which can be explained by a reduction in the amount of light received by the foliage and also by interference with gas exchange for the foliage and stems. In areas suffering from smelter smoke, lichens are conspicuously absent.

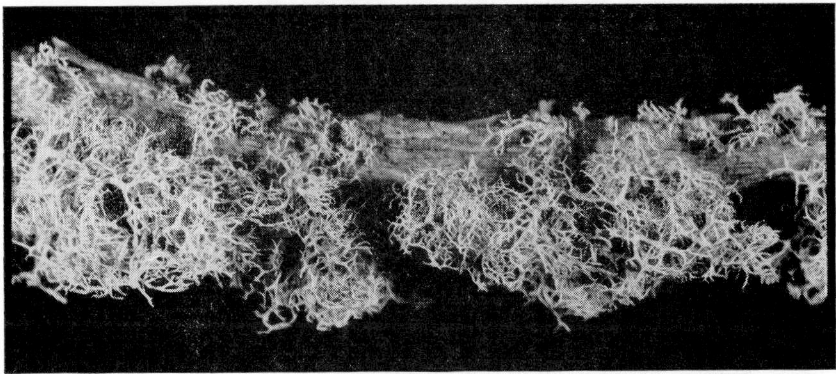

Fig. 136. A fruticose lichen (*Evernia vulpina*) on a dead, decorticated branch of ponderosa pine.

Spanish moss or long moss (*Tillandsia usneoides* L.), the long, pendulous, gray, hairlike or whiskerlike plant that develops so profusely on hardwoods and conifers (Fig. 137), especially on live oak and southern cypress, is neither a lichen nor a moss but belongs to the pineapple family (Bromeliaceae) of the flowering plants (Uphof 1931). Occasionally it is said to kill trees by smothering and reduction of light.

Fig. 137. Spanish moss on southern cypress.

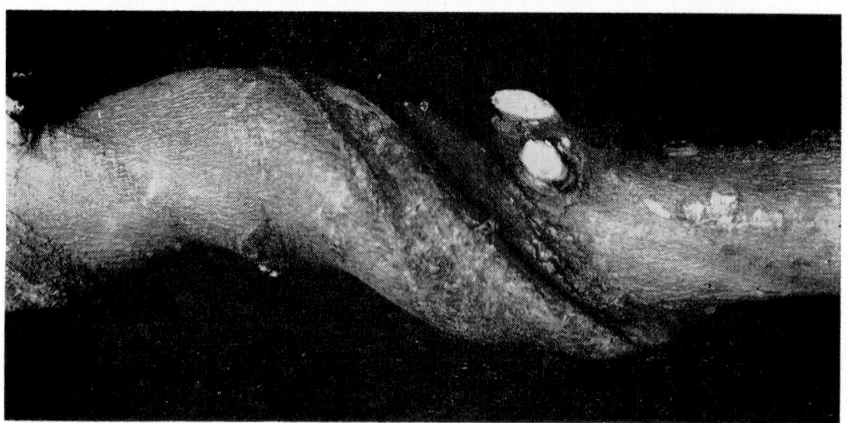

Fig. 138. Main stem of a hardwood sapling deformed by a climber.

CLIMBERS

Climbers have their own root system in the soil but ascend by climbing or leaning on various supports, trees being commonly used. The most injurious of climbers are the twiners, which ascend trees by twining so tightly around the trunk that the tree may be killed by strangulation or at least have the main stem badly deformed (Fig. 138). Climbing bittersweet (*Celastrus scandens* L.) commonly injures hardwoods in this way (Lutz 1943). Even the climbers which do not twine tightly about the trunk may develop so much foliage in the crown of the tree that light is cut off from the leaves of the tree and the weight of the vine may cause branches to break off. During stand-improvement work, the stems of all climbers should be severed above ground level.

REFERENCES

Anderson, R. L., and F. H. Kaufert: The Control of Dwarf Mistletoe on Black Spruce. *Minn. Forestry Notes,* 13:[1]–[2] (1953).

Andrews, S. R.: Dwarfmistletoe of Ponderosa Pine in the Southwest. *U.S. Dept. Agr., Forest Serv., Forest Pest Leaflet* 19:1–4 (1957).

Baldwin, J. T., Jr., and Bernice M. Speese: *Phoradendron flavescens:* Chromosomes, Seedlings, and Hosts. *Am. J. Botany,* 44:136–140 (1957).

Bates, C. G.: Better Seeds, Better Trees. *J. Forestry,* 25:130–144 (1927).

Blumer, J. C.: Mistletoe in the Southwest. *Plant World,* 13:240–246 (1910).

Bray, W. L.: The Mistletoe Pest in the Southwest. *U.S. Dept. Agr., Bur. Plant Ind. Bull.,* 166:1–39 (1910).

Buckland, D. C., and E. G. Marples: Management of Western Hemlock Infested with Dwarf Mistletoe. *Brit. Columbia Lumberman,* 36(5):50–51, 136, 138, 140 (1952).

Cannon, W. A.: The Anatomy of *Phoradendron villosum* Nutt. *Bull. Torrey Botan. Club,* 28:374–390 (1901).

Cannon, W. A.: Observations on the Germination of *Phoradendron villosum* and *P. californicum. Bull. Torrey Botan. Club,* 31:435–443 (1904).

Dowding, Eleanor S.: The Vegetation of Alberta III. The Sandhill Areas of Central Alberta with Particular Reference to the Ecology of *Arceuthobium americanum* Nutt. *J. Ecol.,* 17:82–105 (1929).

Dowding, Eleanor S.: *Wallrothiella arceuthobii,* a Parasite of the Jack-pine Mistletoe. *Can. J. Research,* 5:219–230 (1931).

Dufrenoy, J.: The Parasitism of *Arceuthobium* (*Razoumowskya*) *campylopodum* on *Pinus jeffreyi. Phytopathology,* 26:57–61 (1936).

Ellis, D. E.: Anthracnose of Dwarf Mistletoe Caused by a New Species of *Septogloeum. J. Elisha Mitchell Sci. Soc.,* 62:25–50 (1946).

Fosberg, F. R.: Notes on Mexican Plants. *Lloydia,* 4(4):274–290 (1941).

Fowells, H. A., and G. H. Schubert: Seed Crops of Forest Trees in the Pine Region of California. *U.S. Dept. Agr., Tech. Bull.,* 1150:1–48 (1956).

Gill, L. S.: *Arceuthobium* in the United States. *Trans. Conn. Acad. Arts Sci.,* 32:111–245 (1935).

Gill, L. S.: Dwarfmistletoe of Ponderosa Pine in the Southwest. *Rocky Mt. Forest and Range Expt. Sta. Paper*, 14:1–9 (1954).

Gill, L. S.: Dwarfmistletoe of Lodgepole Pine. *U.S. Dept. Agr., Forest Serv., Forest Pest Leaflet*, 18:1–7 (1957).

Gorrie, R. M.: A Destructive Parasite of the Himalayan Blue Pine. *Indian Forester*, 55:613–617 (1929).

Graser, H. I.: Methods for Control of Mistletoe; Spot Spraying in Winter Proves Effective. *Diamond Walnut News*, 36:10 (1954).

Greenham, C. G., et al.: A Progress Report on Mistletoe Control Investigations. *Australian Forestry*, 15:62–64 (1951).

Harris, J. A., G. J. Harrison, and T. A. Pascoe: Osmotic Concentration and Water Relations in the Mistletoes, with Special Reference to the Occurrence of *Phoradendron californicum* on *Covillea tridentata*. *Ecology*, 11:687–702 (1930).

Harris, J. A., T. A. Pascoe, and I. D. Jones: Note on the Tissue Fluids of *Phoradendron juniperinum* Parasitic on *Juniperus utahensis*. *Bull. Torrey Botan. Club*, 57:113–116 (1930).

Hawksworth, F. G.: Unusual Hosts for Two Southwestern Mistletoes. *U.S. Dept. Agr., Plant Disease Reptr.*, 36:246 (1952).

Hawksworth, F. G.: *Phoradendron juniperinum*, a New Host for *Uredo phoradendri*. *U.S. Dept. Agr., Plant Disease Reptr.*, 37:258 (1953).

Hawksworth, F. G.: Observations on the Age of Lodgepole Pine Tissues Susceptible to Infection by *Arceuthobium americanum*. *Phytopathology*, 44:552 (1954).

Hawksworth, F. G.: Rate of Spread and Intensification of Dwarfmistletoe in Young Lodgepole Pine Stands. *J. Forestry*, 56:404–407 (1958).

Hedgcock, G. G., and N. R. Hunt: Notes on *Razoumofskya campylopoda*. *Phytopathology*, 7:315–316 (1917).

Kimmey, J. W.: Dwarfmistletoes of California and Their Control. *Calif. Forest and Range Expt. Sta., Tech. Paper*, 19:1–12 (1957).

Kimmey, J. W., and J. L. Mielke: Western Dwarfmistletoe on Ponderosa Pine. *U.S. Dept. Agr., Forest Serv., Forest Pest Leaflet*, 40:1–7 (1959).

Korstian, C. H., and W. H. Long: The Western Yellow Pine Mistletoe. *U.S. Dept. Agr. Bull.*, 1112:1–35 (1922).

Kuijt, J.: Dwarf Mistletoes. *Botan. Rev.*, 21:569–626 (1955).

Latham, D. H., et al.: Dodder in Forest Nurseries. *U.S. Dept. Agr., Plant Disease Reptr.*, 22:23–24 (1938).

Long, W. H.: Mistletoe and Smelter Smoke. *Phytopathology*, 12:535–536 (1922).

Lutz, H. J.: Injuries to Trees Caused by *Celastrus* and *Vitis*. *Bull. Torrey Botan. Club*, 70:436–439 (1943).

Mielke, J. L.: Infection Experiments with *Septogloeum gillii*, a Fungus Parasitic on Dwarfmistletoe. *J. Forestry*, 57:925–926 (1959).

Münch, E.: Das Lärchensterben. *Forstwiss. Centr.*, 58:469–494, 537–562, 581–590, 643–671 (1936).

Munns, E. N.: Effect of Fertilization on the Seed of Jeffrey Pine. *Plant World*, 22:138–144 (1919).

Pearson, G. A., and F. H. Wadsworth: An Example of Timber Management in the Southwest. *J. Forestry*, 39:434–452 (1941).

Perez-Llano, G. A.: Lichens—Their Biological and Economic Significance. *Botan. Rev.*, 10:1–65 (1944).

Phillips, J. F. V.: The Influence of *Usnea* sp. (near *barbata*) upon the Supportmg Tree. *Trans. Roy. Soc. S. Africa*, 17:101–107 (1929).

Pomerleau, R.: Le gui de l'épinette noire dans le Québec. *Naturaliste can.*, 69:11–31 (1942).

Romell, L. G.: Hanglavar och tillväxt hos norrländsk gran. [German summary.] *Statens Skogsförsöksanst.* (*Sweden*), *Medd.*, 19:405–421 (1922).

Scholl, R.: Weitere Untersuchungen über Veränderungen der Reaktionslage des Birnbaumes (*Pirus communis* L.) gegenüber der Mistel (*Viscum album* L.). *Phytopathol. Z.*, 28:237–258 (1957).

Trelease, W.: "The Genus *Phoradendron*." Urbana, Ill.: University of Illinois Press (1916).

Tubeuf, C. von: "Monographie der Mistel." Munich: R. Oldenbourg-Verlag (1923).

Uphof, J. C. Th.: *Tillandsia usneoides* als Pflanzenschädling. *Z. Pflanzenkrankh.*, 41:593–607 (1931).

Wagener, W. W.: Mistletoe in the Lower Bole of Incense Cedar. *Phytopathology*, 15:614–616 (1925).

Wagener, W. W.: The Limitation of Two Leafy Mistletoes of the Genus *Phoradendron* by Low Temperatures. *Ecology*, 38:142–145 (1957).

Weir, J. R.: *Wallrothiella arceuthobii*. *J. Agr. Research*, 4:369–378 (1915).

Weir, J. R.: Larch Mistletoe: Some Economic Considerations of Its Injurious Effects. *U.S. Dept. Agr. Bull.*, 317:1–25 (1916*a*).

Weir, J. R.: Mistletoe Injury to Conifers in the Northwest. *U.S. Dept. Agr. Bull.*, 360:1–39 (1916*b*).

Weir, J. R.: Experimental Investigations on the Genus *Razoumofskya*. *Botan. Gaz.*, 66:1–31 (1918).

Wellwood, R. W.: Some Effects of Dwarf Mistletoe on Western Hemlock. *Forestry Chronicle*, 32:282–296 (1956).

York, H. H.: The Anatomy and Some of the Biological Aspects of the "American Mistletoe." *Texas Univ. Bull.*, 120:1–31 (1909).

CHAPTER 16

Stem Diseases: Decay

Decay or rot in wood is caused by fungi. Although bacteria have occasionally been associated with certain forms of decay, the extent to which they can break down solid wood is not yet determined. The fungous hyphae, which cannot be seen by the naked eye unless they occur in a closely woven feltlike mass known as a "mycelial felt" or "sheet" (Fig. 196), feed on the substances composing the cell walls of the wood. They use certain constituents of the cell walls, neglecting others, with the result that the cell walls are broken down, the wood being thus greatly weakened and more or less destroyed. It is the breaking down of the wood and the change in its physical and chemical properties that is termed "decay." Strictly speaking decay cannot be considered as disease, since in living trees it is usually only the dead heartwood that is involved in the process, but occasionally living sapwood is invaded and killed after decay is well advanced in the heartwood. However, the relation of the fungus to the tree can be considered one of parasitism if the dependence of one individual upon any part of another living individual for food is accepted as the basis of parasitism. Fungi require a balance between air and moisture to cause decay. Dry wood cannot decay because there must be at least 15 per cent of moisture in wood based on oven-dry weight before rot can begin, and 25 to 32 per cent (the fiber saturation point) or more is needed for decay of consequence. On the other hand, completely saturated wood cannot decay because air is essential to fungous growth and activity.

Since the classical work of Hartig (1878) first gave a comprehensive understanding of decay, a voluminous literature on the subject has developed because of the tremendous losses in timber stands, in wood during conversion, and in wood in service. Decay may also be beneficial, since wood-destroying fungi are mainly responsible for the disintegration of wood debris, thus adding valuable organic matter to the soil. The more recent general accounts of decay are those by Cartwright and Findlay (1958), Wagener and Davidson (1954), and Findlay (1956).

TYPES OF DECAY

In their effect on wood there are two broad classes of wood-destroying fungi. The first class, causing white rots, decomposes all components of the wood, including lignin. Affected wood is reduced to a spongy mass, to white pockets or streaks of various sizes separated by areas of firm, strong wood, or to a stringy or fibrous condition. The decomposed wood is usually white, but it may be yellow, tan, or even light brown in color. The second class, causing brown rots, decomposes the cellulose and its associated pentosans, leaving the lignin more or less unaffected. The wood is reduced to a carbonous mass in varying shades of brown, which can be powdered between the fingers and to which the name "dry rot" is often loosely applied. Dry wood cannot decay. Brown rot and white rot fungi can be separated by a chemical culture test (page 358). Sometimes the two types work in the same tree, a white rot fungus attacking the heartwood first and then being followed by a brown rot fungus that destroys the remaining relatively sound wood between the white pockets. In Douglas fir *Fomes laricis* occasionally follows *F. pini* in this way. White rots are not likely to follow brown rots (Findlay 1940).

Decays are often described according to their position in the tree as root rot, butt rot or stump rot confined to the base of the tree, trunk rot in the main portion of the bole, and top rot confined to the upper portion of the bole. Such terms must not be accepted literally for any one decay, since it is not uncommon to find rots out of their usual position in the trunk. Again the names "sap rot" and "heart rot" are used to denote the type of wood attacked.

STAGES OF DECAY

There are various stages in decay as wood is changed from sound to completely decayed. In the earliest stages the wood appears to be hard and firm, the only evidence of attack, if any, being a slight to marked color change from the normal. This is known as the "incipient," "initial," "early," "beginning," "first," "primary," "advance," or "invasion" stage. In some cases there is no indication of incipient decay, hyphae extending for several feet longitudinally in advance of the visible evidence of rot. Such a hidden stage can be detected only by a microscopic or cultural diagnosis. Incipient decay is dangerous since it is easily overlooked, and in some types of decay, notably brown rot, the wood is seriously weakened in the incipient stage and should not be used where strength is required. Furthermore, certain decays continue to develop under favorable conditions after the tree is converted, so that placing wood with incipient decay in use where decay durability is essential can result in unusually rapid deterioration. Then, too, the discolorations indicating incipient decay are easily confused with harmless discolorations or color variations, resulting in degrading or rejecting sound wood.

After the incipient stage is passed, the wood becomes more and more noticeably affected until it is finally changed completely in appearance and structure, with the continuity of the wood tissues destroyed. This is known as the "advanced," "late," "mature," "typical," "complete," "ultimate," or "destruction" stage, in which generally the strength of the wood is so reduced that it can be crumbled between the fingers or easily broken. The destruction may even go so far that the heartwood breaks down completely, leaving a large hollow in the tree with only the relatively thin layer of sapwood to serve as support. Occasionally an intermediate stage can be recognized between the incipient and advanced stages.

In a decaying trunk all stages of decay are present, passing vertically up or down the tree from advanced decay, where the fungus has been at work the longest, through incipient decay to sound wood. The same progression may be seen on a cross section passing from the center of the decayed portion of the heartwood to the sapwood. Vertically, incipient decay may extend a few inches to many feet beyond advanced decay, varying with the individual tree and the species of wood-destroying fungus. Horizontally or across the grain the extent of incipient decay is usually limited to 1 or 2 inches.

GROSS CHARACTERS OF DECAY

In the advanced stage of decay, wood is always so markedly changed in its physical properties that its condition is readily apparent to the unaided eye. In the incipient stage, its true condition is easily overlooked and sometimes can be determined only by careful laboratory diagnosis. Consequently decayed wood is often manufactured and used for purposes it cannot fulfill, so that a knowledge of the gross characters of decayed wood is essential for intelligent utilization.

Color. Pronounced color changes from the normal invariably characterize the advanced stage of decay, which is always easy to recognize, although it is not usually possible to identify exactly the fungus causing the rot merely by the appearance of the advanced decay. Discolorations accompany the incipient stage of some decays, more often in the white rots, but color changes may be absent. A color indicating incipient decay in one host may be different or absent in another, although the decay is caused by the same fungus. Furthermore there are numerous color variations in wood arising from other causes that can be confused with discolorations indicating incipient decay (Boyce 1923*a*; Anonymous 1943; Scheffer 1944; Hansbrough and Englerth 1944; Englerth and Hansbrough 1945).

Of course color is a natural characteristic of wood while still in the living tree (Anonymous 1956). For the first few years after formation, wood is white or nearly so, and finally, when the sapwood is transformed to heartwood, a decided color change takes place in most woods. In some, however, the change is negligible so that it is difficult to distinguish heartwood from sapwood. Furthermore there are natural color variations in the heartwood of a species. The heartwood of sweet gum is usually highly figured. In Sitka spruce the heartwood has a faint reddish tinge, slightly distinguishing it from sapwood, but in some trees the heartwood has a pronounced reddish- or brownish-pink color which is quite uniform throughout. The heartwood of sugar, western white, and eastern white pines often becomes a pink, light-red or vinous-red color on drying. The brown heartwood of incense and western red cedars often has a reddish to purplish tinge. Color variations may be connected with changes in the rate of growth. In Douglas fir heartwood that has a distinct reddish or reddish-orange hue, the reddish color may be intensified in long regular bands following a definite group of growth rings, narrower than those on both sides or containing a greater proportion of late wood. The so-called "yellow fir" from the slowly grown, exceedingly narrow-ringed outer layers of old Douglas fir is another example. Occasionally an apparent discoloration in heartwood may result from the failure of the wood to change color uniformly during the transition from sapwood to heartwood.

In hardwoods incipient decay is usually indicated by a shade of brown different from the color range for normal heartwood of the species, or by a bleached appearance of the wood, which has been changed to a cream, straw, or lemon-yellow color. *Fomes igniarius*, which causes a white heart rot in hardwoods, first produces a brownish discolored zone, as does *F. fraxinophilus* in white ash and *F. applanatus* in other hardwoods. In conifers, shades of brown are again evident, which may be preceded or accompanied by pinkish, reddish, or purplish shades. Incipient decay caused by *Fomes pini* in Douglas fir appears as a pronounced reddish-purple or olive-purple discoloration, in white and red spruce as a light purplish gray later deepening to a reddish brown, and in ponderosa pine as a pronounced pink color, which gives way to a red brown. Next the white pockets develop, and in western hemlock the floccosoids, which are white spots or streaks in the wood produced by whitish deposits within the cells, may be confused with the early development of white pockets of decay (Gerry 1943). Incipient decay caused by *Polyporus anceps* and *Fomes annosus* in pines is indicated by pinkish to reddish discolorations, and incipient decay in western white pine caused by *Polyporus tomentosus* var. *circinatus* varies from light to dark reddish brown. All these fungi cause white pocket rots. The incipient stage of brown stringy rot in western hemlock and white and lowland white firs caused by *Echinodontium tinctorium* is first indicated by light-brown or golden-tan spots or larger areas of discolored heartwood. Although the wood appears hard and firm, on drying after manufacture it is brash and frequently falls apart along the growth rings. Darker brown shades in the

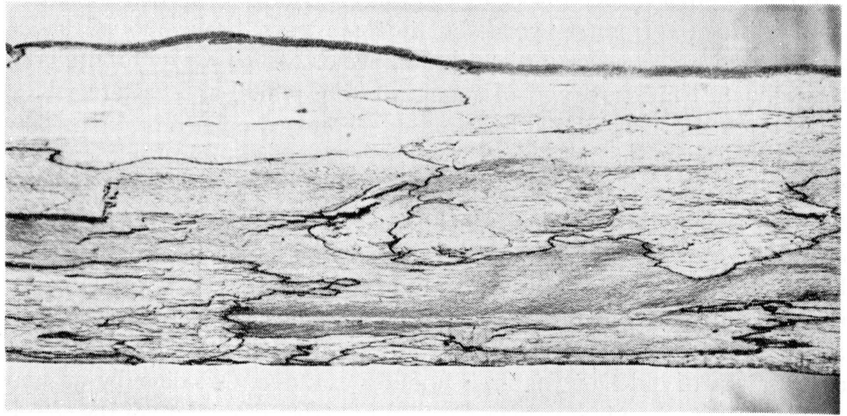

Fig. 139. Zone lines or black zones in decayed wood of maple, probably produced by *Xylaria polymorpha*. (*Photograph by H. F. Marco.*)

brown heartwood of western red cedar usually indicate decay (Findlay and Pettifor 1941). The incipient stages of certain white rots appear as water-soaked areas in the heartwood of freshly felled trees. *Ganoderma lucidum* on eastern hemlock and *Polyporus dryophilus, P. pilotae,* and *Stereum subpileatum* on hardwoods produce this abnormality. However, water-soaked heartwood may be independent of fungous action, which is so for "wetwood" of Scotch pine, Norway spruce, and silver fir in Europe (Lagerberg 1935; Michels 1943), "wetwood" of elms caused by a bacterium (Carter 1945), "glassy" balsam fir (Schrenk 1905), and "watercore" in certain western conifers. Bacteria are associated with the weaker wetwood and dark-colored heartwood of poplars (Clausen and Kaufert 1952; Seliskar 1952; Wallin 1954).

The incipient stages of brown rots are not commonly indicated by marked discoloration of the wood. There are some exceptions. In Douglas fir, *Fomes laricis* usually produces no discoloration in the incipient stage, but occasionally the wood becomes a brilliant purplish-red. In ponderosa pine a pronounced reddish-brown to chocolate-brown color bordered by a pinkish-red zone is evident, making the incipient stage of decay caused by *F. laricis* difficult to distinguish from that caused by *F. pini*.

The incipient stages of other decays are described in the chapter on the various rots.

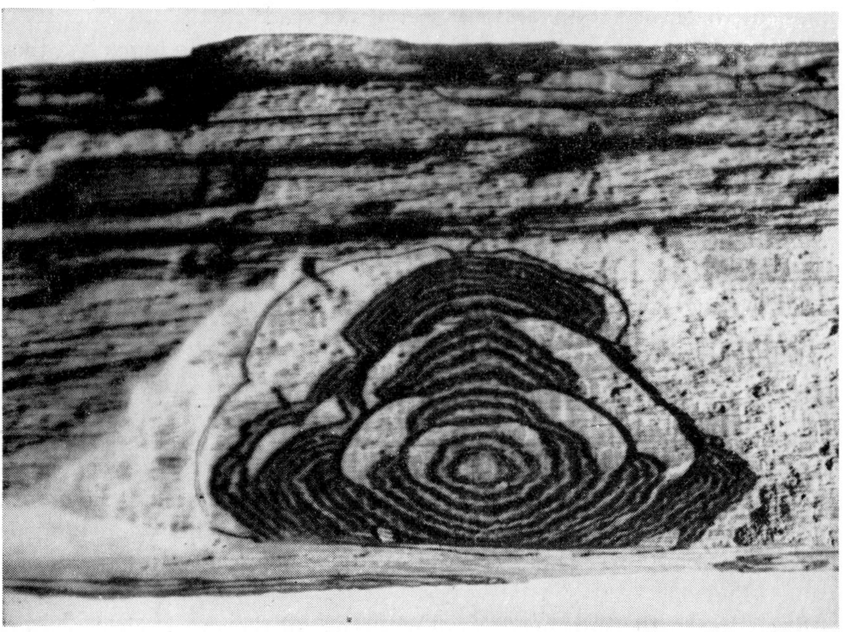

Fig. 140. Zone lines or black zones forming an unusual pattern in decayed wood of maple. The lines were probably produced by *Xylaria polymorpha*. (*Photograph by H. F. Marco.*)

Zone Lines. Narrow zone lines, usually black (Fig. 139) but sometimes dark brown, are a common phenomenon of many decays. They are frequently formed in the white rots but are rare in the brown rots, and they are more common in decayed wood of hardwoods than in conifers. At times black zone lines may be fantastic in pattern (Fig. 140). According to Campbell (1933:126), there are three causes for zone lines, viz., (1) the antagonism of the mycelia of two different fungi occupying the same substratum, (2) the action of the single mycelium of certain fungi, and (3) the production of "wound gum" stimulated by (*a*) natural wounding, (*b*) parasitic invasion, fungal or bacterial, or (*c*) desiccation.

Zone lines associated with decay are composed of dense, gnarled masses of brown, swollen, or bladderlike hyphae occupying the lumina of the various cells in the path of the line and associated with a brown pigment produced by the hyphae. These brown hyphae arise from hyaline mycelium.

Zone lines in decayed wood were formed under controlled conditions whenever the mycelium was exposed to a large amount of air and to a considerable concentration of water, provided atmospheric oxygen was present (Hopp 1938). This accounts for the frequent formation of zone lines just back of freshly cut surfaces of decayed wood.

Although zone lines are often formed by *Fomes pini* in tamarack, they are rarely formed in Douglas fir or hard pines. *Fomes igniarius* and *Xylaria polymorpha* commonly form zone lines in hardwoods, the last named probably producing these lines more abundantly than any other fungus. Zone lines are of little diagnostic value and may even be misleading when a fungus causing the original decay is followed by another forming zone lines, which can happen with *Fomes applanatus* and *Xylaria polymorpha.*

Odor. In general the odor of decayed wood is noticeable only in the advanced stage, then for only a few decays, and usually only from freshly cut surfaces. Decay in Sitka spruce caused by *Polyporus schweinitzii,* in oaks caused by *P. berkeleyi,* and in poplars and willows caused by *Trametes suaveolens* has a distinct odor of anise oil. With decay caused by *T. suaveolens* this odor is retained for some time after the wood is air dried. Decay in ponderosa and sugar pines caused by *Lentinus lepideus* has a characteristic odor consisting of an aromatic resinous and fungous odor. The odors of wood-destroying fungi in culture are of diagnostic value (Badcock 1939).

Strength. In the advanced stage of decay all the mechanical properties of wood are so affected that the wood is usually worthless for any purpose where strength is necessary. Incipient decay may or may not seriously weaken wood, depending on the type of decay. Wood with the incipient stage of certain white rots is little weakened, but in general the action of white rots in this respect is variable. Wood with the incipient stage of brown rots is usually seriously impaired in strength. Toughness or shock resistance is first affected by decay so that the wood breaks abruptly across the grain with a relatively low resistance, indicating little tensile strength. Wood breaking in this way is termed "brash," but there are causes of brashness other than decay, chief among which is density or specific gravity much lower than the average for the species (Koehler 1933). Strength in compression parallel to the grain is affected least in the early stages of decay. Tough wood breaks gradually with continued splintering and with a relatively high resistance. The best test for toughness and brashness in wood is its degree of resistance and manner of breaking as a beam under shock or impact. A rough test can be made by prying up slivers to see if the wood pulls out with a splintery appearance or abruptly. Microscopical changes in decayed Sitka spruce give an approximate idea of the degree of reduction in toughness (Waterman and Hansbrough 1957). Specific gravity cannot be used as a measure of the strength of decayed wood, because, within a species, the strength of decayed wood is less than that of sound wood of equal density, and decayed wood may show considerable loss in strength before there is any appreciable reduction in specific gravity (Scheffer 1936:22). Sitka spruce decayed by *Poria monticola* was seriously weakened when the infected wood had lost only 2 per cent in dry weight (Mulholland 1954).

Among the white rots, Sitka spruce and Douglas fir with incipient decay caused by *Fomes pini* showed little weakening (Scheffer et al. 1941). In jack pine "red stain" caused by an unknown fungus had no appreciable effect on strength, but advanced red rot produced by *Fomes pini* caused marked weakening (Atwell 1948), as it did in Douglas fir (Wood 1955). *Polyporus hispidus*, which produces a white rot, caused serious reduction in toughness of English ash before the decay could be detected visually (Cartwright, Campbell, and Armstrong 1936), and western red cedar with incipient decay probably caused by *Poria weirii* was reduced about 20 per cent in strength (Findlay and Pettifor 1941). Sitka spruce and Douglas fir wood with incipient decay caused by *Polyporus schweinitzii*, *Fomes laricis*, and *F. pinicola*, all of which produce brown rots, were seriously weakened (Scheffer et al. 1941). *Trametes serialis*, which causes a brown pocket rot in logs and lumber of various conifers, almost immediately reduced the strength of Sitka spruce (Cartwright et al. 1931; Armstrong 1935).

Kiln-dried lumber may show collapse, honeycombing (internal collapse), or checking, which sometimes can be explained by the excessive shrinkage of areas with incipient decay during rapid drying, the cell walls of the infected wood being weakened. A slower drying schedule or air seasoning will probably eliminate this defect, although the mycelium of the wood-destroying fungus will not be killed by air seasoning (page 474).

Unless it is definitely known that the incipient stage of a particular decay does not weaken wood, affected wood must not be used when high strength is needed.

Heat Conductivity. Heat conductivity of infected wood is evidently greater than that of sound wood, since fungi in wood in the advanced stage of decay were killed by heat in a shorter time than the same fungi in wood in the incipient stage (Hubert 1924*b*:17).

Water-holding Capacity. Decayed wood absorbs and loses water more rapidly than does sound wood, but the result of alternate periods of wetting and drying is a higher average moisture content for decayed than for sound wood (Scheffer 1936:33).

CHEMISTRY OF DECAY

The decay of wood by fungi is a chemical action on the substances composing wood by enzymes secreted by fungous hyphae. Wood-destroying fungi produce enzymes freely, but the action of the individual enzymes is unknown. The chemistry of decay is complicated and still obscure, since wood itself is such a complex substance that its chemical composition is not yet thoroughly understood (Campbell 1952).

Although the brown rot fungi as a rule attack cellulose and its associated pentosans with little attack on lignin, the white rot fungi attack all components of the wood. The bleaching effect induced by all white rot fungi is probably explained by pigment destruction rather than by lignin destruction, except the bleaching associated with white pocket rots, since the pockets finally contain areas of almost pure cellulose fibers, naturally white in color. Campbell (1932*a*) divides the white rot fungi into three groups on the following chemical bases: (1) early attack on lignin and pentosans and delay in the disorganization of cellulose, (2) early disintegration of cellulose and its associated pentosans, and delayed attack on lignin, and (3) early decomposition of both lignin and cellulose in varying proportions.

The ultimate primary reactions involved in decay are probably oxidation and hydrolysis (Campbell 1932*b*). With *Merulius lacrymans* the decomposition serves to supply sufficient moisture for the fungus to continue decay of wood in buildings without any external source of water supply (Miller 1932). It is possible that other fungi can also supply most of their moisture requirements in this way.

MICROSCOPIC CHARACTERS OF DECAY

The structure of the cell wall of higher plants is not exactly known (Esau 1953). In woody tissues it is believed that the wall is a laminated structure composed of many lamellae differing in chemical and physical properties. Three primary divisions of the wall are recognized: (1) the middle lamella or intercellular material, (2) the primary wall, and (3) the secondary wall, which in turn is usually three-layered, sometimes more. Lignification makes the walls hard and stiff. The middle lamella when first formed is composed of pectic material that later becomes intensely lignified. Whether this lignification is brought about by partial transformation of the pectic compounds or merely by deposit of lignin is unknown. The primary wall is formed by the first layer of cellulose laid down, which becomes lignified after the cell has attained full size. The secondary wall includes all the cell-wall material deposited within the primary wall after the cell has attained full size. It is composed of a cellulose framework that is not altered when lignification occurs, but lignin and other amorphous noncellulose materials permeate the framework. The secondary wall is variable and complex, both structurally and chemically. Under the microscope, unless special preparations are made,

Fig. 141. Photomicrograph of a section of a southern yellow pine decayed by *Fomes pini* showing numerous hyphae and many bore holes in the tracheid walls. (*Reproduced from J. Agr. Research* 29, *no.* 11.)

the middle lamella, primary wall, and immediately adjacent layers of the secondary wall appear as one layer.

Wood-destroying fungi feed on the substances composing the cell walls, but different fungi do not utilize all the components equally, this inequality resulting in different types of decay (Meier 1955). Furthermore the different types of cells may not be disintegrated equally. Either lignin or cellulose may be preferentially attacked, or both may be about equally affected, depending on the species of fungus. The transformation of the various wood compounds into substances nutritively valuable to the fungus depends on enzymes secreted by the mycelium, and these enzymes can affect the wood substance not in direct contact with the hyphae. Hubert (1924*a*:535) and Bavendamm (1936:968; 1951) have given the most complete accounts of the microscopic characters of decay. In decayed wood there are usually two types of hyphae. The younger, which are very small, hyaline, and irregularly branched, are most commonly found in wood with the incipient stage of decay. These hyphae may be so small and scanty that they are hard to detect. The older hyphae are of large size, usually much branched or anastomosed, frequently colored in varying shades of brown, and sometimes encrusted with amorphous decomposition products. Older hyphae are most abundant in wood in the advanced stage of decay, but they may also occur in the incipient stage.

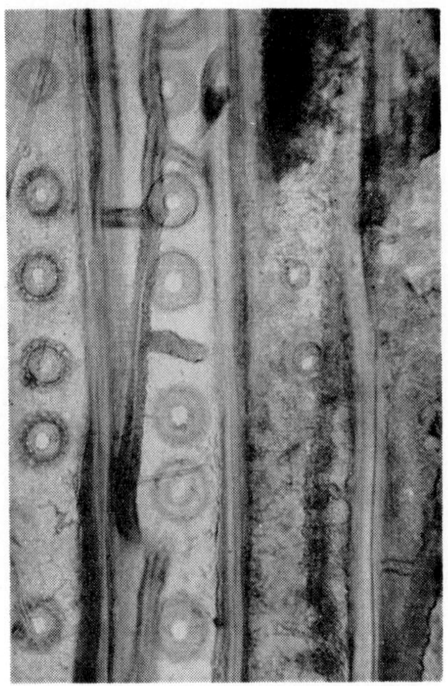

Fig. 142. Photomicrograph of a radial section of western hemlock decayed by *Echinodontium tinctorium*. Both large and small hyphae are present. The two tracheids on the right are so filled with decomposition products that the heavily encrusted hyphae are practically obscured. (*Photograph by A. A. McCready.*)

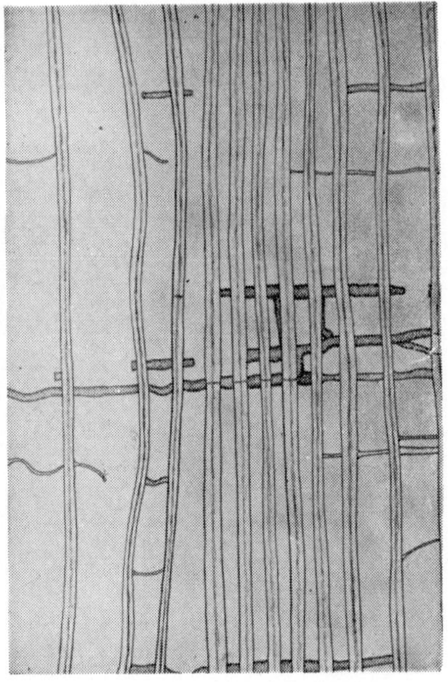

Fig. 143. Radial section of western red cedar decayed by *Poria weirii*. The hyphae cross the tracheids at right angles, only occasionally giving off a branch extending up or down a lumen. None is shown here. The large hyphae are sharply constricted in passing through the walls. (*Drawn by A. A. McCready.*)

Sometimes hyphae cannot be found in badly decayed wood, although there is abundant evidence of their former presence such as bore holes (hyphal holes), shrinkage cracks, decomposition products, and cell walls in various stages of dissolution. This suggests that the hyphae may disintegrate when the wood has been completely broken down and exhausted of its nutritive substances. Hyphae are most readily seen in radial sections of decayed wood, next in tangential sections, and are often difficult or impossible to find in cross sections.

The hyphae wander about in the lumina (cavities) of the cells, usually following in a general way the long axis of the cell (Figs. 141, 142), often closely appressed to the wall, and generally passing from cell to cell directly through the cell walls, or sometimes through the pit membranes. In wood decayed by *Paxillus panuoides* Fr. the hyphae are densely concentrated in the tracheids, becoming aggregated into fine strands (Findlay 1932:345). Occasionally a hypha may cross a number of tracheids at

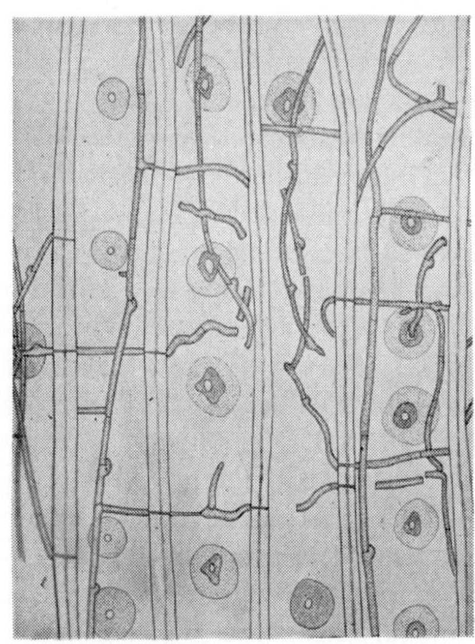

Fig. 144. Radial section of western hemlock decayed by *Echinodontium tinctorium.* The hyphae are sharply constricted in passing through the walls, clamp connections are particularly abundant in the second tracheid from the left, and the bordered pits are partially disintegrated. (*Drawn by A. A. McCready.*)

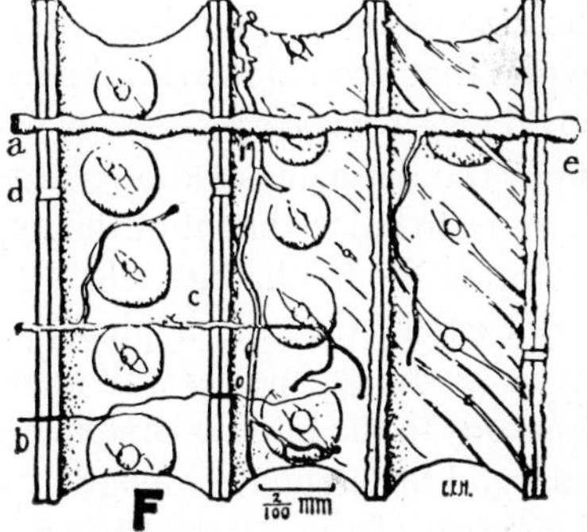

Fig. 145. Radial section from the edge of a rot pocket of incense cedar decayed by *Polyporus amarus.* The tracheid walls show roughened surfaces, diagonal shrinkage cracks are abundant, empty bore holes are evident, and some of the bordered pits have been almost completely destroyed. (*Reprinted by permission from "An Outline of Forest Pathology" by Hubert, published by John Wiley & Sons, Inc.*)

right angles without giving off any or only an occasional branch extending up and down the lumina. With *Poria weirii* in western red cedar[1] this was the characteristic habit in decayed wood from two widely separated localities (Fig. 143). Clamp connections or buckles of the simple or medallion type are often found on the hyphae of the wood destroyers (Fig. 144), since these structures are characteristic of the higher fungi to which most of the wood destroyers belong. Hyphae are sometimes most

[1] McCready, A. A.: A Microscopic Study of Brown Stringy Rot and Yellow Ring Rot. Yale University, Master's Essay (unpublished) (1933).

abundant in the ray cells as in decay caused by *Echinodontium tinctorium.* The hyphae of fungi causing molds and stains are frequent in wood, particularly in sapwood, but they usually pass from cell to cell through the pit membranes, occasionally boring directly through the walls. They are commonly concentrated in the rays and never have clamp connections.

The penetration of hyphae through the cell walls is a chemical action, the cell wall being locally dissolved by enzymes in advance of the actual passage of the hypha (Proctor 1941). In passing through the cell wall all except the finer hyphae are at first markedly constricted but quickly resume normal size when the opposite lumen is reached. Consequently, at first the hyphal hole is small and may remain so, or, as found by Cartwright for *Trametes serialis* (1930:9) in Sitka spruce and by Nutman (1929:53) for *Polyporus hispidus* in ash, after the cell wall is penetrated the size of the hyphal hole may increase to many times the size of the hypha that formed it, with the hypha in the hyphal hole swelling so that it no longer shows constriction. The hyphal hole may enlarge to several times the diameter of the enlarged hypha in it. This indicates that when there is no enlargement of the hyphal hole, enzymes are secreted solely by the tip of the hypha but, when the hyphal hole enlarges, enzyme secretion is not confined to the hyphal tip. The hyphal hole is enlarged first in the layers of the secondary wall adjacent to the lumen, and enlargement then proceeds to the middle lamella; consequently, until the hyphal hole is fully enlarged throughout its length, two cone-shaped openings are formed with the apices of the cones at the middle lamella. This is the so-called "hourglass" type of hyphal hole, which apparently is merely a stage in development.

Elongated cavities, enzymatically produced, following the angle of the cellulose fibrils are formed in the secondary cell walls by microfungi, such as *Chaetomium,* destructive to cellulose and mostly belonging to the Ascomycetes and Fungi Imperfecti (Schacht 1863; Roelofson 1956; Duncan 1960). This type of decay, known as soft rot, is usually in the surface layers of wood in service. Although retaining its shape the wood is extremely soft and brittle when dry.

In addition to hyphae and hyphal holes, there are other manifestations of decay. In the advanced stage, diagonal or spiral shrinkage cracks may appear in the cell walls, particularly in tracheid walls, of wood affected by brown rot (Fig. 145), although there is a possibility that these cracks may appear sometimes in sound wood. These shrinkage cracks may also extend from hyphal holes or pit cavities. Corrosion marks, uneven surfaces and uneven thickness of secondary walls, disintegration of walls, dissolution of the middle lamella, enlargement of the pit cavities, or complete destruction of the bordered pits may occur. Cells may be more or less filled with masses of yellow or brown decomposition products deposited irregularly, which often characterize decay caused by *Echinodontium tinctorium* (Fig. 142); or the ray cells may contain numerous brown globules of a substance resembling wound gum. Resin deposits may occur.

When the mycelium is fine and difficult to detect, staining is essential. Often a single stain, such as safranin or Bismarck brown, may be satisfactory, but sometimes it is necessary to use two stains of contrasting color, one for the wood and the other for the hyphae. Combinations of malachite green and acid fuchsin (Boyce 1918), Bismarck brown and methyl violet (Hubert 1922), and safranin and picroaniline blue (Cartwright 1929) can be used quickly and satisfactorily. Various microchemical tests for lignin and cellulose compounds are useful when examining decayed wood. If the sections are treated with a solution of aniline sulfate, the lignified tissue stains a brilliant yellow, whereas phloroglucin followed by concentrated hydrochloric acid produces a red color. Alcoholic iodine combined with sulfuric acid or zinc chloroidide alone stain cellulose blue. However, these microchemical tests must not be depended upon entirely in diagnosing decay because changes brought about by mechanical rupture cause lignified cell walls to react positively to tests for cellulose (Robinson 1920:54). Furthermore, microchemical tests can be misleading, since the color reaction may depend upon only one component of the compound, which may have been removed or changed by the fungus (Hirt 1932:20). Consequently microchemical tests should be supported by a chemical analysis before interpreting the results.

Until decayed wood has been studied more extensively, caution must be observed in identifying a fungus causing decay from the microscopic characters of the affected wood alone.

CULTURAL CHARACTERS OF FUNGI CAUSING DECAY

Hyphae may be so scanty in the incipient stage of decay that it is extremely difficult to detect the presence of a fungus by microscopic examination, but the mycelium may develop when small bits of the affected wood are placed on a culture medium in a sterile tube or flask and incubated at temperatures of from 70 to 80°F. Malt agar is a particularly suitable medium for wood-inhabiting fungi. Since this and similar media are chemically complex and variable, for exacting tests it is sometimes necessary to use synthetic media (Pieper, Humphrey, and Acree 1917). In general, liquid media are of limited value for wood-destroying fungi. Culture work must be done under sterile conditions to avoid contamination because mixed cultures are worthless in diagnosing decay.

Cultural characters, both macroscopic and microscopic, are extremely useful in identifying wood-inhabiting fungi provided a large number of cultures of definite identity are available for comparison (Nobles 1948). Again two fungi with sporophores and decay so similar that their specific identity is questionable may behave so differently in their cultural reactions to temperature and moisture as to warrant maintaining them as distinct species (Snell, Hutchinson, and Newton 1928). The production of sporophores in culture, which can be facilitated by special methods (Badcock 1943), is a further aid in identification. By inoculating the unknown fungus into sterile wood blocks, typical decay will usually be produced, although this may require several months or longer, whereas cultures on agar will develop in a few days to a few weeks. Fungi associated with white rots secrete oxidase; and when an alcoholic solution of gum guaiac is dropped on the surface of actively growing cultures of wood-destroying fungi on malt agar in Petri dishes or culture tubes, within 2 or 3 minutes a blue color appears in the cultures of white-rotting fungi, but there is no color change in the fungi causing brown rots (Nobles 1958). A further aid to identification is by mating a known with an unknown mycelium in a slide culture (Jorgensen 1954). If anastomosing occurs, the mycelia belong to the same species. However, failure to anastomose does not positively indicate different species.

SPREAD OF DECAY

After decay has continued for some time, the conks, sporophores, fruiting bodies, or fructifications are developed. These, which are usually quite large, appear on the trunk or on the branches as crusts or flat or hoof-shaped brackets, or on the ground nearby coming up from a decayed root as toadstools. In rare cases on living trees the mycelium forming conks issues directly from the heartwood through the sapwood and bark, but ordinarily finds its way out through wounds, or most commonly through dead branch stubs not yet grown over. The conks are annual, living for one season only and being destroyed by birds, rodents, and insects, or perennial, living for a few to many years. The annual conks are usually soft and mushy or somewhat leathery, whereas the perennial ones are hard and corky or woody. The variously colored upper surface may be smooth or rough, uniform or more or less zonate. The zonate character is most common among perennial conks. The undersurface may be smooth or have spines, gills, or tubes, the latter being most common. The mouths of the tubes vary from tiny and perfectly circular to large and irregular in outline. On the smooth surface, on the spines and gills, or in the tubes are borne the microscopically small, simple spores. It has been estimated that a single annual sporophore of *Polyporus squamosus* produced from 50 to 100 billion spores during its life and a perennial sporophore of *Fomes applanatus* liberated 30 billion spores daily for a period of 6 months (Buller 1909:84; White 1919:140). After release from the sporophore and under laboratory conditions, spores have remained viable for 3 weeks to nearly 6 months, depending on the species of wood-destroying fungus (Harrison 1942).

These myriads of spores are carried by air currents, and, upon alighting in a suitable place, they germinate. The hyphae cannot penetrate bark or living sapwood; consequently a spore must light on heartwood or exposed deadwood directly connected with heartwood, for decay to start in a living tree. However, all the dead branches form bridges from the outside to the heartwood, as does the exposed dead sapwood where the bark has been knocked off by injuries such as lightning wounds, breaking branches, falling trees, and many others. Broken tops give direct access to the heartwood, as do fire scars when the fire has eaten deeply enough into the tree. Of all wounds those caused by fire are the most frequent points of entrance for wood-destroying fungi. As soon as a wound occurs, healing over begins; but this is a slow process, requiring years, as does the natural pruning and growing over of dead branches. Meanwhile the tree is exposed to infection. In the more highly resinous conifers, smaller wounds are quickly covered with pitch, which acts as an excellent antiseptic dressing to prevent the growth of hyphae, but large wounds are not protected in this way. In hardwoods plugging of the wood cells by wound gum and tyloses affords some protection (Higgins 1919; Rankin 1933). An unusually effective gum-impregnated zone, impervious to wood-destroying fungi, is quickly formed just below the surface in fire-scarred red gum and persimmon (Hepting and Blaisdell 1936).

After decay has once started, it is usually only a question of time until conks appear and more spores are released to infect other trees. It is apparent that only an infinitesimal number of the spores produced ever cause infection. If any significant portion of them found their mark, practically every tree would be infected, and most overmature trees largely decayed.

RESISTANCE TO DECAY

Little is known about the resistance of living trees to decay of the heartwood, but this probably depends on genetic variability between trees within a species and on variability of the physical condition of the wood.

It is common to find two trees of the same age, of the same size, both free from wounds or wounded in the same way, and growing under exactly the same conditions within a few feet of each other. One will be sound, whereas the heartwood of the other will be completely decayed. It might be assumed that this depends entirely on the chances of infection, but in Douglas fir attacked by *Fomes pini*, where the fungus almost invariably enters through dead branch stubs that have died naturally, such an assumption seems unwarranted. In incense cedar attacked by *Polyporus amarus*, the decay spreads slowly in some trees and rapidly in others (Boyce 1920:21). Moisture content of the heartwood was the factor controlling decay in living subalpine spruce, whereas differences in growth-ring width and specific gravity did not appreciably affect the rate of decay (Etheridge 1958). In young hardwoods infected through fire scars, the decay column was usually confined to the wood laid down up to the time of scarring and rarely extended to wood laid down after scarring (Hepting 1935:20), which indicates some factor of resistance in the subsequently formed wood. Observations on western conifers indicate that heart rot entering through a fire scar will invade wood laid down subsequent to the formation of the scar.

Young stands of defective species are relatively free from decay for years after heartwood formation begins, and it is usually only when stands are mature and overmature that extensive decay is prevalent. This indicates an increase in decay with lessened vigor. However, individuals may be completely decayed years before the age at which the stand as a whole is affected. Furthermore, investigations of the relation of site to decay for a number of species have failed to give a consistent result (Etheridge 1958:187; Riley 1952:728). In aspen in Colorado, decay, largely caused by *Fomes igniarius*, was lowest in site quality 1 and highest in site quality 3 (Davidson, Hinds, and Hawksworth 1959). In western hemlock, decay caused by *Echinodontium tinctorium* decreased as the site index increased, whereas decay caused by *Fomes pini* increased with increasing site index (Foster, Craig, and Wallis 1954:154). In young hardwoods in the Mississippi Delta no relation was found between the rate of diameter growth and the rate of decay (Hepting 1935:21). Evidently, lack of vigor has no fixed relation to decay but varies with species of trees and species of fungi.

The resistance to decay of a species in use has no relation to its resistance as a living tree. The heartwood of western red cedar is highly decay resistant in contact with the soil, but in living trees it suffers considerably from decay. Conversely, ponderosa pine, which is not decay resistant in use, can be classed as a relatively sound species in forest stands. White fir is easily decayed both as a tree and in use, whereas

Port Orford cedar is reasonably free from decay as a living tree and highly decay resistant in use.

RATE OF DECAY

The rate of decay in living trees must be considered from two viewpoints, i.e., in individual trees and in stands. Economically the latter is all-important.

Little is known about the rapidity with which decay develops in individual trees, because this information is difficult to obtain. The only way the volume of decay in a tree can be exactly measured is to fell and open the tree. Then if the year in which infection occurred can be determined, an average figure for yearly increase can be obtained. But it is usually impossible to determine just when decay entered a tree, and, if artificial inoculation is attempted, a few years at least must elapse before a result can be secured. Five years after appropriate rot fungi were inoculated into several species of living northern hardwoods, decay had progressed a maximum average yearly extent of 1.9 feet in bigtooth aspen and a minimum of 0.12 feet in sugar maple (Silverborg 1959). Following careless pruning of young Norway spruce, decay caused by *Stereum sanguinolentum* extended vertically 6.5 to 11.0 feet in a maximum of 5 years (Risley and Silverborg 1958).

In Germany decay in the heartwood of oak caused by *Fomes igniarius* varied from 0.12 to 1.23 feet in yearly vertical progress, with an average of 0.16 to 0.30 foot (Münch 1915:517). There was no significant difference in the upward and downward rate of spread from the point of infection. Decay in incense cedar caused by *Polyporus amarus* was found to vary considerably in rate of development. In general the minimum vertical yearly progress of decay varied from 0.01 to 0.20 foot a year, whereas the maximum ranged from 0.01 to 0.35 foot, although in individual trees the rate could be much slower or faster (Boyce 1920:21). The figures were determined by connecting the focus of decay with a single wound healed over. By determining the age at which the wound occurred and the number of years elapsing before it healed over, the minimum and maximum length of time the decay had been in the tree could be fixed and then the average possible minimum and maximum yearly progress of the rot determined. In young Mississippi Delta hardwoods decay spreads upward from fire scars most rapidly in oaks, 2.3 inches per year, with ashes, red gum, sugarberry (hackberry), and persimmon following in order (Hepting 1935:14). In bottomland oaks decay has spread vertically from 2 to more than 9 inches yearly (Toole 1954, 1956; Toole and Furnival 1957). Conditions governing the wide differences in rate of decay in individual trees are unknown. One factor may be the character of the mycelium, because indications are that wood may be decayed more quickly by a dikaryotic than by a haploid mycelium (Aoshima 1954).

For stands there is a tendency to estimate decay as progressing more rapidly than it actually does, particularly when stumpage is increasing in value and closer utilization is being practised. For incense cedar the maximum increase in decay, based on cubic-foot volume, occurred between 223 and 259 years, when the total increase amounted to 22 per cent or 0.6 per cent annually (Boyce 1920:35). For western white pine the average maximum yearly increase, based on cubic-foot volume, was 0.32 per cent between 131 and 180 years on the bottom site (Weir and Hubert 1919*a*:8). For Douglas fir in western Oregon on site II, decay volume increased 16 per cent or 0.27 per cent yearly from 190 to 250 years and from 285 to 345 years, whereas on site III it increased 24.0 per cent or 0.32 per cent yearly from 25 to 325 years (Boyce and Wagg 1953:51). Douglas fir is a species in which decay has been considered to progress rapidly.

From the foregoing it can be assumed that in general decay of living trees is slow.

PERSISTENCE OF FUNGI CAUSING DECAY

Although when once established all wood-destroying fungi attacking heartwood seem to persist indefinitely in living trees, they vary considerably in their ability to continue development after the trees are dead and after manufacture. *Fomes pini,* so prevalent in living conifers, may continue development to some extent in dead trees or logs, but it has never been found decaying wood in service, although much wood with incipient decay has been used in ties, poles, and buildings. *F. igniarius,* prevalent on both living and dead trees, does not continue its disintegration of wood in service. *Polyporus amarus* will sometimes continue to decay dead incense cedars, standing or down, but when decayed trees are utilized as poles or posts the rot either stops development or continues so slowly as to be of no consequence. The value of pecky cypress for greenhouse benches and other places where high decay durability is essential shows that the fungus stops growth after the trees are manufactured. *Poria monticola* on the other hand, which occurs in living conifers, has been found causing decay of building timbers, the inference being that heartwood with incipient decay from living trees had been used in the buildings. Unless it is definitely known that a fungus will not decay wood in service, it is inadvisable to use timbers with incipient decay where decay durability is important.

CONTROL OF DECAY

Individual trees as components of the forest are a relatively low-value commodity. Only the stand represents value of consequence, so that any attempt to prevent or control decay directly in the individual would be so expensive that by the time the tree was cut the investment in it would be far more than its stumpage value. Consequently, control of decay must be concentrated on reducing financial loss to the minimum in stands of merchantable size already decayed and on preventing significant loss in future stands.

SCALING AND ESTIMATING

A knowledge of the indications of decay in logs and trees is prerequisite to relatively exact scaling and estimating. The importance of accurate scaling and estimating is evident, since the entire lumber industry is largely dependent on them. Logs are bought and sold on board-foot scale alone, timberland changes ownership, and logging operations are based on the amount of timber that can be obtained from a given unit of area as determined by estimating. Consequently, exact estimating and scaling reduce the financial risk in the lumber industry, particularly in exploiting a species with a large amount of defect. When logs are purchased that actually cut out considerably less than scaled, there is a serious loss, particularly since operators always figure on an overrun of the scale. Even more serious is the loss when a mill and roads are constructed and logging equipment set up only to find that because of decay the amount of timber obtained from the area tributary to the mill is so much less than estimated that it is not possible to make a profit throughout the life of the operation.

Most decays afford some indications of their presence in a stand, and some, like the common decay in Douglas fir known as red ring rot caused by *Fomes pini,* rarely occur in an appreciable amount in a Douglas fir without some outward indications of the defect. The most certain indication of decay is the presence of conks, but it is necessary to distinguish between the different kinds of conks to evaluate properly the amount of decay in the tree. Some fungi develop conks abundantly, starting shortly after decay has begun in the heartwood; others develop only occasional or rare conks after extensive decay has occurred. A single conk of one species may normally represent only a few board feet of decay, whereas one of another species may mean that the entire merchantable volume of the tree is destroyed. Other conks are those of fungi that cannot decay heartwood of living trees but grow only on dead sapwood, so that the appearance of these on a wound on a living tree means only a superficial decay of no consequence. Conks of *Polyporus schweinitzii* are usually found, not on the tree but on the ground near it, coming up from a decayed root, yet these conks are sure indications of butt rot in the tree. Most sporophores appearing on the ground have no relation to decay in living trees.

Just as good an indication of decay in a tree as conks themselves are signs that conks have been present or are developing. Fragments of an old conk may be found still attached to the trunk or lying near the butt of the tree, the bark may be discolored at the former point of attachment, or a hole may be left by the rotting branch from which the conk appeared. Swollen knots on Douglas fir and other conifers caused by *Fomes pini* and shot-hole cups on incense cedar caused by *Polyporus amarus* are still more enduring outward records of rot in the heartwood. By chopping into knots the characteristic decay often becomes apparent, or the same color or same substance may be seen as when a conk is cut open. These rotten knots with the same substance as the interior or context of the conk are known as "punk knots" when no swelling of the surrounding wood results. In case there is a swelling, they are termed "swollen knots." Occasionally, decayed knots may be sunken.

Scars of all kinds may be clues to decay. Butt rots are commonly associated with fire scars. It is possible for years afterward to pick out trees with such scars completely healed over because the bark has an abnormal appearance, the regular arrangement of the bark scales being interrupted. Swollen or churn-butted conifers often have butt rot because the abnormal shape is the result of healing over of a basal scar. Broken or dead tops are common points of entrance for decay, and a slight sharp crook in a trunk indicates a broken top of decades ago, long since grown over, but which was an entrance point for decay for a number of years. Large branches broken off may reveal decay in their heartwood, showing that the trunk is also decayed. Pounding on the trunk, although a fairly satisfactory method for determining the presence of decay in small, rather thin-barked timber when the heartwood is completely broken down, has little value in overmature stands of species with bark so thick that it deadens the sound and also makes it impracticable to remove the bark to obtain a solid sounding place. If trees are not too large, decay can also be detected by use of an increment borer or auger. A thulium X-ray unit (Eslyn 1959) and the transmission of ultrasonic energy (Waid and Woodman 1957) have been used experimentally to detect decay in living trees.

In addition to actual indications of decay there are other abnormalities that are often confused with them. One of the most common is burls or galls (page 298). These conspicuous swellings, found on many conifers and some hardwoods in varying degrees, are disorganized masses of woody tissue often formed as a direct response to the irritation caused by an injury, particularly a bruise. A tree much larger than the average in a stand and with many dead limbs in the lower crown is not necessarily decayed. Such wolf trees have merely grown faster than their neighbors, and the branches did not die so soon from lack of light, thus requiring a longer time for the dead branches to break off and for the branch stubs to heal over. Abnormalities in branching, such as branch "fans" on Douglas fir or the "candelabra" branches on incense cedar, are never connected with decay, although abnormal top branching is often the result of dead or broken-off tops through which decay frequently enters. A variation in character of the bark does not indicate decay unless the change is brought about by healing over a scar.

Estimation of cull is more exact when it is based on indicators, since indicator cull factors can be applied to individual tree volumes. Of course, indicator cull factors, being based on averages, are not so satisfactory for estimating single trees or a small number as they are for extensive stands. However, decays in some species have practically no indicators, so that flat cull factors must be determined from cut stands and then applied to uncut stands by diameter classes, age classes, or tree classes based on vigor. Since cull factors increase with increasing age and vigor is broadly related to age in uneven-aged stands, cull factors can be determined by tree classes and then applied to a stand with reasonable accuracy (Kimmey 1954) or by age classes (Morawski, Basham, and Turner 1958). Another approach is to classify living trees of each species as either "suspect," trees having one or more decay indicators, or "residual," trees free from any decay indicators, to examine representative trees in order to determine average cull factors for the two groups, and to apply these factors to the stand (Foster, Thomas, and Browne 1953). Because of possible differences between cull factors over a large region, neither indicator nor flat cull factors for one section should be applied elsewhere without first verifying their applicability.

Western Species. Hardwoods being of little importance in western forests, most information on extent of decay in relation to outward indications deals with conifers (Boyce 1930).

In northwestern California, western Oregon, and southwestern Washington, overmature stands of Douglas fir frequently are highly defective, but progressing northward the amount of decay decreases, until in British Columbia losses are low (Thomas and Thomas 1954). However, stands severely decayed and those lightly decayed may occur in any region.

Most of the decay is caused by *Fomes pini;* the development of conks (Fig. 153), swollen knots (Fig. 156), and infrequent punk knots closely follows the progress of decay in the heartwood. The decay, known as red ring rot or conk rot, extends on an average 7 feet in the heartwood above the highest and below the lowest conk of a series in 140-year-old stands, increasing in a straight-line relationship to 30 feet above and below in 360-year-old stands, whereas for swollen knots and the infrequent punk knots the distances are one-half those for conks (Boyce and Wagg 1953:19). These distances include an allowance for hidden rot, but do not include incipient decay, which extends an average of 3.5 feet beyond advanced decay, because incipient decay is merchantable. For a single conk or for several in a group a deduction is made of 8 feet above and 8 feet below, the equivalent of one 16-foot log. This applies principally to the younger age classes, since single sporophores occur infrequently in older trees. When the stand age is known only approximately, a distance of 12 feet above the highest and 12 feet below the lowest conk in a series can be used for stands 100 to 200 years old, 19 feet in each direction for stands 200 to 300 years old, 26 feet in each direction for stands 300 to 400 years old, and 34 feet in each direction for stands over 400 years old. For swollen knots and punk knots these distances should be halved. Red ring rot can also be detected in fire-killed timber because the dry punky substance of the conks, swollen knots, and punk knots extending into the tree burns readily, resulting in cylindrical holes reaching several inches into the trunk, but it is impossible to tell whether the hole was formerly occupied by a conk or a swollen or punk knot. Since *Fomes pini* infects Douglas fir almost entirely through dead branches, wounds are of no importance as indicators of red ring rot.

A single conk of *Fomes laricis* (Fig. 196) causing brown trunk rot indicates that a tree is unmerchantable. Conks are developed only occasionally, so trees must be scanned for broken branches showing the rot, because a branch with decayed heartwood indicates extensive rot in the trunk. Conks of *Polyporus schweinitzii* (Fig. 194) causing red-brown butt rot may be found on butt wounds or on the ground near by, coming up from a decayed root. The rot is confined to the roots and butt, rarely extending beyond the first 16-foot log. Culling 8 feet of the butt log when a conk is present is a sufficient average allowance for this defect. The majority of fire-scarred Douglas firs become infected with this decay, so culling 8 feet of the butt log of every fire-scarred tree or for a distance of 2 feet beyond the last visible evidence of the scar, if it is over 8 feet long, is the rule. Since the rot usually runs out in the first 16-foot log, the longitudinal extent of the rot column in the butt log in overmature trees can be judged by scalers as follows: if the average diameter of the rot on the butt does not exceed 4 inches, the extent of the rot up the log should be considered as 1 foot for each inch in diameter, if the average diameter is from 5 to 10 inches deduct 1 inch from the diameter and apply the same rule, if 11 to 20 inches deduct 2 inches, if from 21 to 30 inches deduct 3 inches, and if 31 inches or over deduct 4 inches. Conks of *Fomes roseus* (Fig. 198) causing yellow-brown top rot occur so high in the tree that they are usually overlooked, but since the rot either is beyond the merchantable limit or affects only the low-grade top logs this makes no difference. Conks of *Polystictus abietinus* are sometimes found around fire scars, but the decay is confined to dead sapwood, which is mostly slabbed out in sawing, so little loss occurs.

Although cull from decay is not difficult to estimate in Douglas fir because of the preponderance of red ring rot caused by *Fomes pini* with its abundant conks and swollen knots, fructifications of other fungi on other species are developed sparingly or not at all on living trees. A conk of *Echinodontium tinctorium* on the butt 16 feet of the trunk of western hemlock or any of the firs indicates that the first two 16-foot logs are cull; a conk higher up means that the main portion of the bole is unmerchantable, whereas two or more conks some distance apart denote an unmerchantable tree. Rusty knots or rusty branch stubs indicate decay in the heartwood, and the rusty red color of the decayed stubs can be readily exposed with an axe. In big timber this method of detection is principally of value for felled trees. Practically all loss in incense cedar is pocket dry rot caused by *Polyporus amarus*. A single conk or shot-hole cup of this fungus means an unmerchantable tree, but these indicators are developed sparingly. A living tree of any species with a single conk of *Fomes officinalis* is a cull.

However, estimates of decay losses for most species must be based on cull factors developed for various sections or by a careful examination of a large sample of decayed trees as follows: western hemlock in southeastern Alaska (Kimmey 1956), in the Queen Charlotte Islands (Foster and Foster 1951), on Vancouver Island (Buckland, Foster, and Nordin 1949), in the Upper Columbia region of British Columbia (Foster, Craig, and Wallis 1954), in the Kitimat region of British Columbia (Foster, Browne, and Foster 1958), and in western Oregon and Washington (Englerth 1942); Sitka spruce in southeastern Alaska (Kimmey 1956) and in the Queen Charlotte Islands (Bier, Foster, and Salisbury 1946); subalpine spruce in Alberta (Etheridge 1956); Engelmann spruce in the central Rocky Mountains (Hornibrook 1950); Pacific silver fir on Vancouver Island (Buckland, Foster, and Nordin 1949) and in the Kitimat region of British Columbia (Foster, Browne, and Foster 1958); white and red fir in northwestern California (Kimmey 1950, 1957); alpine fir in eastern British Columbia (Bier, Salisbury, and Waldie 1948; Foster 1954) and in the central Rocky Mountains (Hornibrook 1950); ponderosa pine in the Rocky Mountain region (Andrews 1955); ponderosa and sugar pines in California (Kimmey 1950, 1954); lodgepole pine in the central Rocky Mountains (Hornibrook 1950); incense cedar in California (Kimmey 1950, 1954); western red cedar in southeastern Alaska (Kimmey 1956); redwood in California (Kimmey and Hornibrook 1952); black cottonwood in the Middle Fraser region of British Columbia (Thomas and Podmore 1953); and hardwoods in California (Kimmey 1950).

Eastern Species. In the East considerable data are available for estimating the amount of defect in hardwoods but little for conifers.

Conifers. Balsam fir, which usually grows in uneven-aged stands, is so highly defective because of trunk rot caused by *Stereum sanguinolentum* and white stringy butt rot largely caused by *Corticium galactinum* (Basham, Mook, and Davidson 1953) that in older trees more than half the gross volume may be unmerchantable. These fungi rarely fruit on living trees, and there are no other regular indicators of their occurrence. Furthermore, there is considerable variation in the amount of decay in different localities (Davidson 1957). Consequently, estimates of decay must be based on an examination of a representative sample of felled trees in each locality.

Eastern spruces are relatively sound, especially red spruce. Most of the loss is red ring rot caused by *Fomes pini* and red heart rot caused by *Stereum sanguinolentum.* Butt rots are common, caused by a number of fungi, among which *Corticium galactinum, Polyporus circinatus,* and *Poria subacida* are probably most important. The significance or frequency of decay indicators has not been established, so that estimates must be determined by felled samples. In the Maritime Provinces of Canada, cull in white spruce based on board-foot volume increased from 4 per cent at 70 years to 22 per cent at from 150 to 170 years (Davidson and Redmond 1957). Cull also increased with diameter. In red spruce, cull was never more than 5 per cent of the gross board-foot volume. In both species, butt rots caused more loss than trunk rots.

Eastern pines are reasonably sound, most of the cull being red ring rot caused by *Fomes pini,* but conks develop sparingly in relation to total cull. Eastern white pine in Ontario must be judged by age, the percentage of cull by board-foot volume varying from 6.4 per cent in the 60-year age class to 46.8 per cent in the 220-year age class (White 1953). In Maryland, conks of *F. pini* scattered along the bole of Virginia pine meant a cull tree, whereas two or more conks several feet apart indicated one cull log (Fenton and Berry 1956). There were no other indicators. In Michigan, the only method for estimating decay caused by *F. pini* in young stands of jack pine was by increment boring (Varner 1950).

Aspen. This species occurs in extensive, pure, even-aged stands over large areas in the Northern Lake States and to a lesser extent in northern New England, usually following fire. Although the aspen type is considered temporary, it will be a long time before adequate fire protection and proper silviculture relegate this species to its original minor position. Aspen is short-lived and defective. Aspen heartwood is often stained brown or mottled red by Fungi Imperfecti prior to invasion by decay (Basham 1958). *Fomes applanatus* causes a white butt rot seldom extending up more than 2 feet. *Fomes igniarius* causes most of the decay. Advanced decay is usually indicated by conks (Fig. 174) whose development follows rather closely the progress of the rot, so that they may be used with considerable accuracy in estimating the amount of cull as is shown by the following table. A small conk is 1 inch or less and a large conk more than 1 inch in width.

Relation of Conks of Fomes igniarius to the Extent of Rot in Minnesota

Number of conks	Size of conks	Extension of rot, ft.	
		Above conk	Below conk
One	Small	2	2.5
	Large	2.8	5
More than one	Small	3	3
	Large	5	5.5

SOURCE: Horton and Hendee 1934.

However, sporophores of *Armillaria mellea* and *Fomes applanatus* which cause butt rot in aspen are not found often, and therefore decay by these fungi can be overlooked in estimating. To avoid this a rule of thumb, based on rot diameters up to 8 inches, has been developed for aspen in Minnesota (Brown 1934:941). For each stand the average rot diameter for dominant trees 1 foot above ground level is determined by increment boring. The trees should be taken mechanically along a compass line, and about 25 will suffice. The average rot diameter to the nearest inch is multiplied by 3.5 to get the percentage of rot for the individual stand, and these percentages weighted by the gross volume of the individual stands give the average rot percentage for all the stands in the estimating unit.

Northern Hardwoods. In general there is considerable decay in northern hardwoods. Most of the information on decay indicators in relation to cull is from the Adirondack region of New York (Silverborg 1954). A conk of *Fomes applanatus,* a fungus that occurs on many hardwoods, indicates a decay column 6 to 7 feet both above and below the conk. *F. fomentarius,* usually on dead timber, sometimes occurs on living trees. A conk denotes a rot column extending about 4 to 5 feet both above and below the conk, but the decay usually occupies about one-half of the cross-sectional area of the trunk. Decay is often associated with other abnormalities of the bole. In connection with Nectria or Eutypella canker, decay extends no more than 1 foot beyond the vertical limits of the canker. A straight seam or crack in the bole usually results in less than 5 per cent cull, whereas a spiral seam may result in 5 per cent to complete cull, depending on the purpose for which the tree is to be used. An infolding seam indicates a complete cull. Butt scars extending to the ground result in more decay than those beginning 6 inches or more above ground level.

For sugar maple in eastern Ontario cull based on board-foot volume increased from 30 per cent in the 40-year age class to 57 per cent in the 320-year age class (Nordin 1954). For *Hydnum septentrionale* a deduction of 13 feet up and 6 feet down should be made, for *Fomes igniarius* 8 feet each way (a total of 16 feet), for *F. connatus* 2 to 4 feet each way, for a sterile conk of *Polyporus glomeratus* 5 to 7 feet each way, and for *Ustulina vulgaris,* which usually fruits at the butt, 3 to 4 feet up in the Adirondacks and 9 feet up in Ontario. Rot usually extends from 3 to 5 feet above the top of a butt scar, and the same allowance should be made above the top of a butt bulge, whereas for a stem bulge the cull should extend only to the vertical limits of the bulge. Rotten branches 3 inches or less in diameter indicate about 1 linear foot of cull, whereas those over 3 inches indicate 1 to 1½ feet lost.

On red maple a conk of *Fomes connatus* indicates a rot column of from 3 to 5 feet each way from it, and a sterile conk of *Polyporus glomeratus* a column of 6 to 8 feet each way.

Yellow birch is largely decayed by *Fomes igniarius,* a conk of this fungus indicating a decay column 8 to 10 feet each way from it, and by *Poria obliqua,* a sterile conk indicating decay 5 to 7 feet each way. An allowance of 12 to 14 feet should be made above the top of butt scars and butt bulges, and for stem bulges an allowance of 5 to 6 feet above and below the vertical limits of the swelling. Rotten branches 3 inches or less in diameter indicate 1½ to 2 linear feet of cull, whereas those over 3 inches indicate 2 to 5 linear feet.

Beech is highly defective, much of the decay being caused by *Fomes igniarius.* A conk of this fungus indicates a decay column 6 to 7 feet each way from it, for a sterile conk of *Polyporus glomeratus* 5 to 6 feet each way, and for *Ustulina vulgaris* 3 to 4 feet up from the butt. An allowance of 3 to 5 feet should be made above the top of butt scars and butt bulges. Rotten branches 3 inches or less in diameter indicate 1 linear foot of cull, whereas those over 3 inches indicate 1 to 1½ linear feet.

Central Hardwoods. A heart rot of oaks is often caused by *Polyporus hispidus,* the fungus being associated with pronounced trunk cankers (page 292). An average allowance of 1 foot below and 1 foot above the canker was found to be justifiable in estimating the volume loss from decay in affected trees on an area in Connecticut (Sleeth and Bidwell 1937:783).

According to Hepting and Hedgcock (1937) decay in hardwoods in this region is largely connected with wounding, and the vast majority of wounds are basal wounds caused by fire. Regardless of locality, unwounded trees of seedling or seedling-sprout origin remain relatively free from decay to a high age, but unwounded trees of sprout origin are quite likely to have butt rot, the decay entering from the old stump (page 376). Twenty-three wood-destroying fungi were identified on oaks alone, but numerous others also caused decay. There was considerable variation in the amount of decay in trees of the same species, other factors being equal, because of the action of different fungi. One group represented by *Polyporus berkeleyi* and *Armillaria mellea* causes butt rots rarely extending more than a few feet up the bole; another group represented by *Polyporus sulphureus* and *Hydnum erinaceus* may enter through wounds in the butt or anywhere along the trunk and cause extensive rot; and a third group including *Fomes everhartii* and *Polyporus dryophilus* rarely enters through butt scars but usually infects through wounds in the upper part of the trunk and can produce extensive rot. The fungi more commonly causing decay in oaks are *P. sulphureus, P. spraguei, H. erinaceus,* and *A. mellea.*

Unfortunately the production of conks bears little relation to the incidence of heart rot. However, Hepting and Hedgcock (1937) have developed a series of three-variable graphs for oaks and yellowpoplar, so that the cull percentage for stands, if primarily of seedling or seedling-sprout origin, can be approximated by determining the percentage of trees with basal wounds for each diameter class (Fig. 146). Another series of curves based on the width of basal wounds 12 inches above ground level enables the butt-cull volume of oaks to be predicted ahead for 60 years and more, almost immediately after a fire has occurred (Hepting 1941).

Top rot is related to the incidence of wounds and rotten branch stubs

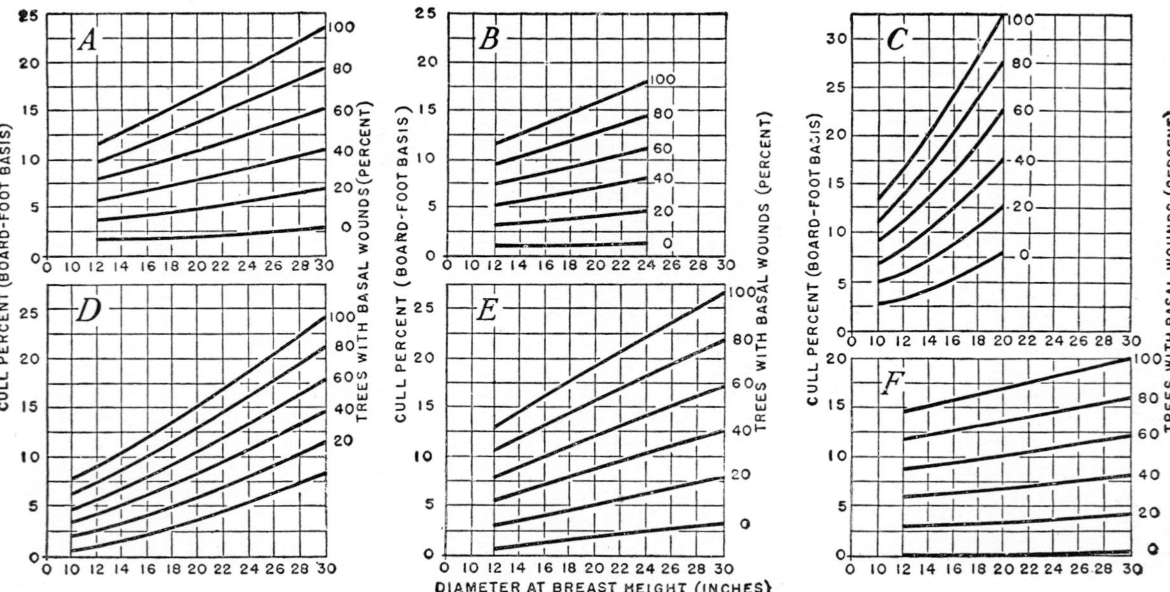

Fig. 146. Relation of cull percentage to tree diameter for tree populations with different percentages of basal wounds. *A*, red oak; *B*, black oak; *C*, scarlet oak; *D*, white oak; *E*, chestnut oak; *F*, yellowpoplar. (*Reprinted from U.S. Dept. Agr. Tech. Bull.* 570.)

on the bole, and five risk classes of oaks have been established, together with the average volume of cull from top rot in trees of each category (Hepting, Garren, and Warlick 1940): Class A—no rotten stubs over 3 inches thick, no large surface wounds, no holes, and less than three blind knots or healed stubs, anywhere on the bole up to 8 feet above merchantable top. Cull less than 5 board feet. Class B—one rotten stub or large surface wound, or three blind knots. Cull about 25 board feet. Class C—two rotten stubs or large surface wounds, or a total of four to five blind knots, rotten stubs, or large wounds. Cull about 65 board feet. Class D—three rotten stubs or large surface wounds, or a total of six to seven blind knots, rotten stubs, or large wounds. Cull about 95 board feet. Class E—four or more rotten stubs or large surface wounds, or a total of eight or more blind knots, rotten stubs, or large wounds. Cull about 220 board feet. A blind knot is an overgrown knot that leaves a scar resembling a navel. Any tree with the top broken out, with a hole in the bole, or bearing a conk is likely to be largely cull. Class A trees are good risks if left for a future cut, the others are poor risks.

Black cherry on the Allegheny Plateau is decayed by a number of fungi, but only *Fomes pinicola* regularly forms conks, so that wounds and decayed branch stubs must be depended upon in judging decay in living trees (Davidson and Campbell 1943). Conks of *F. pinicola* indicate an over-all rot column of 12 feet, 6 feet more or less each way from the conk.

Southern Hardwoods. In the bottom lands of the lower Mississippi River, most decay is butt rot connected with fire scars. The height of hollows and butt bulges, which indicate fire scars, and rot diameter at the stump are related to the extent of rot aboveground in the trunk, so that estimating can be done with more accuracy (Toole 1959). The relationship varies with tree species and applies specifically to bottom-land hardwoods. However, it should be useful elsewhere in the South.

PREVENTION OF WOUNDING

Open wounds are avenues for the entrance of decay into the heartwood, and it is not until a wound is completely grown over that the tree is no longer exposed to infection. Some species will remain relatively sound throughout their life except for decay through wounds, whereas others, where most of the decay enters through branches which have died naturally from lack of light, may contain considerable defect even though largely free from mechanical injury. Douglas fir in western Oregon and Washington typifies the second group (Boyce 1923*b*:8). Wounds are caused by many agencies, some of which can be controlled. Lightning may kill tops or tear narrow strips of bark and wood from trees, but in general lightning scars are only occasional entrance points of decay. Some decay enters through frost cracks, which may also serve to carry decay over a greater length of the bole than normal. Wind, snow, and ice cause broken tops that generally become infected (Heikinheimo 1920; Lagerberg 1919; Roth 1941). Animals and insects cause wounds that serve as infection courts (Baxter and Grow 1955; Buckland and Marples 1952; Redmond 1957).

Man is responsible for many wounds. Blazes that extend through the bark into the wood may become infected (Silvén 1944; Weir 1920), and even insignificant scars made with a timber scribe may give rise to decay (Oppliger 1932). Increment borings sometimes result in decay, particularly in hardwoods, and always result in more or less discoloration of the wood in the vicinity of the hole (Toole and Gammage 1959). After boring, holes in less resinous conifers such as spruces, balsam firs, and hemlocks should be plugged to prevent decay. Untreated plugs of a durable hardwood may be used, but hardwood plugs impregnated with a water-soluble preservative such as sodium fluorate are better. Discoloration and decay often follow the tapping of sugar maples (Scott and Sproston 1950). Logging, particularly machine logging, commonly wounds reserved trees so that serious decay may result in thin-barked species such as western hemlock, Sitka spruce, and true firs (Wright and Isaac 1956). After a stand is opened up, sunscald frequently occurs on western hemlock, and decay infection often follows. Given the age and area of a logging scar on western hemlock or Sitka spruce, the volume of decay accompanying it can be calculated. Also the radial and vertical extent of decay in an individual of western hemlock can be determined and log bucking done accordingly. For sugar maple the width of a logging scar is the most useful factor in predicting the amount of cull in wounded trees; the wider the scar, the more rot. Decay rarely enters pines through healthy turpentine faces, but dry faces, for which fungi seem to be at least partly responsible, are entrance points for decay (True and McCulley 1945; True 1946).

Fire scars are responsible for decay in many species. Sound hardwoods cannot be grown in the East if the trees are fire-scarred (Hepting 1935; Stickel and Marco 1936; Burns 1955). An analysis of over 3,000 oaks, basswoods, and yellowpoplars showed that cull from butt defect was only 1.5 per cent for trees without basal wounds as compared to 15.5 per cent for trees with basal wounds, and that 97 per cent of the butt scars were probably caused by fire (Hepting and Hedgcock 1937:18). Analysis of fire damage in eastern hardwoods will show that the major loss is not the immediate effect of the fire but is the ultimate effect, expressed years later, when the stand is cut with a heavy cull owing to decay that entered through fire scars.

The prevention of wounding, particularly scars caused by fire, will assist materially in reducing the incidence of decay in many species, but in eastern hardwoods adequate fire protection is essential for the production of reasonably sound timber.

PRUNING

With most conifers, clear lumber cannot be produced in the short rotations necessitated by commercial forestry. Clear softwood lumber now largely comes from virgin stands 200 years or more in age, but rotations beyond 50 to 100 years or so are not economically practicable. Consequently artificial pruning of crop trees is necessary, particularly in coniferous plantations, if the expense of establishing the plantation is to be justified. The usual practice is to prune as soon as possible, without checking growth of the tree, to a sufficient height to clear a 16-foot butt log, removing the limbs after they die or pruning both dead and living limbs. Fall is the best time to prune. In pruning to this height, limbs over 1 inch in diameter are rarely encountered. If thicker branches are pruned, decay is likely to occur. Although there are no artificially pruned stands of high age in this country as yet, pruned trees so far examined have shown little evidence of decay entering through the stubs of pruned-off branches, and it seems unlikely that artificial pruning will increase the incidence of decay by comparison with natural pruning (Childs 1956; Hawley and Clapp 1935). Artificially pruned trees should be less subject to heart rot by those wood-destroying fungi which largely infect through dead branches, since, when living branches are cut off, the stub is quickly covered with an antiseptic resin dressing, and when dead branches are removed close to the trunk the stub heals over with relative rapidity, whereas dead branches persist for many years before they drop off naturally. The most effective method for reducing western red rot in ponderosa pine stands under 60 years old is to select small-branched trees for crop trees and prune them (Andrews 1955). Artificial pruning of larger living branches will increase decay hazard because the exposed wood is not completely covered by resin. *Stereum sanguinolentum*, which causes rapid decay of heartwood, has infected pruning wounds on eastern white pine when branches 2 or more inches in diameter were cut off (Sleeth 1938).

Pruning hardwoods has been little practised in this country, but in northern hardwoods no significant difference was found in the amount of decay resulting from dead branch pruning and live branch pruning (Skilling 1958). Less infection followed artificial pruning than natural pruning. In general, large living branches should not be pruned because wounds 2 inches or more in diameter heal slowly. However, in sugar maple branch stubs less than 4 inches in diameter rarely become rotted within 20 years (Hesterberg 1957:24). In oaks only live branches that will leave wounds less than 1.5 inches across should be removed because decay is likely to enter through larger wounds (Roth 1948).

SEEDLING AND SPROUT HARDWOODS

Hardwoods originate in two ways, either as coppice sprouts from a stump or as seedlings. Seedling sprouts, which result from the cutting of a seedling or a small sapling, have the characteristics of seedlings, since they use the root system of the previous tree in its entirety. Trees originating from a stump 2 inches or less in diameter at the ground line are arbitrarily classed as seedling sprouts. Trees of sprout origin have a comparatively high incidence of butt rot entering from the old stump, and the larger the stump the more likelihood of decay attacking the sprout. In hardwood stands, seedlings or seedling sprouts should always be favored over sprouts. Sprout stands should be managed on short rotations and if feasible should ultimately be converted into seedling or seedling-sprout stands if quality timber is desired.

The following practices to keep decay at a minimum apply directly to sprout oak stands (Roth 1956) and less directly to sprout stands of other species:

1. Oak clump sprouts 3 inches dbh or smaller can be thinned in any convenient manner without creating a decay hazard.
2. Clumps composed of sprouts larger than 3 inches can be thinned without creating a decay hazard if the crotch is of the wide U type.
3. In removing surplus sprouts of any diameter, the cut should be made with a saw to minimize bark loosening, and the wound should be kept as small as possible by making a horizontal or slightly sloping cut. Never leave a stub.
4. Clumps of sprouts larger than 3 inches in diameter having a V type crotch should be treated as a unit, either all cut or all left.

The most effective improvement can be made in young stands up to 3 inches dbh and not over 15 years old. Larger and older stands are difficult to improve. Often the number of sprouts in a clump cannot be reduced without leaving large open butt wounds.

Trees on areas that have been repeatedly cut over are likely to be highly defective, so that improvement operations may be difficult and the chances of obtaining a sound stand poor. In trees badly wounded by fire the butt rot hazard is high, so that procedures to reduce decay transmission from stumps and stubs are less likely to reduce losses from butt rot. However, sprout stands arising after severe burns have less decay than those arising after cutting operations without fire, because fire preceding the establishment of a sprout stand kills the stump above ground level, thus forcing sprouts to develop from buds at or below the ground line.

ROTATION

In their youth trees are practically free from decay, but loss increases with advancing age. Decay cannot start until heartwood formation begins, which may be at from 15 to 40 years of age depending on the species. Theoretically, from that time on trees are liable to decay, but with most species it is usually many years later before decay of consequence is evident. It then remains to be decided just when the increase in decay becomes serious, since this determines the maximum age at which the stand can be cut if loss of economic importance is to be avoided. This limiting age has been termed the "pathological rotation" (Meinecke 1916:59), which is in the same category as other so-called "rotations" such as the technical, financial, and so forth. Actually there can be only one rotation, which is determined by the interaction of financial, technical, silvicultural, and pathologic considerations. The age at which decay causes consequential loss will be well beyond the actual rotation set by other considerations except in a few short-lived species such as aspen, which is heavily decayed at an early age. In Minnesota, loss through decay definitely lowers the rotation for aspen stands, and in Utah, although gross periodic increment of aspen trees does not culminate until about 120 years, net periodic increment culminates at about 80 years, owing to the inroads of decay.

The age at which stands must be cut to avoid serious decay has been determined for some species as follows: aspen in Minnesota 40 to 50 years (Schmitz and Jackson 1927); aspen in Utah 80 to 90 years (Meinecke 1929); aspen in northern Ontario 120 years (Black 1951); yellow birch in Nova Scotia 170 years (Stillwell 1955); balsam fir in eastern Canada 60 to 90 years (McCallum 1928); balsam fir in the Lake states 80 years (Kaufert 1935); balsam fir in New England and New York 70 years (Spaulding and Hansbrough 1944); alpine fir and amabilis fir in eastern British Columbia 130 years (Bier, Salisbury, and Waldie 1948); white fir in southern Oregon 150 years (Meinecke 1916); amabilis fir 275 to 325 years and western hemlock 225 to 275 years in western British Columbia (Buckland, Foster, and Nordin 1949); western hemlock in Idaho and Montana 100 to 120 years (Weir and Hubert 1918); western white pine in Idaho and Montana 100 to 120 years (Weir and Hubert 1919*a*); eastern white pine in Ontario 160 to 170 years (White 1953);

incense cedar in California, in the intermediate range 165 years, in the optimum range 210 years (Boyce 1920); Douglas fir in western Oregon and Washington 150 years (Boyce 1932); white spruce 160 years and red spruce 170 years in the Maritime Provinces (Davidson and Redmond 1957); subalpine spruce in Alberta 190 years (Nordin 1956); and Sitka spruce in the Queen Charlotte Islands 250 to 300 years with extremely little decay in trees less than 200 years old (Bier, Foster, and Salisbury 1946:22).

The prevention of decay in the future forest is relatively simple in those species growing in even-aged stands where clear-cutting is practiced, as for example Douglas fir in western Oregon and Washington. The stand is clear-cut to make way for the new stand. Even if decayed trees with conks are reserved for seed trees they will do little if any harm, since under proper management the succeeding stand will in turn be cut before it has attained the age at which there can be appreciable loss through decay, and it is unlikely that decay of the dead heartwood will affect the vital functions of the tree so as to reduce the quality of the seed (Boyce 1927). After about 26 years Douglas firs grown from seed from mother trees decayed by *Fomes pini* and from sound mother trees showed no significant difference when grown on the same sites (Munger and Morris 1936:19; 1942). Removing decayed trees from mature and overmature stands in advance of logging to prevent the spread of decay is not advisable unless it can be done at a profit on the trees removed. The operation would be expensive, it is not necessary for the protection of future stands, and its effect on lessening the increase in decay by reducing new infections in present stands would be doubtful unless the work was done over extensive natural units of area, such as entire watersheds.

In uneven-aged pure or mixed stands, such as the mixed coniferous forest in California, the problem is more difficult at least during the transition period from the overmature virgin stands of the present to the regulated stands of the future. When such stands are handled by either the selection or shelterwood method, trees are reserved that will have passed the age at which they are liable to extensive decay before time for the next cut. Consequently all trees decayed at the time of the first cutting, particularly those with conks, should be removed to reduce the chances of spreading decay to the remaining stand. Although opening up stands by cutting results in drier conditions less favorable to germination and penetration of spores of wood-destroying fungi, yet this influence cannot be relied upon to reduce significantly the number of new infections in reserved trees (Weir and Hubert 1919*b*). Trees reserved for the next cutting should be free from open wounds through which decay might enter. As far as possible scarred trees should be felled, care exercised to avoid wounding the reserve trees during logging, and adequate fire protection given the growing stand.

There are still extensive stands of overmature, defective timber, particularly in the West, which are nature's gift, having cost nothing to produce. Even though the rate of decay in stands is relatively slow (page 362), there comes a time when increase in volume through growth is offset by loss in volume through decay; then each succeeding year brings a decline in net volume until the stand breaks up and is replaced. In aspen in Minnesota this decline begins at about 90 years and in balsam fir in the Lake states at about 100 years. Exceptional stands of defective species may remain relatively sound throughout their life. Although the rate of loss in volume through decay can be determined with some exactness, it is difficult to determine the actual financial loss since so many interrelated factors, other than the mere rate of decay, complicate this figure. The loss in volume is made up in part by growth so that the net volume loss is less than the actual volume increase in decay. On the other hand, there is a quality loss in the older timber that is not replaced, because decay usually is most extensive in the more valuable portion of the trunk. To offset the loss in volume and quality, there is a probable increase in the value of stumpage and consequent closer utilization. The proximity to market has an important bearing, since some operations have a sale for low-grade, partially decayed material that to others is a complete loss. The general condition of the market is significant, because selling timber on a low market may result in greater financial loss than will the inroads of decay over a period of years. The decision as to whether or not defective timber should be cut at a given time is far from dependent on the rate of decay alone; the value increase may more than offset the volume decrease through decay, so that each stand will require individual consideration of the various factors involved.

Yield tables, by which the feasibility of commercial timber production is determined and the rotation is set, demand in addition a knowledge of prospective loss for defective species. Yield tables are based on normal, i.e., fully stocked, stands and gross yield. But the net yield and not the gross yield is important, since the former determines the monetary return when the stand is cut; consequently a yield study in any timber type subject to defect is not complete and is actually misleading unless accompanied by a study of loss.

REFERENCES

Andrews, S. R.: Red Rot of Ponderosa Pine. *U.S. Dept. Agr. Monog.*, 23:1–34 (1955).

Anonymous: "Recognition of Decay and Insect Damage in Timbers for Aircraft and Other Purposes." London: H.M. Stationery Office (1943).

Anonymous: Wood: Colors and Kinds. *U.S. Dept. Agr., Agr. Handbook*, 101:1–36 (1956).

Aoshima, K.: Decay of Beech Wood by the Haploid and Diploid Mycelia of *Elfvingia applanata. Phytopathology*, 44:260–265 (1954).

Armstrong, F. H.: Further Tests on the Effect of Progressive Decay by *Trametes serialis* Fr. on the Mechanical Strength of the Wood of Sitka Spruce. *Forestry*, 9:62–64 (1935).

Atwell, A. E.: Red-stain and Pocket-rot in Jack Pine: Their Effect on Strength and Serviceability of the Wood. *Can. Dept. Mines and Resources, Forest Serv., Forest Products Labs. Circ.*, 63:1–23 (1948).

Badcock, E. C.: Preliminary Account of the Odour of Wood-destroying Fungi in Culture. *Brit. Mycol. Soc. Trans.*, 23:188–198 (1939).

Badcock, E. C.: Methods for Obtaining Fructifications of Wood-rotting Fungi in Culture. *Brit. Mycol. Soc. Trans.*, 26:127–132 (1943).

Basham, J. T.: Decay of Trembling Aspen. *Can. J. Botany*, 36:491–505 (1958).

Basham, J. T., P. V. Mook, and A. G. Davidson: New Information Concerning Balsam Fir Decays in Eastern North America. *Can. J. Botany*, 31:334–360 (1953).

Bavendamm, W.: Erkennen, Nachweis und Kultur der holzfarbenden und holzzersetzenden Pilze. *Handbuch Biol. Arbeitsmethoden*, 12:927–1134 (1936).

Bavendamm, W.: Mikroskopisches Erkennen und Bestimmen von holzbewohnenden und holzzersetzenden Pilzen. *Handbuch Mikroskop Technik*, 5(2):817–844 (1951).

Baxter, D. V., and T. Grow: Unusual Decay in Young Oak Plantations Following Early Rodent Injury. *Mich. Forestry*, 12:1–2 (1955).

Bier, J. E., R. E. Foster, and P. J. Salisbury: Studies in Forest Pathology IV. Decay of Sitka Spruce on the Queen Charlotte Islands. *Can. Dept. Agr. Publ.* 783, *Tech. Bull.*, 56:1–35 (1946).

Bier, J. E., P. J. Salisbury, and R. A. Waldie: Studies in Forest Pathology V. Decay in Fir, *Abies lasiocarpa* and *Abies amabilis*, in the Upper Fraser Region of British Columbia. *Can. Dept. Agr. Tech. Bull.*, 66:1–28 (1948).

Black, R. L.: Poplar Decay Study. *Can. Dept. Agr., Forest Biol. Div., Sci. Serv., Bi-monthly Progr. Rept.*, 7(2):2 (1951).

Boyce, J. S.: Imbedding and Staining of Diseased Wood. *Phytopathology*, 8:432–436 (1918).

Boyce, J. S.: The Dry-rot of Incense Cedar. *U.S. Dept. Agr. Bull.*, 871:1–58 (1920).

Boyce, J. S.: Decays and Discolorations in Airplane Woods. *U.S. Dept. Agr. Bull.*, 1128:1–51 (1923*a*).

Boyce, J. S.: A Study of Decay in Douglas Fir in the Pacific Northwest. *U.S. Dept. Agr. Bull.*, 1163:1–20 (1923*b*).

Boyce, J. S.: Decay and Seed Trees in the Douglas Fir Region. *J. Forestry*, 25:835–839 (1927).

Boyce, J. S.: Decay in Pacific Northwest Conifers. *Yale Univ., Osborn Botan. Lab. Bull.*, 1:1–51 (1930).

Boyce, J. S.: Decay and Other Losses in Douglas Fir in Western Oregon and Washington. *U.S. Dept. Agr. Tech. Bull.*, 286:1–59 (1932).

Boyce, J. S., and J. W. B. Wagg: Conk Rot of Old-growth Douglas-fir in Western Oregon. *Oregon Forest Products Lab.* (*Corvallis*) *Bull.*, 4:1–96 (1953).

Brown, R. M.: Statistical Analyses for Finding a Simple Method for Estimating the Percentage Heart Rot in Minnesota Aspen *J. Agr. Research*, 49:929–942 (1934).

Buckland, D. C., R. E. Foster, and V. J. Nordin: Studies in Forest Pathology VII. Decay in Western Hemlock and Fir in the Franklin River Area, British Columbia. *Can. J. Research C.*, 27:312–331 (1949).

Buckland, D. C., and E. G. Marples: Rodent Damage to Western Hemlock. *Forestry Chronicle*, 28:25–30 (1952).

Buller, A. H. R.: "Researches on Fungi." London: Longmans, Green & Co., Ltd. (1909).

Burns, P. Y.: Fire Scars and Decay in South Missouri Oaks. *Missouri Agr. Expt. Sta. Bull.*, 642:1–8 (1955).

Campbell, A. H.: Zone Lines in Plant Tissues. I. The Black Lines Formed by *Xylaria polymorpha* (Pers.) Grev. in Hardwoods. *Ann. Appl. Biol.*, 20:123–145 (1933).

Campbell, W. G.: The Chemistry of the White Rots of Wood. III. The Effect on Wood Substance of *Ganoderma applanatum* (Pers.) Pat., *Fomes fomentarius* (Linn.) Fr., *Polyporus adustus* (Willd.) Fr., *Pleurotus ostreatus* (Jacq.) Fr., *Armillaria mellea* (Vahl.) Fr., *Trametes pini* (Brot.) Fr., and *Polystictus abietinus* (Dicks.) Fr. *Biochem. J.*, 26:1829–1838 (1932*a*).

Campbell, W. G.: A Chemical Approach to the Study of Wood Preservation. *Forestry*, 6:82–89 (1932*b*).

Campbell, W. G.: The Biological Decomposition of Wood. In L. E. Wise and E. C. John: "Wood Chemistry," 2d ed. New York: Reinhold Publishing Corporation (1952).

Carter, J. C.: Wetwood of Elms. *Illinois Nat. Hist. Survey Bull.*, 23:407–448 (1945).

Cartwright, K. St. G.: A Satisfactory Method of Staining Fungal Mycelium in Wood Sections. *Ann. Botany*, 43:412–413 (1929).

Cartwright, K. St. G.: A Decay of Sitka Spruce Caused by *Trametes serialis*, Fr. *Gt. Brit. Dept. Sci. Ind. Research, Forest Prods. Research Bull.*, 4:1–26 (1930).

Cartwright, K. St. G., W. G. Campbell, and F. H. Armstrong: The Influence of Fungal Decay on the Properties of Timber I—The Effect of Progressive Decay by *Polyporus hispidus*, Fr., on the Strength of English Ash (*Fraxinus excelsior* L.). *Proc. Roy. Soc. (London) B.*, 120:76–95 (1936).

Cartwright, K. St. G., and W. P. K. Findlay: "Decay of Timber and Its Prevention." London: H. M. Stationery Office (1958).

Cartwright, K. St. G., et al.: The Effect of Progressive Decay by *Trametes serialis* Fr., on the Mechanical Strength of the Wood of Sitka Spruce. *Gt. Brit. Dept. Sci. Ind. Research, Forest Prods. Research Bull.*, 11:1–18 (1931).

Childs, T. W.: Pruning and Occurrence of Heart Rot in Young Douglas-fir. *U.S. Dept. Agr., Forest Serv., Pacific Northwest Forest and Range Expt. Sta. Research Note*, 132:1–5 (1956).

Clausen, V. H., and F. H. Kaufert: Occurrence and Probable Cause of Heartwood Degradation in Commercial Species of *Populus*. *J. Forest Products Research Soc.*, 2(4):62–67 (1952).

Davidson, A. G.: Studies in Forest Pathology XVI. Decay of Balsam Fir, *Abies balsamea* (L.) Mill., in the Atlantic Provinces. *Can. J. Botany*, 35:857–874 (1957).

Davidson, A. G., and D. R. Redmond: Decay of Spruce in the Maritime Provinces. *Forestry Chronicle*, 33:373–380 (1957).

Davidson, R. W., and W. A. Campbell: Decay in Merchantable Black Cherry on the Allegheny National Forest. *Phytopathology*, 33:965–985 (1943).

Davidson, R. W., T. E. Hinds, and F. G. Hawksworth: Decay of Aspen in Colorado. *U.S. Dept. Agr., Forest Serv., Rocky Mt. Forest and Range Expt. Sta., Sta. Paper*, 45:1–14 (1959).

Duncan, Catherine G.: Wood-attacking Capacities and Physiology of Soft-rot Fungi. *U.S. Dept. Agr., Forest Serv., Forest Prods. Lab. Rept.*, 2173:1–28 (1960).

Englerth, G. H.: Decay of Western Hemlock in Western Oregon and Washington. *Yale Univ., School Forestry Bull.*, 50:1–53 (1942).

Englerth, G. H., and J. R. Hansbrough: The Significance of the Discolorations in Aircraft Lumber: Noble Fir and Western Hemlock. *U.S. Dept. Agr., Forest Pathol., Spec. Release*, 24:1–10 (1945).

Esau, Katherine: "Plant Anatomy." New York: John Wiley & Sons, Inc. (1953). (Chapter 3: The Cell Wall, pp. 35–73.)

Eslyn, W. E.: Radiographical Determination of Decay in Living Trees by Means of the Thulium X-ray Unit. *Forest Sci.*, 5:37–47 (1959).

Etheridge, D. E.: Decay of Subalpine Spruce on the Rocky Mountain Forest Reserve in Alberta. *Can. J. Botany*, 34:805–816 (1956).

Etheridge, D. E.: The Effect on Variations in Decay of Moisture Content and Rate of Growth in Subalpine Spruce. *Can. J. Botany*, 36:187–206 (1958).

Fenton, R. H., and F. H. Berry: Heart Rot of Virginia Pine in Maryland. *Northeast. Forest Expt. Sta., Forest Research Notes*, 56:1–4 (1956).

Findlay, W. P. K.: A Study of *Paxillus panuoides* Fr., and Its Effects upon Wood. *Ann. Appl. Biol.*, 19:331–350 (1932).

Findlay, W. P. K.: Studies in the Physiology of Wood-destroying Fungi. III. Progress of Decay under Natural and under Controlled Conditions. *Ann. Botany*, 4:701–712 (1940).

Findlay, W. P. K.: Timber Decay—A Survey of Recent Work. *Forestry Abstr.*, 17:317–327, 477–486 (1956).

Findlay, W. P. K., and C. D. Pettifor: Dark Coloration in Western Red Cedar in Relation to Certain Mechanical Properties. *Empire Forestry J.*, 20:64–72 (1941).

Foster, R. E.: Decay of Alpine Fir in the Upper Fraser Region. *Can. Dept. Agr., Sci. Serv., Forest Biol. Div., Bi-monthly Progr. Rept.*, 10(5):3 (1954).

Foster, R. E., J. E. Browne, and A. T. Foster: Studies in Forest Pathology XIX. Decay of Western Hemlock and Amabilis Fir in the Kitimat Region of British Columbia. *Can. Dept. Agr. Publ.*, 1029:1–37 (1958).

Foster, R. E., H. M. Craig, and G. W. Wallis: Studies in Forest Pathology XII. Decay of Western Hemlock in the Upper Columbia Region, British Columbia. *Can. J. Botany*, 32:145–171 (1954).

Foster, R. E., and A. T. Foster: Studies in Forest Pathology VIII. Decay of Western Hemlock on the Queen Charlotte Islands, British Columbia. *Can. J. Botany*, 29:479–521 (1951).

Foster, R. E., G. P. Thomas, and J. E. Browne: A Tree Decadence Classification for Mature Coniferous Stands. *Forestry Chronicle*, 29:359–366 (1953).

Gerry, Eloise: Western Hemlock "Floccosoids" (White Spots or Streaks). *U.S. Dept. Agr., Forest Serv., Forest Prods. Lab. Mim.*, 1392:1–3 (1943).

Hansbrough, J. R., and G. H. Englerth: The Significance of the Discolorations in Aircraft Lumber: Sitka Spruce. *U.S. Dept. Agr., Forest Pathol., Spec. Release*, 21:1–14 (1944).

Harrar, E. S.: Defects in Hardwood Veneer Logs: Their Frequency and Importance. *U.S. Dept. Agr., Forest Serv., Southeast. Forest Expt. Sta. Paper*, 39:1–45 (1954).

Harrison, C. H.: Longevity of the Spores of Some Wood-destroying Hymenomycetes. *Phytopathology*, 32:1096–1097 (1942).

Hartig, R.: "Die Zersetzungserscheinungen des Holzes der Nadelholzbäume und der

Eiche in forstlicher botanische und chemischer Richtung." Berlin: Springer-Verlag (1878).

Hawley, R. C., and R. T. Clapp: Artificial Pruning in Coniferous Plantations. *Yale Univ., School Forestry Bull.*, 39:1–36 (1935).

Heikinheimo, O.: Suomen lumituhoalueet ja nüden metsät. [German summary.] *Inst. Quaest. Forest. Finland Commun.*, 3:1–134, 1–17 (1920).

Hepting, G. H.: Decay Following Fire in Young Mississippi Delta Hardwoods. *U.S. Dept. Agr. Tech. Bull.*, 494:1–32 (1935).

Hepting, G. H.: Prediction of Cull Following Fire in Appalachian Oaks. *J. Agr. Research*, 62:109–120 (1941).

Hepting, G. H., and Dorothy J. Blaisdell: A Protective Zone in Red Gum Fire Scars. *Phytopathology*, 26:62–67 (1936).

Hepting, G. H., K. H. Garren, and P. W. Warlick: External Features Correlated with Top Rot in Appalachian Oaks. *J. Forestry*, 38:873–876 (1940).

Hepting, G. H., and G. G. Hedgcock: Decay in Merchantable Oak, Yellow Poplar, and Basswood in the Appalachian Region. *U.S. Dept. Agr. Tech. Bull.*, 570:1–29 (1937).

Hesterberg, G. A.: Deterioration of Sugar Maple Following Logging Damage. *U.S. Dept. Agr., Forest Serv., Lake States Forest Expt. Sta., Sta. Paper*, 5:1–58 (1957).

Higgins, B. B.: Gum Formation with Special Reference to Cankers and Decays of Woody Plants. *Georgia Expt. Sta. Bull.*, 127:23–59 (1919).

Hirt, R. R.: On the Biology of *Trametes suaveolens* (L.) Fr. *N.Y. State Coll. Forestry Syracuse Univ., Tech. Publ.*, 37:1–36 (1932).

Hopp, H.: The Formation of Colored Zones by Wood-destroying Fungi in Culture. *Phytopathology*, 28:601–620 (1938).

Hornibrook, E. M.: Estimating Defect in Mature and Overmature Stands of Three Rocky Mountain Conifers. *J. Forestry*, 48:408–417 (1950).

Horton, G. S., and C. Hendee: A Study of Rot in Aspen on the Chippewa National Forest. *J. Forestry*, 32:493–494 (1934).

Hubert, E. E.: A Staining Method for Hyphae of Wood-inhabiting Fungi. *Phytopathology*, 12:440–441 (1922).

Hubert, E. E.: The Diagnosis of Decay in Wood. *J. Agr. Research*, 29:523–567 (1924*a*).

Hubert, E. E.: Effect of Kiln Drying, Steaming, and Air Seasoning on Certain Fungi in Wood. *U.S. Dept. Agr. Bull.*, 1262:1–20 (1924*b*).

Jorgensen, E.: A Method for the Study of Mycelial Anastomoses. *Friesia*, 5:75–79 (1954).

Kaufert, F. H.: Heart Rot of Balsam Fir in the Lake States, with Special Reference to Forest Management. *Minn. Agr. Expt. Sta. Tech. Bull.*, 110:1–27 (1935).

Kimmey, J. W.: Cull Factors for Forest-tree Species in Northwestern California. *U.S. Dept. Agr., Forest Serv., Calif. Forest and Range Expt. Sta., Forest Survey Release*, 7:1–30 (1950).

Kimmey, J. W.: Cull and Breakage Factors for Pines and Incense-cedar in the Sierra Nevada. *U.S. Dept. Agr., Forest Serv., Calif. Forest and Range Expt. Sta., Forest Research Notes*, 90:1–4 (1954).

Kimmey, J. W.: Cull Factors for Sitka Spruce, Western Hemlock and Western Redcedar. *U.S. Dept. Agr., Forest Serv., Alaska Forest Research Center, Sta. Paper*, 6:1–31 (1956).

Kimmey, J. W.: Application of Indicator Cull Factors to White and Red Fir Stands in the Sierra Nevada. *U.S. Dept. Agr., Forest Serv., Calif. Forest and Range Expt. Sta., Forest Research Notes,* 127:1–5 (1957).

Kimmey, J. W., and E. M. Hornibrook: Cull and Breakage Factors and Other Tree Measurement Tables for Redwood. *U.S. Dept. Agr., Forest Serv., Calif. Forest and Range Expt. Sta., Forest Survey Release,* 13:1–28 (1952).

Koehler, A.: Causes of Brashness in Wood. *U.S. Dept. Agr. Tech. Bull.,* 342:1–39 (1933).

Lagerberg, T.: Snöbrott och toppröta hos granen. [German summary.] *Statens Skogsförsöksanst. (Sweden), Medd.,* 16:115–162 (1919).

Lagerberg, T.: Barrträdens vattved. [English résumé.] *Svenska Skogsvårdsför. Tidsskr.,* 33:177–264 (1935).

Lockhard, C. R., J. A. Putnam, and R. D. Carpenter: Log Defects in Southern Hardwoods. *U.S. Dept. Agr. Handbook,* 4:1–37 (1950).

McCallum, A. W.: Studies in Forest Pathology I. Decay in Balsam Fir (*Abies balsamea* Mill.). *Can. Dept. Agr. Bull.,* 104(N.S.):1–25 (1928).

Meier, H.: Über den Zellwandabbau durch Holzvermorschungspilze und die submikroskopische Struktur von Fichtentracheiden und Birkenholzfasern. *Holz Roh- u. Werkstoff,* 13:323–338 (1955).

Meinecke, E. P.: Forest Pathology in Forest Regulation. *U.S. Dept. Agr. Bull.,* 275:1–63 (1916).

Meinecke, E. P.: Quaking Aspen: A Study in Applied Forest Pathology. *U.S. Dept. Agr. Tech. Bull.,* 155:1–34 (1929).

Michels, P.: Der Nasskern der Weisstanne. *Holz,* 6:87–99 (1943).

Miller, V. V.: "Points on the Biology and Diagnosis of House Fungi." [In Russian.] Leningrad: State Forestal Technical Publishing Office (1932). [*Rev. Appl. Mycol.,* 12:257–259 (1933).]

Morawski, Z. J. R., J. T. Basham, and K. B. Turner: A Survey of a Pathological Condition in the Forests of Ontario 1958. *Ontario Dept. Lands and Forests, Div. Timber, Forest Resources Inventory, Cull Studies Rept.,* 25:1–96 (1958).

Mulholland, J. R.: Changes in Weight and Strength of Sitka Spruce Associated with Decay by a Brown-rot Fungus, *Poria monticola. J. Forest Products Research Soc.,* 4:410–416 (1954).

Münch, E.: Versuche über Baumkrankheiten. *Naturw. Z. Forst- u. Landwirtsch.* 8:389–408, 425–447 (1910).

Münch, E.: Untersuchungen über Eichenkrankheiten. *Naturw. Z. Forst- u. Landwirtsch.,* 13:509–522 (1915).

Munger, T. T., and W. G. Morris: Growth of Douglas Fir Trees of Known Seed Source. *U.S. Dept. Agr. Tech. Bull.,* 537:1–40 (1936).

Munger, T. T., and W. G. Morris: Further Data on the Growth of Douglas-fir Trees of Known Seed Source. *Pacific Northwest Forest Expt. Sta.* (1942). (Mimeographed.)

Nobles, Mildred K.: Studies in Forest Pathology. VI. Identification of Cultures of Wood-rotting Fungi. *Can. J. Research C. Botan. Sci.,* 26:281–431 (1948).

Nobles, Mildred K.: A Rapid Test for Extracellular Oxidase in Cultures of Wood-inhabiting Hymenomycetes. *Can. J. Botany,* 36:91–99 (1958).

Nordin, V. J.: Studies in Forest Pathology XIII. Decay in Sugar Maple in the Ottawa-Huron and Algoma Extension Forest Region of Ontario. *Can. J. Botany,* 32:221–258 (1954).

Nordin, V. J.: Heart Rots in Relation to the Management of Spruce in Alberta. *Forestry Chronicle,* 32:79–84 (1956).

Nutman, F. J.: Studies of Wood-destroying Fungi I. *Polyporus hispidus* (Fries). *Ann. Appl. Biol.*, 16:40–64 (1929).

Oppliger, F.: Stammbeschädigung durch Reisserstriche. *Schweiz. Z. Forstw.*, 83:59–61 (1932).

Pieper, E. J., C. J. Humphrey, and S. F. Acree: Synthetic Culture Media for Wood-destroying Fungi. *Phytopathology*, 7:214–220 (1917).

Proctor, P.: Penetration of the Walls of Wood Cells by the Hyphae of Wood-destroying Fungi. *Yale Univ., School Forestry Bull.*, 47:1–31 (1941).

Rankin, W. H.: Wound Gums and Their Relation to Fungi. *Natl. Shade Tree Conf. Proc.*, 9:111–115 (1933).

Redmond, D. R.: Infection Courts of Butt-rotting Fungi in Balsam Fir. *Forest Sci.*, 3:15–21 (1957).

Riley, C. G.: Studies in Forest Pathology IX. *Fomes igniarius* Decay of Poplar. *Can. J. Botany*, 30:710–734 (1952).

Risley, J. H., and S. B. Silverborg: *Stereum sanguinolentum* on Living Norway Spruce Following Pruning. *Phytopathology*, 48:337–338 (1958).

Robinson, W.: The Microscopical Features of Mechanical Strains in Timber and the Bearing of These on the Structure of the Cell-wall in Plants. *Phil. Trans. Roy. Soc. London, Ser. B*, 210:49–82 (1920).

Roelofson, P. A.: Eine mögliche Erklarung der typischen Korrosionfiguren der Holzfasern bei Moderfäule. *Holz Roh- u. Werkstoff*, 14:208–210 (1956).

Roth, E. R.: Top Rot in Snow-damaged Yellow Poplar and Basswood. *J. Forestry*, 39:60–62 (1941).

Roth, E. R.: Healing and Defects Following Oak Pruning. *J. Forestry*, 46:500–504 (1948).

Roth, E. R.: Decay Following Thinning of Sprout Oak Clumps. *J. Forestry*, 54:26–30 (1956).

Schacht, H.: Ueber die Veränderung durch Pilze in abgestorbenen Pflanzenzellen. *Jahrb. wiss. Botan.*, 3:442–483 (1863).

Scheffer, T. C.: Progressive Effects of *Polyporus versicolor* on the Physical and Chemical Properties of Red Gum Sapwood. *U.S. Dept. Agr. Tech. Bull.*, 527:1–45 (1936).

Scheffer, T. C.: Diagnostic Features of Some Discolorations Common to Aircraft Hardwoods. *U.S. Dept. Agr., Forest Pathol., Spec. Release*, 19:1–5 (1944).

Scheffer, T. C., et al.: The Effect of Certain Heart Rot Fungi on the Specific Gravity and Strength of Sitka Spruce and Douglas-fir. *U.S. Dept. Agr. Tech. Bull.*, 779:1–24 (1941).

Schmitz, H., and L. W. R. Jackson: Heartrot of Aspen with Special Reference to Forest Management in Minnesota. *Minn. Agr. Expt. Sta. Tech. Bull.*, 50:1–43 (1927).

Schrenk, H. von: Glassy Fir. *Missouri Botan. Garden Ann. Rept.*, 16:117–120 (1905).

Scott, W. W., and T. Sproston, Jr.: A *Valsa* Associated with the Decay and Discoloration Following the Tapping of Sugar Maple Trees. [Abstract.] *Phytopathology*, 40:24–25 (1950).

Seliskar, C. E.: Wetwood Organism in Aspen, Poplar Is Isolated. *Farm and Home Research* (*Colo. Agr. Expt. Sta.*), 2(6):6–11, 19–20 (1952).

Silvén, F.: Stämpelröta hos gran i Norrland. ["Blaze Rot" in Norway Spruce in Norrland.] *Norrlands Skogsvårdsförbund Tidsskr.*, 1944(2):135–158 (1944). [*Forestry Abstr.*, 9:507–508 (1948).]

Silverborg, S. B.: Northern Hardwoods Cull Manual. *State Univ. N.Y. Coll. Forestry at Syracuse Bull.*, 31:1–45 (1954).

Silverborg, S. B.: Rate of Decay in Northern Hardwoods Following Artificial Inoculation with Some Common Heartrot Fungi. *Forest Sci.*, 5:223–228 (1959).

Skilling, D. D.: Wound Healing and Defects Following Northern Hardwood Pruning. *J. Forestry*, 56:19–22 (1958).

Sleeth, B.: Pruning Wounds as an Avenue of Entrance for *Stereum sanguinolentum* in Northern White Pine Plantations. *U.S. Dept. Agr., Allegheny Forest Expt. Sta., Tech Note*, 22:1–3 (1938).

Sleeth, B., and C. B. Bidwell: *Polyporus hispidus* and a Canker of Oaks. *J. Forestry*, 35:778–785 (1937).

Snell, W. H., W. G. Hutchinson, and K. H. N. Newton: Temperature and Moisture Relations of *Fomes roseus* and *Trametes subrosea*. *Mycologia*, 20:276–291 (1928).

Spaulding, P., and J. R. Hansbrough: Decay in Balsam Fir in New England and New York. *U.S. Dept. Agr. Tech. Bull.*, 872:1–30 (1944).

Stickel, P. W., and H. F. Marco: Forest Fire Damage Studies in the Northeast III. Relation Between Fire Injury and Fungal Infection. *J. Forestry*, 34:420–423 (1936).

Stillwell, M. A.: Decay of Yellow Birch in Nova Scotia. *Forestry Chronicle*, 31:74–83 (1955).

Thomas, G. P., and D. G. Podmore: Studies in Forest Pathology XI. Decay in Black Cottonwood in the Middle Fraser Region, British Columbia. *Can. J. Botany*, 31:675–692 (1953).

Thomas, G. P., and R. W. Thomas: Studies in Forest Pathology. XIV. Decay of Douglas Fir in the Coastal Region of British Columbia. *Can. J. Botany*, 32:630–653 (1954).

Toole, E. R.: Rot and Cankers on Oak and Honeylocust Caused by *Poria spiculosa*. *J. Forestry*, 52:941–942 (1954).

Toole, E. R.: *Polyporus hispidus* on Southern Bottomland Oaks. *Phytopathology*, 45:177–180 (1955).

Toole, E. R.: Heartrot in Bottomland Red Oaks Two Years after Inoculation. *U.S. Dept. Agr., Plant Disease Reptr.*, 40:823–826 (1956).

Toole, E. R.: Decay after Fire Injury to Southern Bottom-land Hardwoods. *U.S. Dept. Agr., Forest Serv., Tech. Bull.*, 1189:1–25 (1959).

Toole, E. R., and G. M. Furnival: Progress of Heart Rot Following Fire in Bottomland Red Oaks. *J. Forestry*, 55:20–24 (1957).

Toole, E. R., and J. L. Gammage: Damage from Increment Borings in Bottomland Hardwoods. *J. Forestry*, 57:909–911 (1959).

True, R. P.: Some Fungi Are Agents of Dry Face. *AT-FA J.*, 8(8):11–15, 20 (1946).

True, R. P., and R. D. McCulley: Defects above Naval Stores Faces Are Associated with Dry Face. *Southern Lumberman*, 171(2153):200–203, 204 (1945).

True, R. P., E. H. Tryon, and J. F. King: Cankers and Decays of Birch Associated with Two Poria Species. *J. Forestry*, 53:412–415 (1955).

Varner, R. W.: The Rennerfelt Method for Determining the Occurrence of Decay and Its Application to Michigan Forests. *Papers Mich. Acad. Sci.*, 34(I):79–82 (1950).

Wagener, W. W., and R. W. Davidson: Heart Rots in Living Trees. *Botan. Rev.*, 32:61–134 (1954).

Waid, J. S., and M. J. Woodman: A Non-destructive Method of Detecting Diseases in Wood. *Nature (London)*, 180:47 (1957).

Wallin, W. B.: Wetwood in Balsam Poplar. *Minn. Forestry Notes*, 28:1–2 (1954).

Waterman, Alma D., and J. R. Hansbrough: Microscopical Rating of Decay in Sitka Spruce and Its Relation to Toughness. *Forest Prods. J.*, 7:77–84 (1957).

Weir, J. R.: Note on the Pathological Effects of Blazing Trees. *Phytopathology*, 10:371–373 (1920).

Weir, J. R., and E. E. Hubert: A Study of Heart-rot in Western Hemlock. *U.S. Dept. Agr. Bull.*, 722:1–37 (1918).

Weir, J. R., and E. E. Hubert: A Study of the Rots of Western White Pine. *U.S. Dept. Agr. Bull.*, 799:1–24 (1919*a*).

Weir, J. R., and E. E. Hubert: The Influence of Thinning on Western Hemlock and Grand Fir Infected with *Echinodontium tinctorium*. *J. Forestry*, 17:21–35 (1919*b*).

White, J. H.: On the Biology of *Fomes applanatus* (Pers.) Wallr. *Trans. Roy. Can. Inst.*, 12:133–174 (1919).

White, L. T.: Studies in Forest Pathology. X. Decay of White Pine in the Timagami Lake and Ottawa Valley Areas. *Can. J. Botany*, 31:175–200 (1953).

Wood, L. W.: Properties of White-pocket Douglas-fir Lumber. *U.S. Dept. Agr., Forest Serv., Forest Prods. Lab. Rept.*, 2017:1–15 (1955).

Wright, E., and L. A. Isaac: Decay Following Logging Injury to Western Hemlock, Sitka Spruce, and True Firs. *U.S. Dept. Agr. Tech Bull.*, 1148:1–18 (1956).

CHAPTER 17

The Rots

There are so many rots or decays caused by such a large number of wood-destroying fungi that it is impossible to do more than briefly discuss the more common and important ones. The rots are not grouped according to the organisms causing them but according to the type of decay. A few have already been considered in the chapter on root diseases. Some rots have not yet been connected with the causal organism. The distribution of many of the fungi causing them is not yet throughly known. For a complete list of the wood-destroying fungi the reader must refer to monographs of the various families containing wood destroyers, particularly of the family Polyporaceae, where detailed taxonomic descriptions of the conks, fruiting bodies, fruit bodies, fructifications, or sporophores may be found. Although some wood destroyers are quite specific as to hosts and condition of the wood attacked, in general they are sufficiently adaptable so that it is not unusual to find species normally on conifers occasionally occurring on hardwoods, or to find a slash destroyer invading the heartwood of a living tree, or vice versa. Although there are a great many fungi that can decay the wood of dead trees, the number that can cause extensive decay in the dead heartwood of living trees is smaller, and only a few can kill and decay living sapwood.

Since in many cases the presence of conks bears a definite relation to the amount of decay in living trees, it is important for timber estimators and scalers to know the more common conks. Even if typical conks are not produced, abortive or sterile conks are just as valuable provided the fungus producing them is known. It is also desirable to know whether a fungus causing decay in a living tree can continue destruction of wood in service, because if it can, wood in any stage of decay should be rejected for use where resistance to rot is necessary.

Modern knowledge of wood-destroying fungi is based on the classical papers of Hartig (1874, 1878). All wood-destroying fungi, with occasional exceptions, belong to a few families of the class Basidiomycetes, so that to avoid frequent repetition of family names the genera are listed here under the families: family Agaricaceae—*Collybia, Lentinus, Pholiota, Pleurotus, Schizophyllum;* Hydnaceae—*Echinodontium, Hydnum;* Polyporaceae—*Daedalea, Fistulina, Fomes, Ganoderma, Lenzites, Merulius, Polyporus, Poria, Trametes;* Thelephoraceae—*Coniophora, Hymenochaete, Peniophora,* and *Stereum.*

RED RING ROT

Red ring rot, ring scale, red heart, white speck, honeycomb, conk rot, or pecky rot, a white pocket rot, is caused by the ring scale fungus, *Trametes pini* (Thore) Fr., now commonly called *Fomes pini* (Thore)

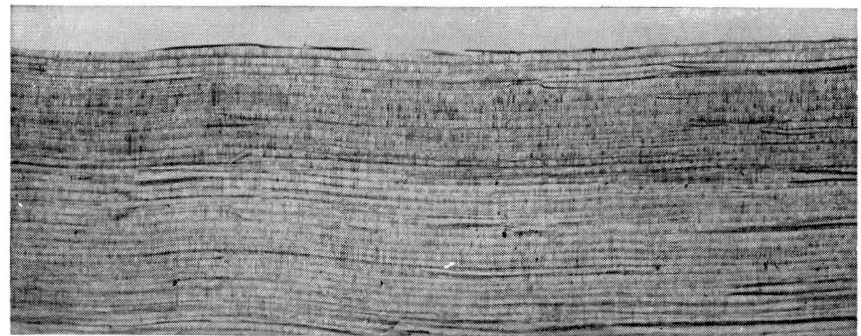

Fig. 147. Radial view of Douglas fir wood showing the incipient stage of red ring rot caused by *Fomes pini.* The dark-colored band at the top is incipient decay.

Lloyd by American mycologists. The fungus has a widespread distribution in the North Temperate Zone (Haddow 1938*a*), where its ravages have caused it to be extensively investigated (Haddow 1938*b*; Boyce and Wagg 1953). In North America it attacks all conifers with the possible exceptions of bigtree, redwood, southern cypress, cypresses (*Cupressus* spp.), and *Tumion.* The reports of it on hardwoods are doubtful. It is particularly severe on Douglas fir, larches, pines, and spruces.

Damages. Losses resulting from red ring rot far exceed those from any other decay. In Douglas fir stands of western Oregon and Washington where the loss from decay may be 50 per cent or more of the gross board-foot volume in some areas, with an average for the region of 17 per cent, red ring rot comprises about 81 per cent of the loss. Heavy losses from this rot also occur in eastern white pine.

The Decay. The incipient stage appears as a discoloration of the heartwood. In Douglas fir it is a pronounced reddish-purple or olive-purple discoloration (Figs. 147, 148); in white and red spruce a light purplish gray later deepening to a reddish brown; in western white pine a pink to reddish; and in ponderosa pine a pronounced pink, which gives way to red-brown. The incipient stage is often termed "red heart," "firm red rot," or "firm red stain," because the wood is usually firm and tough.

Fig. 148. End of a Douglas fir log with red ring rot caused by *Fomes pini.* The upper portion of the log shows the discoloration characteristic of the incipient stage and the lower portion the ring scale characteristic of the advanced stage. (*Photograph by A. H. Hodgson.*)

The wood may be heavily infiltrated with resin. In western red cedar the incipient stage is indicated by a narrow bluish to reddish-brown band along the lateral margins of the advanced stage. Strength tests in Douglas fir and Sitka spruce have shown no significant weakening of wood in the incipient stage.

The advanced stage appears as few to many elongated, somewhat spindle-shaped, more or less pointed pockets or cavities parallel to the grain and separated by apparently sound wood, sometimes resin filled (Fig. 149). In the pockets the wood is reduced to a white soft fibrous mass of cellulose. In some cases the pockets may be empty. In others the pockets may merge so that the rotted wood is a mass of white fibers with no pockets apparent. Brown to black fine zone lines may sometimes be distributed irregularly through the decayed wood, particularly in

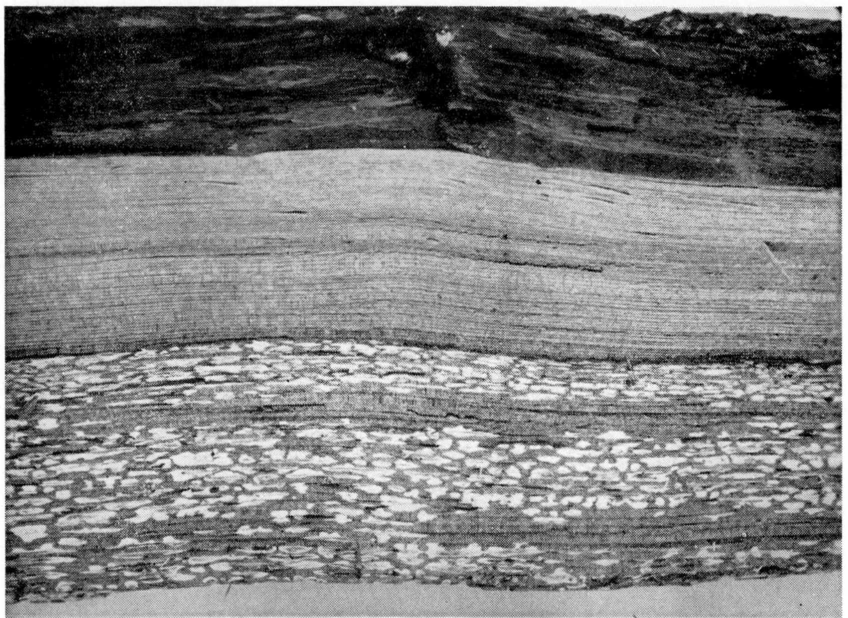

Fig. 149. Radial view of Douglas fir wood showing the advanced stage of red ring rot (at the bottom) caused by *Fomes pini,* the incipient stage, sapwood and bark.

western larch, tamarack, and red spruce. Seen in cross section on the end of a log the pockets may be scattered rather uniformly throughout the affected wood (Fig. 150), or they may be in the form of ring scale (Figs. 148, 151), which quite commonly results from the rot destroying early wood more readily than late wood and also being confined to definite groups of annual rings. The ring scale character is common in Douglas fir, hard pines, and western larch but is not so frequent in soft pines and spruces.

In general the decay is confined to heartwood, but in Douglas fir, western larch, and lowland white fir it may attack living sapwood after decay of the heartwood is well advanced. In lowland white fir, sapwood and inner bark are killed, resulting in the formation of elliptical, sunken, cankerlike areas on the trunk within which conks develop profusely (Owens 1936:147). The rot is not limited to a special portion of the bole, except that it is not frequent in the butt. The decay may continue development after an infected tree dies or is cut, but the fungus is of no importance as a destroyer of slash. Rot has never been known to develop

Fig. 150. End of a sugar pine log with the heartwood completely affected with red ring rot caused by *Fomes pini.* The rot is fairly uniform. (*Photograph by G. G. Hedgcock.*)

in wood in service. Microscopically the decay has been studied by Hartig (1874:46).

Indications of Decay. The perennial sporophores are extremely variable. In shape they vary from thin shell-shaped to bracketlike and hoof-shaped with some tendency for the shape to be uniform on a host species but with sufficient variation so this cannot be relied upon. In size they range from 1 or 2 inches in width to more than 1 foot, with the average ranging from 4 to 8 inches. The upper surface is rough, dull grayish or brownish black, and with approximately concentric furrows parallel to the lighter brown margin. When a conk is actively growing the margin is velvety and light golden brown in color. The undersurface is a grayish-brown or rich-brown color with the tube mouths varying from small and almost circular to large and irregular or daedaloid (Fig. 152). Conks from Scotch pine in Europe and from Douglas fir in the western United States would hardly be recognized as belonging to the same species. On living trees the conks nearly always issue from knots or branch stubs

Fig. 151. End of a ponderosa pine log with the heartwood destroyed by red ring rot caused by *Fomes pini*. This illustrates the "ring scale" condition.

(Fig. 153), rarely from wounds, and in lowland white fir they come out directly through the sapwood and bark. On dead trees, standing or down, they frequently issue directly through the bark, particularly in red spruce and balsam fir. They may then be crustlike instead of bracketlike. A thin annual type of conk that is said to appear anywhere on the bark of balsam firs, larches, and spruces in eastern North America and in Europe

Fig. 152. Sporophore or conk of *Fomes pini* from Douglas fir showing the rough black upper surface and the large irregular mouths of the tubes on the undersurface.

Fig. 153. Conks of *Fomes pini* on a living Douglas fir. (*Photograph by G. G. Hedgcock.*)

Fig. 154. End of a Douglas fir log showing a swollen knot of *Fomes pini.* (*Photograph by J. S. Boyce and R. E. McArdle.*)

has been named *F. pini* var. *abietis* Karst. The interior of the conks is yellow-brown in color and is corky or punky.

Conks are the most reliable indications of decay. In Douglas fir in western Oregon and Washington no more than a few board feet of wood are decayed before conks or swollen knots appear. In addition punk knots are excellent indications of decay. These knots are filled with the same yellow-brown punky substance composing the interior of the conk, indicating either points from which old conks have fallen or the beginning of a new conk. The punk knot may be overgrown by sapwood and new bark so that it is necessary to chop through solid wood before the punk knot is reached. These are called "blind conks." In fire-killed timber the punk knots burn out leaving cylindrical holes reaching several inches into the trunk. Swollen knots result from the living wood tissue trying to grow over the knot where a conk is forming (Fig. 154), and in many cases on Douglas fir the appearance is accentuated by the growing conk actually pushing out the thick, dead outer bark in a characteristic manner. The swelling resulting from the attempt to heal over a punk knot is marked when contrasted with the appearance of a normal, sound knot when grown over (Fig. 155). Swollen knots are easily recognized on

Fig. 155. End of a sound Douglas fir log showing normal knots grown over. (*Photograph by J. S. Boyce and R. E. McArdle.*)

living trees (Figs. 156, 157). Swollen knots have also been found on Sitka spruce, red spruce, various pines, and western larch. On western white pine in addition to the characteristic punk knots, there are at some knots peculiar ram's-horn-like projections up to 6 or 8 inches long that show definite annual ridges of growth. They are abortive conks composed of the familiar brown punky substance mixed with resin. In western

Fig. 156. Swollen knots of *Fomes pini* on a living Douglas fir.

Fig. 157. Swollen knots of *Fomes pini* on two living western larches. (*Photograph by A. S. Rhoads.*)

white pine in northern Idaho, red ring rot will extend 2 feet toward the top from a punk knot and 4 feet toward the butt. Punk knots in Engelmann spruce and alpine fir are often sunken. Heavy resin flow from the base of dead branch stubs is a common indication of red ring rot in red spruce.

Nearly all infections, regardless of host, occur through dead branch stubs, with open wounds rarely becoming infected. As a consequence, scars are of little value as indicators of red ring rot, and accordingly protecting trees from wounding has little value in preventing infection.

RED ROT

Red rot is a white pocket rot of conifers caused by *Polyporus anceps* Pk., also known as *P. ellisianus* (Murr.) Sacc. & Trott. The fungus is most frequent on and destructive to ponderosa pine (Andrews 1955). This is by far the most important fungus decaying ponderosa pine slash and to that extent it is beneficial, but it also attacks living trees, causing

Fig. 158. Radial or edge-grain view of the advanced stage of red ray rot in ponderosa pine caused by *Polyporus anceps.*

heart rot. The losses caused by it are not serious in the Pacific Coast region, but in the Rocky Mountain region, particularly in Arizona and New Mexico, it is not uncommon for 20 per cent or more of the overmature trees in ponderosa pine stands to show red rot in some portion of the bole, and immature trees are also decayed.

The Decay. At first this rot was confused with red ring rot caused by *Fomes pini,* but it is distinctly different (Fig. 158). On the end of a log of living ponderosa pine, heartwood in the incipient stage shows as reddish to reddish-brown discolored areas that may be roughly circular in outline or radiate outward from the center of the log, like spokes from the hub of a wheel. Red ring rot does not have this appearance in cross section. Other points of difference in longitudinal section are (1) incipient decay of red rot is reddish to reddish brown, that of red ring rot pink, reddish to purplish. (2) In the advanced stage the white pockets in red rot are not well defined and have squarish ends and a tendency to run together; in red ring rot the pockets are sharply defined with more or less pointed ends. (3) Wood with red rot finally becomes much disintegrated and crumbles readily when handled; with red ring rot it always remains firm. (4) Mycelium of red rot when growing more or less ex-

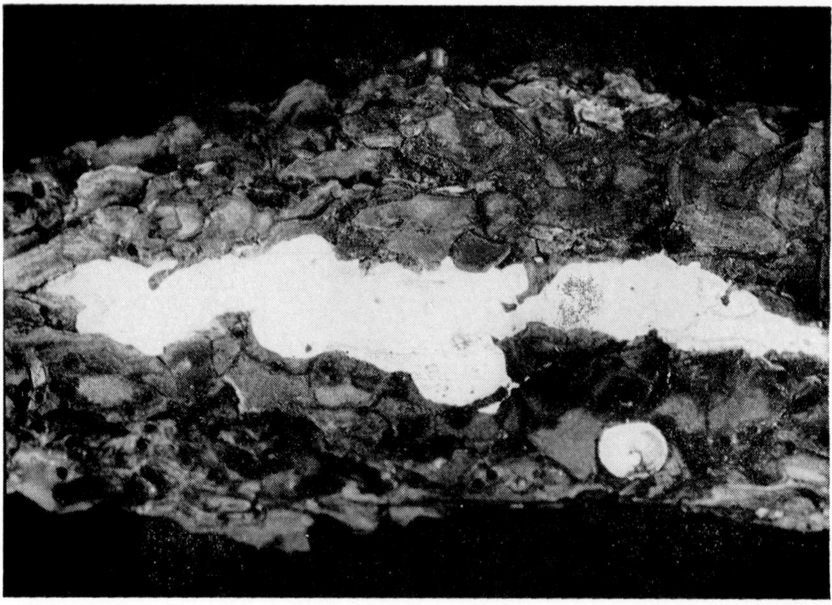

Fig. 159. Sporophore of *Polyporus anceps* on the bark of a ponderosa pine log.

posed to the air is white; that of red ring rot turns brown. (5) Zone lines are never formed in red rot; they sometimes are in red ring rot. (6) In red rot black flecks may occur in the white pockets, in red ring rot they never do. The decay is not limited to any one portion of the bole except that it is much less frequently found in the butt than elsewhere.

On ponderosa pine slash the fungus causes a rapid and destructive decay of the sapwood similar to that in the heartwood of living trees. The heartwood is left largely intact. Pieces of slash less than 4 inches in diameter inside bark are rarely attacked. On slash the fungus produces a wet, soggy rot even in dry situations. Apparently, during dry periods it lives on its own metabolic water.

Indications of Decay. The rot is most difficult to detect in living trees, since conks rarely develop on them and then only on dead branches. There are no swollen knots. Infection occurs through dead branches.

The white annual conks appear on the underside of logs or fallen trees as flat crusts on the bark, rarely showing a bracket tendency (Fig. 159). On slash the fungus begins growth as a cottony layer of mycelium between the bark and sapwood. This cottony layer develops into a firm white mycelium felt that attaches the bark firmly to the sapwood. *P. anceps* will not attack decorticated wood, because such wood is too dry on the outside for the fungus to become established.

This fungus is harmful in causing a heart rot of living trees and beneficial as a rapid acting sap rot of slash. Future losses can be reduced by proper pruning (page 375).

SIMILAR ROTS

Polyporus tomentosus Fr. var. *circinatus* (Fr.) Sartory & Maire causes red root and butt rot in living conifers (Gosselin 1944). It causes much butt rot in western white pine in northern Idaho, the decay there usually being associated with fire scars, and in eastern white pine in Ontario. Eastern spruces are particularly susceptible; trees may be killed by it. Infection appears to take place through lateral roots. Possibly the fungus enters its host as a mycorrhizal component (page 120).

In the incipient stage of this rot, decayed wood is firm but dark reddish brown in color, whereas the incipient stage of red ring rot caused by *Fomes pini* with which red root rot may be confused is pinkish, reddish, or purplish but not generally so dark. In the advanced stage the wood is characterized by more or less elliptical white pockets, the firm wood between the pockets being brown in color (Fig. 160). The dark-yellow to light-brown conks on trees are sessile and applanate, but on the ground have a central stalk (Fig. 160).

Fomes annosus (page 108) causes a white pocket rot of conifers that can be confused with the decays caused by *F. pini* and *Polyporus tomentosus* var. *circinatus*.

Hydnum abietis, the fir hydnum, causes long-pocket rot of the heartwood of living lowland white fir, alpine fir, and western hemlock in the Pacific Northwest (Hubert 1931:305). In the incipient stage the wood is discolored a faint yellowish or brownish with scattered darker spots. The advanced stage is characterized by elongated pockets separated by firm, reddish-brown wood. The pockets may be empty or may contain white fibers. The corallike, white to cream, soft annual sporophores, which range from 4 to 10 inches wide by 6 to 12 inches high, are produced on old logs or dead fallen trees, but they occasionally appear on living trees at wounds. Infection occurs through wounds and dead branches.

F. juniperinus (v. Schr.) Sacc. & Syd., the juniper fomes, causes juniper pocket rot of the heartwood of living junipers, particularly eastern red cedar and western juniper (Von Schrenk 1900). The incipient decay is first visible as a light-yellowish discoloration, which becomes more pro-

Fig. 160. Advanced decay in the heartwood of western white pine caused by *Polyporus tomentosus* var. *circinatus* and a conk from the ground.

nounced, the affected wood becoming whitish and soft, until finally the advanced stage is characterized by large hollow vertical pockets lined with yellowish-white fibers. The pockets are commonly placed one above the other, and, when they join, long tubes result. The perennial, hoof-shaped conks are developed sparingly and blend somewhat in color with the bark (Fig. 161). When young, the upper surface is brown with numerous cracks, but later it becomes deep brown to almost black and very rough. The undersurface is light brown, with a smooth, light yellow-brown margin. The circular mouths of the tubes are easily seen with the naked eye. The interior of the conk is a pronounced reddish brown. A tree with a single conk as a rule is completely unmerchantable.

F. nigrolimitatus (Romell) Egel. (*F. putearius* Weir) causes a white pocket rot of several western conifers, commonly Douglas fir and western larch (Weir 1914). The fungus is usually found on dead down trees or slash, and it occurs on living western red cedar (Buckland 1946). The advanced decay somewhat resembles that caused by *F. pini* except that the white pockets are always larger, sometimes attaining a length of 2 inches and varying in breadth according to the structure of the host (Fig.

Fig. 161. Conks of *Fomes juniperinus* from western juniper—a young one on the left, an old one on the right.

162). The perennial thick brown crustlike sporophores are found on the underside of down trees or logs and are often a foot or more in length. The interior of the conk has thin horizontal black lines between the older layers of tubes.

Polyporus abietinus Dicks. ex Fr., the purple conk, causes a white pocket rot known as "pitted sap rot" or "hollow pocket rot" confined to dead coniferous sapwood (Baxter 1948). It is common on dead timber on slash, and on logs in storage throughout the United States. It may attack the dead sapwood of wounds on living trees. In the incipient stage the wood is faint yellowish to tannish in color and slightly softened. In the advanced stage the wood is honeycombed with small pockets that at first may be filled with whitish fibers, but later the pockets are empty (Fig. 163). In this stage the wood is spongy or corky. The small, thin annual conks are shelflike or sometimes crustlike with a hairy, zoned,

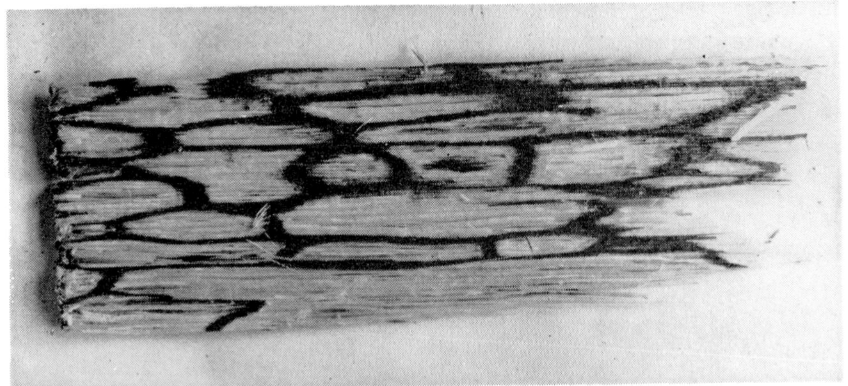

Fig. 162. White pocket rot in western larch caused by *Fomes nigrolimitatus.*

grayish upper surface and a lavender to purplish undersurface that turns light brown with age. The mouths of the tubes are regular to irregular in shape and conspicuous. The conks develop abundantly along bark crevices or on decorticated wood.

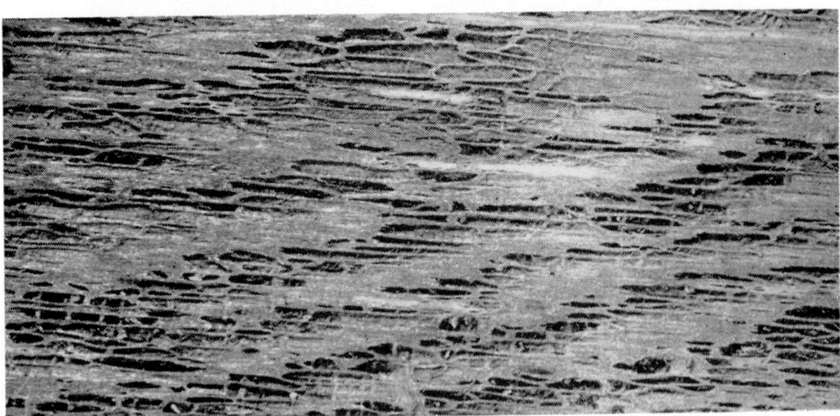

Fig. 163. Advanced stage of pitted sap rot in red fir caused by *Polyporus abietinus.*

Polyporus pargamenus Fr. causes a white pocket rot of the dead sapwood of hardwoods that is practically identical with pitted sap rot of conifers caused by *P. abietinus* (Rhoads 1918). *P. pargamenus* is abundant in the eastern half of the United States, but it is rare in the West.

Rarely it is found on conifers. Conks of the two fungi are so similar that they can be distinguished only by an expert (Fig. 164).

Polyporus dryophilus Berk. causes a white pocket rot, termed "piped rot," of oaks and occasionally of poplars (Hedgcock and Long 1914).

Fig. 164. Conks of *Polyporus pargamenus* on the dead portion of a living chestnut oak. (*Photograph by H. F. Marco.*)

The fungus is widespread in Europe and the United States, where it is particularly destructive in the Southwest and on the Pacific Coast. In oaks the incipient decay first appears as a darkened water-soaked area in the heartwood. The advanced decay appears as long narrow white pockets running together longitudinally and separated by strips of firm brown wood (Fig. 165). The decay is confined to heartwood of living trees, although it may sometimes continue to develop slowly for some years after an infected tree dies. The rot column is usually in the main and upper portion of the bole. In poplars there are indications that the fungus kills living sapwood after decay is well advanced in the heartwood. The large brown annual conks are shelf-shaped or hoof-shaped, the latter being usual, and they are characterized by a hard granular sandstonelike core. The conks commonly issue from knots or occasionally from wounds.

Stereum subpileatum Berk. & Curt. causes a white pocket rot of oaks known as "honeycomb heart rot" (Long 1915). It is found from Virginia, Ohio, and Missouri south to Arkansas and Florida. In the incipient stage the wood has a water-soaked appearance, whereas in the advanced stage it is characterized by white, oval to circular pockets from which finally

Fig. 165. Piped rot in the heartwood of California black oak caused by *Polyporus dryophilus*. Sound sapwood is also shown.

the cellulose disappears leaving the empty pockets separated by thin walls—a honeycomb appearance. In living trees the decay is usually in the butt, entering through fire scars. More frequently the fungus attacks dead trees, where it invades both sapwood and heartwood. The thin perennial shelving grayish conks occur only on dead trees or dead areas of living trees.

Polyporus ludovicianus (Pat.) Sacc. & Trott. on oaks and red gum and *P. zonalis* Berk. on oaks both cause white pocket rots of living trees in the South (Overholts 1938).

Stereum frustulosum (Pers.) Fr. causes a white pocket rot of oaks known as "partridge wood" (Von Schrenk and Spaulding 1909:60; Cartwright and Findlay 1936:18). The rot, which affects dead trees or the heartwood of living trees, first appears as a dark-brown discoloration of the wood, but later small spindle-shaped pockets lined with white mycelium develop. The resupinate fructifications on the surface of the decaying wood appear like sheets of cracked mud, the individual pieces varying from $1/16$ to $1/4$ inch in diameter.

Hymenochaete rubiginosa (Schrad.) Lev. causes a white pocket rot of dead decorticated American chestnut and more rarely of decorticated oak (Brown 1915). The decay is similar to partridge wood caused by *S. frustulosum,* but the pockets are smaller. The thin brown annual fructifications are either half-circular brackets in clusters or resupinate crusts with a diameter up to 2 inches. They appear only on decorticated wood.

Polyporus croceus (Pers.) Fr. (*P. pilotae* Schw.) causes a white pocket or piped butt rot of living oaks, American chestnut, and eastern chinquapin, commonly entering through fire scars, or in sprouts from the old stump (Long 1913). Sometimes decay occurs in the upper part of the tree, entering through dead branches. Incipient decay appears water-soaked and brownish, whereas advanced decay is characterized by small lens-shaped pockets with white linings in the early wood of the annual rings. The annual conks are shelflike, soft, watery, and buff to orange colored. They attain a width of 6 inches and develop on either living trees or dead fallen wood.

Stereum gausapatum Fr. causes a white pocket or white mottled rot of oaks in the eastern United States (Davidson 1934; Herrick 1939). *S. spadiceum* Fr. causes an identical rot of the heartwood of English oak (Cartwright and Findlay 1936:15). In the United States the decay is particularly prevalent in sprout stands, the rot invading the heartwood of the sprouts from the old stumps. In the incipient stage white lines appear that usually follow the early wood vertically, although they may spread throughout the growth rings. Next, most of the early wood is rotted, and in the advanced stage all the wood is light colored and brittle. The small thin crustlike to slightly shelf-shaped conks have a tobacco-colored upper surface and a smooth, snuff-brown undersurface. They develop abundantly on old stumps, cut stubs, and slash.

Fomes applanatus (Pers.) Gill., the shelf fungus or artist's conk, causes a white mottled rot (White 1919; Englerth 1942:16). It is widely distributed, mostly on hardwoods but to some extent on conifers, particularly in the Pacific Northwest. In general it is mostly on dead timber, but it does attack living trees, entering through wounds and destroying the heartwood for a few feet in the lower bole, although the rot may extend for 12 to 15 feet or more in the trunk. Living sapwood may be killed after appreciable decay of the heartwood has occurred. Hardwoods in storage and use, particularly mine timbers, are decayed.

In the incipient stage the wood is somewhat bleached with the bleached area encircled by a dark-brown band $1/4$ inch or more in width. Wood of western hemlock in the incipient stage is violet to lilac in color. In the advanced stage the wood is whitish to cream in color, soft, and spongy, with fine black zone lines through it. On the radial face the wood has a mottled appearance because of horizontal streaks or irregular patches of white mycelium or bleached wood (Fig. 166). The hard woody shelf-like perennial conks may attain a width of 2 or more feet (Fig. 167). The upper surface is smooth, zoned, and grayish or grayish black, whereas the undersurface is white when fresh but becomes yellowish with age. If the fresh undersurface is bruised it immediately turns brown, making the conk a favorite medium for amateur etchers. The interior of the conk is dark red-brown, soft, and corky. The characteristic spores

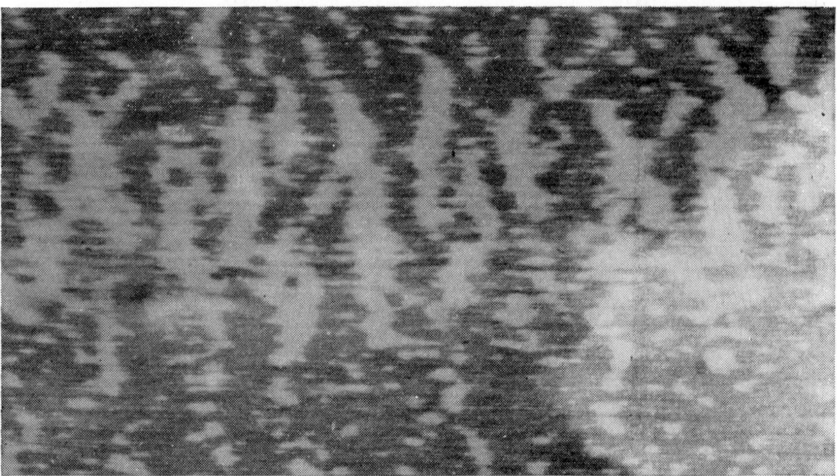

Fig. 166. White mottled rot in black cottonwood caused by *Fomes applanatus*. (*Photograph by A. S. Rhoads.*)

are large and yellowish in color and have truncated apices. Conks are produced abundantly on dead timber and less frequently on living trees, where they always issue from wounds.

The tinder fungus, *F. fomentarius* (Fr.) Kickx., which is widely distributed, causes a white mottled rot of birches, beech, poplars, and other hardwoods (Hilborn 1942). Although it decays chiefly sapwood and heartwood of dead timber, it occasionally causes heart rot of living trees and also attacks living sapwood (page 293).

Wood in the incipient stage of decay is brownish in color, firm, and seemingly not altered. Wood with advanced decay is yellowish white, soft, and spongy, with narrow dark-brown to black zone lines, and often with frequent small radial cracks filled with yellowish mycelium, which give the wood a mottled appearance. The decay usually begins in the upper part of the bole and progresses downward. Infection of living trees takes place through branch stubs and wounds. The hard perennial hoof-shaped conks up to 8 inches wide are profusely produced on dead trees. They have a smooth, grayish-black or grayish-brown, concentrically

Fig. 167. Conks of the shelf fungus (*Fomes applanatus*) on black poplar. (*Photograph by A. S. Rhoads.*)

zoned upper surface, with a grayish to brown undersurface. The mouths of the tubes are small and circular in outline. The interior of the conk is dark brown and punky with the tubes usually white stuffed or white encrusted. Except for the smooth upper surface, the conks are easy to confuse with those of *F. igniarius*.

F. fraxinophilus (Pk.) Sacc., the ash fomes, causes a white mottled rot of the heartwood of white and green ashes (Von Schrenk 1903*a*;

Baxter 1925:553). It may also occur on dead trees. The incipient decay appears first as a brownish discoloration in which white spots soon develop. In the advanced stage the wood is straw colored to white, soft, and crumbly, with longitudinal white areas on the radial face. The perennial hard woody shelf-shaped to somewhat hoof-shaped conks have a roughened dark-brown or black upper surface and a brownish undersurface. The conks may attain a width of 1 foot or more. Their development follows closely the progress of decay in the heartwood, the first ones appearing when the wood has decayed only a short distance from the point of infection. The fungus usually enters through dead branch stubs.

F. johnsonianus (Murr.) Lowe causes a soft white laminated rot of the heartwood of ash, which in the incipient stage is a white mottled rot (Davidson, Lombard, and Campbell 1959). The brown, usually resupinate conks occur on the underside of down logs. The fungus is also found on other hardwoods.

Polyporus squamosus (Huds.) Fr., the scaly or saddleback fungus so prevalent in Europe (Buller 1906), occasionally causes a white mottled rot of the heartwood of living hardwoods in this country (Von Schrenk and Spaulding 1909:38). The decay is usually limited to the immediate vicinity of wounds, and there are indications that living sapwood can be invaded after decay is established in the heartwood. The fungus also develops on stumps and down trees. The somewhat fleshy annual conks, which issue from wounds on living trees, are whitish to dingy yellowish in color with a short, thick, excentric or lateral stalk. The upper surface is covered with broad appressed scales. The sporophore is commonly 18 or more inches wide.

P. adustus (Willd.) Fr., the scorched conk, causes a white mottled or cubical rot of dead hardwoods, usually confined to sapwood, and is most common on poplars and red gum (Baxter 1949:21). It causes considerable loss of red gum logs and lumber in storage yards. The annual thin fleshy to corky conks, up to 3 inches wide, sometimes occurring in clusters, are smoky gray in color.

P. borealis Fr. causes a white mottled rot of the heartwood of living red spruce and eastern hemlock (Atkinson 1901:202). The large fleshy white annual conks, with a shaggy upper surface, usually occur in clusters, overlapping one another.

Fomes robustus Karst. (*F. hartigii*) causes a white heart rot of western hemlock (Englerth 1942:17). It also occurs on eastern hemlock. Incipient decay in western hemlock is a reddish-purple to brown discoloration that extends longitudinally an average of 7.3 feet beyond the advanced decay. On the radial face, advanced decay has a laminated appearance because the springwood, which is bleached, is more decayed than the summerwood, which retains its normal color. On the tangential surface light-buff to brown horizontal streaks often appear, giving the wood a mottled appearance. Brown zone lines are common. The rot is usually limited to the side of the tree where infection took place. The rot extends longitudinally in the inner sapwood, then spreads into the heartwood and outer sapwood, usually killing the cambium so that a canker results. The brown perennial conks generally develop as thick crusts on the undersurface of branch stubs, although they occasionally appear directly on the trunk, then being hoof-shaped. The behavior of the fungus on eastern hemlock is similar. The fungus decays other conifers and also hardwoods, particularly oaks (Baxter 1952*b*).

Lentinus tigrinus Fr. causes a white mottled butt rot of living hardwoods, commonly associated with fire scars. It is one of the most important decay fungi in the Mississippi Delta hardwoods. The toadstool fructification is white, with a cap that is depressed in the center and more or less covered with blackish-brown hairy scales. The fructifications rarely develop on living trees.

Polyporus berkeleyi Fr. causes a white rot of the heartwood in the butt of living oaks and other hardwoods throughout the East, which is known as string and ray rot (Long 1913; Weir 1913). In the West it is also associated with decay in Douglas fir, Pacific silver fir, and western larch. The decay is confined to the roots and butt, not extending more than 15 feet up the trunk. In oaks incipient decay appears as yellowish white, relatively firm wood. The advanced decay at first is yellowish brown and crumbly, but with the vessels, the cells that immediately adjoin the vessels, and the rays intact so that the rays are interwoven with long flat strings of wood. The decayed wood has an anise-oil odor. Finally the entire mass of rotted wood decomposes leaving a hollow in the butt of the tree, lined with a white or cream-colored layer of mycelium. The whitish to light-buff sporophores (Fig. 168), shaped like those of *P. schweinitzii,* arise either just at the base of the tree from the root collar or on the ground some distance away, coming up from a decayed root. The fungus most commonly enters through fire scars.

P. fissilis Berk. causes a soft ropy whitish or cinnamon-colored butt rot of the heartwood of living oaks and red gum in the South (Overholts 1938). The rays in the decayed wood are persistent. The white fructifications become yellowish on drying. The undersurface is slightly lavender. In addition, sterile conks form as soft watery mucilaginous or soapy white masses in crevices near the fructifications.

P. frondosus Fr. causes a butt rot of the heartwood of living oaks and American chestnut (Long 1913). The decay is usually connected with fire scars. The incipient decay appears water-soaked and reddish with white lines running upward into it from the advanced decay. The latter is at first whitish and later straw colored, with the rays in oaks almost unaffected. The annual sporophores, which come up through the ground from decayed roots or appear just at the base of the tree, are buff to smoky gray with a branched stem giving rise to many pilei. The entire fructification forms a somewhat globose mass up to 2 feet in diameter.

Fig. 168. A conk of *Polyporus berkeleyi* from a root of western larch.

Pleurotus ostreatus Jacq., the oyster mushroom, causes a white, flaky rot of the sapwood and heartwood of various hardwoods (Learn 1912). The light-colored advanced decay is surrounded by a narrow brown zone of incipient decay. The fleshy annual conks are shelving structures, either sessile or with a short, stout, excentric stalk (Fig. 169). The upper surface is smooth, white or grayish, and the gills on the undersurface extend onto the stalk. Infection usually occurs through open wounds.

Fomes everhartii (Ell. & Gall.) v. Schr., the charred fomes, causes a soft white or yellow flaky heart rot of living hardwoods, particularly oaks, and the fungus can also attack living sapwood (Baxter 1925:545). The decay is usually confined to the lower trunk. Invasion of the sapwood results in gnarly swellings on the trunk of oak (page 293). Advanced decay is characterized by narrow dark-brown zone lines, which on a cross section of a decayed stem are frequently concentrically arranged. The flaky character occurs because decay is more rapid between the rays. The perennial hard woody shelf-shaped conks, up to 1 foot in width, have

Fig. 169. Sporophores of the oyster mushroom (*Pleurotus ostreatus*) on the trunk of a living yellowpoplar. (*Photograph by R. P. Marshall.*)

a yellowish-brown upper surface, which becomes grayish black or black, rough, cracked, and concentrically grooved with age, although the margin still remains brown. The undersurface is reddish brown. Sterile black clinkerlike conks are sometimes produced on beech similar to those produced by *Poria obliqua* on birches (Hirt 1930).

Poria andersonii (E. & E.) Neumann causes a white heart rot of living hardwoods, especially oaks, similar to that caused by *F. everhartii* (Campbell and Davidson 1939*a*). The fructifications develop under the bark of dead trees and down logs.

P. subacida (Pk.) Sacc., the white root conk, causes a soft spongy white rot known as "feather rot," "spongy root rot," or "stringy butt rot" (Baxter 1935; 1949:20). The fungus is found on dead coniferous and hardwood trees and slash and in addition causes a root and butt rot of the heartwood of living conifers. The decay rarely extends higher than 6 to 10 feet in the trunk. Feather rot is the most common butt and root rot of balsam fir, causing considerable loss and predisposing infected trees to wind throw. In balsam fir infection apparently largely occurs through the roots.

Fig. 170. Advanced stage of feather rot caused by *Poria subacida* in the heartwood of balsam fir.

Incipient decay is first apparent as a faint darkening of the wood. Advanced decay has irregularly shaped pockets in the springwood that run together forming masses of white fibers (Fig. 170). At this stage the annual rings separate readily. Small black flecks may appear in the pockets or in the masses of white fibers. Finally the heartwood is reduced to a water-soaked mass of white fibers. In dead material the sapwood is also decayed. The annual crustlike white to straw-colored or cinnamon-buff conks (Fig. 171) may form sheets several feet in extent on the underside of fallen trees. On living trees they usually develop on the underside of root crotches or exposed roots. A yellow form on western hemlock has been called *P. colorea* (Englerth 1942:20).

Odontia bicolor (A. & S.) Bres. causes a stringy white butt rot of conifers and hardwoods, characterized by white mycelial felts and black flecks (Nobles 1953). The dusty, cartridge-buff, crustlike fructifications have short fragile teeth.

Corticium galactinum (Fr.) Burt causes a stringy yellow butt and root rot of conifers and hardwoods (White 1951). The crustlike, cream-colored sporophores on roots and logs may grow out over adjacent soil and debris.

WHITE TRUNK ROT

White trunk rot or white heart rot is caused by the false tinder fungus, *Fomes igniarius* (L.) Gill. (Spaulding 1951). The fungus is practically world wide, attacking a great variety of hardwoods, but it is not found on conifers. Aspens, beeches, and birches are particularly susceptible hosts. Verrall (1937) recognizes three fairly distinct groups within the species, designated as the aspen type, the birch type, and the type from

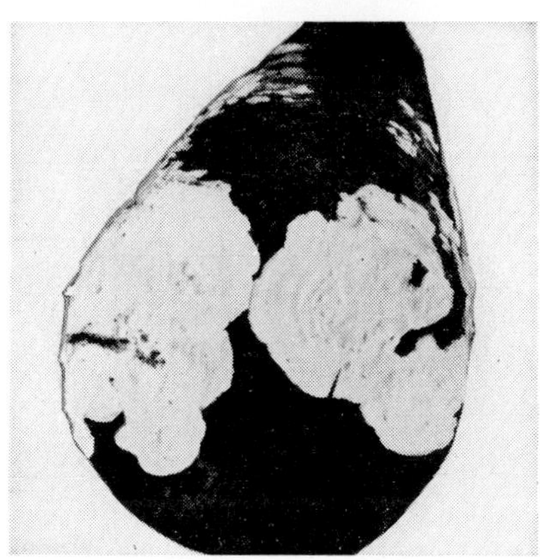

Fig. 171. Conks of *Poria subacida* on an old conifer log. (*Photograph by R. C. Lorenz.*)

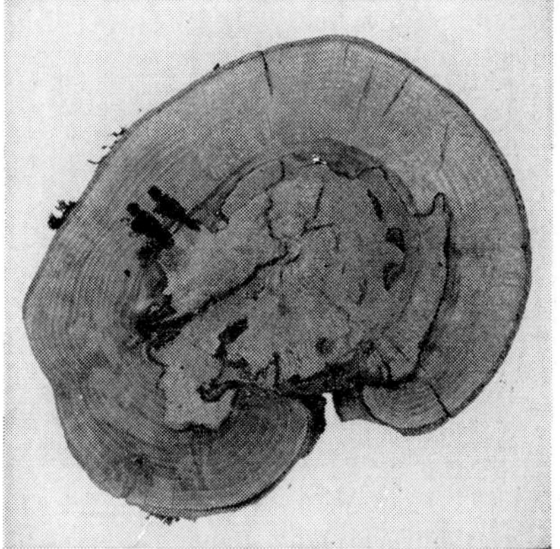

Fig. 172. Cross section of a living beech showing white trunk rot caused by the false tinder fungus (*Fomes igniarius*). Black zone lines are abundant in the decayed wood. (*Reproduced from U.S. Dept. Agr., Bur. Plant Ind. Bull.* 149.)

miscellaneous hosts. These groups differ in the morphology of their conks, in their cultural characteristics, in their rate of decaying some wood species, and in their reaction to zinc chloride.

Damage. The false tinder fungus causes more loss than any other wood destroyer of hardwoods. It is particularly severe on aspen (Riley 1952). Beech is severely affected in the Adirondack Mountains, and in many stands it is difficult to find a sound tree.

Decay. Incipient decay appears as yellowish-white spots, streaks, or larger areas in the heartwood, the whole usually surrounded by a yellowish-green to brownish-black zone. In the advanced stage the wood is light in weight, soft, whitish, and rather uniform in texture, with little or no tendency to crack, but with fine black lines running throughout (Fig. 172). On the cross section of a stem the black lines are seen to have an irregularly concentric arrangement. In aspen three stages of decay have been recognized (Schmitz and Jackson 1927:10). In the incipient stage the wood is faintly colored from light pink to straw-brown; in the intermediate stage it is colored from straw to chocolate-brown, but it is still hard and firm; and in the advanced stage is included all soft, punky wood irrespective of color. Wood in the incipient and intermediate stages is utilized for some purposes but in the advanced stage is always rejected.

Fig. 173. Conk of the false tinder fungus (*Fomes igniarius*) on a living beech.

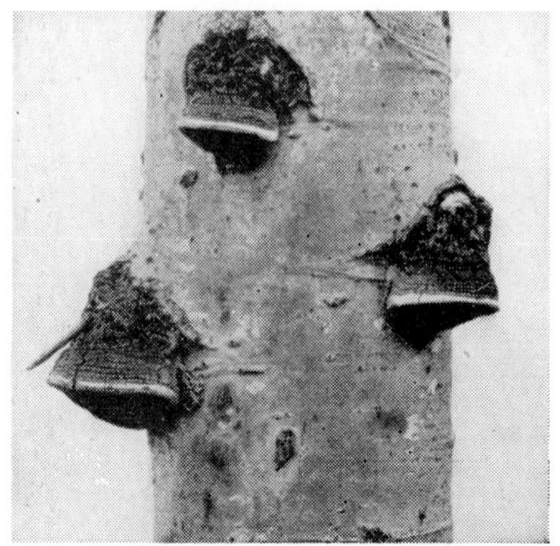

Fig. 174. Conks of the false tinder fungus (*Fomes igniarius*) on a living aspen. (*Reproduced from U.S. Dept. Agr. Yearbook for* 1907.)

In general the decay is confined to heartwood, but in yellow birch the fungus sometimes kills living sapwood, causing cankers on the trunk (page 293). Rot is not confined to any particular section of the bole but is more often found centered in the main portion of the trunk rather than in the butt or top. After an infected tree dies, the fungus continues its development, decaying the dead sapwood as well as the heartwood. It is a common destroyer of dead trees and cull logs in the forest, but it has not been found decaying timber in storage yards or in service.

Indications of Decay. The hard woody thick perennial conks are usually hoof-shaped, but sometimes shelflike, and they may attain a width of 8 or more inches (Figs. 173, 174). The upper surface is grayish black to black, dull or shiny, at first smooth, but becoming rough and cracked with age. The undersurface is brown with the mouths of the tubes small and circular in outline. The context or interior of the conk is rusty brown and shows many layers of tubes, those in the older layers being conspicuously stuffed with white. Normally one tube layer is formed each year. The conks usually issue from knots or occasionally from wounds. On dead down trees the conks are sometimes thick crusts. A single conk of average size usually indicates a rot column of 15 linear feet or more in the heartwood, except in aspen where the development of conks follows rather closely the progress of decay in the heartwood (page 369). Typical punk knots may appear at branch stubs, showing the rusty-brown punky substance when chopped into. Infection occurs through branch stubs and open wounds.

Fig. 175. A sterile, clinkerlike conk, probably of *Poria obliqua,* on paper birch. (*Reproduced from U.S. Dept. Agr., Bur. Plant Ind. Bull.* 149.)

Fig. 176. Conk of *Fomes connatus* in an old wound on the trunk of a red maple.

SIMILAR ROTS

Poria obliqua (Pers.) Bres. causes a heart rot in living birches similar to that caused by *Fomes igniarius* (True, Tryon, and King 1955). It occurs occasionally on American beech and *Ostrya.* Black warty rough clinkerlike sterile conks develop on infected trees (Fig. 175), issuing from knots and wounds; and on yellow birch they often appear in the center of conspicuous cankers presumably caused by the fungus killing the sapwood. The inconspicuous short-lived fruiting conks break through the bark after the tree dies. A sterile conk indicates a rot column of 15 linear feet or more in the heartwood.

F. geotropus Cke., which causes a soft white rot of hardwoods, often in connection with fire scars, is particularly serious in the Mississippi Delta forests. In southern magnolia incipient decay is grayish black with distinct greenish and reddish tinges (Johnson and Edgerton 1936). The upper surface of the thick woody conks, which may attain a width of 10 inches, is straw colored and marked with numerous concentric ridges. The undersurface is rose colored when young but becomes light brown with age.

Armillaria mellea (page 104) causes a soft white to light-yellow rot that may be stringy in conifers. In western hemlock the incipient stage has a faint purplish color (Englerth 1942:22).

Fomes connatus (Weinm.) Gill. causes a soft straw-colored to yellowish-brown rot, somewhat stringy, of the heartwood of living hardwoods, particularly maples, that may finally disintegrate the wood completely, leaving a hollow (Baxter 1948:205; Spaulding 1939). The small perennial conks, which rarely attain a width of 6 inches, are usually somewhat hoof-shaped, corky to woody, and white to yellowish, and the upper surface is usually covered with moss or algal growth. Unless the conks are dissected, they appear annual rather than perennial. The fungus may enter the tree through a wound or branch stub but generally fruits on a basal seam or scar (Fig. 176). The conks indicate a limited extent of decay (page 370).

F. rimosus (Berk.) Cke. causes a spongy yellow heart rot of living black locust throughout the range of the tree, and it occasionally attacks mesquite (Von Schrenk 1901; 1914). In black locust, infection occurs through older branches and through tunnels made by the locust borer. The decay extends outward from the center of the heartwood in a series of radial lines along the larger rays. Finally the heartwood is transformed into a soft spongy yellow or brown mass. Red-brown mycelial felts characterize the decay. The perennial hard woody conks, which may be a foot wide, are bracket- to hoof-shaped and at first uniformly brown, but with age the upper surface becomes black, furrowed, and cracked while the undersurface is a dull red-brown. The interior of the conk is bright yellowish brown. Conks on a tree indicate extensive decay.

Polyporus robiniophilus (Murr.) Lloyd (*Trametes robiniophila*) occasionally causes a soft white rot of living black locust in the northern part of the range of the tree (Kauffman and Kerber 1922). The annual shelflike conks, up to 6 inches wide, are white and watery, but firm, when fresh and are brownish and corky when dry.

Trametes suaveolens (L.) Fr. causes a soft white rot of the heartwood of living willows and occasionally poplars (Hirt 1932). It also destroys the sapwood of dead trunks. The incipient decay is characterized by fine bleached lines extending vertically through the heartwood beyond the advanced decay, which is white, dry, and corky to punky and has a characteristic anise odor. Rot usually begins in the lower trunk and progresses upward. It seldom goes below the root collar. The annual leathery to corky conks, up to 6 inches or more wide, are white when young but turn gray to yellowish with age. When fresh, they have a strong anise odor.

Polyporus hispidus (Bull.) Fr. causes a spongy white rot of the heartwood of living hardwoods, especially black ash and oaks in the eastern United States (Stewart 1951; Toole 1955). In England it attacks white ash (Nutman 1929). The decay apparently does not develop in dead trees. The decayed heartwood, usually in the upper portion of the trunk, in the incipient stage appears slightly bleached and in the advanced stage is reduced to a soft spongy yellowish or whitish mass. The annual shelflike conks, up to 10 inches in width, have a dark-brown coarse velvety to hairy upper surface and a golden-brown undersurface. With age they become uniformly dark brown. They usually appear on the upper portion of the trunk issuing from branch stubs, frost cracks, or other wounds. In oaks, particularly of the black oak group, the conks are often associated with large irregular trunk cankers (page 292).

P. glomeratus Peck causes a light-brown spongy heart rot of living maples and American beech. The advanced decay is surrounded by firm dark wood (Good and Nelson 1951). Obtuse-elongated or flattened, dark-brown to black, roughened sterile conks form on beech often at unhealed branch stub openings (Campbell and Davidson 1939*b*). They rarely protrude more than 3 inches. A canker develops on the trunk around the sterile conks. Ridges and fluted areas may precede conk development and canker formation. On red maple, cankers develop in connection with old branch stubs, but protruding sterile conks are rare. The yellowish-brown fruiting conks form on the underside of down trees or logs.

Ptychogaster cubensis Pat. (Fungi Imperfecti) causes a white to yellow-brown heart rot of living oaks and southern wax-myrtle (Davidson, Campbell, and Weber 1942). The conidial fructifications are brown hairy masses developing at branch stubs or wounds.

Hydnum erinaceus Fr., the hedgehog fungus, occasionally causes a white rot of the heartwood of living oaks (Von Schrenk and Spaulding 1909:44). In the advanced stage the soft spongy white rot may completely decompose, leaving large hollows, which are usually lined with yellowish mycelium. The soft annual conk is white at first and then turns yellowish or brownish with age. It is globular in shape, with a hairy top and long slender teeth or spines on the lower surface (Fig. 177).

H. septentrionale Fr. occasionally causes a soft spongy white rot of the heartwood of living maples and other hardwoods (Baxter 1925:550). In the incipient stage there is a zone of brownish discolored wood surrounding the white rot area, and fine black zone lines are usually present in the advanced decay. The large soft soggy annual creamy-white conks occur in bracketlike clusters on the trunk (Fig. 178). The clusters measure up to 2 feet or more in width and are 12 to 16 inches deep. The upper surface is almost plain, slightly scaly, with the slender teeth or spines on the undersurface attaining a length of ½ inch.

Collybia velutipes Curt. causes a soft, spongy white rot of the sapwood of living hardwoods (Biffen 1898). The conks, which are small toadstools with central stems whose bases are covered with dark-brown velvety

Fig. 177. Conks of the hedgehog fungus (*Hydnum erinaceus*) growing from a fire scar on black oak. (*Photograph by R. C. Lorenz.*)

Fig. 178. Conks of *Hydnum septentrionale* on the trunk of a sugar maple. (*Photograph by F. Kaufert and R. C. Lorenz.*)

hairs, issue in clusters from wounds. The upper surface of the cap is yellow or brownish as are the gills on the undersurface.

Polyporus obtusa Berk. causes a spongy, creamy-white heart rot of hardwoods, particularly oaks, that may be locally destructive since the rot advances into the living sapwood, finally killing the tree (Riley 1947). The large, corky-fibrous, cream-colored, bracketlike conks have somewhat labyrinthine tube mouths.

Daedalea unicolor (Bull.) Fr. causes a heart rot of living hardwoods, particularly maples, associated with cankers (page 293). In the incipient stage decayed wood is yellow; later it becomes white and soft. The hairy upper surfaces of the small corky conks, which commonly occur in clusters, vary from brown to gray.

D. confragosa (Bolt.) Fr. causes a soft white rot of many hardwoods, being particularly prevalent as a slash destroyer in eastern hardwood forests. It may be found on conifers. It sometimes occurs on living trees in the vicinity of wounds, especially on willows. The annual leathery to rigid conks are shelf-shaped and may attain a width of 6 inches. Occasionally they completely encircle a small dead stem. The upper surface is grayish to brownish, smooth, and concentrically zoned, and the mouths of the tubes on the undersurface are elongated and wavy in outline. Occasionally the under surface is gill-like.

Polyporus gilvus (Schw.) Fr., the rusty conk, causes a white rot of dead sapwood of many hardwoods and occasionally it occurs on conifers. It is prevalent on dead trees, stumps, and slash, and on living trees it commonly decays dead sapwood around wounds, occasionally invading the heartwood (Hirt 1928). In the incipient stage the wood appears slightly bleached, and, as the decay progresses, the wood becomes a light-cream color, is brittle when dry, and tends to split along the rays and tangentially along the limits of the annual rings. The cracks fill with brown matted mycelium. The small annual yellowish-brown to reddish-brown leathery to corky conks develop in profusion.

Polyporus tulipiferae (Schw.) Overh. causes a white rot of hardwoods usually on dead sapwood of slash or timbers in service (Baxter 1950). Occasionally it occurs on coniferous wood. The white fructifications are crustlike to somewhat reflexed.

Polyporus versicolor L. ex Fr., the rainbow conk, causes a soft white spongy rot of dead sapwood (Jay 1935). The rotted wood is of uniform texture, much reduced in weight, and brittle when dry. Throughout the United States this is the most common fungus on hardwood slash, and it is occasionally found on coniferous wood. It also frequently attacks wood in storage and in service. It decays the heartwood of living hardy catalpa, entering through wounds and dead branches, and in sprouts entering from the old stump. It continues to decay the wood after infected trees have been felled. The thin tough leathery annual conks, up to 2 inches or more in width, have a hairy or velvety upper surface varying from white through brown to grayish black in color, with more or less conspicuous multicolored zones.

P. hirsutus Wulf. ex Fr., the hairy conk, causes a sap rot of dead hardwoods practically identical with that caused by *P. versicolor*. It is rare on conifers. It also attacks wood in storage and in service. The fungus is said to kill the sapwood of living trees. The conks are similar in size and shape to those of *P. versicolor*, with a hairy grayish, yellowish, or brownish upper surface that although zonate is not multicolored. Several other species of *Polyporus* are known to attack dead sapwood and heartwood of living trees, usually entering through wounds (Weir 1923).

Stereum hirsutum (Willd.) Fr., which causes a soft uniform white sap rot of hardwood slash, is widespread throughout the United States, and the fungus is also abundant in other parts of the world (Ward 1898). In South Africa it causes a heart rot of blue gum (*Eucalyptus globulus* Labill.), also killing the living sapwood (Bottomley and Carlson 1920). The small thin leathery crustlike to somewhat shelf-shaped sporophores have a cream-buff to grayish hairy upper surface and a buff to smoke-gray smooth undersurface.

S. *murraii* (B. & C.) Burt causes a white or light-brown sweet-smelling heart rot of living hardwoods, particularly yellow birch. Cankers are commonly associated with this decay (page 293). The thin, largely crustlike, but sometimes reflexed fructifications, which develop on the cankers, have a pale olive-buff fruiting surface.

Peniophora gigantea (Fr.) Massee causes a superficial sap rot of both conifer and hardwood slash and wood products in storage. It often appears as a fluffy white mycelial growth on a wood surface. The fructifications are smooth, white to creamy or smoky-gray, flat waxy to leathery patches, which on drying are easily separated from the wood.

Polyporus lucidus Leys. ex Fr., *P. tsugae* (Murr.) Overh., and *P. oregonensis* (Murr.) Kauffman, the varnish or lacquer conks, cause a soft spongy white rot with numerous black spots scattered throughout. These fungi are usually on dead timber, affecting both sapwood and heartwood, although *P. lucidus* can be pathogenic (Pirone 1957). *P. lucidus* occurs on eastern hardwoods, *P. tsugae* on eastern conifers, commonly on eastern hemlock (West 1919), and *P. oregonensis* on western conifers, particularly on balsam firs and western hemlock, causing some decay in living trees of the latter. The large annual conks with a reddish shiny lacquerlike upper surface and usually with a short thick lateral stalk are conspicuous on logs, stumps, and standing or fallen dead trees (Fig. 179).

Polyporus volvatus Pk., the pouch fungus, causes a superficial soft grayish rot of the sapwood of coniferous slash and dead trees. Under laboratory conditions the fungus has caused fairly rapid decay of coniferous wood (Schmitz 1923). The white to tan leathery globose annual conks develop in profusion on dead standing and down trees, issuing from insect burrows in the bark. The brown tube layer is completely covered by a leathery membrane continuous with the upper surface. A small opening forms in this membrane through which beetles can pass to distribute the spores.

Schizophyllum commune Fr. is abundant on hardwood slash and dead parts of living trees, causing a superficial decay of the sapwood. The small thin sometimes lobed conks, from ½ to 2 inches wide, are fan-shaped with a grayish-white downy upper surface and with brownish forked gills on the undersurface. The gills are split along the lower edge. The fungus is common on fruit trees, sometimes attacking living sapwood (Essig 1922; Putterill 1922), and in South Africa it has caused withering of planted acacia about 3 years old (Ledeboer 1946).

Daldinia vernicosa (Schw.) Ces. & de Not. (Ascomycetes, Xylariaceae) is frequently found on dead parts of living hardwoods and particularly on dead, fire-scorched trees. In English ash the fungus causes a white mottled rot with black specks or dark lines (Cartwright and Findlay 1942:230). The fructifications containing the sunken perithecia are hemispherical in shape, black in color, and carbonaceous in texture.

Fig. 179. Varnish or lacquer conks (*Polyporus tsugae*) on a stump of eastern hemlock. (*Photograph by R. P. Marshall.*)

Ustulina vulgaris Tul. (Ascomycetes, Sphaeriaceae), the coal fungus, causes a brittle white heart rot with prominent black zones in the butts of living hardwoods (Campbell and Davidson 1940; Wilkins 1943). It is prevalent on sugar maple sprouts in the Northeast. The black carbonaceous crustose fructifications develop on large stumps and logs. On American beech the fructifications appear abundantly on flat cankered areas.

BROWN STRINGY ROT

Brown stringy rot of the heartwood of living conifers in the western United States and Canada is caused by the Indian paint fungus, *Echinodontium tinctorium* E. & E. (Boyce 1930; Thomas 1958; Weir and Hubert 1918). Its hosts are balsam firs, hemlocks, Engelmann spruce, and Douglas fir, its occurrence on the last two hosts being rare. It is most prevalent on lowland white fir, white fir, and western hemlock. Although the fungus may continue development after the death of an infected tree, it is of no importance as a destroyer of dead material.

Damage. In the aggregate the damage is great. Losses of 25 per cent or more of the gross board-foot volume are not uncommon in mature stands of white fir and lowland white fir. In northern Idaho losses in western hemlock often amount to 30 per cent or more in a given stand, but in western Oregon and Washington losses in this species are minor, the fungus being limited to high elevations and poor sites.

Fig. 180. A broken white fir log with the heartwood completely destroyed by brown stringy rot caused by *Echinodontium tinctorium*. (*Photograph by G. G. Hedgcock.*)

Decay. In white fir the first indication of incipient decay is light-brown or golden-tan spots or larger areas of discoloration in the light-colored heartwood, although the wood from 1 to several feet in advance of this discoloration is invaded by hyphae. The discoloration may be accompanied by small but clearly distinct radial burrows resembling, somewhat, shallow insect burrows. Next, rusty-reddish streaks appear following the grain. The wood still seems strong and only slightly softened if at all, but actually it is so weakened that when sawed into boards it may fall apart along the growth rings after drying. The discoloration usually extends from 2 to 6 feet beyond the advanced stage. The light-brown color intensifies, the wood softens, shows a decided tendency to separate along the early wood in the annual rings, and finally becomes brown with pronounced rusty-reddish streaks, fibrous, and stringy (Fig. 180). The butt of a decayed tree frequently becomes hollow. In western hemlock the rot is essentially the same.

Fig. 181. Conks of the Indian paint fungus (*Echinodontium tinctorium*) on a living white fir. (*Photograph by A. S. Rhoads.*)

Brown stringy rot is not confined to any definite portion of the bole, and in older trees it frequently extends the entire length of the heartwood in the trunk running into the roots and larger branches. In cross section the rot is usually circular in area, commonly destroying all the heartwood and leaving only a shell of sapwood, but the decay does not encroach on the living sapwood.

Indications of Decay. The hard woody hoof-shaped perennial conks range from a few inches to 1 foot or more in width. The upper surface is black, dull, rough, and cracked, and the undersurface is grayish, level, and thickly set with hard coarse spines. The interior or context is a striking rust-red or brick-red color, which is also often seen in wood in the advanced stage of decay exposed to the air. The Indians made a pigment from the substance of the conks, hence the common name. The conks are formed on the underside of dead branch stubs (Fig. 181), and normally the rusty color runs from the decayed heartwood through these stubs or knots. The conks are commonly located on the middle or lower middle section of the trunk and are developed quite abundantly,

Fig. 182. Advanced stage of red heart rot in balsam fir caused by *Stereum sanguinolentum.*

although not every decayed tree bears conks. A single conk indicates extensive decay. Swollen knots are not formed, but punk knots, called "rusty knots" owing to their rust-red color that reaches to within an inch or so of the end of the branch stub, are common. The significance of conks and rusty knots in estimating is discussed on page 367.

Infection occurs through branch stubs and wounds. In a decayed tree with a frost crack, the rot usually extends the length of the crack.

RED HEART ROT

Red heart rot, a decay of conifers, is caused by the bleeding Stereum, *Stereum sanguinolentum* A. & S. The fungus has been found killing planted conifers in the West (page 113), decaying the heartwood of eastern white pine after entering through large pruning scars (page

375), and in Scandinavia causing a serious decay in living trees and stored pulpwood of Norway spruce. It is widespread in North America on coniferous slash, and it occasionally decays dead sapwood of wounds on living trees. The fungus causes a severe heart rot of living balsam fir (McCallum 1928:7). Spruces are also affected.

Decay. Balsam fir heartwood in the incipient stage of decay appears water-soaked and reddish brown in color, and is firm. On the end of a log in the most characteristic form, there are rays extending out from the main body of the rot, somewhat similar to red ray rot caused by *Polyporus anceps*. In most cases the decay is a solid circular mass or is in irregular patches. Thin white mycelial felts appear early. The advanced decay is light in weight, light brown in color, dry, and friable in texture with a tendency to stringiness (Fig. 182). The rot occurs throughout the main and top portions of the bole, rarely invading the butt.

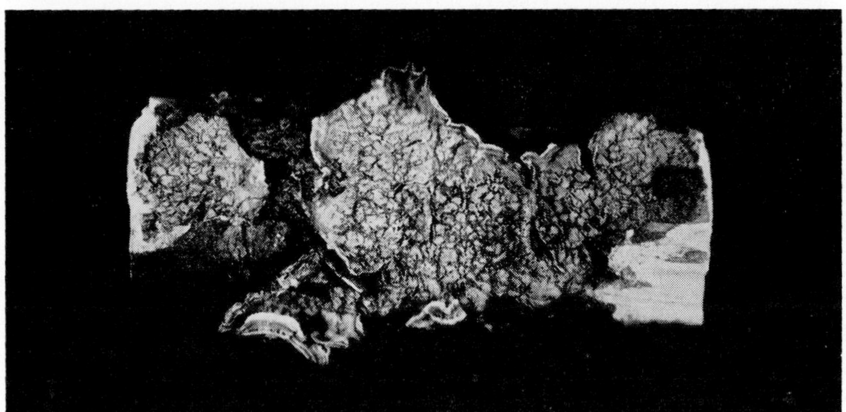

Fig. 183. Conk of the bleeding Stereum (*S. sanguinolentum*) on a dead branch of eastern hemlock.

Indications of Decay. The thin annual leathery conks are for the most part crustlike, with a narrow bracketlike margin (Fig. 183). They are small, seldom exceeding 2 inches in width. The upper surface is silky and pale olive-buff in color, whereas the smooth undersurface bleeds readily when wounded and dries to a grayish-brown color. The sporophores do not occur upon living trees except occasionally on dead branches, but they appear in profusion on dead wood.

Stereum pini (Schleich. ex Fr.) Fr. decays the heartwood of living pines, particularly lodgepole pine, and occasionally Douglas fir and white spruce (Nobles 1956). In cross section the incipient stage of decay is a reddish-brown discoloration in the heartwood with short radiating extensions around the periphery, like incipient decay of S. *sanguinolentum* and *Polyporus anceps*. The advanced stage, known only in cultures, is a yellow, fine, stringy rot. The small, thin, whitish crustlike fructifications appear on dead branches.

YELLOW RING ROT

Yellow ring rot or yellow laminated rot of the heartwood of living and dead western red cedar and other conifers is caused by *Poria weirii* Murr. (Baxter 1953). The fungus causes a severe root and butt rot of Douglas fir (page 113).

Fig. 184. Radial section through the butt and root collar of a western red cedar showing yellow ring rot in the heartwood caused by *Poria weirii*. The crustlike conk can be seen on the left-hand side of the trunk extending from about level with the ruler down the trunk to just beyond the root crotch. (*Photograph by E. E. Hubert.*)

Yellow ring rot is a serious decay of western red cedar in the Intermountain Region but is of minor consequence on the Pacific Coast. In the incipient stage there first appears a yellowish discoloration in the brown heartwood, which remains firm. Next, the yellowish color intensifies, and the wood softens. In the advanced stage the affected wood

Fig. 185. Stump of western red cedar showing the advanced stage of yellow ring rot in the heartwood caused by *Poria weirii*. (*Photograph by A. S. Rhoads.*)

has a yellowish or brownish discoloration and is soft, and the annual rings separate, giving the wood a laminated appearance on the radial face (Fig. 184). Sometimes small patches or a thin web of dark-brown hyphae develop between the lamina, and the wood shows long striations or thin, closely packed pockets. Finally the decomposed wood is reduced to a brown stringy or crumbly mass, and a hollow may form (Fig. 185). Usually the rot is confined to the butt, extending up 6 to 12 feet, but it may run higher. The rot column is cone-shaped at its upper limit.

The conks are light- to dark-brown flattened crusts with stratifications or layers, showing that they are perennial. The substance of the conk is dark brown and the undersurface is light brown with the mouths of the tubes numerous and small. The conks on living trees are commonly on the root crotches (Fig. 184) and partially obscured by the duff. They are more abundant on the undersides of down trees or logs, developing there in long layers. The decay commonly enters through fire scars, which, together with other basal wounds, indicate that a tree has been exposed to infection.

SIMILAR ROTS

Poria albipellucida Baxter causes white ring rot of the butts of western red cedar, which is prevalent on the Pacific Coast but is rarely, if ever, found on living trees in the Intermountain Region (Buckland 1946). The fungus also occurs on redwood (Kimmey 1958). In cedar the decay is similar to yellow ring rot, except that in the incipient stage a bluish or darker reddish marginal discoloration may appear, and in the advanced stage there are white mycelial flecks between the lamina, with the wood surface showing minute elongate pits. The late advanced stage is a typical ring rot, rarely stringy. The thin, white, crustlike sporophores rarely occur on living trees. In redwood, incipient decay is usually dark brown but sometimes almost black; then it is called "black heart."

Pleurotus ulmarius Bull., the elm fungus, causes a brown rot of the heartwood of living hardwoods, particularly white elm (Learn 1912). The rot may also extend into the sapwood. The affected wood separates readily along the annual rings. The large fleshy annual sporophores have a long excentric stalk. The upper surface is smooth and varies from white to yellow or brown. The gills on the undersurface are notched at the point of attachment to the stalk. The conks usually issue from crotches and pruning wounds.

Pholiota adiposa Fr., the yellow cap fungus, causes a brown mottled rot of the heartwood of living hardwoods and to a lesser extent of certain conifers (Hubert 1931:428). Basswood, birches, poplars, and balsam firs are the more common hosts. It may be that other species of *Pholiota* also cause brown mottled rot.

In conifers the incipient decay appears as light-yellowish areas of discoloration in the heartwood. These areas darken and brownish streaks may appear, giving wood in the advanced stage a mottled appearance

(Fig. 186). If the rotted wood is sawed or split along the grain, narrow yellowish to tan-colored strands of mycelium may pull out, leaving narrow irregular channels or pits in the wood. Trees with decay of long standing may become hollow-butted. In basswood the incipient stage appears as a reddish-brown discoloration, which fades into whitish areas

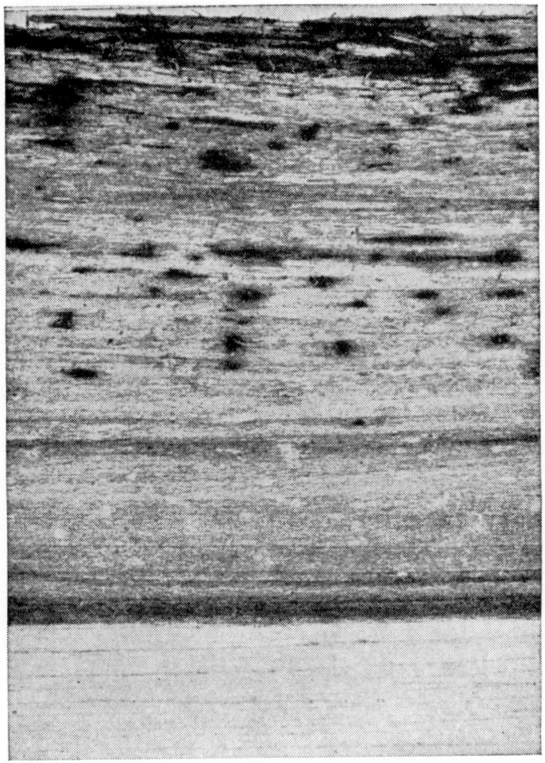

Fig. 186. The incipient and advanced stages of brown mottled rot in basswood caused by *Pholiota adiposa*. The light- and dark-colored patches are both shown in the advanced stage at the top. (*Reprinted by permission from "An Outline of Forest Pathology" by Hubert, published by John Wiley & Sons, Inc.*)

with brown streaks. Small white patches next appear. In the advanced stage the wood is whitish with tan to brownish streaks, and the small white patches become yellowish to rusty red or brown.

The soft annual conks, which are mushroomlike with yellow central stems and caps with a sticky, yellowish, slightly scaly upper surface, appear on the trunks, stumps, or old logs in clusters (Fig. 187). The undersurface of the cap has yellowish to brownish gills.

POCKET DRY ROT OF INCENSE CEDAR

Pocket dry rot, pecky cedar, pin rot, or incense cedar dry rot is a brown cubical pocket rot of the heartwood of living incense cedar caused by *Polyporus amarus* Hedgc. (Boyce 1920; Wagner and Bega 1958). The fungus may continue development in dead trees that were decayed while living, but it does not develop in wood in service. Cedar posts with this decay are as durable as sound posts. The decay is found throughout the range of incense cedar, and heavy losses result.

Fig. 187. Sporophores of *Pholiota adiposa* on the end of a decayed red fir log.

Fig. 188. End of an incense cedar log showing pocket dry rot in the heartwood caused by *Polyporus amarus.*

Decay. The pockets, which are elongated in the direction of the grain, vary in length from ½ to about 1 foot. They are several times longer than broad, with rounded or somewhat pointed ends. In the incipient stage the wood in the pockets is at first firm and faintly yellowish brown. This color deepens slightly, and the wood becomes somewhat soft. The incipient stage extends but a short distance vertically beyond the advanced stage, and a distance of 2 feet beyond the last visible evidence will usually exclude all decay. The incipient stage is only faintly apparent, occurring as it does in pockets with the color in the very earliest stage differing but slightly, when at all, from normal wood. An occasional pocket may occur several feet in advance of the main body of decay. The wood, even in an incipient pocket, is weakened, although the wood between pockets does not appear to be affected. The purplish coloration of the heartwood so often found in freshly cut incense cedar is not an incipient stage of decay. Cedar colored in this way is sound.

In the advanced stage the pockets are filled with a dark-brown carbonaceous crumbly mass, typical of brown cubical rots, and the line of separation between the sound and decayed wood is sharp. In some pockets small cobweblike or feltlike masses of white mycelium occur.

Fig. 189. Conks of *Polyporus amarus* on a living incense cedar. (*Photograph by J. L. Mielke.*)

Fig. 190. A shot-hole cup of *Polyporus amarus.* This is an enduring record of the place originally occupied by a conk and shows the knothole in the bark from which the conk issued.

The pockets are separated from each other by what appears to be sound wood, although occasionally streaks of straw-colored or brownish wood may extend vertically between two pockets. This is especially noticeable between young pockets.

On the ends of logs the pockets are circular, elliptical, or somewhat irregular in shape, with firm, normal-appearing wood separating them (Fig. 188). The rot column may extend from the butt to the upper limit of the heartwood.

Indications of Decay. The annual conks during most years are scarce. Issuing from knots, they usually appear in the summer or autumn. They are bracket-shaped with rounded tops (Fig. 189) or half bell-shaped.

When young, the conks are soft and mushy with a smooth, tan-colored top and a bright sulfur-yellow undersurface with numerous small tubes sometimes exuding clear yellow drops of sweet liquid. Later the conks become tough and cheesy, turning brown, and finally become hard and dry. However, they are usually so quickly destroyed by insects, birds, and squirrels that the last stage is rarely reached. Although conks are not common, an accurate indication of the place formerly occupied by a conk is supplied by a shot-hole cup, which appears as a cup-shaped depression below a knot. The depression, which is riddled with numerous fine holes caused by insects which infest the conk, is created by birds chopping out the bark to get the insects (Fig. 190). At first shot-hole cups have the color of the freshly opened bark of the tree but later become weathered and gray with age. A conk or a shot-hole cup always indicates that the bole is unmerchantable from ground level to a height of 10 to 50 feet above the conk or cup. Fire scars and branch stubs are the common points of entrance for pocket dry rot.

OTHER BROWN CUBICAL POCKET ROTS

Stereum sp. probably causes the brown pocket rot of the heartwood of living southern cypress known as peckiness (Davidson, Toole, and Campbell 1959). The decay (Fig. 191) closely resembles pocket dry rot of incense cedar caused by *Polyporus amarus*. Although the rot is widespread in cypress stands, it does not continue to develop in wood products. This is evident because pecky cypress lumber is highly decay resistant.

Poria asiatica (Pilát) Overh. causes a brown cubical butt rot and a brown cubical pocket trunk rot of living western red cedar (Buckland 1946). It is responsible for the greatest decay loss to this tree throughout British Columbia. Wood in the incipient stage, extending several feet beyond the advanced stage, is straw colored to pale yellow-brown, soft, and "dead" in appearance. Wood in the advanced stage is light brown to brown, brittle, and breaks into small cubes along the shrinkage cracks. The surfaces of the cubes may be covered with white mycelium. In the butt the advanced decay is piped or tubular in form, or a solid mass of decay. Higher in the trunk the two types of butt rot either develop into typical pocket rot or continue as an unbroken column. The thin white crustlike sporophores rarely if ever occur on living trees.

Polyporus basilaris Overh. causes a brown cubical pocket rot of heartwood in the butts of living Monterey cypress (Bailey 1941). The pockets are sharply delimited from adjacent firm wood. In the advanced stages of decay the pockets coalesce to form masses of rot with thin strips of firm wood scattered through the mass. The small conks, variable in shape, are imbricate where well developed; tough, leathery, and dirty grayish when fresh; but hard, rigid, and grayish brown to blackish when dry.

Poria sequoiae Bonar causes a brown pocket rot in the heartwood of redwood that occurs throughout the bole (Kimmey 1958). Incipient decay appears as a dark brown discoloration, sometimes almost black. Then it is termed "black heart." In the final stage, pockets become so numerous or individually so large that they may comprise an almost solid mass of rot. Mycelium felts may occur in the shrinkage cracks. The annual sporophores are small, thin, white crusts found in crevices and remote corners of cavernous fire scars on dead trees and on the ends of logs.

Fig. 191. End of a log of southern cypress showing peckiness of the heartwood probably caused by *Stereum* sp.

Lentinus kauffmanii Smith causes a brown pocket rot of the heartwood of living Sitka spruce on the Pacific Coast (Bier and Nobles 1946). The fungus may continue to develop in dead trees for 50 years. The decay is most prevalent as a root and butt rot, but it may also occur in the main trunk. In the incipient stage the wood in the pockets is a faint brown color. In the advanced stage the broadly lens-shaped elongated pockets are filled with a brown crumbly mass that forms cubes. The line of demarcation between the sound and decayed wood is sharp. The wood between pockets is firm. Finally, the pockets may become so numerous that they merge, forming a solid rot column. Isolated pockets often occur at some distance from the main body of rot. The small inconspicuous mushroomlike conks develop only on infected wood that is exposed to the air, and therefore they do not occur on living trees. There are no known external indications of decay in living trees.

Trametes serialis Fr. causes a brown cubical rot of conifers, rarely occurring on hardwoods, that tends to be pocketlike. It decays wood in service, and on Sitka spruce and Douglas fir it causes heart rot in living trees. The annual white leathery conks are usually flat but occasionally partially shelflike. They occur on dead material, rarely on living trees. This species and the following one with which it has been confused (Nobles 1943) are similar in habit and as far as known cause identical decays.

Poria monticola Murr. (*P. microspora* Overh.) causes a brown cubical rot of conifers. On Sitka spruce and Douglas fir it commonly causes a heart rot of living trees as well as decaying wood in service. Incipient decay first appears as streaks or elongated patches of faint yellowish-brown discoloration. At first glance it would appear as though drops of oil had fallen on the wood. A marked purple discoloration may accompany the incipient decay. The yellowish-brown areas enlarge, and pockets of brown cubical rot develop, the margins of the pockets not being sharply delimited. Finally the pockets may merge so that a large mass of brown cubical rot is formed. The paper-thin, small evanescent flat white conks rarely occur on standing trees.

Fomes subroseus (Weir) Overh., the rose-colored conk, often misnamed *Trametes carnea* Nees and for many years confused with *F. roseus*, causes a brown pocket rot of the heartwood of living eastern red cedar and of fruit trees (Baxter 1951). The fungus occurs commonly on dead coniferous timber and wood in service but rarely on hardwood material. In red cedar the pockets are from 1 inch to several feet in length, filled with roughly cubical pieces of brown, charcoallike decayed wood, and separated by firm, hard wood. The annual conks, which appear on cedar in the holes around dead branch stubs, are about ½ inch wide and irregular in shape, but they can be recognized by the rose-colored undersurface. On dead conifers where the conks develop normally they are shelflike, up to 3 inches in width, and have a smooth pinkish to rose-colored upper surface (which becomes grayish or black with age), as well as a rose-colored undersurface.

Lenzites saepiaria (Wulf.) Fr., the slash conk, which has a world-wide distribution, causes a brown pocket rot of dead coniferous sapwood, occasionally also decaying the outer layers of heartwood (Baxter 1952*a*). Sometimes it is found on hardwoods. It rarely occurs on living trees. The fungus is one of the most common destroyers of coniferous slash, but it also causes extensive damage by decaying exposed timber in service, particularly poles, posts, and ties.

The incipient stage of the decay is evident as elongated patches of yellowish to yellowish-brown discoloration. The advanced stage is characterized by elongated pockets filled with brown, friable wood that breaks up into small cubical pieces and crushes easily. On cross section the long axis of the pockets is radial. Mycelium felts in varying shades of yellow and brown are common in the pockets. As decay progresses the pockets tend to merge, becoming large. On the ends of timbers the rot is first noticeable as dark reddish-brown areas, more or less irregular in outline, but when the timbers are opened, pockets of decay are found extending in from these darkened areas for 1 or 2 feet.

The annual conks are usually long, narrow, shelflike formations issuing through cracks in the bark or season checks on decorticated wood. The velvety upper surface varies from yellowish-red to dark reddish brown in color but on weathering becomes grayish black. The undersurface has light- to dark-brown rather thick gills with occasional cross connections. *Trametes americana* Overh. causes the same type of decay, and the conks are the same except that the undersurface has tubes. The mouths of the tubes are angular, labyrinthiform, or so elongated that a gill-like appearance results. Conks of the two fungi may be intermingled, arising from the same underlying decayed wood.

L. trabea (Pers.) Fr., usually on broadleaf but occasionally on coniferous wood, causes a brown cubical rot, which tends to be pocketlike and which is quite similar to decay caused by *L. saepiaria.* It occurs on slash and is extremely destructive to wood in service such as stadium seats, sash, doors, and building timbers. In color and shape the conks are similar to *L. saepiaria,* but the undersurface usually has elongated pores and is never entirely gilled.

Polyporus leucospongia Cke. & Hk. is a western alpine species on large decorticated coniferous slash and wood in service, causing a brown carbonizing cubical rot of both sapwood and heartwood (Baxter 1954). The cream to light-buff conks are spongy.

RED-BROWN BUTT ROT

A brown cubical rot of the heartwood of living trees known as "red-brown butt rot," which is also found in the roots, is caused by the velvet top fungus, *Polyporus schweinitzii* Fr. (Boyce 1930; Childs 1937; Hiley 1919:126). The fungus may continue development in dead trees, logs, and stumps, but it is not important as a destroyer of slash nor does it decay wood in service. The fungus is widespread throughout Europe, Asia, and North America, attacking many conifers, but it is rare on hardwoods. Douglas fir, pines, and spruces are common hosts in the United States.

Damage. This fungus causes the most serious butt rot of conifers in the United States both as to the number of trees attacked and the volume of wood destroyed. Losses are more important than the actual

Fig. 192. Heartwood of the butt of a Douglas fir completely destroyed by red-brown butt rot caused by *Polyporus schweinitzii.* (*Photograph by G. G. Hedgcock.*)

Fig. 193. Conks of the velvet top fungus (*Polyporus schweinitzii*) in an old wound on the butt of a western white pine. The upper conk is fresh and growing, as indicated by the light-colored margin; the lower ones are old, dried up, and withered. (*Photograph by J. R. Weir.*)

volume of wood destroyed indicates because it is the valuable butt logs that are damaged, and furthermore infected trees are predisposed to wind throw. In general old trees suffer, but the fungus can be parasitic on young trees, since planted conifers have been killed by it (page 114), and it has also attacked the roots of 1- and 3-year-old eastern white pine seedlings growing in pots (Wean 1937), but it did not attack hardwood seedlings (Berk 1948).

Decay. There may be no visible evidence of incipient decay, or affected wood may show a light-yellowish to pale reddish-brown discoloration in the shape of spires running ahead of the advanced decay for a few inches to several feet. This color intensifies and the wood becomes soft and cheesy. In the advanced stage the wood is yellow-brown to red-brown in color (depending on the host), brittle, breaks into large cubes on drying, and can be easily crumbled into a fine powder (Fig. 192). In the shrinkage cracks, thin crustlike layers of mycelium that are easily mistaken for resinous crusts are formed.

The rot is confined to the heartwood of the roots and butt. It usually extends from 2 to 8 feet or so above ground level in the trunk, although in large trees of Douglas fir on the Pacific Coast it has been known to extend 90 feet or more upward. Rarely it may be found as a trunk rot not connected with the butt, being localized around a dead branch stub. The rot column is normally conical from the base of the tree upward and in cross section is usually circular. For scaling and estimating Douglas fir on the Pacific Coast with this decay see page 366.

Fig. 194. A conk of the velvet top fungus (*Polyporus schweinitzii*) on the ground near the base of an infected Douglas fir. (*Photograph by H. J. Rust.*)

Indications of Decay. The annual conks, which develop most abundantly during moist weather in the late summer and fall, are to be found on the butts of infected trees issuing from old wounds or on the ground near by, coming up from a decayed root. On the tree the conk is a thin bracket, and frequently two or more brackets grow one above the other (Fig. 193). On the ground, when viewed from above, the conk is circular in shape and sunken in the center, tapering to a short thick stalk (Fig. 194). When fresh the upper surface is velvety, concentrically zoned, and reddish-brown in color with a light yellow-brown margin. The undersurface is dirty green and turns red-brown when bruised. The mouths of the tubes are large and irregular in outline. The substance of the conk is moist and cheesy. When the conks are old and dried they become corky and have a deep red-brown or blackish-brown color. Punk knots are rare, and swollen knots are unknown.

Infection is largely through basal wounds, particularly fire scars, so in any fire-scarred coniferous stand in which this fungus is present, the

Fig. 195. Conks of *Polyporus balsameus* on the butt of a small, living balsam fir. (*Photograph by H. G. Eno.*)

incidence of butt rot will be high. It has been assumed that the fungus spreads through the soil to infect roots and that infection may also occur by contact of a decayed root with a sound one. These ideas need substantiation.

BALSAM BUTT ROT

Balsam butt rot or brown butt rot, a brown cubical butt rot of living balsam fir, is caused by the balsam conk, *Polyporus balsameus* Pk. The fungus is prevalent on dead trees from Minnesota eastward to the New England states and eastern Canada. As a butt rot in balsam fir this decay is secondary in importance to feather rot caused by *Poria subacida*, but it is an important factor in the rapid deterioration of balsam fir stands killed by the spruce budworm, *Choristoneura fumiferana* (Clem.).

Decay. In the incipient stage affected wood is yellowish to light buff in color and is somewhat softer than sound wood. In the advanced stage the wood is brown in color, brittle, breaks into large cubes along shrinkage cracks, and is easily crushed to a clay-colored powder. Thin white mycelial felts appear in the shrinkage cracks. Advanced decay is quite similar to rot caused by *P. schweinitzii;* consequently in the absence of conks the two decays cannot be separated with certainty except by cultural characteristics. Generally brown butt rot caused by *P. balsameus* is lighter in color than that caused by *P. schweinitzii.*

In living trees the decay is a root and butt rot of heartwood, although it may be able to encroach on living sapwood. The rot column in more seriously decayed trees may extend from 6 to 12 feet above ground level, but generally it does not extend for more than 3 to 4 feet.

Indications of Decay. The annual fleshy-tough shelflike conks, rigid when dry, up to 2 inches wide, have a pale-brown upper surface distinctly marked with concentric zones and a white undersurface. They commonly develop in clusters, overlapping one another (Fig. 195). They are usually quickly destroyed by insects so that conks are present for a short time only during the growing season. They may be found on the trunk, on exposed roots, or on the ground at a point where a decayed root is near the surface. Apparently infection is largely through wounds so that trees with basal scars, particularly fire scars and frost cracks, are commonly infected. Resin flow from exposed roots and from root crotches commonly accompanies this decay. Some diseased trees are swollen at the butt.

BROWN TRUNK ROT

A brown cubical rot of the heartwood of living trees is caused by the quinine fungus, *Fomes laricis* (Jacq.) Murr. also often called *F. officinalis* (Boyce 1930; Faull 1916; Martinez 1944). The fungus may continue development in dead trees, logs, and stumps, but it is not important as a destroyer of slash. It occasionally decays wood in service, but only when the timber was sawn from infected heartwood. The conks are important enough for medicinal purposes in Europe so that the possibility of cultivating them artificially has been investigated (Borzini 1941). The fungus is widespread in Europe and North America, confined to conifers. It is rare in the East and common in the West, where its principal hosts are Douglas fir, ponderosa and sugar pines, and western larch. In eastern Canada, eastern white pine is attacked. In the West the decay causes considerable loss to its principal hosts.

Decay. The incipient stage may be almost imperceptible, or it may appear as a faint yellowish to reddish-brown discoloration, the wood softening as the discoloration deepens. Occasionally in Douglas fir incipient decay is indicated by a brilliant and intense purplish discoloration, but in most decayed Douglas firs this unusual color is lacking. In

Fig. 196. An old chalky conk of the quinine fungus (*Fomes laricis*) and decayed wood showing the white mycelial felts. Both from Douglas fir. (*Photograph by E. E. Hubert.*)

this host incipient decay may extend 8 feet or more beyond advanced decay but the average is about 3½ feet. In ponderosa pine the incipient stage appears as a red-brown or pronounced brown discoloration in the pale-lemon to light orange-brown heartwood, bounded by a narrow band of pronounced pink or red color, particularly noticeable on a radial or tangential surface. The discolored wood seems to be firm and strong, but it is seriously weakened. Incipient decay may extend for many feet beyond advanced decay.

In the advanced stage the wood is a brown mass, easily crumbled between the fingers. It is more or less separated into irregular cubes by cracks in shrinking. The cracks are commonly occupied by conspicuous white mycelial felts (Fig. 196), although in weathered material or decay of long standing the felts may disappear. In Douglas fir these felts may completely fill large cracks or old wind shakes, becoming $\frac{1}{4}$ inch thick, 1 foot or more wide, and several feet long. The felts have a bitter flavor. This decay almost exactly resembles that caused by *Polyporus sulphureus,* but the mycelial felts differ microscopically (Faull 1916:198). The rot is confined to heartwood of the main and upper portions of the bole, occasionally appearing in the butt.

Indications of Decay. The only certain indications of this decay are the perennial, long cylindrical pendulous or roughly hoof-shaped conks (Fig. 196). They do not develop abundantly, none appearing on many decayed trees and rarely more than one appearing on a single tree. The conks may attain considerable age and large size. When they develop in the cylindrical shape they may be 2 feet or more in length, or as hoof-shaped brackets they may be 10 to 18 inches or more wide. The substance of a conk is white, soft, and cheesy when young, and rather crumbly and chalky when old and dry, with an intensely bitter flavor. The upper surface is chalky white or brownish, rough and zoned, whereas the undersurface is white with small tube mouths. The conks issue through knots or old wounds, a single conk commonly indicating an unmerchantable tree. No swollen or punk knots are formed.

Infection occurs through wounds or branch stubs in the upper part of the tree and spreads throughout the trunk. Occasionally infection occurs through fire scars. Broken tops are a frequent point of entrance for this decay; consequently in a stand where the fungus is known to occur, broken-topped trees are probably infected. Large broken-off branches in which the decay can be seen indicate extensive rot in the bole.

OTHER BROWN CUBICAL ROTS

Polyporus sulphureus (Bull.) Fr., the sulfur fungus, causes a brown cubical rot of the heartwood of living hardwoods and conifers, occasionally killing living sapwood (Atkinson 1901:208; Von Schrenk and Spaulding 1909:37). It may develop in dead trees, logs, and stumps, but it is not an important slash destroyer. It decays chestnut poles in the Northeast. The fungus is widespread in the United States, oaks being a common host among hardwoods; balsam firs, Douglas fir, and spruces are the conifers most commonly affected. The sulfur fungus causes the most serious brown rot of hardwoods, but the loss resulting does not compare with that from white trunk rot caused by *Fomes igniarius.* In general damage to conifers is small, except that it causes heavy losses to red fir in California.

The decay is so similar to brown trunk rot caused by *F. laricis* that a separate description of it is unnecessary. In conifers it is often a butt

Fig. 197. Conks of the sulfur fungus (*Polyporus sulphureus*) on the trunk of a living butternut. (*Photograph by R. P. Marshall.*)

rot, whereas in hardwoods the decay is less localized. The annual, shelf-like conks are conspicuous and frequently overlap one another in clusters of from 2 to 20 or more (Fig. 197). When fresh they are soft, fleshy, and moist, with a bright orange-red upper surface, brighter at the margin, and a brilliant sulfur-yellow undersurface. When old they are hard, brittle, and chalky or dirty white in color. Infection occurs through dead branch stubs and wounds.

Poria cocos (Schw.) Wolf causes a brown root and butt rot of conifers and hardwoods, and also decays dead timber (Baxter 1949; Davidson and Campbell 1954; Fritz 1954). Large oblong to subglobose sclerotia with a light-brown surface and a white interior, known as tuckahoes, are developed in the soil by this fungus. The white crustlike fructifications form on the surface of the sclerotia.

F. roseus Fr., the rose-colored conk, causes a brown cubical rot in the

Fig. 198. Cross section of a decayed top of a living Douglas fir showing yellow-brown top rot caused by *Fomes roseus* and the upper surface of a conk. Diameter inside bark of the stem is 3.5 inches.

heartwood of various living conifers, particularly Douglas fir, known as "brown top rot" because the decay usually originates in the upper part of the bole (Boyce 1930:27). The fungus occurs more commonly on dead trees and wood in service. The relation of *F. roseus* to *F. subroseus* (page 434) as to the type of rot caused needs further investigation. In living trees no color change is visible in the heartwood first invaded by the mycelium, but later on discoloration appears. It is first evident as a faint brownish or yellowish-brown color, the outer limits of which are sometimes marked by a zone of greenish discoloration. In the advanced stage the wood is yellowish to reddish brown in color, soft, and breaks into irregular cubes. The perennial woody bracketlike conks (Fig. 198), ranging up to 6 inches wide, have a hard rough zoned black upper surface and a delicate rose-colored undersurface. Infection occurs principally through broken tops and dead branch stubs.

F. pinicola (Swartz) Cke., the red belt fungus, which is widespread, causes a brown cubical rot of the sapwood and heartwood of dead trees known as "brown crumbly rot," many conifers and some hardwoods being affected (Mounce 1929). The fungus is the most common destroyer of dead coniferous timber. It can be responsible for the rapid deterioration of merchantable timber killed by wind throw, insects, and

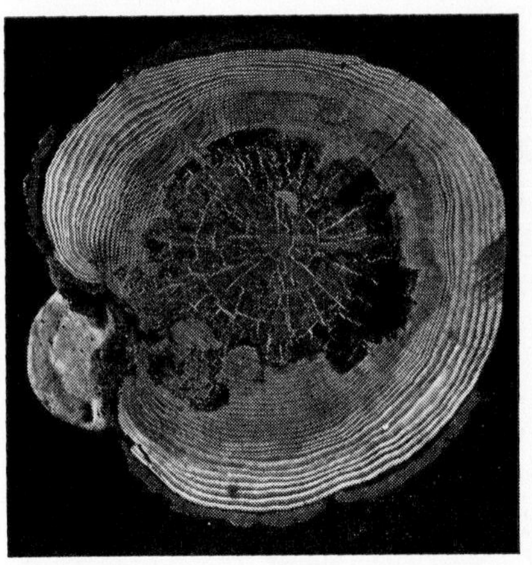

Fig. 199. Cross section of a lowland white fir with advanced stage of decay in the heartwood caused by *Fomes pinicola*. The rot is cubical in character with white mycelial felts filling the shrinkage cracks. The undersurface of a conk attached to the trunk is also shown. (*Reproduced from U.S. Dept. Agr. Bull.* 658.)

Fig. 200. Conks of the red belt fungus (*Fomes pinicola*) on a dead Douglas fir. The light-colored margin is prominent.

other agencies (Von Schrenk 1903*b*). It also causes heart rot in living trees. Inoculations of living trees showed no strains of the fungus specialized as to host (Hirt and Eliason 1938).

The incipient stage of decay appears as a faint yellowish to brownish discoloration. Wood in the advanced stage is reduced to a yellowish-brown or occasionally a light reddish-brown crumbly mass with a tendency to break into cubes (Fig. 199). Prominent mycelial felts develop in the shrinkage cracks. The hard woody perennial shelf- to hoof-shaped conks may attain a width of 2 feet, but on an average they vary from 2 to 10 inches (Fig. 200). The upper surface is rather smooth, zoned, gray to black in color and commonly has a wide red margin. When fresh, the undersurface is white to yellowish but on drying becomes cream to brownish.

Lentinus lepideus Fr., the scaly cap, which has a world-wide distribution, causes a brown cubical rot of coniferous wood, particularly pines. It is of most importance as a destroyer of wood used in contact with the ground such as foundation timbers, ties, and paving blocks. It causes some decay of slash, developing in stumps and old logs, but it rarely attacks limbs or small tops. The conks are a familiar sight on stumps in cutover ponderosa pine stands. Occasionally it decays the heartwood of living pines, the rot not being limited to any particular portion of the bole (Wagener 1929).

Incipient decay is indicated by a pale-yellowish to brownish discoloration. In the advanced stage the wood is reddish-brown, brittle, breaks into large cubes on drying, and crumbles readily under pressure. Thin white mycelial felts occupy the shrinkage cracks. Before drying, decayed wood has a sharp aromatic resinous odor. The annual tough-fleshy buff-colored conks are of the mushroom type with a central scaly stalk having a dark brown to black base (Fig. 201). The upper surface of the thick circular cap, from 2 to 6 inches in diameter, is covered with brownish to blackish scales, and the gills on the undersurface are notched close to the point of attachment to the stalk. On drying the sporophores become very hard.

Fistulina hepatica (Huds.) Fr., the beefsteak fungus, causes a brown cubical rot of the heartwood of living oaks known as "brown oak," "red oak," or "foxiness" (Cartwright 1937; Cartwright and Findlay 1936:10; Davidson 1935). It also occurs on chestnut. The wood may or may not be reduced in strength depending on the degree of infection, but in either case its value for paneling and furniture construction is enhanced in England.

The incipient stage of decay appears at first as a streaky brown discoloration of the heartwood, but gradually the heartwood of the entire trunk is affected and becomes a rich-brown color. When the brown wood contains black patches it is known as tortoise-shell oak. The decay develops so slowly that the advanced stage, in which the wood is deep reddish-brown with a tendency to cubical cracking, is seldom reached. The wood never becomes so soft or cracks so much as that decayed by the other brown cubical rots.

The annual conks with or without thick lateral stalks, found on trunks of living trees or stumps of oaks and on stumps of American chestnut, are always shelflike, 3 to 8 inches broad, thick, and deep purplish red or brown (Thomas 1937). The soft red juicy flesh, which is variegated with brighter streaks, exudes a red bloodlike juice when broken. The free or separate cylindrical tubes on the undersurface appear like toothed projections, particularly in the early stages.

Polyporus spraguei Berk. & Curt., causes a brown friable rot of the

Fig. 201. Sporophore of the scaly cap (*Lentinus lepideus*) from a western larch railroad tie.

heartwood of the butt and roots of living American chestnut, oaks, and persimmon (Weir 1927). The fungus continues development on stumps and dead trees as long as any heartwood is left. The annual, slightly bracket-shaped conks, up to 6 inches in width, are fleshy-tough and watery when fresh and rigid when dry. The upper surface is somewhat rough and white to grayish in color, whereas the undersurface is whitish.

P. betulinus (Bull.) Fr., the birch conk, causes a brown cubical rot in birches (Von Schrenk and Spaulding 1909:51; Macdonald 1937). The fungus occurs throughout the range of birches and is confined to dead trees, standing or down, attacking sapwood. The decayed wood is yellowish brown, cracking radially and tangentially with thin white mycelial felts forming in the cracks. It is light in weight and powders readily under pressure. The corky annual conks (Fig. 202) are shelf- to hoof-

Fig. 202. Conks of *Polyporus betulinus* on the trunk of a dead yellow birch. (*Photograph by R. C. Lorenz.*)

shaped with a smooth grayish upper surface, and a yellow-white undersurface changing to brown with age. The incurved margin projects below the undersurface. The brown tube layer separates readily from the other part of the conk. The conks are from 2 to 10 or more inches in width and usually develop abundantly on decayed trees.

Daedalea quercina (L.) Fr., which has a world-wide distribution, causes a brown cubical rot of dead timber and exposed wood in service such as poles, posts, and bridge timbers. Occasionally it decays the heartwood of living trees in the immediate vicinity of butt wounds. Oaks and American chestnut are its usual hosts in the United States. In the advanced stage the wood is reduced to a yellow-brown friable mass with a tendency to break into small cubes. The conks are corky and shelf-shaped, ranging up to 7 inches wide, with a grayish to almost black smooth upper surface and a cream-colored to brownish undersurface (Fig. 203). The mouths of the tubes are large, elongated, and irregular in shape. The conks are more or less perennial, beginning growth again when conditions are favorable.

Stereum fasciatum Schw. causes a brown crumbly rot of hardwoods, attacking both sapwood and heartwood. It is found on slash, and it also commonly decays ties. The thin leathery annual shelflike imbricate conks have a clay-colored upper surface when young that becomes grayish to slate colored with age. The undersurface is light brown and smooth. In the Gulf and South Atlantic states it is largely replaced by *S. lobatum* (Kunze) Fr., a similar-appearing fungus.

Fig. 203. Conks of *Daedalea quercina* on the end of an oak tie. (*Reproduced from U.S. Dept. Agr. Bull.* 510.)

Coniophora cerebella Pers. (*C. puteana*) causes a brown cubical rot of conifers and occasionally of hardwoods. It occurs on slash, but it is of most importance as a destroyer of building timbers. It sometimes decays living trees (Macdonald 1939). The annual crustlike fleshy fructifications are usually a little over 2 inches in diameter, or they may be elongated. The surface is smooth or slightly wavy and olive to brownish in color with a whitish margin.

DRY ROT

Dry rot or building rot, a brown cubical rot which is widespread in this country, is caused by the building Poria, *Poria incrassata* (B. & C.) Curt. (Humphrey 1923; Baxter 1940). It decays both coniferous and broadleaf wood. The fungus is confined to structural timbers in buildings or stored lumber, rarely attacking poles, posts, ties, or slash in the forest.

Building Poria is an outstanding destroyer of coniferous timber in buildings throughout the United States, particularly in the Southern and Pacific Coast states where the mild humid climate is so favorable to it.

Decay. In the advanced stage the decay is a typical brown cubical rot with mycelial felts frequently filling the shrinkage cracks (Fig. 206). Mycelial felts appear in cracks between boards, or even on exposed surfaces if there is sufficient moisture. Irregular vinelike or rootlike rhizomorphs, dirty gray to brownish in color, and sometimes attaining the

Fig. 204. Sporophore of the building Poria (*P. incrassata*) on a Douglas fir structural timber. (*Photograph by L. O. Overholts.*)

thickness of a wrist, are formed abundantly, enabling the fungus to transport water long distances from the ground to the second or third floor of a building. The rhizomorphs may also develop as flattened veins embedded in sheets of mycelium. This ability to supply water to moisten the wood makes the fungus particularly destructive since dry wood cannot decay.

Indications of Decay. The first indications are often paperlike fan-shaped sheets of mycelium that appear on wood in damp places. Decayed timbers may sound hollow when tapped or may crack to expose the whitish mycelium. Thick mycelial masses may protrude along the bottom of baseboards or at other cracks where timbers are joined. The annual leathery crustlike sporophores vary in color from orange, through olivaceous to deep purplish black depending on their age and location (Fig. 204). Often all these colors may be represented in order from the margin to the center of a single fructification. The tubes vary from shallow in poorly developed sporophores to a centimeter deep in well-developed ones.

Merulius lacrymans (Wulf.) Fr., the dry rot fungus or tear fungus (Lohwag 1955), is the European counterpart of *Poria incrassata* which is unknown there. *M. lacrymans* is rare in the United States, possibly because it is adapted to moderate temperatures, most reports of its occurrence being based on decay caused by *P. incrassata.* The two fungi produce exactly similar decays on coniferous wood so that the rots cannot be identified unless sporophores are present. In addition to the water supply obtained by rhizomorphs, *M. lacrymans* is said to provide moisture adequate for its activity by decomposition of the wood (page 352). Its annual spongy-fleshy, crustlike sporophores are orange when fresh and light brown when dry. The tubes are reduced to wide, shallow pits, giving the surface a wrinkled appearance. Large drops of water commonly exude from the sporophores. Several species of *Merulius* are found on slash in the forest, *M. americanus* Burt and *M. tremellosus* Schrad. being common.

REFERENCES

Andrews, S. R.: Red Rot of Ponderosa Pine. *U.S. Dept. Agr. Monog.*, 23:1–34 (1955).

Atkinson, G. F.: Studies of Some Shade Tree and Timber Destroying Fungi. *Cornell Univ. Agr. Exp. Sta. Bull.*, 193:199–235 (1901).

Bailey, H. E.: The Biology of *Polyporus basilaris. Bull. Torrey Botan. Club,* 68:112–120 (1941).

Baxter, D. V.: The Biology and Pathology of Some of the Hardwood Heart-rotting Fungi. *Am. J. Botany,* 12:522–576 (1925).

Baxter, D. V.: Some Resupinate Polypores from the Region of the Great Lakes. VI. *Papers Mich. Acad. Sci.,* 20(1934):273–281 (1935).

Baxter, D. V.: Some Resupinate Polypores from the Region of the Great Lakes. XI. *Papers Mich. Acad. Sci.,* 25:145–170 (1940).

Baxter, D. V.: Some Resupinate Polypores from the Region of the Great Lakes. XVIII. *Papers Mich. Acad. Sci.,* 32(1946):189–211 (1948).

Baxter, D. V.: Some Resupinate Polypores from the Region of the Great Lakes. XIX. *Papers Mich. Acad. Sci.,* 33(1947):9–30 (1949).

Baxter, D. V.: Some Resupinate Polypores from the Region of the Great Lakes. XX. *Papers Mich. Acad. Sci.,* 34(1948):41–56 (1950).

Baxter, D. V.: Some Resupinate Polypores from the Region of the Great Lakes. XXI. *Papers Mich. Acad. Sci.,* 35(1949):43–59 (1951).

Baxter, D. V.: Some Resupinate Polypores from the Region of the Great Lakes. XXII. *Papers Mich. Acad. Sci.,* 36(1950):69–83 (1952*a*).

Baxter, D. V.: Some Resupinate Polypores from the Region of the Great Lakes. XXIII. *Papers Mich. Acad. Sci.*, 37(1951):93–110 (1952*b*).

Baxter, D. V.: Some Resupinate Polypores from the Region of the Great Lakes. XXIV. *Papers Mich. Acad. Sci.*, 38(1952):115–123 (1953).

Baxter, D. V.: Some Resupinate Polypores from the Region of the Great Lakes. XXV. *Papers Mich. Acad. Sci.*, 39(1953):125–138 (1954).

Berk, S.: Inoculation Experiments with *Polyporus schweinitzii. Phytopathology,* 38:370–377 (1948).

Bier, J. E., and Mildred K. Nobles: Brown Pocket Rot of Sitka Spruce. *Can. J. Research C. Botan. Sci.*, 24:115–120 (1946).

Biffen, R. R.: On the Biology of *Agaricus velutipes,* Curt. (*Collybia velutipes,* P. Karst.). *J. Linnean Soc. London, Botany,* 34:147–162 (1898).

Borzini, G.: Primo contributo allo studio delle possibilita di una coltivazione artificiale del 'Fomes officinalis' (Will.) Fr. *Boll. staz. patol. vegetale,* 21:221–234 (1941) [*Rev. Appl. Mycol.*, 25:523 (1946)].

Bottomley, A. M., and K. A. Carlson: Parasitic Attack on *Eucalyptus globulus.* A Note on *Stereum hirsutum* in Plantations in the Transvaal. S. *Africa Dept. Agr. and Forestry Bull.*, 3:1–8 (1920).

Boyce, J. S.: The Dry-rot of Incense Cedar. *U.S. Dept. Agr. Bull.*, 871:1–58 (1920).

Boyce, J. S.: Decay in Pacific Northwest Conifers. *Yale Univ., Osborn Botan. Lab. Bull.*, 1:1–51 (1930).

Boyce, J. S., and J. W. B. Wagg: Conk Rot of Old-growth Douglas-fir in Western Oregon. *Oregon Forest Products Lab.* (*Corvallis*) *Bull.*, 4:1–96 (1953).

Brown, H. P.: A Timber Rot Accompanying *Hymenochaete rubiginosa* (Schrad.) Lév. *Mycologia,* 7:1–20 (1915).

Buckland, D. C.: Investigations of Decay in Western Red Cedar in British Columbia. *Can. J. Research C. Botan. Sci.*, 24:158–181 (1946).

Buller, A. H. R.: The Biology of *Polyporus squamosus Huds.*, a Timber Destroying Fungus. *J. Econ. Biol.*, 1:101–138 (1906).

Campbell, W. A., and R. W. Davidson: *Poria andersonii* and *Polyporus glomeratus,* Two Distinct Heart-rotting Fungi. *Mycologia,* 31:161–168 (1939*a*).

Campbell, W. A., and R. W. Davidson: Sterile Conks of *Polyporus glomeratus* and Associated Cankers on Beech and Red Maple. *Mycologia,* 31:606–611 (1939*b*).

Campbell, W. A., and R. W. Davidson: *Ustulina vulgaris* Decay in Sugar Maple and Other Hardwoods. *J. Forestry,* 38:474–477 (1940).

Cartwright, K. St. G.: A Reinvestigation into the Cause of "Brown Oak," *Fistulina hepatica* (Huds.) Fr. *Brit. Mycol. Soc. Trans.*, 21:68–83 (1937).

Cartwright, K. St. G., and W. P. K. Findlay: "The Principal Rots of English Oak." *Gt. Brit. Dept. Sci. Ind. Research, Forest Prods. Research.* London: H. M. Stationery Office (1936).

Cartwright, K. St. G., and W. P. K. Findlay: Principal Decays of British Hardwoods. *Ann. Appl. Biol.*, 29:219–253 (1942).

Childs, T. W.: Variability of *Polyporus schweinitzii* in Culture. *Phytopathology,* 27:29–50 (1937).

Davidson, R. W.: *Stereum gausapatum,* Cause of Heart Rot of Oaks. *Phytopathology,* 24:831–832 (1934).

Davidson, R. W.: Forest Pathology Notes 2. *Fistulina hepatica* Causing "Brown Oak." *U.S. Dept. Agr., Plant Disease Reptr.*, 19:95–96 (1935).

Davidson, R. W., and W. A. Campbell: *Poria cocos,* a Widely Distributed Wood-rotting Fungus. *Mycologia,* 46:234–237 (1954).

Davidson, R. W., W. A. Campbell, and G. F. Weber: *Ptychogaster cubensis,* a Wood Decaying Fungus of Southern Oaks and Waxmyrtle. *Mycologia,* 34:142–153 (1942).

Davidson, R. W., Frances F. Lombard, and W. A. Campbell: A White Heart Rot of *Fraxinus* Caused by *Fomes johnsonianus* (Murr.) Lowe. *U.S. Dept. Agr., Plant Disease Reptr.,* 43:1148–1149 (1959).

Davidson, R. W., E. R. Toole, and W. A. Campbell: A Preliminary Note on the Cause of "Pecky" Cypress. *U.S. Dept. Agr., Plant Disease Reptr.,* 43:806–808 (1959).

Englerth, G. H.: Decay of Western Hemlock in Western Oregon and Washington. *Yale Univ., School Forestry Bull.,* 50:1–53 (1942).

Essig, F. M.: The Morphology, Development, and Economic Aspects of *Schizophyllum commune* Fries. *Univ. Calif. (Berkeley) Publs. Botany,* 7:447–498 (1922).

Faull, J. H.: *Fomes officinalis* (Vill.) a Timber-destroying Fungus. *Trans. Roy. Can. Inst.,* 11:185–209 (1916).

Fritz, Clara W.: Fructifications of *Poria cocos* (Schw.) Wolf on Poplar. *Can. J. Botany,* 32:545–548 (1954).

Good, H. M., and C. D. Nelson: A Histological Study of Sugar Maple Decayed by *Polyporus glomeratus* Peck. *Can. J. Botany,* 29:215–223 (1951).

Gosselin, R.: Studies on *Polystictus circinatus* and Its Relation to Butt-rot of Spruce. *Farlowia,* 1:525–568 (1944).

Haddow, W. R.: On the Classification, Nomenclature, Hosts and Geographical Range of *Trametes pini* (Thore) Fries. *Brit. Mycol. Soc. Trans.,* 22:182–193 (1938*a*).

Haddow, W. R.: The Disease Caused by *Trametes pini* (Thore) Fries in White Pine (*Pinus strobus* L.) *Trans. Roy. Can. Inst.,* 22:21–80 (1938*b*).

Hartig, R.: "Wichtige Krankheiten der Waldbäume." Berlin: Springer-Verlag (1874).

Hartig, R.: "Die Zersetzungserscheinungen des Holzer der Nadelholzbäume und der Eiche in forstlicher botanische und chemischer Richtung." Berlin: Springer-Verlag (1878).

Hedgcock, G. G., and W. H. Long: Heart-rot of Oaks and Poplars Caused by *Polyporus dryophilus. J. Agr. Research,* 3:65–77 (1914).

Herrick, J. A.: The Biology of *Stereum gausapatum. Ohio State Univ., Abstr. Diss. Ph. D.,* 28:167–173 (1939).

Hilborn, M. T.: The Biology of *Fomes fomentarius. Maine Agr. Expt. Sta. Bull.,* 409:161–214 (1942).

Hiley, W. E.: "The Fungal Diseases of the Common Larch." New York: Oxford University Press (1919).

Hirt, R. R.: The Biology of *Polyporus gilvus* (Schw.) Fr. *N.Y. State Coll. Forestry Syracuse Univ., Tech. Publ.,* 22:1–47 (1928).

Hirt, R. R.: *Fomes everhartii* Associated with the Production of Sterile Rimose Bodies on *Fagus grandifolia. Mycologia,* 22:310–311 (1930).

Hirt, R. R.: On the Biology of *Trametes suaveolens* (L.) Fr. *N.Y. State Coll. Forestry Syracuse Univ., Tech. Publ.,* 37:1–36 (1932).

Hirt, R. R., and E. J. Eliason: The Development of Decay in Living Trees Innoculated with *Fomes pinicola. J. Forestry,* 36:705–709 (1938).

Hubert, E. E.: "An Outline of Forest Pathology." New York: John Wiley & Sons, Inc. (1931).

Humphrey, C. J.: Decay of Lumber and Building Timbers Due to *Poria incrassata* (B. & C.) Burt. *Mycologia,* 15:258–277 (1923).

Jay, B. A.: A Study of *Polystictus versicolor. Bull. Misc. Inform. Kew,* 1934:409–424 (1935).

Johnson, H. W., and C. W. Edgerton: A Heart Rot of Magnolia Caused by *Fomes geotropus. Mycologia,* 28:292–295 (1936).

Kauffman, C. H., and H. M. Kerber: A Study of the White Heart-rot of Locust, Caused by *Trametes robiniophila. Am. J. Botany,* 9:493–508 (1922).

Kimmey, J. W.: The Heart Rots of Redwood. *U.S. Dept. Agr., Forest Serv., Forest Pest Leaflet,* 25:1–7 (1958).

Learn, C. D.: Studies on *Pleurotus ostreatus* Jacq. and *Pleurotus ulmarius* Bull. *Ann. Mycol.,* 10:542–556 (1912).

Ledeboer, M. S. J.: *Schizophyllum commune* as a Wound Parasite: A Warning to Wattle Growers. *J. S. African Forestry Assoc.,* 13:39–40 (1946).

Lohwag, K.: "Erkenne und Bekämpfe den Hausschwamm und seine Begleiter." Vienna and Munich: Verlag Georg Fromme Co. (1955).

Long, W. H.: Three Undescribed Heart-rots of Hardwood Trees, Especially of Oak. *J. Agr. Research,* 1:109–128 (1913).

Long, W. H.: A Honeycomb Heart-rot of Oaks Caused by *Stereum subpileatum. J. Agr. Research,* 5:421–428 (1915).

McCallum, A. W.: Studies in Forest Pathology I. Decay in Balsam Fir (*Abies balsamea,* Mill.). *Can. Dept. Agr. Bull.,* 104 (n.s.):1–25 (1928).

Macdonald, J. A.: A Study of *Polyporus betulinus* (Bull.) Fries. *Ann. Appl. Biol.,* 24:289–310 (1937).

Macdonald, J. A.: *Coniophora puteana* (Schum.) Karst. on Living *Sequoia gigantea. Ann. Appl. Biol.,* 26:83–86 (1939).

Martínez, J. B.: El Agárico blanco ('Ungulina officinales' Pat.) en los Cedrales del Marruecos espanol. [English summary.] *Anales Madrid jard. botan,* 4:75–115 (1944).

Mounce, Irene: Studies in Forest Pathology II. The Biology of *Fomes pinicola* (Sw.) Cooke. *Can. Dept. Agr. Bull.,* 111 (n.s.):1–75 (1929).

Nobles, Mildred K.: A Contribution toward a Clarification of the *Trametes serialis* Complex. *Can. J. Research C. Botan. Sci.,* 21:211–234 (1943).

Nobles, Mildred K.: Studies in Wood-inhabiting Hymenomycetes. I. *Odontia bicolor. Can. J. Botany,* 31:745–749 (1953).

Nobles, Mildred K.: Studies in Wood-inhabiting Hymenomycetes III. *Stereum pini* and Species of *Peniophora* Sect. Coloratae on Conifers in Canada. *Can. J. Botany,* 34:104–130 (1956).

Nutman, F. J.: Studies of Wood-destroying Fungi I. *Polyporus hispidus* (Fries). *Ann. Appl. Biol.,* 16:40–64 (1929).

Overholts, L. O.: Notes on Fungi from the Lower Mississippi Valley. *Bull. Torrey Botan. Club,* 65:167–180 (1938).

Owens, C. E.: Studies on the Wood-rotting Fungus *Fomes pini.* I. Variations in Morphology and Growth Habit. II. Cultural Characteristics. *Am. J. Botany,* 23:144–149, 235–248 (1936).

Pirone, P. P.: *Ganoderma lucidum,* a Parasite of Shade Trees. *Bull. Torrey Botan. Club,* 84:424–428 (1957).

Putterill, V. A.: The Biology of *Schizophyllum commune* Fries with Special Reference to Its Parasitism. *S. Africa Dept. Agr., Sci. Bull.,* 25:1–35 (1922).

Rhoads, A. S.: The Biology of *Polyporus pargamenus* Fries. *N.Y. State Coll. Forestry Syracuse Univ., Publ.,* 11:1–197 (1918).

Riley, C. G.: Heart Rot of Oaks Caused by *Polyporus obtusus. Can. J. Research. C. Botan. Sci.,* 25:181–184 (1947).

Riley, C. G.: Studies in Forest Pathology. IX. *Fomes igniarius* Decay of Poplar. *Can. J. Botany,* 30:710–734 (1952).

Schmitz, H.: Notes on Wood Decay—I. The Wood Destroying Properties of *Polyporus volvatus. J. Forestry,* 21:502–503 (1923).

Schmitz, H., and L. W. R. Jackson: Heartrot of Aspen with Special Reference to Forest Management in Minnesota. *Minn. Agr. Expt. Sta. Tech. Bull.,* 50:1–43 (1927).

Schrenk, H. von: Two diseases of Red Cedar Caused by *Polyporus juniperinus* n. sp. and *Polyporus carneus* Nees. *U.S. Dept. Agr., Div. Veg. Physiol. and Pathol. Bull.,* 21:1–22 (1900).

Schrenk, H. von: A Disease of the Black Locust (*Robinia pseudacacia* L.). *Missouri Botan. Garden Ann. Rept.,* 12:21–31 (1901).

Schrenk, H. von: A Disease of White Ash Caused by *Polyporus fraxinophilus. U.S. Dept. Agr., Bur. Plant Ind. Bull.,* 32:1–20 (1903*a*).

Schrenk, H. von: The "bluing" and the "Red Rot" of the Western Yellow Pine, with Special Reference to the Black Hills Forest Preserve. *U.S. Dept. Agr., Bur. Plant Ind. Bull.,* 36:1–40 (1903*b*).

Schrenk, H. von: Two Trunk Diseases of the Mesquite. *Ann. Missouri Botan. Garden,* 1:243–252 (1914).

Schrenk, H. von, and P. Spaulding: Diseases of Deciduous Forest Trees. *U.S. Dept. Agr., Bur. Plant Ind. Bull.* 149:1–85 (1909).

Spaulding, P.: Spongy White Rot of Hardwoods. *Mass. Forest and Park Assoc., Tree Pest Leaflet,* 38:[1–4] (1939).

Spaulding, P.: White Trunk Rot of Hardwoods (Caused by *Fomes igniarius*). *Soc. Am. Foresters, New England Sect., Tree Pest Leaflet,* 20 (rev.):[1–2] (1951).

Stewart, D. M.: Heart Rot of Black Ash in Minnesota. *Phytopathology,* 41:569–570 (1951).

Thomas, G. P.: Studies in Forest Pathology XVIII. The Occurrence of the Indian Paint Fungus, *Echinodontium tinctorium* E. & E., in British Columbia. *Can. Dept. Agr. Publ.,* 1041:1–30 (1958).

Thomas, W. S.: The Beefsteak Mushroom. *J. N.Y. Botan. Garden,* 38:201–207 (1937).

Toole, E. R.: *Polyporus hispidus* on Southern Bottomland Oaks. *Phytopathology,* 45:177–180 (1955).

True, R. P., E. H. Tryon, and J. F. King: Cankers and Decays of Birch Associated with Two Poria Species. *J. Forestry,* 53:412–415 (1955).

Verrall, A. F.: Variation in *Fomes igniarius* (L.) Gill. *Minn. Agr. Expt. Sta. Tech. Bull.,* 117:1–41 (1937).

Wagener, W. W.: *Lentinus lepideus* Fr.: A Cause of Heart Rot of Living Pines. *Phytopathology,* 19:705–712 (1929).

Wagener, W. W., and R. V. Bega: Heart Rot of Incense-cedar. *U.S. Dept. Agr., Forest Serv., Forest Pest Leaflet,* 30:1–7 (1958).

Ward, H. M.: On the Biology of *Stereum hirsutum* (Fr.). *Phil. Trans. Roy. Soc. London, Ser. B.,* 189:123–134 (1898).

Wean, R. E.: The Parasitism of *Polyporus schweinitzii* on Seedling *Pinus strobus. Phytopathology,* 27:1124–1142 (1937).

Weir, J. R.: Some Observations on *Polyporus berkeleyi. Phytopathology,* 3:101–103 (1913).

Weir, J. R.: Two New Wood-destroying Fungi. *J. Agr. Research,* 2:163–165 (1914).

Weir, J. R.: The Genus *Polystictus* and Decay of Living Trees. *Phytopathology,* 13:184–186 (1923).

Weir, J. R.: Butt Rot in *Diospyrus virginiana* Caused by *Polyporus spraguei. Phytopathology,* 17:339–340 (1927).

Weir, J. R., and E. E. Hubert: A Study of Heart-rot in Western Hemlock. *U.S. Dept. Agr. Bull.,* 722:1–37 (1918).

West, E.: An Undescribed Timber Decay of Hemlock. *Mycologia,* 11:262–266 (1919).

White, J. H.: On the Biology of *Fomes applanatus* (Pers.) Wallr. *Trans. Roy. Can. Inst.,* 12:133–174 (1919).

White, L. T.: Studies of Canadian Thelephoraceae VIII. *Corticium galactinum* (Fr.) Burt. *Can. J. Botany,* 29:279–296 (1951).

Wilkins, W. H.: Studies in the Genus *Ustulina* with Special Reference to Parasitism VI. A Brief Account of Heart Rot of Beech (*Fagus sylvatica* L.) Caused by *Ustulina. Brit. Mycol. Soc. Tans.,* 26:169–170 (1943).

CHAPTER 18

Deterioration of Dead Timber

Entire timber stands are often killed by various agencies. Once trees are dead, the sapwood of all species and the heartwood of most deteriorate rapidly so that within a few years the material does not pay the cost of salvage. The sapwood is first stained by fungi and damaged by boring insects and season checks. Stained sapwood always lowers the lumber grade, and if insect infestation is severe enough, the sapwood may be unmerchantable. Next the sapwood is attacked by wood-destroying fungi, and when it is decayed, the heartwood is invaded by either the same or different fungi. Destruction of the heartwood progresses more slowly but is nevertheless certain, except that heartwood of cedars and junipers will remain sound for many years, sometimes indefinitely. The heartwood of down Douglas fir in the moist forests of the Pacific Northwest also resists decay for a long time. Although blight-killed chestnut deteriorated slowly, by 1957 it could no longer be salvaged profitably. The younger the timber, the more rapidly it deteriorates because of its smaller size and the greater proportion of sapwood to heartwood. Preventing deterioration of dead timber is impossible from the practical standpoint, and, to avoid loss, the timber must be utilized as quickly as possible after death. Delay results in rapidly increasing cull until finally the timber is worthless. Sound wood from dead trees is just as durable and as strong when in use as sound wood from living trees.

INSECT-KILLED TIMBER

Extensive killing of timber by bark beetles and defoliators occurs periodically. Although hardwoods in the East have suffered somewhat from gypsy and brown-tail moths, it is in coniferous forests that the most serious depredations have occurred. Killing of pines in the West by *Dendroctonus* spp. takes place sporadically over extensive areas, lodgepole, western white, and ponderosa pines especially being subject to recurring epidemics.

Ponderosa pine killed by bark beetles, or felled to control such outbreaks, after two seasons of exposure was no longer salvable (Boyce 1923; Patterson 1930; Von Schrenk 1903). By the end of the first season blue stain was well developed throughout the sapwood, resulting in degrade of the lumber. Bark-beetle-killed Pacific silver fir in western Washington deteriorated annually 5 per cent or more of the total cubic-foot volume up to 4 years after death, and pulp from trees dead more than 2 years was decreased in quantity and quality (Wright, Coulter, and Gruenfeld 1956). Engelmann spruce at high elevations killed by bark beetles remained sound for at least 20 years, and losses were gradual over a considerably longer period because low temperatures delayed decay for about 3 years until the wood became too dry to rot (Mielke 1950).

Balsam fir is sometimes extensively killed by the spruce budworm *Christoneura fumiferana.* The dead timber deteriorates rapidly, five fungi being consistently associated with the deterioration in a more or less regular succession pattern (Basham 1959). Although timber salvaged within 1 year after death will produce a satisfactory yield of pulp both in quantity and in quality, timber dead 3 years or longer is not salvable (Basham 1951). Western hemlock killed by the looper (*Lambdina fiscillaris lugubrosa*) also deteriorates rapidly so that sawlog volume is reduced beyond the point of economic recovery by the fifth year (Engelhardt 1957).

Spruce stands in northern New England and eastern Canada are often killed by spruce budworm, sawfly (*Neodiprion*), or eastern spruce beetle (*Dendroctonus piceaperda*). Spruces do not deteriorate so rapidly as balsam fir. The cause of death is an important factor in the rate of deterioration from sap rot (Riley and Skolko 1942). On the Gaspé Peninsula in Quebec it was determined that the loss in merchantable cubic-foot volume for trees killed by beetle was nil in 1 to 2 years after death, 1.1 per cent in 2 to 3 years, 3.9 per cent in 5 to 6 years, and 8.7 per cent in 7 to 8 years. The figures for the same periods for sawfly-killed trees were 2.4, 5.7, 43.6, and 48.3 per cent respectively. Trees killed by beetle had loose bark much of which was subsequently removed by woodpeckers, consequently the wood dried and decay was retarded. On the other hand, the bark remained relatively intact on trees killed by sawfly, the wood stayed moist, and decay was accelerated. In sawfly-killed trees the rate of decay lessened after 5 to 6 years because much of the bark was then removed by woodpeckers. Red and blue stain of sapwood was prevalent during the first season. Sap rot was largely caused by *Fomes pinicola* and

Polyporus abietinus. In northern Maine spruce killed by beetle deteriorated at about the same rate, and trees were still salvable even a dozen years after death (Cary 1917).

FIRE-KILLED TIMBER

The rate of deterioration of fire-killed stands of Douglas fir in western Oregon and Washington depends on the growth type of the timber, i.e., young growth (red fir) from 60 to 250 years old and from 10 to 50 inches dbh, intermediate growth (bastard fir) from 200 to 400 years old and from 30 to 70 inches dbh, and old growth (yellow fir) mostly over 400 years old and from 40 to 100 inches dbh or more (Kimmey and Furniss 1943). Salvage operations have been carried on for about 1 to 2 years after a fire in young growth, for 4 to 7 years in intermediate growth, and for 5 to 10 years or more in old growth. Young growth averaged about 50 per cent deterioration in 3 to 4 years, intermediate growth 50 per cent in 10 to 15 years, and old growth 50 per cent in 15 to 20 years. Old-growth trees of average size did not become completely deteriorated until 60 years or more after death. The actual loss by direct burning is seldom more than 3 per cent of the gross volume. Insect infestation and partial blue-staining of the sapwood occur the first year after a fire. By the end of the third year the sapwood is no longer usable because of insect borings and decay by fungi, principally *Fomes pinicola* and *Polyporus abietinus.* Deterioration of the heartwood then proceeds at a slower rate. Heartwood is decayed mainly by *F. pinicola,* and somewhat by *F. laricis* and *Lenzites saepiaria.* Loosening of the bark begins about the third year, and from the sixth to tenth years it falls off in masses. Fuel value of the wood is not affected by fire killing at least for many years, nor is strength, provided the material is free from insect borings, decay, and checks. Fast-grown heartwood with wide growth rings is much less resistant to insects or fungi than slow-grown narrow-ringed dense material. Fire-killed western hemlock and balsam firs deteriorate much more rapidly than Douglas fir, usually being unsalvable 5 years after death regardless of size or age. Most of the heartwood of western red cedar remains sound in standing dead trees for 65 years or more. In California, large white fir is salvable for about 3 to 4 years, ponderosa and Jeffrey pines for about 5 years, sugar pine for approximately 10 years, and Douglas fir from 15 to 20 years (Kimmey 1955).

However, in most regions, because of the smaller size of the timber, fire-killed stands must be cut within a year or two. In northern Idaho fire-killed western white pine logged more than 2 years after death is seldom handled at a profit (Bradner and Anderson 1930). In Quebec the loss from sap rot in fire-killed pulpwood stands of spruce, balsam fir, and jack pine where the burns were severe did not exceed 3.5 per cent of the merchantable volume in the first 5 years after death (Skolko 1947). However, where the burns were light or moderate, the losses in spruce and balsam fir amounted to from 12 to 28 per cent in 2 years. The sapwood of jack pine became heavily stained the first year. In Ontario after 2 years considerable deterioration occurred in fire-killed jack, red, and eastern white pines (Basham 1957, 1958). The faster deterioration than in Quebec may possibly be explained by the fact that the Ontario fires occurred in the spring and early summer and the Quebec fires in the late summer and fall, or that conditions for the 5 years following the fires were moister in Ontario than in Quebec.

WIND-THROWN TIMBER

There are various factors that make trees liable to wind throw or wind fall, the most important being deep butt scars, wet soil, shallow soil, opening up of stands by cutting, shallow root system, shape and height of the tree, heavy snow, sleet or ice, and decay of the roots and trunk. Hardwoods are not often wind thrown, but conifers suffer severely, especially shallow-rooted species such as balsam fir. Fungi are important in two ways: (1) they predispose trees to wind fall, and (2) they decay the trees after they are down, causing heavy losses unless the timber can be salvaged quickly.

Root and butt rots are most serious in predisposing trees to wind fall, but trunk rots also play a part, and it is not unusual to find decayed trees broken off at considerable heights above the ground. Fungi predisposing trees to wind fall are particularly serious on cutover areas where part of the stand is reserved for seed trees or to form the nucleus of the next cut. The opening up of the stand makes the reserved trees liable to wind fall, and if in addition they are decayed, the chances of loss are greatly increased. Consequently, reserved trees on cutover areas should be free from decay, particularly root and butt rots, if they must last for some years. Ordinarily the wood-destroying fungi in the heartwood of blown-down trees are of little consequence in subsequent deterioration, other fungi causing the major loss.

Complete or almost complete wind throw over extensive areas has occurred occasionally in western Oregon and Washington, affecting western hemlock (Fig. 205), Pacific silver fir, Sitka spruce, Douglas fir, and western red cedar, with individual trees ranging up to 117 inches dbh and with the average for the several species varying from 23 to 47 inches (Buchanan and Englerth 1940; Childs and Clark 1953). Trees with butt rot were broken off a few feet above ground level, and sound trees were uprooted. For the first three seasons of exposure most of the loss was occasioned by blue stain and ambrosia-beetle borings in the sapwood. A little decay of the sapwood appeared the second season, from the fourth season on sap rot was severe, and by the end of the sixth season the sapwood of all species had been practically destroyed. In addition, considerable heartwood of western hemlock, Pacific silver fir, and Sitka spruce had been decayed.

Fig. 205. A stand of second-growth western hemlock composed of large poles and standards wind thrown on the Olympic Peninsula in 1921. This timber, containing a large percentage of sapwood, deteriorated rapidly.

Western hemlock and Pacific silver fir, irrespective of size, were destroyed by decay within 8 years after being blown down and Sitka spruce within 15 years. Douglas fir heartwood decays rather slowly, so that, 15 years after being blown down, trees over 30 inches dbh still contained considerable merchantable volume. Western red cedar heartwood is highly decay resistant, so that, after 15 years, decay loss in this species did not exceed the original volume of sapwood, and apparently the heartwood would remain sound for many years. Wind-thrown trees in the spruce-hemlock and Douglas fir old-growth types of the Washington and Oregon coast must be salvaged within 2 years after a blowdown to avoid any loss from decay. *Fomes pinicola* and *F. applanatus* caused most of the decay in all species, except western red cedar, the sapwood of which was rotted by *Polyporus cuneatus* (Buchanan 1940). The loss in pulpwood of balsam fir 3 to 4 years after wind throw in Newfoundland was less than 1 per cent (Davidson and Newell 1953).

Once timber is wind thrown, deterioration by sap stain, checking, boring insects, and decay begins. The smaller the tree the more rapid the damage, making quick salvage essential, since the only practical way to prevent loss is by immediate logging. Wind fall in conifers may result in infestation of the standing living timber on and around the wind-fall area by the large bark-beetle population developing in the down trees. Although such epidemics usually subside quickly after reaching their peak, yet considerable standing timber may be killed. This additional damage usually can be prevented if logging is completed within two seasons after wind fall occurs. Little is known about the deterioration of wind-thrown hardwoods because they are rarely subject to extensive wind fall.

LOGS

Logs are usually removed from the logging area promptly, but occasionally it may be necessary to leave them on the ground in the woods or store them at the mill for several weeks or even months before they can be converted, thus subjecting them to the risk of stain and decay of the sapwood. If logs can be stored continuously in water, no loss will occur, but adequate water storage is rarely available. In logging southern hardwoods it is commonly necessary to store logs for some time. The warm, humid climate of the region favors rapid deterioration, particularly of red gum, which may begin to stain after 1 week and to decay after 3 weeks. Consequently, chemical treatment within 24 hours after cutting is necessary (Anonymous 1949).

Stain and decay in logs can be retarded by spraying the ends and debarked areas of the sides with 2 per cent pentachlorophenol in No. 2 fuel oil solution or any standard lumber dip (page 508) in three times normal concentration in water. For insect, stain, and decay control use No. 2 fuel oil solution containing about 2 per cent pentachlorophenol and 0.5 per cent gamma isomer of benzene hexachloride. The logs must be completely covered with the solution. These chemicals can be applied with an ordinary hand-operated garden sprayer, but a power sprayer is necessary for logs in piles. To prevent end checking, any good moisture-resistant coating can be put on after spraying.

SLASH

The disposal of slash after logging operations is always a problem in those regions where the demand for wood is not sufficient to make possible the utilization of slash and where the slash is a serious fire hazard or does not permit adequate restocking. The usual method of disposal on public lands in the United States has been piling and burning or occasionally broadcast burning, but there has been a tendency to overdo slash burning, applying the method arbitrarily to types where its value was doubtful or where it was actually harmful. Slash burning robs the soil of organic matter and usually damages advance reproduction to some degree. Wherever slash disposal by burning is not essential, the method of disposal in relation to the rate of decay becomes important. The method used should be one that will permit adequate restocking and at the same time return the slash as organic matter to the soil in the shortest time, thus reducing fire hazard.

From the standpoint of forest sanitation only, burning is the best method of slash disposal to reduce the incidence of wood-destroying fungi, provided cull logs and tops are destroyed or charred as well as branches. Stumps should also be charred since they support many wood-rotting fungi (Käärik and Rennerfelt 1957). Fungi attacking living trees rarely develop on charred wood. Some fungi causing extensive decay in living trees will persist and develop conks on the cull logs and tops after the decayed trees have been girdled or felled, thus being a source of infection to the reserve trees on the cutover area and to the surrounding stand (Hubert 1920; Spaulding 1934; Wright 1934; Gill and Andrews 1956). Others, such as *Stereum sanguinolentum* causing red heart rot in balsam fir and *Polyporus anceps* causing red rot in ponderosa pine, develop conks almost exclusively on slash from which living trees are infected. In ponderosa pine forests of the Southwest it is the practice to pile brush with the aim of charring large limbs, tops, and cull logs in contact with the ground, thus preventing fruiting of *P. anceps*. However, the usual method of piling and burning slash does not include cull logs and tops, the material on which most of the wood destroyers fruit, whereas broadcast burning which does destroy or char the larger slash is so destructive, except in a few forest types, that the damage resulting must far outweigh any benefit from the reduction in conks of wood-destroying fungi. It is unlikely that in most forest types the infection of living trees by fungi from slash will be serious enough to warrant the increased expense of disposing of large cull material on cutover areas, since generally the fungi causing most of the decay in living trees are not abundant even on large slash. However, it is advisable to leave large cull logs or down trees as much as possible in the open to facilitate drying out and check sporophore production. It has been concluded that the sporadic and brief insect epidemics which occasionally arise from slash in the United States do not cause enough damage to warrant special methods of slash disposal (Craighead et al. 1927).

Ponderosa Pine. Slash disposal in ponderosa pine forests has been fully discussed elsewhere, showing that no one method of disposal will meet all conditions in a given stand, so that diversified disposal comprising piling and burning, scattering, and leaving slash as it falls will all be necessary (Munger and Westveld 1931; Pearson and McIntyre 1935). Slash should not be piled and left unburned because it will probably require 40 or 50 years for the piles to rot completely (Long 1915). Lopped slash (*i.e.*, with branches severed from tops and cull logs) decays more slowly than unlopped, since lopping causes slash to dry out sooner. Ponderosa pine slash is largely decayed by *Polyporus anceps, P. abietinus,* and *Lenzites saepiaria,* the first causing a wet, soggy, rapid white rot of the sapwood which will not burn unless dry and then is quickly consumed without producing living cinders (Gill and Andrews 1956). *P. abietinus* decay has similar burning qualities. *L. saepiaria* causes a dry, brittle, slow, brown charcoallike rot of the sapwood that is dangerous during fire, because it produces living cinders that can be carried long distances by wind. *P. anceps* invades cull logs and tops down to 4 or 5 inches in diameter, but severed branches of even larger diameter are not attacked by it. However, if still attached to the trunk, even smaller branches are readily decayed for some distance out by *P. anceps.* Furthermore, cull logs and tops from which the branches have been cut are decayed more slowly by *P. anceps* because it does not decay sapwood from which the bark has been removed or become loosened. Consequently, if the bark is loosened by a bark-beetle infestation, decay is considerably delayed. *P. anceps* disintegrates ponderosa pine slash in from 8 to 15 years, depending on moisture and temperature conditions, but *L. saepiaria* requires from 20 to 30 years. Slash of Douglas fir, lodgepole pine, western larch, and white fir when associated with ponderosa pine deteriorates at a slower rate because it is largely decayed by wood destroyers other than *P. anceps.*

Shortleaf Pine. Investigations in Arkansas demonstrated that this slash when piled decays slowly, requiring 3 to 6 years longer to disintegrate than when lopped and scattered or left unlopped (Long 1917; Siggers 1935). When slash piles are burned, unfavorable soil conditions that greatly delay seedling establishment over 10 to 15 per cent of the cutover area are created beneath the piles. There is no significant advantage in lopping and scattering, except for pulling tops away from advance growth or isolated seed trees, since slash left as it has fallen decays rapidly enough so that the additional fire hazard persists only up to the third or fourth year after logging. Most of the decay is caused by *Lenzites saepiaria* and *Polyporus abietinus*. The rotting of oak slash in Arkansas followed the same course in the same time as shortleaf pine slash; *Stereum* spp. causing the decay (Long 1917).

Eastern White Pine in the Northeast. This slash is decayed by many fungi (Spaulding and Hansbrough 1944). The slash is usually left in irregular piles or windrows of varying compactness depending on the density of the stand cut. Under average conditions, most of the slash is well rotted about 17 years after cutting. Under the most favorable conditions for decay, this period may be shortened by about 3 years. On the other hand, unfavorable conditions, i.e., too wet or too dry situations, will retard decay for years. On wet ground, slash should be left as little as possible in contact with the soil to prevent waterlogging. On very dry ground, slash should be piled compactly in order to prevent rapid drying, which also retards decay, but not so much as waterlogging. If piling is too costly, lopping will be of some value. On medium situations the slash can be left more or less as it falls, but with as much contact with the soil as possible.

Eastern Hemlock in the Northeast. This slash decays in the same general way but a little more rapidly than eastern white pine slash.

Hardwoods in the Northeast. This slash decays rapidly so that in about 5 years the abnormal fire hazard has passed. Consequently burning is rarely justified. The slash is disintegrated about 15 years after cutting (Spaulding and Hansbrough 1944). The slash can best be left as it has fallen, since the cost of lopping is scarcely warranted by the increased rate of decay in certain species only, unless such species comprise a major portion of the slash. Furthermore lopping increases the possibility of smothering seedlings. The effect of lopping varies somewhat in relation to the durability of the wood, the less durable species showing the least difference between lopped and unlopped slash because the wood rots quickly in any case. Aspen, paper birch, and poplars are in this category. Yellow birch slash waterlogs so readily under usual conditions that lopping is of no benefit and on wet soil is harmful. Waterlogging retards decay. Basswood and beech waterlog often enough so that lopping is of questionable value. Chestnut, white oak, and white ash will hardly waterlog under any conditions, and decay will be somewhat hastened by lopping.

Balsam Fir and Red Spruce in the Northeast. This slash decays slowly, constituting an abnormal fire hazard for 8 to 12 years, requiring

21 years or more to decompose completely, crushing and killing advance growth and preventing the establishment of new reproduction for 15 to 20 years (Westveld 1931:50). Consequently dense slash should be burned as logging progresses. Beech, birches, and maples are commonly associated with these conifers, but the hardwood slash decays two to three times as fast. Abnormal fire hazard is gone at the end of 3 or 4 years, and in 7 or 8 years the hardwood slash disappears. Consequently it is unnecessary to burn it.

Hardwoods and Eastern Hemlock in the Lake States. On clear-cut areas where the bulk of the slash is left in large piles distributed aimlessly over the area with scattered branches between the piles, hardwood-hemlock slash is an extra-acute fire hazard for at least 8 years, jack pine slash for 7 years, and other coniferous slash for from 12 to 15 years (Zon and Cunningham 1931:17). By the end of the eighth year in hardwood-hemlock slash, scattered branches are well disintegrated and piles have decayed and settled, and in 15 or 16 years the piles are thoroughly rotted and broken down, so that the fire hazard is completely gone. If there is no hemlock in the slash, decay is more rapid. Slash deteriorates much more rapidly on selectively cut areas because there is not so much of it, the slash is more scattered and consequently decays more rapidly, and the reserve stands afford protection against drying. On such cuttings where the wood is utilized down to 3 inches in diameter and no hemlock is present, the additional fire hazard persists for only 2 or 3 years, and in 4 years the slash is well decayed. Even when hemlock is mixed with hardwoods, the slash deteriorates sooner on selectively cut than on clear-cut areas. Whether lopping might hasten decay has not been determined.

Sitka Spruce, Western Hemlock, and Douglas Fir. West of the summit of the Cascade Mountains in Oregon and Washington the two principal forest types are the spruce-hemlock and the Douglas fir. The usual method of slash disposal is broadcast burning. Fire hazard on unburned slashings is exceedingly high for the first few years after logging. In both types disintegration of needles and twigs required 10 to 12 years after logging, although on very dry sites in the Douglas fir type the majority of needles and twigs remained intact longer (Childs 1939). One-third to one-half of the sapwood of tops and cull logs on unburned slashings in both types was usually decayed within 5 years after logging, but from then on decay progressed more slowly. The decay of slash was too slow to be of value in fire control during the period of high hazard, because decay did not cause a marked reduction in the inflammability of slash until after coniferous reproduction or brush became high enough to shade the slash sufficiently to prevent drying. Although a few of the slash-inhabiting fungi also attack living trees, conks were rare on slash old enough for the reproduction to have become liable to decay, so that slash is not likely to cause an appreciable increase in decay in succeeding stands.

REFERENCES

Anonymous: Prevention of Deterioration in Stored Southern Hardwood Logs. *U.S. Dept. Agr., Forest Serv., Southern Forest Expt. Sta.* (1949). (Mimeographed.)

Basham, J. T.: The Pathological Deterioration of Balsam Fir Killed by the Spruce Budworm. *Pulp Paper Mag. Can.,* 52(9):120–122, 125–126, 128, 130, 133–134 (1951).

Basham, J. T.: The Deterioration by Fungi of Jack, Red, and White Pine Killed by Fire in Ontario. *Can. J. Botany,* 35:155–172 (1957).

Basham, J. T.: Studies in Forest Pathology XVII. The Pathological Deterioration of Fire-killed Pine in the Mississagi Region of Ontario. *Can. Dept. Agr. Publ.,* 1022:1–38 (1958).

Basham, J. T.: Studies in Forest Pathology XX. Investigations of the Pathological Deterioration in Killed Balsam Fir. *Can. J. Botany,* 37:291–326 (1959).

Boyce, J. S.: The Deterioration of Felled Western Yellow Pine on Insect-control Projects. *U.S. Dept. Agr. Bull.,* 1140:1–7 (1923).

Bradner, M., and I. V. Anderson: Fire-damaged Logs—What Is the Loss? *Timberman,* 31(7):38–44 (1930).

Buchanan, T. S.: Fungi Causing Decay in Wind-thrown Northwest Conifers. *J. Forestry,* 38:276–281 (1940).

Buchanan, T. S., and G. H. Englerth. Decay and Other Volume Losses in Windthrown Timber on the Olympic Peninsula, Wash. *U.S. Dept. Agr. Tech. Bull.,* 733:1–29 (1940).

Cary, A.: Effect of Insect Attack on Spruce Timber. *Am. Lumberman,* 2201:44 (1917).

Childs, T. W.: Decay of Slash on Clear-cut Areas in the Douglas Fir Region. *J. Forestry,* 37:955–959 (1939).

Childs, T. W., and J. W. Clark: Decay of Windthrown Timber in Western Washington and Northwestern Oregon. *U.S. Dept. Agr., Div. Forest Pathol., Spec. Release,* 40:1–20 (1953).

Craighead, F. C., et al.: The Relation of Insects to Slash Disposal. *U.S. Dept. Agr. Circ.,* 411:1–12 (1927).

Davidson, A. G., and W. R. Newell: Pathological Deterioration in Wind-thrown Balsam Fir in Newfoundland. *Forestry Chronicle,* 29:100–107 (1953).

Engelhardt, N. T.: Pathological Deterioration of Looper-killed Western Hemlock on Southern Vancouver Island. *Forest Sci.,* 3:125–136 (1957).

Gill, L. S., and S. R. Andrews: Decay of Ponderosa Pine Slash in the Southwest. *Rocky Mt. Forest and Range Expt. Sta. Research Notes,* 19:1–2 (1956).

Hepting, G. H., and E. R. Roth: The Fruiting of Heart-rot Fungi on Felled Trees. *J. Forestry,* 48:332–333 (1950).

Hubert, E. E.: The Disposal of Infected Slash on Timber-sale Areas in the Northwest. *J. Forestry,* 18:34–56 (1920).

Käärik, A., and E. Rennerfelt: Investigations on the Fungal Flora of Spruce and Pine Stumps. *Statens Skogsforskningsinst.* (*Sweden*), *Medd.,* 47(7):1–88 (1957).

Kimmey, J. W.: Rate of Deterioration of Fire-killed Timber. *U.S. Dept. Agr. Circ.*, 962:1–22 (1955).

Kimmey, J. W., and R. L. Furniss: Deterioration of Fire-killed Douglas-fir. *U.S. Dept. Agr. Tech. Bull.*, 851:1–61 (1943).

Long, W. H.: A New Aspect of Brush Disposal in Arizona and New Mexico. *Soc. Am. Foresters Proc.*, 10:383–398 (1915).

Long, W. H.: Investigations of the Rotting of Slash in Arkansas. *U.S. Dept. Agr. Bull.*, 496:1–14 (1917).

Mielke, J. L.: Rate of Deterioration of Beetle-killed Engelmann Spruce. *J. Forestry*, 48:882–888 (1950).

Munger, T. T., and R. H. Westveld: Slash Disposal in the Western Yellow Pine Forests of Oregon and Washington. *U.S. Dept. Agr. Tech. Bull.*, 259:1–57 (1931).

Patterson, J. E.: Milling Beetle-killed Pine Timber. *Timberman*, 31(3): 162–172 (1930).

Pearson, G. A., and A. C. McIntyre: Slash Disposal in Ponderosa Pine Forests of the Southwest. *U.S. Dept. Agr. Circ.*, 357:1–28 (1935).

Riley, C. G., and A. J. Skolko: Rate of Deterioration in Spruce Killed by the European Sawfly. *Pulp Paper Mag. Can.*, 43:521–524 (1942).

Schrenk, H. von: The "Bluing" and the "Red Rot" of the Western Yellow Pine, with Special Reference to the Black Hills Forest Preserve. *U.S. Dept. Agr., Bur. Plant Ind. Bull.*, 36:1–35 (1903).

Siggers, P. V.: Slash-disposal Methods in Logging Shortleaf Pine. *Southern Forest Expt. Sta. Occas. Papers*, 42:1–5 (1935).

Skolko, A. J.: Deterioration of Fire-killed Pulpwood Stands in Eastern Canada. *Forestry Chronicle*, 23:128–144 (1947).

Spaulding, P.: Persistence of Heart-rotting Fungi in Girdled Trees. [Abstract.] *Phytopathology*, 24:17–18 (1934).

Spaulding, P., and J. R. Hansbrough: Decay of Logging Slash in the Northeast. *U.S. Dept. Agr. Tech. Bull.*, 876:1–22 (1944).

Westveld, M.: Reproduction on Pulpwood Lands in the Northeast. *U.S. Dept. Agr. Tech. Bull.*, 223:1–52 (1931).

Wright, E.: Survival of Heart Rots in Down Timber in California. *J. Forestry*, 32:752–753 (1934).

Wright, E., W. K. Coulter, and J. J. Gruenfeld: Deterioration of Beetle-killed Pacific Silver Fir. *J. Forestry*, 54:322–325 (1956).

Zon, R., and R. N. Cunningham: Logging Slash and Forest Protection. *Univ. Wisconsin Agr. Expt. Sta. Research Bull.*, 109:1–36 (1931).

CHAPTER 19

Deterioration of Forest Products: Decay

The losses caused by wood-destroying fungi in wood after it has been converted for the many purposes for which it is used are enormous, but these losses can be greatly reduced by proper air seasoning, kiln drying, and preservative treatment (Peck 1956; Hunt and Garratt 1953). Damage to wood during seasoning, storage, and service has been a source of difficulty since time immemorial, being particularly serious in the days of wooden ships when the specifications of the marine departments of European nations reflected attempts to get the most durable wood for war vessels. The course of naval history has been influenced by decay, the difficulties encountered before Nelson's victory at Trafalgar being a fascinating story (Albion 1926). These losses, both direct and indirect, have led to the development of a specialized branch of forest pathology sometimes termed "products pathology" or "wood pathology." In this book it is impossible to do more than touch upon the important aspects of the field in such a way as to enable the teacher or practitioner to develop the subject in detail. General information on decay has been given in Chap. 16. Recent work on decay of forest products has been summarized by Findlay (1956).

RESISTANCE OF WOOD TO DECAY

Where decay may shorten its period of service, wood must be properly selected, cared for, and used, to avoid serious losses. The choice of species is important as is also the choice of wood of proper quality within a species. Fungi require a certain balance of moisture, air, and temperature in order to develop in wood. Fungi may remain dormant in wood for several years and then resume development when conditions become favorable. Sound wood or that in which fungi have been killed is constantly exposed to new infection. Some fungi decay wood quickly, others slowly, consequently, depending on the fungus at work as well as other conditions, the usefulness of a piece of wood can be destroyed in a few months, or it may take years. In addition to Basidiomycetes and some of the larger Ascomycetes, molds, such as species of *Chaetomium*, cause decay of wood in service known as soft rot (page 356).

Knowledge of the relative decay resistance of wood in service is essential for the proper selection of species for various uses. The best available grouping of woods according to decay resistance is the following one based on actual service records and miscellaneous information (Anonymous 1950). *High:* catalpas, cedars, chestnut, bald cypress, junipers, black locust, red mulberry, Osage orange, mesquite, redwood, black walnut, and Pacific yew. *Intermediate:* Douglas fir, honey locust, western larch, chestnut oak, white oak, eastern white pine, southern yellow pine, and sassafras. Honey locust is classed as intermediate but is nearly as decay resistant as some of the species in the highest group. *Low:* aspens, basswood, cottonwood, balsam firs, and willows. Ashes, beech, birches, hemlocks, sugar maple, northern red oak, and spruces are included in this group but may be nearly as high in decay resistance as those in the next higher group.

The foregoing rating must be understood as an approximate average only, from which specific cases may vary considerably, and as having application only where the wood is used under conditions favoring decay. Under such conditions the sapwood of all North American species will usually rot quickly, natural decay resistance being a property of the heartwood only. In a few species such as spruces and white firs the sapwood is often difficult to distinguish from the heartwood. In general both the heartwood and sapwood of such species are not decay resistant. Included or internal sapwood is not decay resistant (MacLean and Gardner 1958).

Service records in sufficient number enable an approximation to be made of the decay resistance of a species of wood in service, provided the decay-resistance figure is used in comparison with the same conditions as were present in the original service tests. The natural decay resistance of any species varies greatly according to the conditions under which it is exposed, and this fact and the number of years required are the principal difficulties in controlled outdoor service tests. This has led to testing the natural decay resistance of wood in sterile jars, flasks, or tubes under controlled laboratory conditions to get an approximation of the relative decay resistance of various species in from a few months to a year. Loss in weight is the criterion generally used, requiring minimum equipment and a simple procedure (McNabb 1958). Since the size and shape of the test blocks affect the rate of weight loss from decay, some standardization is necessary to compare results between different laboratories (Findlay 1953). Other criteria suggested as being faster have been various strength, chemical, and electrical tests; hygroscopicity; sound transmission; and dimensional change; but after an evaluation of all methods so far tried, Hartley (1958) concluded that the weight-loss method was still the most useful. Several fungi should be used in such tests because different species of wood react differently to different fungi.

Although laboratory tests are quicker, they are not true tests under actual service conditions; and although they give information on relative decay resistance, they do not tell absolute decay resistance. Because there is such a difference between the results of outdoor service tests and laboratory experiments, in order to avoid confusing the two types of data Hubert has suggested that the length of service of wood with respect to decay should be termed "decay durability" and the relative resistance of wood to decay under controlled laboratory conditions should be termed "decay resistance."

FACTORS IN THE RESISTANCE OF WOOD TO DECAY

There are various factors contributing to the decay resistance of wood in service, some of which are inherent in the wood; others are dependent on the conditions of use.

Chemicals. Sapwood is not resistant to decay even in those species in which heartwood is highly resistant, because the heartwood contains large amounts of extractives poisonous to fungi, which are formed during the change from sapwood to heartwood. The theory that the substances present in sapwood such as starch, sugars, and nitrogenous materials furnish food for wood-destroying fungi, thus encouraging their growth, is not a necessary explanation for the rapid decay of sapwood, for the difference in decay resistance of heartwood and sapwood of decay-resistant species is explained by the presence of substances in the heartwood that retard decay rather than to the presence of substances in the sapwood that promote it.

Among the decay-resistant woods, chestnut, oaks, black locust, and red mulberry contain tannin, whereas Osage orange and black walnut contain large quantities of soluble coloring matter. A toxic water extractive is present in redwood. Western red cedar heartwood contains the extractive thujaplicin, which is about as fungicidal as sodium pentachlorophenol (Rennerfelt 1948), and a water-soluble phenolic toxin (Roff and Atkinson 1954). Some cedars and southern cypress contain toxic oils, but others are dependent for their decay resistance on water-soluble extractives. In ponderosa pine, which is a relatively non-decay-resistant wood, there is little difference in the toxicity of the water extractives from sapwood and heartwood, and there is a loss of certain toxic volatile materials resulting from the temperatures used in kiln drying (Anderson 1931). Hot-water-soluble extractives from Douglas fir, black locust, and white oak in the order given were decreasingly toxic to two wood-decay fungi, whereas in service black locust is highly decay resistant, while white oak and Douglas fir are intermediate (Waterman 1946). Numerous constituents of coniferous heartwood have been tested for fungicidal activity (Rennerfelt 1956; Rennerfelt and Nacht 1955). Toxicity tests of extractives give only a general indication of the relative decay resistance of the wood naturally containing the toxins. For the species so far investigated (Sherrard and Kurth 1933; Cartwright 1941, 1942; Scheffer and Hopp 1949; Scheffer 1957; Zabel 1948), heartwood from different positions in the same tree, from different trees of the same species, from different varieties of a species, and from different localities but not from different regions is known to vary in decay resistance. European larch in Switzerland varies in decay resistance according to altitude, the more susceptible wood coming from trees at the lowest and highest elevations (Gäumann 1948). The variation in different parts of the same tree is explained by the uneven distribution of toxic extractives, decay resistance generally decreasing from the outer heartwood toward the center and from the butt toward the top. However, in white oaks the heartwood in the upper part of the trunk is most resistant. In the same species the central heartwood decays about twice as fast as the outer heartwood. The differences between individuals, varieties, and localities may depend upon differences in composition of the toxic extractives. In selecting the most decay-resistant wood the part of the tree from which it is taken must be considered. Certain woods give off volatile substances that stimulate the growth of some fungi (Findlay 1956:321).

Resin seems to be essentially nontoxic, but it is valuable as a waterproofing substance in wood, preventing penetration of wood-destroying fungi (Verrall 1938).

The more rapid decay of timber in rich than in poor soils may in part be explained by the greater supply of nitrogen in rich soils resulting in a greater fungous population, and by the infiltration of nitrogenous material into the timber making it less resistant to decay, since decay under laboratory conditions has been accelerated by first treating the wood with nitrogenous substances (Findlay 1934; Schmitz and Kaufert 1936; Hungate 1940). Furthermore, wood treated with urea to accelerate seasoning has proved more susceptible to decay than untreated wood in laboratory tests (Kaufert and Behr 1942). Wood impregnated with fire-retarding ammonium salts will probably resist decay better than untreated wood (Scheffer and Van Kleeck 1945).

Specific Gravity. Based on experience with wood in service it has long been accepted that, within a species, the higher the specific gravity or density of the wood, the greater is its decay durability. This has been determined as applicable to the heartwood of European larch (Björkman 1944). Garren (1939) found that the higher the specific gravity of loblolly pine sapwood, the more resistant it was to decay by *Polyporus abietinus.* However, Richards (1950) found no relation between decay resistance and specific gravity of sapwood from a single log each of loblolly pine and sweet gum. After investigating leached western red cedar and considering the results of others, Southam and Ehrlich (1943) concluded that, although for a single species of wood the higher the specific gravity the greater the initial decay resistance may be, this tendency is nullified or may even be reversed as decay proceeds, and thus is of little practical value. Thorough investigation is needed of woods naturally free from toxic substances to determine the relation between specific gravity and decay. Between species, specific gravity is not a measure of relative durability, since chemical constituents of the heartwood are such an important factor. Western red cedar with an average specific gravity of 0.33 is a very durable wood, but Sitka spruce with an average of 0.40 and western white pine with an average of 0.38 are not durable. Coast Douglas fir with an average of 0.48, although somewhat decay durable, does not compare with cedar in this respect. No correlation was found between density or ring width and decay in Scotch pine (Rennerfelt 1947), or in English oak (Jørgensen 1953). Since the relation between specific gravity and decay is still unsettled, where decay resistance must be considered in the use of wood it seems advisable to select material of better than average specific gravity if possible.

Water and Air. Wood must be moist before spores of wood-destroying fungi can germinate on it. Spores of *Lenzites saepiaria* on shortleaf pine germinated sparsely until the relative humidity of the air was above 95 per cent. This humidity was high enough for the wood to reach fiber saturation point so that free moisture appeared as a film on the wood surface (Zeller 1920:68). Furthermore there must be a proper balance between water and air in wood before it can decay, an amount of air equivalent to more than 20 per cent of the volume of a piece of wood being necessary (Snell 1929:545). Although wood can decay at a moisture content as low as 15 per cent based on oven-dry weight (Bavendamm and Reichelt 1938), wood must usually reach fiber saturation point, which varies from 25 to 32 per cent, before decay becomes severe (Snell, Howard, and Lamb 1925). An excess of water in wood also prevents decay. For those fungi tested by Snell on Sitka spruce wood, with a specific gravity of 0.34, it was found that the upper limit of optimum growth was at a moisture content of 150 per cent, whereas the inhibition point of decay was from 190 to 200 per cent. For southern yellow pine sapwood, with a specific gravity of 0.44, the figures were 100 and 145 to 150 per cent, respectively; for Douglas fir, of specific gravity 0.54, they were 70 and 110 to 120 per cent; and for southern yellow pine heartwood, of specific gravity 0.70, they were 50 and 75 to 80 per cent. With increasing specific gravity the moisture content at which decay is inhibited decreases. Since air is necessary for wood-destroying fungi, the more wood substance in a given volume of wood the less water it will hold before the air supply is reduced below the minimum requirement. Wood destroyers also differ in their moisture requirements (Ferdinandsen and Buchwald 1937; Theden 1941). The difficulties encountered in maintaining uniform moisture and air conditions throughout the necessarily long test period, when investigating moisture-decay relations in wood, have been discussed by Haasis (1932), and an apparatus for this purpose has been described by Hatfield (1931).

In this country nothing is known about the subsequent decay of timbers from wood that has been immersed in water during rafting, driving, or pond storage. Decay will not progress during immersion, but wood-destroying fungi in red pine sapwood have survived 38 weeks of water-soaking (Schmitz and Kaufert 1938). It will probably require a long time to kill hyphae in wood by immersion. In Germany it has been found that coniferous timbers after removal from the water are quickly overrun by mold fungi (*Trichoderma* spp.) that are not destructive to wood but which for some time inhibit or retard decay. Thus floated wood is more desirable for building construction than wood subjected to dry transportation, since those molds do not so actively attack the latter (Falck 1931). Furthermore floated wood before it dries out can be effectively treated by coating the surface with a water-soluble preservative such as sodium fluoride paste.

Decay can be prevented if wood is kept air dry, which is the usual condition for furniture and for building timbers not in contact with the ground or with some other continuous source of moisture. However, if the moisture content of dry wood is raised for any extensive period decay can occur. Water-soaked wood will not decay; consequently water storage for logs and timbers is effective.

Temperature. By means of tests with agar cultures it has been established that wood-destroying fungi differ in their temperature requirements for growth, the optimums varying from 20 to 36°C, whereas the maximum temperatures permitting growth varied from less than 6° to as much as 16° above the optimum (Cartwright and Findlay 1934; Humphrey and Siggers 1933). For example, *Merulius lacrymans* grew best at 20°C and *Lenzites saepiaria* at 36°C. Consequently temperature must be a strong factor in determining which fungi will be most active in decaying a piece of wood. The assumption that the temperature requirements of a fungus on wood can be predicted by tests on agar is probably sufficiently correct for practical purposes, yet in some cases it is not true, since the optimums for *Lentinus tigrinus, Polyporus vaporarius,* and *Schizophyllum commune* on wood were lower than on agar (Lindgren 1933; Gäumann 1939).

High or low temperatures will kill the mycelium of fungi in wood, but the temperatures occurring under usual service conditions are not sufficiently extreme to be effective. Wood-destroying fungi have remained alive in air-dry wood for over 9 years (Findlay and Badcock 1954). They differ considerably in their reaction to heat. Moist heat is much more effective than dry heat (Snell 1923). According to Hubert (1924), "The temperatures of commercial kiln runs, excepting temperatures below 120°F, are effective in sterilizing infected wood up to and including 4 by 4 inches square. Pieces 6 by 6 inches and 8 by 8 inches square were sterilized by treatment at 130°F for 9 hours at saturated atmosphere and by steam-pressure treatment." Usual commercial kiln runs require from 3 to 40 or more days at temperatures varying from 95 to 180°F and at relative humidities ranging from 100 down to 30.

Steaming wood, particularly under pressure, increases its susceptibility to decay during the subsequent air-seasoning period, according to laboratory tests (Chapman 1933; Scheffer and Lindgren 1936), indicating the need for air seasoning such wood by the best methods for rapid drying (page 481). After steamed wood has once been dried there is no evidence that it is less decay resistant than unsteamed wood. In some cases heating wood to temperatures of 250°F increases its resistance to fungus attack, possibly as a result of chemical changes or reduced hygroscopicity whereby sufficient water is prevented from entering the cell-wall structure to support decay (Buro 1954).

Ordinary commercial preservative treatments in which the temperatures of the preservative are above 150°F will kill decay fungi in green coniferous wood (Chidester 1939). A temperature of 150°F maintained for 75 minutes is lethal, the time necessary decreasing with increased temperature until only 5 minutes is needed at 212°F. Since these are internal temperatures it requires varying periods for a piece of wood to heat through, depending on its size. The time necessary can be determined from curves (MacLean 1952:58). Internal temperatures lower than 150°F are not practical, since it required 12 hours at 140°F and 24 hours at 122° F to kill.

However, after fungi in wood are killed by heat the material is always subject to reinfection and decay when moisture and temperature conditions become suitable.

Season of Cutting. For Norway spruce, silver fir, and European beech in Switzerland it was found that, during the period of air drying, wood cut in the growing season was less resistant to decay than that cut in the dormant season, the variation being ascribed to a difference in colloidal condition of the cell walls (Gäumann 1938). Accordingly he recommends not felling, if decay resistance is important, from the beginning of the circulation of the sap until ring formation is complete, November and December being the months during which decay resistance of the wood is highest, and May and June lowest. However, after a year's exposure to air there was no difference in decay resistance no matter when the wood was cut. No differences were observed in the rate of decay of beech sapwood (*Fagus crenata*) from trees cut at various times of the year in Japan (Nagai, Aoshima, and Hayashi 1955).

American beech and white birch cut in March decayed more rapidly when stored in the woods than when cut in July and September. White spruce cut in September decayed more than when it was cut in March or July. These differences were apparent over a 6-year period (Hilborn and Steinmetz 1949). Hence where it is important, the effect of time of felling upon the rate of decay should be determined for the tree species in question.

From the practical standpoint, if felled wood cannot be quickly removed from the forest or if it must be used in the green condition, autumn or winter felling is to be preferred, since the wood will have an opportunity to dry somewhat before warm weather, when fungi causing stain and decay are most active. Removing bark is valuable to retard decay. Under the best piling conditions even peeled timbers in warm, humid climates may decay somewhat before they become seasoned.

Position of the Grain. Laboratory tests showed no difference between edge-grained and flat-grained Douglas fir and ponderosa pine in resistance to decay caused by *Lenzites saepiaria, Fomes pinicola,* and several other wood-destroying fungi (Schmitz 1924).

DECAY OF PULPWOOD AND PULP

Decay in pulpwood and pulp can be considered in three categories: (1) rot in pulpwood from living, decayed trees because the wood was not culled properly when the trees were cut, (2) rot in stored pulpwood that was free of decay when the trees were cut, and (3) rot in stored pulp.

Pulpwood is obtained from both conifers and hardwoods. When partially decayed trees are cut into pulpwood, decayed heartwood may be included, and this generally results in a reduced yield and quality of pulp (Glennie and Schwartz 1950). Brown rots are more damaging than white rots. Wood with white rot caused by *Fomes pini* yields satisfactory sulfite pulp. Norway spruce wood with the incipient stage of white rot caused by *Fomes annosus* yields satisfactory chemical pulp (Björkman et al. 1949). Wood with white rot should not be used for mechanical (groundwood) pulp.

It is usual to store pulpwood for some time before it is converted into pulp. Consequently, proper storage is necessary to prevent serious losses from decay (Björkman 1958*a*, 1958*b*; Fritz 1954). In the eastern United States the principal fungi causing storage decay of hardwoods are *Polyporus hirsutus, P. versicolor, P. adustus,* and *Stereum hirsutum,* the first three being prevalent on aspen wood. *Lenzites saepiaria, Peniophora gigantea, Polyporus abietinus, Stereum sanguinolentum, Fomes roseus,* and *F. pini* are prevalent on coniferous wood, the last named undoubtedly being introduced as a heart rot from living trees, whereas the others attack the wood subsequent to cutting.

Pulpwood should be removed from the forest quickly, and before storage it should be peeled, although rough southern pine pulpwood showed less deterioration than peeled pulpwood for short storage periods in Louisiana (Lindgren 1951). At the mills wood is stored either in water or on the ground in huge conical piles of short wood or in long, closely ricked piles which may be over 30 feet high. Wood stored in water is safe from decay, provided it is completely immersed, but usually most of the wood floats so that with long storage the portion out of the water dries out sufficiently to decay. Conical piles afford excellent conditions for decay since the piles are usually established directly on the ground, thus allowing direct fungous infection from the soil. Only the outer wood dries out, leaving the wood in the interior of the pile subject to rapid deterioration. Such conical piles of water-transported wood can be protected from fire and decay by a system of overhead sprays to keep the wood water-soaked. Piling wood in ricks is preferable to conical piling, but the ricks must be raised from the ground on concrete skids or on concrete piers supporting wooden stringers pressure-treated with creosote. Pressure-treated wooden blocks may be substituted for concrete piers. The stringers should be 12 to 18 inches above the ground, and the ricks should be well separated by open spaces. The wood should be used in rotation, and the storage yard kept clean of all wood debris. Fungicides applied to log ends immediately after felling reduce decay during storage (Björkman 1953). Surface applications of 3 to 5 per cent solutions of ammonium bifluoride and sodium fluoride are more effective than chlorinated phenol and organic mercury solutions in reducing decay in peeled and rough southern pine pulpwood (Lindgren and Harvey 1952). Fluorides, while inhibiting growth of wood-rotting fungi, favor growth of molds such as *Trichoderma* which, in turn, increase the permeability of the wood to fluids.

Chemical pulp is ordinarily used immediately or within a few months, but mechanical pulp is often stored in the open, in closed unheated sheds, or in damp basements for 6 months to a year or even longer. Deterioration during storage is often serious, being caused both by a great variety of molding or staining fungi and by some wood destroyers. The molds do not affect the strength of pulp, but they discolor it and commonly bind the pulp particles together so that the molded material does not beat up well, producing a lumpy speckled paper. The wood destroyers make the wood fibers so brittle that they are broken up in beating and much of the pulp is lost in the waste water. Sodium pentachlorophenate at the rate of 2 to 4 pounds per ton of dried pulp, sprayed on the lap as it is being formed in the wet machine, will prevent deterioration of both chemical and mechanical pulp in storage (Nason, Shumard, and Fleming 1940). The application of 0.1 to 0.6 pound of sodium pentachlorophenate per ton of ground pulp by continuous feed at the grinders will reduce slime formed by microorganisms. Since manufacturing conditions and fungous flora differ, each mill should make its own tests to determine the kind and amount of chemical most effective.

DECAY OF FUEL WOOD

There can be a rapid loss in the value of fuel wood that is piled on the ground in the open. Generally the consumer does not realize that the calorific or heating value of the wood has been considerably reduced by decay. This loss is shown by the reduction in the calorific value of split and round paper birch, beech, and red maple in Maine piled on the ground in an open field and in the forest (Hilborn 1936). Split wood deteriorated more rapidly than round, wood stored in the forest more rapidly than that in the open field, and in general white birch deteriorated most rapidly, followed by beech and red maple in order. Based on the heating value of sound, dry wood cut in the winter, the monetary value of split red maple piled in the forest was reduced 74 per cent by the fourth autumn and round paper birch 89 per cent.

DECAY OF STRUCTURES

Buildings. In the United States much decay in buildings is due to the building poria (*Poria incrassata*), which causes the well-known dry rot

(Verrall 1954). The fungus causes a brown cubical rot (Fig. 206), and, because it can transport water for considerable distances, once it has established a source of moisture supply it can decay seemingly dry wood. *Merulius lacrymans,* which is rather rare in this country, behaves the same way. This fungus, so prevalent in northern Europe, probably finds the high summer temperatures and the greater heat maintained in our buildings during the colder months unfavorable to it. Other fungi decaying building timbers are *Coniophora cerebella, Fomes roseus, F. subroseus, Lentinus lepideus, Lenzites saepiaria, L. trabea, Peniophora gigan-*

Fig. 206. Brown cubical rot of a structural timber caused by the building Poria (*Poria incrassata*). White mycelial felts are prominent on the decayed wood, and a few small rhizomorphs are evident.

tea, Trametes serialis, and *Odontia spathulata* (Silverborg 1953). All these fungi cause brown rots of the charcoallike type. White rots are rarely, if ever, found in building timbers.

Decay in buildings and other permanent structures is largely preventable (Anonymous 1952). Two general types of structures are subject to decay, i.e., those in which moisture conditions are usually low such as dwellings, office buildings, and most factories, and those in which moisture conditions are high such as textile weave sheds, nursery packing sheds, canning factories, pulp mills, and creameries. In the first type of structures decay is the result of faulty design, faulty construction, the use of wet timber, or lack of sanitary conditions after construction. In the second type of structures decay is more difficult to prevent, particularly in roofs because of the condensation of moisture on the underside; consequently under such conditions the most advanced methods of construction with decay-resistant durable woods only should be used.

Although most cases of decay are usually in buildings where proper precautions have not been taken, there are great numbers of buildings, particularly houses, in which precautions have been neglected without fungous damage resulting. Nevertheless, the following principles should be applied to prevent decay: (1) Carefully inspect all lumber and timber to be used for construction for signs of dry rot, particularly the brown-colored rots. A rot occurring in the heartwood of living trees may be present in odd pieces of lumber and develop later in the wood of buildings. Kiln-dried lumber is sterilized lumber and should carry no infection provided it has been properly stored and handled to keep it dry. Blue-stained material should be given special attention, since discoloration develops under conditions favoring decay. Durable grades of wood with a large percentage of heartwood should be insisted upon in timber and lumber used in the lower portions of buildings. (2) Nothing but well-seasoned wood should be used. Wet lumber or timber is readily attacked by dry-rot fungi and will not always dry out rapidly enough in a building to prevent damage. The building poria will rarely remain alive longer than a month in air-dry wood (Scheffer and Chidester 1943). (3) Wood should not be placed in contact with moist soil or where moisture can reach it unless the lumber, timber, or wood blocking has been well treated with a wood preservative. (4) Both the sleepers and the subfloor of a wooden basement floor should be treated with preservatives before they are placed over concrete, stone, or brick bases. In addition, ventilation is essential between the base and the wood floor to carry off the accumulating moisture. (5) The ends of girders, joists, and stringers should not be embedded in stone, concrete or brick walls or piers unless ample ventilation is provided or unless the timbers are properly treated against decay. Untreated floor sleepers should not be embedded in concrete. (6) Ample cross ventilation is necessary in basements and beneath floors to remove moisture from the timbers and from moist soils, foundations and footings. Buildings without basements should be on foundations of adequate height with 1 square foot of ventilator area for each 25 linear feet of wall so placed as to prevent dead air pockets. For small buildings with one ventilator in each wall the openings should be at the corners. If adequate ventilation cannot be provided in basementless houses on wet sites, excessive moisture in joists or solid sills can be reduced by covering the soil under the house with roll roofing (Moses 1954). (7) All wood forms should be removed from foundation walls, piers, and footings, and all trim and odd pieces of lumber left after construction should be removed from beneath the building. Dry rot develops readily in this material and may spread to the foundation and floor timbers, particularly if pieces of boards or sticks with one end in contact with the ground are left leaning against joists. (8) Leakage of water through basement walls sunk below the ground level should be prevented by adequately waterproofing the outer faces of such foundation walls and providing drainage for the water in the soil. (9) Where roof decay may be expected due to high humidities in the building, care should be taken to provide sufficient insulation and suitable heat distribution to prevent the condensation of moisture on the roof timbers. (10) In all wet situations where moisture is difficult or impossible to control only naturally durable or treated wood should be used. (11) In buildings where humidities are high it is advisable to use sashes and frames treated with a toxic water repellent (page 486). (12) Proper design of buildings, including a good roof overhang, will reduce the accumulation of water in exterior woodwork (Verrall 1957*a*).

After decay has become established in a building, the source of moisture must be found and removed. All that may be needed to arrest the progress of decay is the maintenance of good ventilation so the wood will dry out. If repairs are necessary, cut a foot or more beyond the visible rot so that wood with incipient decay is eliminated. New green untreated lumber should not be fastened to old decayed wood because then the new material can become infected immediately so that decay may be much more rapid than it was originally.

Some fungi causing decay in buildings are sensitive to heat. Maintaining the temperature of a building at 115°F for 1½ days or more may kill certain fungi and will assist in drying out the wood for a time, thus retarding the activities of others. In some buildings it is feasible, by installing new heating coils, to lower the humidity sufficiently to dry out the building timbers to a point where they will not decay.

Boats. Warm climate, fresh water, leaks, lack of ventilation, and poor material such as sapwood, heartwood of species with a low decay resistance, or unseasoned wood promote decay in boats (Anonymous 1953; Savory 1954). Sometimes decay has been accelerated because the wood was infected before it went into construction. Boats in fresh water are more susceptible than those in salt water. The reaction between iron fastenings and moist heartwood of some hardwoods may accelerate decay in the iron-stained wood (Krause 1954). The most decay-vulnerable parts of boats are stems and transoms, waterways and frameheads, beam ends, and bulwark stanchion ends. Decay can be largely prevented by using heartwood of decay-resistant species at critical points, by using seasoned wood not infected before assembly, by preventing leaks, particularly of fresh water, and by providing and maintaining good ventilation in all parts of the boat. If it is necessary to use sapwood and install tight ceiling, the timber for critical parts of the hull should be impregnated with a suitable preservative, by either pressure or hot and cold bath. Salt packing may occasionally be useful. The addition of a fungicide to the bilgewater may protect wooden members in the bilge area (Scheffer 1953).

In repairing wooden boats, obviously rotting timbers and all wood to within a foot beyond the visibly infected area should be removed. The surrounding wood should be swabbed for a distance of several feet with one of the chlorinated phenols or phenyl mercury oleate (0.5 per cent concentrations), removing any paint before swabbing. The seasoned replacement timbers should be treated with one of these preservatives, preferably by the hot and cold bath method, or at least by swabbing.

The more important fungi causing decay in wooden boats are *Poria monticola* (*P. microspora*), *P. xantha*, *P. carbonica*, and *Lenzites saepiaria* on softwoods, and *Poria oleracea*, *Daedalea quercina*, and *Stereum frustulosum on hardwoods* (Davidson, Lombard, and Hirt 1947).

Balsa is highly susceptible to decay. When used in life floats it should be dipped in a 5 per cent solution of chlorinated phenol for 5 minutes (Rhoads 1948).

Aircraft. Decay in wooden aircraft, which occurs infrequently, has always originated from faulty design, careless construction, or improper maintenance (Boyce and Hepting 1943; Hepting 1944). Aircraft must be constructed so as to keep water from getting into the interior as far as possible, and drain holes must be placed so that water that occasionally does find its way in will drain out quickly. There should be no place in an aircraft where water can remain for any length of time.

DECAY OF STORED LUMBER

Stored lumber is decayed by a number of fungi including those which occur on pulpwood and building timbers. The losses include the actual loss from decay during storage and the indirect loss, often more serious, when lumber infected during storage is placed in structures and decay continues so that the timber fails prematurely. Nearly all these losses result from improper storage conditions that can be corrected (Fritz 1929; Humphrey 1917).

Storage yards should be located on well-drained ground, removed from floods, high tides, and standing water. Yards should be sanitary. All debris should be collected and burned, including decayed foundation or other timbers in service or stored lumber that has become infected. In yards filled to a considerable depth with sawdust and other woody debris, a heavy surfacing with cinders, slag, soil, or similar material is needed.

Lumber should be piled not less than 18 to 24 inches above the ground in humid climates, to ensure ventilation beneath the piles. Weeds should be kept down. Foundation piers should be of concrete, brick, or similar material, or of wood pressure-treated with creosote or other oil. Treated horizontal skid timbers are advisable. As a rule, lumber should not be close piled in the open but should be separated by crossers or stickers 1 inch thick. When stickers are not in use they should be piled on sound foundations and kept dry. If heartwood of decay-resistant species is used for stickers, the danger of decay is decreased. Lateral spacing is also advisable, and the piles should be roofed. Storage sheds should be tightly roofed, but the siding should not run down below the bottom of the foundation sills. Only thoroughly dry lumber should be stored in close piles under cover. Stock that has become infected should never be sold for permanent construction purposes.

Temporary protection against stain and decay in green or partly dried lumber during shipment or storage can be obtained by dipping it in solutions of chlorinated phenols and organic mercurials alone or in mixtures and in combination with borax and soda (Verrall and Mook 1951; Scheffer 1958).

The use of methyl bromide fumigation for the sterilization of lumber already infected with fungi has been suggested where heat-treatment is undesirable or impractical (Cartwright, Edwards, and McMullen 1953).

DECAY OF POSTS, POLES, PILING, MINE TIMBERS, AND TIES

These and similar products usually decay rapidly if untreated because they are generally used in contact with the soil, where conditions for decay are unusually favorable. In some cases these products are cut from living trees already infected with heart rot, the incipient decay being overlooked. In general such rots do not continue development or develop so slowly after a member is in service that other wood-destroying fungi cause the ultimate destruction. For example, there is no record of the incipient stage of red ring rot caused by *Fomes pini*, so common in conifers, ever developing further in wood in service. In fact, this fungus is known to die out in railroad ties in service (Fritz and Atwell 1941). Such material can be used with relative safety, particularly if the wood is given a preservative treatment with oil under pressure and at a high temperature. However, since most decay of these products is caused by brown rot fungi, it is advisable to reject material with incipient brown rot although even here there may be exceptions. Pocket dry rot of incense cedar caused by *Polyporus amarus* does not develop or develops so slowly in wood in service that posts with the advanced stage of this decay can be used.

More commonly these products are infected during storage by wood-destroying fungi that continue to decay the members after they are placed in service. Early decay caused by *Peniophora gigantea* in southern pine poles, piling, and posts in storage can be detected by applying a solution of Alizarine Red S to freshly cut end surfaces. The decayed areas turn yellow, whereas the sound wood appears red (Lindgren 1955). Timbers should not be piled directly on the ground, and during seasoning they should be piled so as to permit free circulation of air all around the piles and if possible around the individual timbers. Where feasible, winter cutting should be practiced because fungi are less active at this season and seasoning will be slower, thus preventing excessive checking. Checks are the main points of entrance for wood-destroying fungi. Timbers should be peeled immediately after trees are felled. A coating of oil paint should never be applied until the material is thoroughly dry, because this will prevent complete seasoning and favor decay. Rail transport is better than water transport, although the latter can be safely used, provided the material is properly piled to ensure rapid drying after it is removed from the water. Seasoning timbers to be used as posts, poles, ties, and similar products does not increase their decay resistance over unseasoned timbers placed in service immediately after cutting. In the Southwest, Utah white oak posts set green and unpeeled were more decay resistant than those seasoned before setting (Long 1941).

Poles. Among the fungi decaying poles in service the most important are *Lenzites saepiaria* and *Lentinus lepideus* on conifers and *Daedalea quercina* on American chestnut in the Northeastern States. *Polyporus sulphureus* (Fig. 207) and *P. spraguei* attack chestnut poles to some extent. In dry climates the decay of poles is near the ground line, but in warm, humid climates such as the South, decay often occurs in the sides and tops of poles many feet above the ground. Poles of decay-resistant species containing internal sapwood are probably liable to more rapid decay (Englerth and Scheffer 1954).

Piling. Piling in water is subject to decay during all but the winter months, unless it is completely submerged. However, if not completely submerged then there is a moisture gradient in the piles that at some point above water level reaches the proper balance of moisture and air for the development of wood-destroying fungi (Lagerberg 1928). Even if completely submerged, wood in brackish water or in sea water of normal salinity is subject to slow decomposition by certain marine fungi causing soft rot (Barghoorn and Linder 1944). Soft rot is less important in fresh water.

Fig. 207. Conks of the sulfur fungus (*Polyporus sulphureus*) on a telephone pole of American chestnut. (*Photograph by R. P. Marshall.*)

Mine Timbers. Because of high humidity, mine timbers are under the most unfavorable conditions to resist decay, and in deep mines this is combined with a relatively even temperature throughout the year (Humphrey 1922; Hornor, Tufft, and Hunt 1925). Depth influences decay only by its effect on temperature and moisture conditions (Fritz 1942). Mine timbers decay rapidly, records on a return air course in one mine showing that the normal life of 10- to 14-inch round, peeled Douglas fir was but 4 to 8 months. Often when crushing occurs, the timbers have first been weakened by decay. Of course, in temporary tunnels or shafts decay resistance is not important, but elsewhere the only way to secure satisfactory service is by preservative treatment of the timber. Fungi decaying mine timbers are often different from those attacking posts, poles, and ties aboveground. White rots as well as brown rots are found in mine timbers (Spaulding 1910), white rot fungi such as *Armillaria mellea, Hydnum erinaceus, Fomes applanatus,* and *Fomes annosus* being prevalent.

Ties. The fungi that decay ties in service are largely the same as, although fewer in number than, those attacking ties in storage (Richards 1938). Consequently it is important to select and store ties properly so that they will not be already infected when placed in the track, because if untreated, infected ties decay more rapidly than sound ones (Roth 1943). *Lenzites saepiaria* and *Lentinus lepideus* are the most serious destroyers of coniferous ties, especially in the West, although others often found are *Fomes roseus, Lenzites trabea, Polyporus abietinus, Peniophora gigantea,* and *Schizophyllum commune.* Decay caused by the last two named, once considered inconsequential, is known to reduce the strength of wood (Richards and Chidester 1940). On hardwood ties a number of fungi occur, but *Lenzites trabea, Polystictus versicolor, Stereum fasciatum,* and *S. lobatum* are probably the most common.

DECAY OF PLYWOOD AND GLUES

The decay resistance of plywood is determined by the least resistant species of wood composing it. Plywood is quite subject to soft rot. If there is sufficient moisture present, plywood will delaminate before it can decay because molds and bacteria readily digest water-resistant glues such as soybean and casein (Duncan 1957). The resistance of these glues can be increased by the addition of 5 per cent sodium trichlorophenate (Kaufert 1943). Phenolic resin adhesives are extremely resistant to fungi and bacteria, and consequently it is advisable to use plywood bonded with one of them in damp situations.

PRESERVATION

Satisfactory long-time service can be secured from many woods placed in contact with the soil, or in other positions where high decay resistance is a requisite, only by preservative treatment. Although some species have extremely resistant heartwood, most do not, and the sapwood of all has little resistance to decay. Round timbers always have more or less sapwood, whereas squared timbers (sawed or hewed) may have most, if not all, of the sapwood removed. Ultimately all wood used where high decay resistance is required will be treated, because the supply of resistant woods is steadily decreasing, as is also the quality of the woods of intermediate decay resistance such as the southern yellow pines. When the old virgin timber is gone and young second growth becomes the chief source of supply, preservative treatment will be essential for practically all timbers to be used in exposed situations.

For information on methods of preservation and toxicity of various chemicals the reader is referred to other publications (Hunt and Garratt 1953; MacLean 1952; Verrall 1957*b*). The ideal preservative, not yet developed, would be one that would not leach from the wood and yet when the wood became moist would form a solution in water poisonous to wood-destroying fungi, whether the preservative was originally impregnated as a water or an oil solution. It is evident that these two requirements are not entirely compatible. Creosote and especially coal-tar creosote is now most widely used. A solution of bichloride of mercury in water is probably the most effective wood preservative, but it is little employed in this country because it is a deadly poison to animals and highly corrosive to metals. In the United States at least, zinc chloride is regarded as the standard water-soluble preservative. The various chemicals used for the purpose of protecting wood may be applied by any one of a number of different processes, but the most effective methods all make use of pressure to secure adequate penetration and absorption.

For exposed wood to be used off the ground such as porch flooring, steps, rails, shutters, and window sashes and frames, toxic water repellents are effective. With kerosene for the dilutent, 5 per cent pentachlorophenol, 0.2 per cent phenyl mercury oleate, and 4.5 per cent copper naphthenate (0.5 per cent copper) have protected southern yellow pine sapwood and heartwood immersed for 15 minutes or longer (Verrall 1946). Two brush treatments, 1 hour apart, with 5 per cent pentachlorophenol or 18 per cent copper naphthenate (2 per cent copper) were also effective. Dip or brush treatments with these chemicals are not efficient on green wood; air-dry wood must be used. Wood so treated can be painted. Paint is not a preservative. In fact, it may be necessary to add 2 to 3 per cent of tetrachlorophenol to paint to prevent discoloration by molding (Turner 1942). Although paint retards the entrance of water into the wood, yet once the wood has become wet, paint retards drying, and therefore under some conditions paint may actually favor decay.

A chemical may be toxic to wood-destroying fungi and yet be of little value when impregnated into wood because it leaches out readily or has other serious drawbacks. As yet it has not been possible to correlate laboratory tests of the toxicity of a preservative with the actual protection it will afford wood in service, since there is no satisfactory short-cut method of determining permanence. For example, *Poria radiculosa* is not able to decay freshly creosoted southern pine sapwood, although it decays creosoted poles that have been in service for at least 7 years (Hudson 1952). However, laboratory tests are valuable to indicate whether or not a chemical is worth trying as a preservative (Colley 1953; Duncan 1958). Such tests are conducted by growing wood-destroying fungi under sterile conditions on nutrient agar containing varying amounts of preservative, or on sawdust, wood discs, or wood blocks impregnated with the preservative. Toxicity tests should be standardized as to both the substrates and fungi employed.

REFERENCES

Albion, R. G.: "Forests and Sea Power." Cambridge, Mass.: Harvard University Press (1926).

Anderson, B. A.: The Toxicity of Water-soluble Extractives of Western Yellow Pine to *Lenzites sepiaria*. *Phytopathology* 21:927–940 (1931).

Anonymous: Factors That Influence the Decay of Untreated Wood in Service and Comparative Decay Resistance of Different Species. *U.S. Dept. Agr., Forest Serv., Forest Prods. Lab. Rept.,* R68 (rev.):1–5 (1950).

Anonymous: Prevention and Control of Decay in Dwellings. *U.S. Dept. Agr., Forest Serv., Forest Prods. Lab. Tech. Note* 251 (rev.): 1–6 (1952).

Anonymous: Decay Studies in Wooden Boats and Ships. *U.S. Dept. Agr., Div. Forest Pathol.* (1953). (Mimeographed.)

Barghoorn, E. S., and D. H. Linder: Marine Fungi: Their Taxonomy and Biology. *Farlowia,* 1:395–467 (1944).

Bavendamm, W., and H. Reichelt: Die Abhängigkeit des Wachstums holzzersetzender Pilze vom Wassergehalt des Nahrsubstrates. *Arch. Mikrobiol.,* 9:486–544 (1938).

Björkman, E.: Om röthärdigheten hos lärkvirke. [English summary.] *Norrlands Skogsvårdsförbunds Tidskr.,* 1944:18–45 (1944).

Björkman, E.: The Occurrence and Significance of Storage Decay in Birch and Aspen Wood with Special Reference to Experimental Preventive Measures. *Kgl. Skogshögskol. Skrifter,* 16:53–90 (1953).

Björkman, E. Lagringsröta och blanad i skogslagrad barr- och lövmassaved. (Storage Decay and Blue Stain in Forest-stored Pine, Spruce, Birch and Aspen Pulpwood.) [English summary.] *Kgl. Skogshögskol. Skrifter,* 29:1–128 (1958*a*).

Björkman, E.: Stockblanad och langringsröta i tall- och grantimmer vid olika avverkningstid och behandling i samband med flottning. (Log Blue Stain and Storage Decay in Pine and Spruce Timber with Special Reference to Felling Time and Treatment During Floating.) [English summary.] *Kgl. Skogshögskol. Skrifter,* 30:1–62 (1958*b*).

Björkman, E., et al.: Om rötskador i granskog och deras betydelse vid framställning av kemisk pappersmassa och silkemassa. [English summary.] *Kgl. Skogshögskol. Skrifter,* 4:1–73 (1949).

Boyce, J. S., and G. H. Hepting: Decay of Wood in Aircraft. *U.S. Dept. Agr., Forest Pathol., Spec. Release,* 12:1–4 (1943).

Buro, A.: Die Wirkung von Hitzbehandlungen auf die Pilzresistenz von Kiefer und Buchenholz. *Holz Roh- u. Werkstoff.,* 12:297–304 (1954).

Cartwright, J. B., D. W. Edwards, and M. J. McMullen: Sterilization of Timber with Methyl Bromide. *Nature (London),* 172:552–553 (1953).

Cartwright, K. St. G.: The Variability in Resistance to Decay of the Heartwood of Home-grown Western Red Cedar (*Thuja plicata* D. Don.) and Its Relation to Position in the Log. *Forestry,* 15:65–75 (1941).

Cartwright, K. St. G.: The Variability in Resistance to Decay of the Heartwood of Home-grown European Larch, *Larix decidua,* Mill. (*L. europaea*) and Its Relation to Position in the Log. *Forestry,* 16:49–51 (1942).

Cartwright, K. St. G., and W. P. K. Findlay: Studies in the Physiology of Wood Destroying Fungi. II. Temperature and Rate of Growth. *Ann. Botany,* 48:481–495 (1934).

Chapman, A. D.: Effect of Steam Sterilization on Susceptibility of Wood to Blue-staining and Wood-Destroying Fungi. *J. Agr. Research,* 47:369–374 (1933).

Chidester, Mae S.: Further Studies on Temperatures Necessary to Kill Fungi in Wood. *Proc. Am. Wood-Preservers' Assoc.,* 35:319–324 (1939).

Colley, R. H.: The Evaluation of Wood Preservatives. *Bell System Tech. J.,* 32:120–169, 425–505 (1953).

Davidson, R. W., Frances F. Lombard, and R. R. Hirt: Fungi Causing Decay in Wooden Boats. *Mycologia,* 39:313–327 (1947).

Duncan, Catherine G.: Effects of Moisture on Bacterial Weakening of Casein-bonded Plywood. *U.S. Dept. Agr., Forest Serv., Forest Prods. Lab. Rept.,* 2077:1–4 (1957).

Duncan, Catherine G.: Studies of the Methodology of Soil-block Testing. *U.S. Dept. Agr., Forest Serv., Forest Prods. Lab. Rept.,* 2114:1–126 (1958).

Englerth, G. H., and T. C. Scheffer: Tests of Decay Resistance of Four Western Pole Species. *U.S. Dept. Agr., Forest Serv., Forest Prods. Lab. Rept.,* 2006:1–8 (1954).

Falck, R.: Ueber den Einfluss des Flossens auf die Widerstandsfähigkeit des Bauholzes gegen Trockenfäule und über den Holzschutz durch Schimmelbefall und Diffusionstränkung. *Mitt. Forstw. u. Forstwiss.,* 2:480–485 (1931).

Ferdinandsen, C., and N. F. Buchwald: Nogle Undersøgelser over Tømmersvampe med saerlight Hensyn til deres Fugtighedskrav. *Dansk Skovfor. Tidsskr.,* 22:685–715 (1937). [*Rev. Appl. Mycol.,* 17:641 (1938).]

Findlay, W. P. K.: Studies in the Physiology of Wood-destroying Fungi. I. The Effect of Nitrogen Content upon the Rate of Decay of Timber. *Ann. Botany,* 48:109–117 (1934).

Findlay, W. P. K.: Influence of Sample Size on Decay Rate of Wood in Culture. *Timber Technol. and Machine Woodworking,* 61:160–162 (1953).

Findlay, W. P. K.: Timber Decay—A Survey of Recent Work. *Forestry Abstr.,* 17:317–327, 477–486 (1956).

Findlay, W. P. K., and E. C. Badcock: Survival of Dry Rot Fungi in Air-dry Wood. *Timber Technol. and Machine Woodworking,* 62:137–138 (1954).

Fritz, Clara W.: Stain and Decay in Lumber-seasoning Yards with Special Reference to Methods of Prevention. *Can. Dept. Interior, Forest Serv. Circ.,* 27:1–15 (1929).

Fritz, Clara W.: Does Depth Influence Rate of Decay in Mine Timbers? *Can. Mining J.*, 63:719–720 (1942).

Fritz, Clara W.: Decay in Poplar Pulpwood in Storage. *Can. J. Botany*, 32:799–817 (1954).

Fritz, Clara W., and E. A. Atwell: Decay in Red-stained Jack Pine Ties under Service Conditions. *Can. Dept. Mines and Resources, Dominion Forest Serv., Forest Products Labs. Div. Circ.*, 58:1–19 (1941).

Garren, K. H.: Studies on *Polyporus abietinus*. III. The Influence of Certain Factors on the Rate of Decay of Loblolly Pine Sapwood. *J. Forestry*, 37:319–323 (1939).

Gäumann, E.: Der Einfluss der Fällungszeit auf die Dauerhaftigkeit des Fichten-, Tannen- und Buchenholzes. *Schweiz. Z. Forstw.*, 89:177–197 (1938).

Gäumann, E.: Ueber die Wachsteumsund Zerstörungsintensität von *Polyporus vaporarius* und von *Schizophyllum commune. Angew. Botan.*, 21:59–69 (1939).

Gäumann, E.: Der Einfluss der Meereshöhe auf die Dauerhaftigkeit des Lärchenholzes. *Schweiz. Anst. forstl. Versuchsw. Mitt.*, 25:327–393 (1948).

Glennie, D. W., and H. Schwarz: Review of the Literature on Decay in Pulpwood, Its Measurement and Its Effect on Wood Properties and Pulp Quality. *Can. Dept. Mines and Resources, Forest Serv., Forest Prods. Labs. Div. Mimeographed Rept.*, 0–153 (1950).

Haasis, F. W.: A Study of Laboratory Methods for Investigating the Relation between Moisture Content of Wood and Fungal Growth. *Phytopathology*, 22:71–84 (1932).

Hartley, C.: Evaluation of Wood Decay in Experimental Work. *U.S. Dept. Agr., Forest Prods. Lab. Rept.*, 2119:1–60 (1958).

Hatfield, I.: Control of Moisture Content of Air and Wood in Fresh-air Chambers. *J. Agr. Research*, 42:301–305 (1931).

Hepting, G. H.: Preventing Decay in Wood Aircraft. *Aero Digest*, 44:126, 128, 142, 213 (1944).

Hilborn, M. T.: The Calorific Value of Decayed Cordwood. *Phytopathology*, 26:905–914 (1936).

Hilborn, M. T., and F. H. Steinmetz: Relation of Time of Cutting to Rate of Decay of Beech, Birch, and Spruce under Natural Storage Conditions. (Abstract.) *Phytopathology*, 39:9 (1949).

Hornor, R. R., H. E. Tufft, and G. M. Hunt: Mine Timber, Its Selection, Storage, Treatment, and Use. *U.S. Bur. Mines Bull.*, 235:I–X, 1–118 (1925).

Hubert, E. E.: Effect of Kiln Drying, Steaming, and Air Seasoning on Certain Fungi in Wood. *U.S. Dept. Agr. Bull.*, 1262:1–20 (1924).

Hudson, M. S.: *Poria radiculosa*, a Creosote Tolerant Organism. *J. Forest Products Research Soc.* 2(2):73–74 (1952).

Humphrey, C. J.: Timber Storage Conditions in the Eastern and Southern States with Reference to Decay Problems. *U.S. Dept. Agr. Bull.*, 510:1–42 (1917).

Humphrey, C. J.: Decay of Mine Timbers. *Proc. Am. Wood-Preservers' Assoc.*, 18:213–222 (1922).

Humphrey, C. J., and P. V. Siggers: Temperature Relations of Wood-Destroying Fungi. *J. Agr. Research*, 47:997–1008 (1933).

Hungate, R. E.: Nitrogen Content of Sound and Decayed Coniferous Woods and Its Relation to Loss in Weight During Decay. *Botan. Gaz.*, 102:382–392. (1940).

Hunt, G. M., and G. A. Garratt: "Wood Preservation." New York: McGraw-Hill Book Company, Inc., (1953).

Jørgensen, E.: Fra Bregentveds Egeskove. Modstandsevne imod Svampeangreb hos bredringet og smalringet Egekaerne. (From the Bregentved Oak Woods. Resistance to Fungal Attack of Broad- and Narrow-ringed Oak Heartwood.) [English summary.] *Dansk Skovfor. Tidsskr.*, 38:452–466 (1953).

Kaufert, F. H.: Increasing the Durability of Casein Glue Joints with Preservatives. *U.S. Dept. Agr., Forest Serv., Forest Prods. Lab. Rept.*, 1332: 1–18 (1943).

Kaufert, F. H., and E. A. Behr: Susceptibility of Wood to Decay: Effect of Urea and Other Nitrogenous Compounds. *Ind. Eng. Chem., Ind. Ed.*, 34:1510–1515 (1942).

Krause, R. L.: Iron Stain from Metal Fastenings May Accelerate Decay in Some Woods. *J. Forest Products Research Soc.*, 4:103–111 (1954).

Lagerberg, T.: Rötskador å barrvirke i vatten. [English summary.] *Svenska Skogsvårdsför. Tidskr.*, 26:617–658 (1928).

Lindgren, R. M.: Decay of Wood and Growth of Some Hymenomycetes as Affected by Temperature. *Phytopathology*, 23:73–81 (1933).

Lindgren, R. M.: Deterioration of Southern Pine Pulpwood During Storage. *Proc. Forest Products Research Soc.*, 5:169–180 (1951).

Lindgren, R. M.: Color Test for Early Storage Decay in Southern Pine. *U. S. Dept. Agr., Forest Serv., Forest Prods. Lab. Rept.*, 2037:1–4 (1955).

Lindgren, R. M., and G. M. Harvey: Decay Control and Increased Permeability in Southern Pine Sprayed with Fluoride Solutions. *J. Forest Products Research Soc.*, 2(5):250–256 (1952).

Long, W. H.: The Durability of Untreated Oak Posts in the Southwest. *J. Forestry*, 39:701–704 (1941).

MacLean, H., and J. A. F. Gardner: Distribution of Fungicidal Extractives in Target Pattern Heartwood of Western Red Cedar. *Forest Prods. J.*, 8:107–108 (1958).

MacLean, J. D.: Preservative Treatment of Wood by Pressure Methods. *U.S. Dept. Agr. Handbook*, 40:I–IV, 1–160 (1952).

McNab, H. S., Jr.: Procedure for Laboratory Studies on Wood Decay Resistance. *Proc. Iowa Acad. Sci.*, 65:150–159 (1958).

Moses, C. S.: Condensation and Decay Prevention under Basementless Houses. *U.S. Dept. Agr., Forest Serv., Forest Prods. Lab. Rept.*, 2010:1–9 (1954).

Nagai, Y., K. Aoshima, and Y. Hayashi: Laboratory Test on the Durability of Beech Wood Felled at Different Seasons. [Japanese, with English summary.] (*Tokyo*) *Govt. Forest Expt. Sta. Bull.*, 77:21–23 (1955).

Nason, H. R., R. S. Shumard, and J. D. Fleming: Microbiology of Pulp and White Water Systems. *Paper Trade J.*, 110(13):30–36 (1940).

Peck, E. C.: Air Drying of Lumber. *U.S. Dept. Agr., Forest Serv., Forest Prods. Lab. Rept.*, 1657 (rev.):1–21 (1956).

Rennerfelt, E.: Några undersökningar över olika rötsvampars förmåga att angripa splintoch Kärnved hos tall. (Some Investigations over the Capacity of Some Decay Fungi to Attack Sapwood and Heartwood of Scots Pine.) [English summary.] *Statens Skogsforskningsinst.* (*Sweden*) *Medd.*, 36(9):1–25 (1947).

Rennerfelt, E.: Investigations of Thujaplicin, a Fungicidal Substance in the Heartwood of *Thuja plicata* D. Don. *Physiol. Plantarum*, 1:245–254 (1948).

Rennerfelt, E.: The Natural Resistance to Decay of Certain Conifers. *Friesia*, 5:361–365 (1956).

Rennerfelt, E., and Gertrud Nacht: The Fungicidal Activity of Some Constituents from Heartwood of Conifers. *Svensk Botan. Tidskr.*, 49:419–432 (1955).

Rhoads, A. S.: Decay in Balsa Life Floats and the Value of Brush and Drip Treatments with Chlorinated Phenol for its Prevention. *J. Am. Soc. Naval Engrs.*, 60:565–595 (1948).

Richards, C. Audrey: Defects in Cross Ties, Caused by Fungi. *Cross Tie Bull.*, 19:3–31 (1938).

Richards, C. Audrey, and Mae S. Chidester: The Effect of *Peniophora gigantea* and *Schizophyllum commune* on Strength of Southern Yellow Pine Sapwood. *Proc. Am. Wood-Preservers' Assoc.*, 36:24–31 (1940).

Richards, D. B.: Decay Resistance and Physical Properties of Wood. *J. Forestry*, 48:420–422 (1950).

Roff, J. W., and J. M. Atkinson: Toxicity Tests of a Water-soluble Phenolic Fraction (Thujaplicin-free) of Western Red Cedar. *Can. J. Botany*, 32:308–309 (1954).

Roth, E. R.: Effect of Invisible Decay on Deterioration of Untreated Oak Ties and Posts. *J. Forestry*, 41:117–121 (1943).

Savory, J. G.: Prevention of Decay of Wood in Boats. *Gt. Brit. Dept. Sci. Ind. Research, Forest Prods. Research Bull.*, 31:1–18 (1954).

Scheffer, T. C.: Treatment of Bilgewater to Control Decay in the Bilge Area of Wooden Boats. *J. Forest Products Research Soc.*, 3(3):72–78, 95 (1953).

Scheffer, T. C.: Decay Resistance of Western Red Cedar. *J. Forestry*, 55:434–442 (1957).

Scheffer, T. C.: Control of Decay and Sap Stain in Logs and Green Lumber. *U.S. Dept. Agr., Forest Serv., Forest Prods. Lab. Rept.*, 2107:1–13 (1958).

Scheffer, T. C., and Mae S. Chidester: Significance of Air-dry Wood in Controlling Rot Caused by *Poria incrassata*. *Southern Lumberman*, 166(2091):53–55 (1943). *U.S. Dept. Agr., Forest Pathol., Spec. Release*, 17:1–4 (1943).

Scheffer, T. C., and H. Hopp: Decay Resistance of Black Locust Heartwood. *U.S. Dept. Agr. Tech. Bull.*, 984:1–37 (1949).

Scheffer, T. C., and R. M. Lindgren: The Effect of Steaming on the Durability of Unseasoned Sap-gum Lumber. *J. Forestry*, 34:147–153 (1936).

Scheffer, T. C., and A. Van Kleeck: The Decay Resistance of Wood Impregnated with Fire-retarding Ammonium Salts. *Proc. Am. Wood-Preservers' Assoc.*, 41:1–7 (1945).

Schmitz, H.: Notes on Wood Decay II. The Comparative Resistance to Decay of Edge-grained and Flat-grained Douglas Fir and Western Yellow Pine Lumber. *Idaho Forester*, 6:12–15 (1924).

Schmitz, H., and F. Kaufert: The Effect of Certain Nitrogenous Compounds on the Rate of Decay of Wood. *Am. J. Botany*, 23:635–638 (1936).

Schmitz, H., and F. Kaufert: Studies in Wood Decay. VII. How Long Can Wood-destroying Fungi Endure Immersion in Water? *Proc. Am. Wood-Preservers' Assoc.*, 34:83–87 (1938).

Sherrard, E. C., and E. F. Kurth: The Distribution of Extractive in Redwood: Its Relation to Durability. *U.S. Dept. Agr., Forest Serv., Forest Prods. Lab.*, R 988:1–6 (1933).

Silverborg, S. B.: Fungi Associated with the Decay of Wooden Buildings in New York State. *Phytopathology*, 43:20–22 (1953).

Snell, W. H.: The Effect of Heat upon the Mycelium of Certain Structural-timber-destroying Fungi within Wood. *Am. J. Botany*, 10:399–411 (1923).

Snell, W. H.: The Relation of the Moisture Contents of Wood to Its Decay. III. *Am. J. Botany*, 16:543–546 (1929).

Snell, W. H., N. O. Howard, and M. V. Lamb: The Relation of Moisture Contents of Wood to Its Decay. *Science*, 62:377–379 (1925).

Southam, C. M., and J. Ehrlich: Decay Resistance and Physical Characteristics of Wood. *J. Forestry*, 41:666–673 (1943).

Spaulding, P.: Fungi of Clay Mines. *Missouri Botan. Garden Ann. Rept.*, 21:189–195 (1910).

Theden, Gerda: Untersuchungen über die Feuchtigkeitsansprüche der wichtigsten in Gebäuden auftretenden holzzerstörenden Pilze. *Angew. Botan.*, 23:189–253 (1941).

Turner, M. B.: Mold Problems of the Paint Industry. *Am. Paint J.*, 26:18–24 (1942).

Verrall, A. F.: The Probable Mechanism of the Protective Action of Resin in Fire Wounds on Red Pine. *J. Forestry*, 36:1231–1233 (1938).

Verrall, A. F.: Progress Report on Tests of Soak and Brush Preservative Treatments for Use on Wood off the Ground. *Southern Lumberman*, 173(2168):36–38 (1946).

Verrall, A. F.: Preventing and Controlling Water-conducting Rot in Buildings. *U.S. Dept. Agr., Forest Serv., Southern Forest Expt. Sta. Occas. Paper*, 133:1–14 (1954).

Verrall, A. F.: How to Prevent Fungus Damage to Wood Structures. *Forest Prods. J.*, 6(1):15A–17A (1957*a*).

Verrall, A. F.: Absorption and Penetration of Preservatives Applied to Southern Pine Wood by Dips or Short-period Soaks. *U.S. Dept. Agr., Forest Serv., Southern Forest Expt. Sta. Occas. Paper*, 157:1–31 (1957*b*).

Verrall, A. F., and P. V. Mook: Research on Chemical Control of Fungi in Green Lumber, 1940–51. *U.S. Dept. Agr. Tech. Bull.*, 1046:1–60 (1951).

Waterman, Alma M.: The Effect of Water-soluble Extractives from the Heartwood of Tropical American Woods on the Growth of Two Wood-decay Fungi. *Trop. Woods*, 88:1–11 (1946).

Zabel, R. A.: Variations in the Decay Resistance of White Oak. *N.Y. State Coll. Forestry Syracuse Univ. Tech Publ.*, 68:1–53 (1948).

Zeller, S.M.: Humidity in Relation to Moisture Imbibition by Wood and to Spore Germination on Wood. *Ann. Missouri Botan. Garden*, 7:51–74 (1920).

CHAPTER 20

Deterioration of Forest Products: Stains

The many stains mostly in the sapwood of logs, lumber, and other wood products cause large losses to the lumber industry (Scheffer and Lindgren 1940; Verrall 1945; Findlay 1959). There are other discolorations in wood that may be confused with stains (Boyce 1923). Although stains do not reduce the strength of wood for most purposes, the unsightly discolorations cause heavy degrade, and it is low-grade lumber that frequently cannot be sold or can be disposed of only at a price below the cost of production. It is doubtful if consumers will ever accept stained wood for all purposes where it is just as suitable as unstained wood. Since conditions favoring stain and mold also favor decay, stained wood should be closely scrutinized for evidences of rot.

There are two general classes of stains: those caused by chemical reactions and those caused by fungi.

CHEMICAL STAINS

Sapwood of many species is subject to discolorations varying widely in appearance but fundamentally the same, resulting from chemical action (Bailey 1910). Sapwood is rich in organic compounds and also contains certain soluble ferments that facilitate the oxidation of such compounds. Under favorable temperature conditions, when green sapwood is exposed to the air, the oxidizing enzymes act on the organic compounds in the sapwood. The result of this oxidation process is discoloration of the sapwood, with the colored substance most noticeable upon microscopic examination in those cells mainly concerned with the storage and transportation of food. Hot humid weather is most favorable for such staining whereas cool dry weather retards or prevents it entirely. Logs immersed in water are not affected. Light is not necessary for the reaction. Discoloration is frequently confined to the immediate surface layer so that it surfaces off when the wood is planed. The wood is not weakened. The superficial chemical stains that occur on sawed lumber can be prevented if necessary by dipping the green sap boards into boiling water for a few minutes as they come green from the saw. The chemical stains that penetrate deeply into logs, bolts, or lumber will require longer immersion in boiling water or treatment with live steam. Darkening of wood often occurs during kiln drying (Kollman, Keylwerth, and Kübler 1951).

Stains of Hardwoods. The sapwood of birches, black cherry, and maples stains a reddish-yellow or rusty color. The sapwood of sweet gum often assumes a reddish-brown shade. The wood of alders becomes intensely red or red-brown on freshly cut surfaces, often within an hour or so after the surface is exposed. If the wood of red alder dries white, the red color will appear upon the addition of water in the presence of air, provided the temperature is favorable. The sapwood of freshly cut persimmon often stains throughout a pronounced dark brown to deep gray, particularly if seasoning is retarded, but the discoloration can be prevented in material up to 2 inches in diameter by steaming at atmospheric pressure for 30 minutes or at 15 pounds pressure for 15 minutes (Scheffer and Chapman 1934). Thicker stock will probably require longer steaming. The sapwood of certain red oaks stains a light- to dark-gray color during air drying that can be prevented by preheating the wood at 200°F for 15 to 45 minutes (Clark 1957). An interior discoloration resembling light interior blue stain may develop in the sapwood of tupelo gum, evergreen magnolia, and sweet bay during kiln drying or air drying. Freshly cut woods rich in tannin, such as oak and chestnut, often stain a bluish or blackish gray after contact with metal saws, veneer knives, or green chains. This iron tannate stain can be prevented by dipping in alkaline solutions to counteract the acid condition of the wood.

Mineral stain or streak is a serious discoloration of sugar maple and other hardwoods that develops only in the living tree and seemingly only in the sapwood (Scheffer 1954). There is some confusion as to what discolorations are included under mineral stain. On sawed material of sugar maple the stain appears either as individual lenticular deep-olive or greenish-black streaks paralleling the grain or as mass discolorations. The discolored wood appears to contain carbonates in abnormal quantities and is harder than normal. During seasoning, the stained wood has a tendency to check. The cause of mineral stain is unknown, but organisms have not been consistently associated with it. The discolorations may be initiated by injuries caused by sapsuckers, insects, and other agencies. Brown heart of sugar maple may be a form of mineral stain (Good, Murray, and Dale 1955). Discolorations in yellowpoplar ranging from light yellow through greens, reds, purples, blues, and browns, commonly referred to as blue butt or mineral stain, develop in the wood of living trees following wounding, apparently from oxidation (Roth 1950).

Stains of Conifers. The most serious chemical stain on conifers is brown stain (Fig. 208), which develops during air drying or kiln drying (Anonymous 1951; Millet 1952). The discoloration appears as a buff to dark-brown stain on ponderosa, sugar, and western white pines. On eastern white pine it is sometimes termed coffee stain. On sugar pine the discoloration frequently occurs just below the surface of sawed boards and consequently escapes notice until after planing. The stain is variable in intensity, rarely shows flecking, is often superficial or just beneath the surface, and has a whitish margin near the ends, edges, or surfaces of

Fig. 208. Brown stain of sugar pine caused by a chemical reaction.

boards and near the juncture of sapwood and heartwood. Chemical brown stain is most common in sapwood but also occurs in heartwood. The discoloration appears in lumber during seasoning but not in logs.

Brown stain can be confused with kiln burn, a very dark browning, scorching, or charring of kiln-dried lumber, resulting from too high temperatures in the kiln. Kiln burn is more uniform in color, and the wood is softer. Machine burn is an easily recognized brown discoloration found in finished stock caused by scorching or burning when the progress of wood through a high-speed machine is somewhat retarded. Water stain, a yellowish to dark-brownish discoloration, appears on the lighter colored woods when water has dripped on them. It can be readily planed off. Brown sap stain caused by a fungus may also be confused with chemical brown stain (page 497).

Stain can be minimized by cutting the logs into lumber soon after felling the trees and by not solid-piling the fresh lumber for more than a day. Slowed kiln drying during the early part of the schedule may prevent or reduce the stain, since it is less prevalent and severe on air-dried stock.

Redwood lumber may develop a black stain and noble fir a reddish-brown stain during drying (Davis 1943; Luxford and Krone 1942). A superficial reddish-brown stain occurs in the sapwood on the ends of incense cedar logs.

FUNGOUS STAINS

The stains responsible for the greatest losses in wood are those caused by fungi. They are usually sap stains, although they may occasionally occur in the heartwood. Discolorations of the heartwood often indicate the incipient stage of decay (page 346), and certain ones, such as brown oak (page 445), which somewhat resemble stains, are known to be caused in the living tree by slow-acting wood-destroying fungi. The prevalent red heart of paper birch is regarded as largely caused by a mold fungus, *Torula ligniperda* (Campbell and Davidson 1941). The dark heartwood of western red cedar contains microorganisms, although no decay results (Eades and Alexander 1934). Light-colored heartwood is free from microorganisms, but both are practically identical in strength. A purple stain in the heartwood of living lodgepole pine and white spruce is associated with *Coryne sarcoides* (Jacq.) Tul., but this fungus is also widespread in the wood of living trees without causing stain or decay (Etheridge 1957). Sap stains, especially blue stain, may develop in the sapwood of dying trees, but commonly they appear in dead trees, logs, or wood products.

Staining fungi may be roughly divided into two groups: (1) those known as "molds," which develop largely on the surface of wood causing discolorations that can be easily planed off, and (2) those which penetrate wood causing discolorations that cannot be planed off. The stain is produced either in some unknown manner or by an actual color solution secreted by the hyphae that stains the wood. Discoloration is usually most intense in the rays, because the bulk of the food material is found in the cavities of the ray cells. The staining fungi attack the cell walls to a negligible extent or not at all; consequently the lack of food, and in some cases lack of moisture, explains why heartwood is so rarely invaded.

MOLDS

Molds, which appear as fluffy whitish to grayish growths, or as blue-green, green, yellowish, or reddish powdery discolorations on the surface of wood, cause superficial stains that can be readily planed off lumber, even though the mycelium can penetrate deeply into wood. Spores, which are produced abundantly, are usually responsible for the discoloration. The common black mold (*Mucor* spp. and *Rhizopus* spp.) sometimes develops during prolonged kiln drying at low temperatures and high humidities. The fluffy mycelium may become so luxuriant that the spaces between the tiers of boards are filled up, preventing air circulation. Ultimately the mycelium will dry up, but the wood surfaces covered by it will surface check because they did not dry evenly with the other parts of the board. Green molds are most frequent on lumber, usually species of *Penicillium, Trichoderma,* and *Gliocladium* being responsible. *Aspergillus* spp. commonly causes black discolorations. Basket veneers made from wood soaked in hot water or steamed until thoroughly saturated are extremely susceptible to molding, both sapwood and heartwood being affected. While soaking the logs in vats, soluble elements leach from the sapwood into the water and then impregnate the heartwood, and the result is subsequent staining and molding of heartwood, which is ordinarily immune (Hedgcock 1933). Molds develop abundantly on pulp (page 477). For commercial fiberboard, the addition of 2 per cent pentachlorophenate to the fiber furnish and adhesive appreciably reduces molding and strength loss (Englerth 1946). Some fungicides favor the development of molds on green sapwood (Verrall 1949).

Tests of loblolly pine sapwood with green mold caused by *Trichoderma lignorum* (Tode) Harz showed that its strength was insignificantly reduced (Chidester 1942). Trichoderma mold increases the permeability of Douglas fir and southern yellow pines to oil- and waterborne preservatives and to rainwater (Lindgren and Wright 1954). Molds tend to retard the development of decay (page 473). Chemical control of blue stain fungi (page 508) is also effective against most mold fungi. Mold can be removed from staves that cannot be chemically treated by brushing the staves mechanically between steel brushes (Baxter 1930).

MINOR STAINS

Many of these stains are bright colored in shades of red, purple, and yellow. They often occur with blue stain, and most of them can be controlled by the same methods. Some of them seem to develop in wood wetter than that in which only blue stain occurs.

Brown Sap Stain. Brown sap stain is confined to sapwood and is common on jack, ponderosa, and red pines, and it occurs occasionally on sugar and western white pines (Fritz 1952). The stain is chocolate brown in color, occurs in streaks, and has numerous darker flecks appearing in it (Fig. 209). The stain apparently occurs in logs that have been stored for considerable time, and it has not been found developing in lumber during air seasoning. Stained wood is not weakened. The stain is caused by *Cytospora* sp. (Fungi Imperfecti, Phomaceae).

Brick-red Stain. Brick-red stain of Sitka spruce, Douglas fir, and eastern white pine, which is caused by *Ascocybe grovesii* Wells (Ascomycetes, Endomycetaceae), appears as brick-red spots or large areas on rough lumber, varying in intensity from faint to pronounced (Davidson and Lombard 1954). It is common and widespread in the Sitka spruce region. It appears in logs and cants in the immediate vicinity of shakes, seams, and checks. In rough lumber it frequently develops where wet

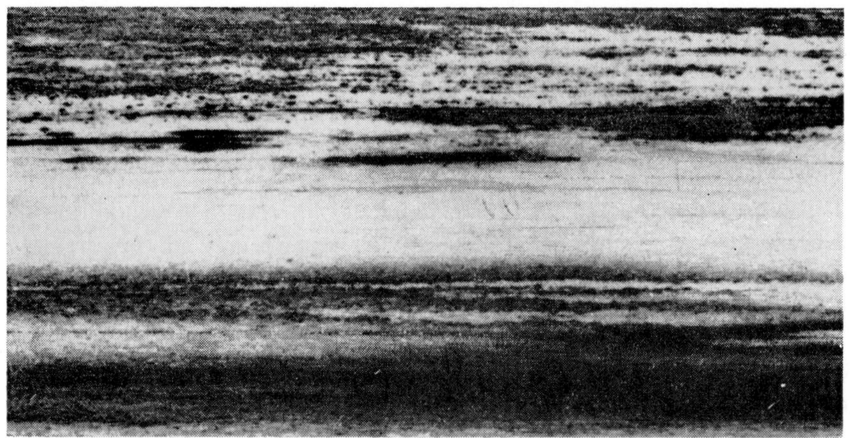

Fig. 209. Brown sap stain of ponderosa pine caused by *Cytospora* sp.

boards or boards and stickers are in contact a long time before drying. The stain may be superficial or deep.

Pink Stain. Pink stain, caused by *Geotrichum* sp. (Fungi Imperfecti, Moniliaceae), appears as a jasper-pink or light-red discoloration of both heartwood and sapwood of southern yellow pine (Chidester 1940). The fungus has also been found in stained heartwood of oak and southern cypress. The stain has been produced in culture in both heartwood and sapwood of several conifers and hardwoods. Other pink, red, orange, and purple stains are produced by species of *Penicillium* and *Fusarium*.

Yellow Stain of Hardwoods. Yellow stain of hardwoods, caused by a species of the *Penicillium divaricatum* group (Ascomycetes, Aspergillaceae), appears on birch, hickory, maple, and oak (Hubert 1929:22). It can develop on logs, lumber, or other wood products during storage. The first symptom may be a moldy appearance of the stock, but later the stain appears pale yellow in color, usually in streaks or irregular areas, and it is most readily seen on freshly surfaced wood. The discoloration often penetrates deeply and cannot be surfaced off. The principal loss caused is the discoloration of the finished product, there being no weakening of the wood.

Red Stain of Boxelder. This stain is confined to living trees, occurring in the heartwood and to some extent in the sapwood of all parts of boxelder from the roots to the smaller branches. It is so prevalent that it has been popularly used as a reliable character for identifying the wood. The color varies from a light coral-red to helebore-red or carmine. The stain is caused by *Fusarium negundi* Sherb. (Toole 1955).

Green Stain of Slash. A brilliant, verdigris-green stain of decorticated, decayed slash is often found in moist situations. This discoloration is caused by *Chlorosplenium* (*Chlorociboria*) *aeruginosum* (Oed.) de Not. (Ascomycetes, Helotiaceae), the fungus being prevalent on hardwoods but infrequent on conifers. The stain is of no economic importance, but it is striking enough to attract attention frequently (Martinez 1943:61, col. pl. III).

Grayish-olive Stain of Hardwoods. This stain, caused by *Lasiosphaeria pezizula* (B. & C.) Sacc. (Ascomycetes, Sphaeriaceae), is common on stored ties or logs of beech, red gum, black gum, and persimmon (Hubert 1921). The discoloration somewhat resembles blue stain, but it is grayish olive instead of bluish gray. Both sapwood and heartwood are affected, the stain usually extending in from 1 to 4 inches on the cut ends of ties or logs. The old hyphae in the wood are olive colored, stout, and twisted and have short cells. The fungus is extremely resistant to heat, probably because it forms spores within the wood tissues.

Grayish-black Stain. *Torula ligniperda* (Willk.) Sacc. (Fungi Imperfecti, Dematiaceae) is associated with various discolorations in hardwoods (Siggers 1922). It has been found causing dark-tan to grayish-black streaks in white ash, dark-colored streaks in red gum and basswood, brownish-black areas in eastern hemlock, and pinkish areas in yellowpoplar. It has also been associated with green to gray to brown discolorations connected with wounds and branch stubs in sugar maple. The fungus occurs in both heartwood and sapwood. It is readily recognized in wood by chains of dark spores that occur frequently in the cell cavities.

BLUE STAIN

This discoloration of wood, known the world over, is caused by a number of fungi. The stain can occur in the sapwood of practically all woods, but conifers are most subject to it, although hardwoods are also

Fig. 210. Blue stain on an eastern white pine board.

Fig. 211. End of a ponderosa pine log showing the sapwood completely blue-stained.

affected (Campbell 1959). White and yellow pines, red gum, and yellow-poplar are especially susceptible. Blue stain may appear in the sapwood when a tree is dying (page 315), because some blue stain fungi can become parasitic on living trees if the moisture content of the sapwood is reduced and the air content raised (Münich 1907–1908:297), but stain more commonly develops in dead trees, logs, lumber, and other wood products throughout the process of manufacture until the wood is dry. If wood has dried without staining, it may discolor if it again becomes moist, but it probably will not stain so readily. Mechanical pulp is subject to severe staining. Occasionally with woods such as Sitka spruce (Hansbrough and Englerth 1944:8), ponderosa pine, and southern pine, the heartwood is partially invaded. Blue stain may also develop in the heartwood of living trees, being prevalent in northern white cedar (Christensen and Kaufert 1942). A blue stain of yellow birch, primarily of the sapwood and localized in the vessels or pores, is termed "black line stain" (Hansbrough 1945). It is presumably caused by *Pullularia pullulans* (de By.) Berkhaut (Campbell 1959:118). The annual loss from blue-stained wood, which results from reduction in grade rather than outright rejection, at one time was estimated at over 10 million dollars in the United States alone.

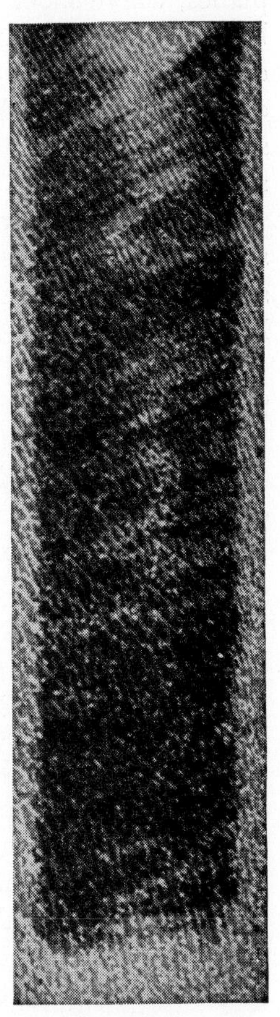

Fig. 212. Interior blue stain in a ponderosa pine board 1½ inches thick. (*Reprinted by permission from "An Outline of Forest Pathology" by Hubert, published by John Wiley & Sons, Inc.*)

Effect on Wood. The grayish, dark-blue, or blackish discoloration may appear as spots, streaks, or irregular areas on the surface of a board or end of a log (Figs. 210, 211). If conditions remain favorable for the fungus the entire sapwood becomes completely blued. In a thick board that has dried unevenly, the central area of the board may blue although the outer portion and surface remain clear (Fig. 212). This is termed "interior blue stain."

The strength of wood is not significantly reduced by blue stain, except in toughness, where the reduction may amount to 25 per cent or more (Chapman and Scheffer 1940; Pettifor and Findlay 1946; Thunell 1952; Scheffer and Lindgren 1940:28). Consequently, blue-stained wood can be used with confidence for structural timbers with reasonably high factors of safety. It is not advisable to use blued wood for small, highly stressed members with low factors of safety, as in certain parts of airplanes, where shock resistance is important. Not all blue stain fungi have been investigated, and therefore strength reduction may be expected in certain cases. *Botryodiplodia theobromae* Pat. causing blue stain in obeche, a tropical African wood, brought about a serious reduction in strength properties, behaving like a typical white rot fungus (Findlay and Pettifor 1939).

Although there are no exact records of the decay resistance of blue-stained wood in service, stained wood in laboratory tests has been slightly less decay resistant than unstained (Björkman 1947; Scheffer and Lindgren 1940:31). The difference was of no practical importance. Although blued wood is popularly supposed to be more difficult to impregnate with preservatives than unstained wood, blued southern yellow pine in carefully controlled experiments absorbed more creosote than unstained wood (Lindgren and Scheffer 1939). Other investigators obtained the same general result (Bellman and Francke-Grossmann 1952). The high moisture content of stained sapwood, not the presence of blue stain fungi, has been responsible for reports of its resistance to impregnation. Stained wood dries more rapidly than unstained (Scheffer 1941).

Blued pulpwood yields a darker colored paper that is generally objectionable. However, the pulp can be bleached to a normal color by the use of additional bleaching powder. Sulfite pulps from stained spruce chips required 7 per cent more bleaching powder and from southern pine chips 9 per cent more to obtain the same whiteness as pulp from unstained wood (Sutermeister 1922; Chidester, Bray, and Curran 1938).

Blue-stained wood can be used for most purposes, provided the discoloration itself is not objectionable, although this factor is of no consequence if wood is to be painted. Although no exact tests are available, experience indicates that paintability and gluing properties of stained and unstained wood are comparable. Since conditions favoring blue stain also favor decay, stained wood should be carefully examined for indications of rot.

Fungi Causing Stain. Blue stain in wood is caused by a number of fungi formerly belonging to the genera *Ceratostomella, Endoconidiophora,* and *Ophiostoma* of the Ascomycetes, but now included in the genus *Ceratocystis* (Hunt 1956), and *Alternaria, Cadophora, Diplodia, Discula, Graphium, Hormodendron, Hormonema, Leptographium, Sclerophoma, Sphaeropsis,* and *Trichosporium* of the Fungi Imperfecti (Lagerberg, Lundberg, and Melin 1927; Nisikado and Yamauti 1933–

1935; Verrall 1939, 1941*a*). In the United States *Ceratocystis* spp. are the most common cause of blue stain in logs and lumber, *C. pilifera* (Fr.) C. Moreau being prevalent in lumberyards throughout the country. Groundwood pulp is stained and molded by some of the foregoing and by many other fungi (Goidanich et al. 1938; Melin and Nannfeldt 1934; Rennerfelt 1941; Robak 1932).

The mycelium of *Ceratocystis* spp. concentrates in the rays and wood parenchyma, because it is there that the food material utilized by these fungi largely occurs (Figs. 213, 214). Hyphae develop sparingly in vessels, tracheids, and other wood cells. Although these fungi live mainly on substances in the cell cavities, they do attack cell walls to a limited

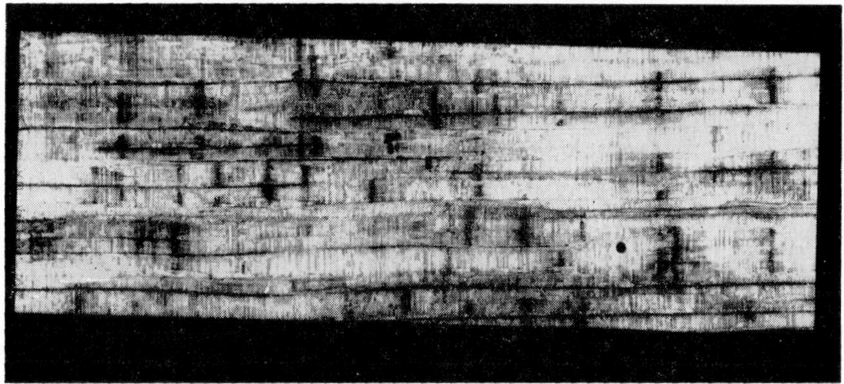

Fig. 213. Radial or edge-grain view of blue-stained eastern white pine showing discoloration concentrated in the rays.

extent (Scheffer and Lindgren 1940:25). Tracheid or fiber cell walls in close contact with hyphae may be thinner than those without such contact. Hyphae often pass directly through the walls by means of bore holes, usually being considerably restricted at the point of passage. Generally the hyphae pass from cell to cell through the simple or bordered pits, the pit membranes offering little resistance. Hyphae sometimes occur in the middle lamella, which may explain reduced shock resistance so frequent in blued wood (Liese 1953). In the rays considerable destruction may occur. The slightly lignified walls between the ray cells may be decomposed completely, while the walls separating the ray cells from the neighboring wood cells may become very thin, thus leaving only an outer framework of the ray. The young hyphae are fine and colorless, but with age they quickly become stout and brown in color. How these brown hyphae cause a blue color in wood is unknown, since microscopic examination of the blued wood shows no blue tint. The various theories have been summarized by Howard (1922:11).

The fructifications or perithecia of *Ceratocystis* under a hand magnifier appear as stiff black hairs or bristles 1 to 2 mm long, with a bulbous base that may be partially or completely embedded in the wood (Fig. 215). They develop on the surface of rough lumber and on the wood

Fig. 214. Blue-stained eastern white pine showing hyphae of *Ceratocystis* sp. concentrated in the rays. (*Drawn by J. W. Kimmey.*)

under the loosened bark, on the bark next to the wood, or in insect galleries in the bark or wood of dead trees and logs. On lumber the perithecia often develop so abundantly as to form a dark hairy coating (Fig. 216). Each perithecium contains a number of asci, each ascus having eight colorless, more or less cylindrical ascospores. Under favorable moisture conditions the ascospores ooze from the ends of the perithecial beaks in white sticky masses. The dissemination of staining fungi is by means of air currents, insects, milling machinery, rain water, and transportation of infected wood (Verrall 1941*b*). Infection of dead trees and logs with bark intact is largely by ambrosia beetles and bark beetles (Leach 1940:217; Mathiesen-Käärik 1953).

Conditions Affecting Staining. Fungi causing blue stain must have a sufficient food supply, which is almost invariably present in the sap-

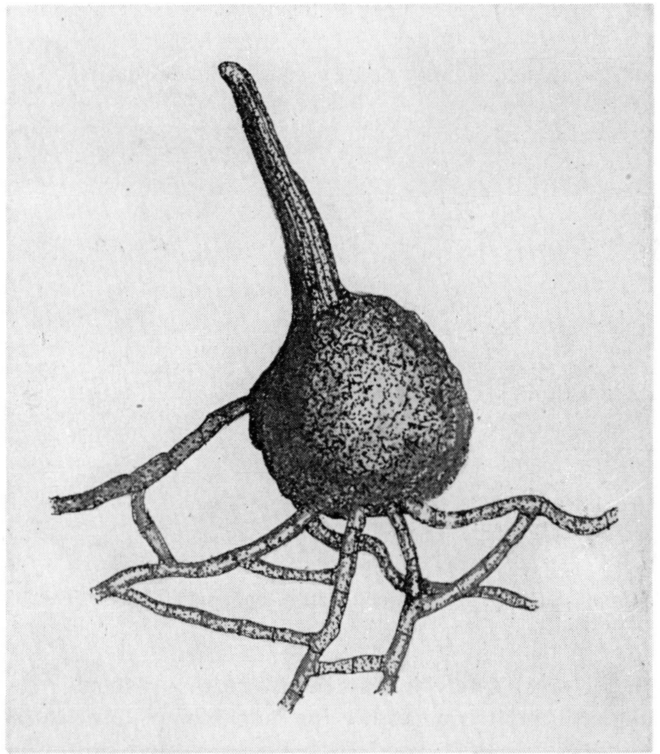

Fig. 215. A perithecium and hyphae of *Ceratocystis minor* (Hedgc.) Hunt from ponderosa pine. Much enlarged.

wood, a sufficient moisture supply, and warm weather. During humid weather in the warm periods of spring, summer, and early autumn when wood dries slowly, staining is most severe. Green lumber fresh from the saw, bulk piled for storage or shipped in closed cars or the holds of vessels, stains severely. Blue stain fungi develop most rapidly when the temperature is from about 75 to 85°F (Lindgren 1942). During warm weather the average temperature within a pile of unseasoned lumber is several degrees lower than that of the surrounding air. At about 40°F and below and at about 95°F and above, growth practically ceases.

The mycelium can be killed by artificially high temperatures. For shortleaf pine sapwood infected with *Ceratocystis pilifera,* the lower limiting moisture content for development of the fungus was 24 per cent based on oven-dry weight of the wood (Lindgren 1942), and there was visible staining only in wood that was at or above fiber saturation point (27 + per cent). This figure agrees approximately with a lower limiting moisture content of 28 per cent previously determined for Scotch pine in Europe. For practical purposes 20 per cent moisture content based on oven-dry weight of the wood should be set as the lower limit for staining to

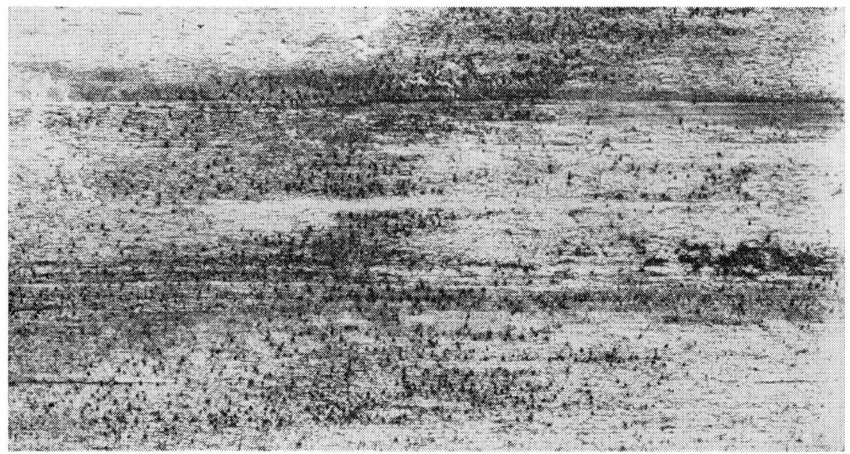

Fig. 216. *Ceratocystis* sp. fruiting abundantly on rough lumber. The black specks are the perithecia.

provide a factor of safety in case moisture requirements vary for different fungi. Since air is necessary for fungi, completely saturated wood does not discolor. Frozen wood (Schmid 1935) and frozen pulp (Rennerfelt 1937) on thawing are more susceptible to staining than unfrozen material.

Timber cut in the autumn and winter is less liable to staining than timber cut in the spring and summer. Experiments with Scotch pine and Norway spruce did not show that nutritively spring- and summer-felled timber was any more favorable for blue stain fungi than winter-felled wood (Lagerberg, Lundberg, and Melin 1927:616). The advantage of winter-felled timber is only that it has a chance to dry somewhat before warm weather arrives to promote the activity of blue stain fungi.

Certain species such as the white and yellow pines are peculiarly susceptible to staining, but the reason for this has not been determined. It may have to do with the acidity of the sapwood or the character of the substances available for food. Either because of too low a moisture content or lack of suitable food material, most likely the latter, blue stain fungi rarely invade heartwood.

Control. This has been exhaustively discussed by Scheffer and Lindgren (1940). Prevention, not cure, is the only control for blue stain, because once discoloration has occurred it cannot be eradicated. Dry wood or wood completely saturated with moisture cannot stain, but since it is impractical to keep wood water-soaked except for limited periods during storage, the most effective method for preventing stain is rapid drying.

In most regions piling green lumber properly as it comes from the saw will prevent staining during all times of the year except for short periods unusually favorable to the development of blue stain fungi. Lumber may be end-racked, that is, the pieces leaned crosswise on edge against a horizontal support, for from 3 to 7 days, which usually reduces the surface moisture content sufficiently to prevent staining except during periods of weather particularly favorable to blue stain fungi. End-racking is more expensive than chemical treatment. Before warping sets in, the stock should be piled in the yard in narrow, elevated, well-spaced, and properly roofed piles. Narrow dry stickers, preferably chemically treated, should be used to separate the tiers of boards, and there should be lateral spacing between boards. Broad stickers induce stain, and stickers not chemically treated are more likely to carry infection. When the stock is reduced to 20 per cent moisture content or less, based on oven-dry weight, it will not stain, but air drying does not kill mycelium in the wood; consequently if it again becomes moist, discoloration can occur. *Ceratocystis* sp. is known to have remained alive for 7 years in air-dried wood (Scheffer and Chidester 1948).

If lumber is immediately kiln dried as it comes green from the saw, bluing is effectively prevented. In addition fungi in lumber up to and including 4 by 4 inches square are killed by the temperatures of commercial kiln runs excepting temperatures below 120°F (Hubert 1924). Pieces up to and including 8 by 8 inches square can be sterilized at 130°F for 9 hours in a saturated atmosphere. Steam-pressure treatment is also effective and quicker.

Where the best methods of air seasoning will not prevent blue stain and it is impossible for mills to kiln-dry any or all of their stock, lumber can be protected against stain and decay for several months, even though it is bulk piled during storage or transport, by dipping or spraying the lumber for from 10 to 30 seconds in chemical solutions as it comes green from the saw, but no chemical will prevent stain if the wood is already infected before dipping. The following chemical solutions have been successful (Verrall and Mook 1951; Verrall and Scheffer 1949).

Product	*Pounds in 100 gal of water*
Borax	32
Dowicide G (sodium pentachlorophenate)	7
Dowicide H (sodium tetrachlorophenate)	6
Lignasan (ethyl mercuric phosphate)	2
Permatox 10s (sodium pentachlorophenate plus borax)	10
Santobrite (sodium pentachlorophenate)	7
Mixture 1	
Lignasan	1
Dowicide G or Santobrite	4
Mixture 2	
Lignasan	½
Dowicide G or Santobrite	2
Borax	6

Borax alone is usually effective on hardwoods only. In the South, Dowicide H is recommended for hardwoods only. The two mixtures are particularly valuable when conditions are severe. These products, except borax, are irritating to the skin if not handled properly. Care is especially needed with the chlorinated phenols. To be effective the solutions must be maintained at full strength.

Dipping is simply an aid in preventing stain and decay during air seasoning, so that it is still necessary to follow the methods of yard piling previously outlined. Bulk piling should be avoided if possible. Lumber should preferably be dipped within 24 hours after it comes from the saw because more than 2 days' delay before dipping can result in severe staining.

If the lumber is cut from infected logs, dipping will not prevent further discoloration. Where logs cannot be immediately sawed into lumber the same chemicals, but in concentrations three or more times as strong as used for lumber, should be sprayed on the ends and barked areas of the freshly cut logs. Borax alone, at the rate of 32 pounds per 100 gallons of water, is effective for hardwood logs.

To prevent molding and staining of wood boxes and crates promoted by damp conditions of storage and aggravated by the use of unseasoned lumber, dipping the panels in a 2 per cent aqueous solution of sodium pentachlorophenate is reasonably effective (Scheffer 1946).

A laboratory method of treating unseasoned discs of sapwood with antistain chemicals and then inoculating the discs with sap-staining fungi enables the prediction with reasonable accuracy and promptness of the possible results of large-scale applications on merchantable lumber (Hatfield et al. 1950).

After wood is blued, there is no successful method of bleaching it. Blue-stained yellowpoplar can be restored to normal color by dipping in a solution of 50 pounds of sodium phosphate to 60 gallons of water. The wood bleaches to a depth of 1/16 to 1/8 inch, but the natural color is only temporarily restored. Blued pulp can be successfully bleached (page 502).

REFERENCES

Anonymous: Chemical Brown Stain in Pine. *U.S. Dept. Agr., Forest Serv., Forest Prods. Lab. Tech. Note,* 254 (rev.):1–2 (1951).

Bailey, I. W.: Oxidizing Enzymes and Their Relation to "Sap Stain" in Lumber. *Botan. Gaz.,* 50:142–147 (1910).

Baxter, D. V.: A Brush Treatment of Moldy Staves. *Phytopathology,* 20:575–582 (1930).

Bellman, H., and Francke-Grossmann, Helene: Versuche zur Trankfähigkeit verblauten Kiefernsplintholzes. I. Tränkung mit öligen Imprägniermitteln. *Holz Roh- u. Werkstoff,* 10:456–463 (1952).

Björkman, E.: On the Development of Decay in Building-timber Injured by Blue Stain. *Svensk Papperstidn.,* 50:49–52 (1947).

Boyce, J. S.: Decays and Discolorations in Airplane Woods. *U.S. Dept. Agr. Bull.,* 1128:1–51 (1923).

Campbell, R. N.: Fungus Sap-stains of Hardwoods. *Southern Lumberman,* 199:115–120 (1959).

Campbell, W. A., and R. W. Davidson: Red Heart of Paper Birch. *J. Forestry,* 39:63–65 (1941).

Chapman, A. D., and T. C. Scheffer: Effect of Blue Stain on Specific Gravity and Strength of Southern Pine. *J. Agr. Research.,* 61:125–133 (1940).

Chidester, G. H., M. W. Bray, and C. E. Curran: Characteristics of Sulphite and Kraft Pulps from Blue-stained Southern Pine. *Paper Trade J.,* 106(14):43–46 (1938).

Chidester, Mae S.: A Pink Stain of Wood Caused by a Species of *Geotrichum. Phytopathology,* 30:530–533 (1940).

Chidester, Mae S.: The Effect of a Mold, *Trichoderma lignorum,* on Loblolly Pine Sapwood. *Proc. Am. Wood-Preservers' Assoc.,* 38:134–138 (1942).

Christensen, C. M., and F. H. Kaufert: A Blue-Staining Fungus Inhabiting the Heartwood of Certain Species of Conifers. *Phytopathology,* 32:735–737 (1942).

Clark, J. W.: A Gray Non-fungus Seasoning Discoloration of Certain Red Oaks. *Southern Lumberman,* 194(2418):35–38 (1957).

Davidson, R. W., and Frances Lombard: Brick Red Stain of Sitka Spruce and Other Wood Substrata. *Phytopathology,* 44:606–607 (1954).

Davis, S.: This Much We Know—We Think. *West Coast Lumberman,* 70(10):44, 46 (1943).

Eades, H. W., and J. B. Alexander: Western Red Cedar: Significance of Its Heartwood Colorations. *Can. Dept. Interior, Forest Serv. Circ.,* 41:1–17 (1934).

Englerth, G. H.: Deterioration of Fiberboard by Molds. *U.S. Dept. Agr., Forest Pathol. Spec. Release,* 29:1–7 (1946).

Etheridge, D. E.: Comparative Studies of *Coryne sarcoides* (Jacq.) Tul. and Two Species of Wood-destroying Fungi. *Can. J. Botany,* 35:595–603 (1957).

Findlay, W. P. K.: Sap-stain of Timber. *Forestry Abstr.*, 20:1–7, 167–174 (1959).

Findlay, W. P. K., and C. B. Pettifor: Effect of Blue Stain on the Strength of Obeche (*Triplochiton scleroxylon*). *Empire Forestry J.*, 18:259–267 (1939).

Fritz, Clara W.: Brown Stain in Pine Sapwood Caused by *Cytospora* sp. *Can. J. Botany*, 30:349–359 (1952).

Goidanich, G., et al.: "Ricerche sulle alterazioni e sulla conservazione della pasta di legno destinata alla fabbricazione della carta esequite nella R. Stazione di Patologia Vegetale di Roma." [French, English, and German summaries.] Rome: Tipografia del Senato dell Dott. G. Bardi (1938).

Good, H. M., P. M. Murray, and H. M. Dale: Studies on Heartwood Formation and Staining in Sugar Maple, *Acer saccharum* Marsh. *Can. J. Botany*, 33:31–41 (1955).

Hansbrough, J. R.: The Significance of Black Line Stain in Yellow Birch Propeller Lumber. *U.S. Dept. Agr., Forest Pathol. Spec. Release*, 23:1–4 (1945).

Hansbrough, J. R., and G. H. Englerth: The Significance of the Discolorations in Aircraft Lumber: Sitka Spruce. *U.S. Dept. Agr., Forest Pathol. Spec. Release*, 21:1–13 (1944).

Hatfield, I., et al.: The Unseasoned Wood Disc Method for Evaluating Sap-stain Control Chemicals. *Phytopathology*, 40:653–663 (1950).

Hedgcock, G. G.: The Prevention of Woodstaining in Basket Veneers. *J. Forestry*, 31:416–420 (1933).

Howard, N. O.: The Control of Sap-stain, Mold, and Incipient Decay in Green Wood with Special Reference to Vehicle Stock. *U.S. Dept. Agr. Bull.*, 1037:1–55 (1922).

Hubert, E. E.: Notes on Sap Stain Fungi. *Phytopathology*, 11:214–224 (1921).

Hubert, E. E.: Effect of Kiln Drying, Steaming, and Air Seasoning on Certain Fungi in Wood. *U.S. Dept. Agr., Dept. Bull.*, 1262:1–20 (1924).

Hubert, E. E.: Sap Stains of Wood and Their Prevention. *U.S. Dept. Com., Wood Utilization* (1929).

Hunt, J.: Taxonomy of the Genus *Ceratocystis*. *Lloydia*, 19:1–58 (1956).

Kollman, F., R. Keylwerth, and H. Kübler: Verfärbungen des Vollholzes und der Furniere bei der kunstlichen Holztrocknung. *Holz Roh- u. Werkstoff*, 9:382–391 (1951).

Lagerberg, T., G. Lundberg, and E. Melin: Biological and Practical Researches into Blueing in Pine and Spruce. *Svenska Skogsvårdsför. Tidskr.*, 25:145–272, 561–739 (1927).

Leach, J. G.: "Insect Transmission of Plant Diseases." New York: McGraw-Hill Book Company, Inc. (1940).

Liese, W.: Elektronenmikroskopische Untersuchungen an verblauten Kiefernholz. *Deut. Botan. Ges. Ber.*, 66:427 (1953).

Lindgren, R. M.: Temperature, Moisture, and Penetration Studies of Wood-staining Ceratostomellae in Relation to Their Control. *U.S. Dept. Agr. Tech. Bull.*, 807:1–35 (1942).

Lindgren, R. M., and T. C. Scheffer: Effect of Blue Stain on the Penetration of Liquids into Air-dry Southern Pine Wood. *Proc. Am. Wood-Preservers' Assoc.*, 35:325–336 (1939).

Lindgren, R. M., and E. Wright: Increased Absorptiveness of Molded Douglas-fir Posts. *J. Forest Products Research Soc.*, 4:162–164 (1954).

Luxford, R. F., and R. H. Krone: Chemical Stain in Noble Fir as Related to Strength. *U.S. Dept. Agr. Forest Serv., Forest Prods. Lab. Mimeograph*, 1391:1–6 (1942).

Martinez, J. B.: "La investigacion de las alteraciones micologicas de la madera. 1.er fasciculo." [French, English, and German summaries.] Madrid: Consejo Superior de Investigaciones Cientificas (1943).

Mathiesen-Käärik, A.: Eine Übersicht über die gerwöhnlichsten mit Borkenkäfern assoziierten Bläuepilze in Schweden und enige für Scheweden neue Bläuepilze. *Statens Skogsforskningsinst. (Sweden), Medd.*, 43(4):1–74 (1953).

Melin, E., and J. A. Nannfeldt: Researches into the Blueing of Ground Wood-pulp. *Svenska Skogsvårdsför. Tidskr.*, 32:397–616 (1934).

Millet, M. A.: Chemical Brown Stain in Sugar Pine. *J. Forest Products Research Soc.*, 2:232–236 (1952).

Münch, E.: Die Blaufäule des Nadelholzes. *Naturw. Z. Land- u. Forstwirtsch.*, 5:531–573; 6:32–47, 297–323 (1907–1908).

Nisikado, Y., and K. Yamauti: Contributions to the Knowledge of the Sap Stains of Wood in Japan. *Ber. Ōhara Inst. landwirtsch Forsch. Kurashiki, Japan,* 5:501–538; 6:467–490, 539–560 (1933–1935).

Pettifor, C. B., and W. P. K. Findlay: Effect of Sap-stain on the Tensile Strength of Corsican Pine Sapwood. *Forestry,* 20:57–61 (1946).

Rennerfelt, E.: Undersökningar över svampinfektionen i slipmassa och dess utveckling däri. [English summary.] *Svenska Skogsvårdsför. Tidskr.*, 35:47–159 (1937). [*Pulp Paper Mag. Can.*, 38:561–568 (1937).]

Rennerfelt, E.: The Development of the Fungus Flora in Wet Mechanical Pulp, Manufactured at Different Temperatures and Stored under Different Conditions. *Göteborgs Högskolas Årsskr.*, 47(22):1–47 (1941).

Robak, H.: Investigations Regarding Fungi on Norwegian Ground Wood Pulp and Fungal Infection at Wood Pulp Mills. *NYT Mag.*, 71:185–330 (1932).

Roth, E. R.: Discolorations in Living Yellow Poplar Trees. *J. Forestry,* 48:184–185 (1950).

Scheffer, T. C.: Drying Rates of Bluestained and Bright Lumber. *Southern Lumberman,* 162(2039):46–48 (1941). *Wood Products,* 46:23–25 (1941).

Scheffer, T. C.: Chemical Dipping Treatments for Controlling Molding and Staining of Wood Boxes and Crates. *U.S. Dept. Agr., Forest Pathol. Spec. Release,* 28:1–11 (1946).

Scheffer, T. C.: Mineral Stain in Hard Maples and Other Hardwoods. *U.S. Dept. Agr., Forest. Serv., Forest Prods. Lab. Rept.,* R1981:1–2 (1954).

Scheffer, T. C., and A. D. Chapman: Prevention of Interior Brown Stain in Persimmon Sapwood During Seasoning. *Hardwood Record,* 72(11):17 (1934).

Scheffer, T. C., and Mae S. Chidester: Survival of Decay and Blue-stain Fungi in Air Dry Wood. *Southern Lumberman,* 177(2225):110–112 (1948).

Scheffer, T. C., and R. M. Lindgren: Stains of Sapwood and Sapwood Products and Their Control. *U.S. Dept. Agr. Tech. Bull.,* 714:1–123 (1940).

Schmid, W.: Ueber Verpilzung von Feuchtholzschliff. *Papier-Fabr.,* 23:380–382, 387–389 (1935). [*Rev. Appl. Mycol.,* 15:270–271 (1936).]

Siggers, P. V.: *Torula ligniperda* (Willk.) Sacc., a Hyphomycete Occurring in Wood Tissue. *Phytopathology,* 12:369–374 (1922).

Sutermeister, E.: The Use of Rotten and Stained Wood for Making Sulphite Pulp. *Pulp Paper Mag. Can.,* 20:513–514 (1922).

Thunnell, B.: Einwirkung der Bläue auf die Festigkeitschaften der Kiefer. *Holz Roh- u. Werkstoff,* 10:362–365 (1952).

Toole, E. R.: Red Stain of Boxelder. *U.S. Dept. Agr., Plant Disease Reptr.,* 39:66–67 (1955).

Verrall, A. F.: Relative Importance and Seasonal Prevalence of Woodstaining Fungi in the Southern States. *Phytopathology,* 29:1031–1051 (1939).

Verrall, A. F.: Fungi Associated with Stain in Chemically Treated Green Lumber *Phytopathology,* 31:270–274 (1941*a*).

Verrall, A. F.: Dissemination of the Fungi That Stain Logs and Lumber. *J. Agr. Research,* 63:549–558 (1941*b*).

Verrall, A. F.: The Control of Fungi in Lumber During Air-seasoning. *Botan. Rev.* 11:398–415 (1945).

Verrall, A. F.: Some Molds on Wood Favored by Certain Toxicants. *J. Agr. Research,* 78:695–703 (1949).

Verrall, A. F., and P. V. Mook: Research on Chemical Control of Fungi in Green Lumber, 1940–51. *U.S. Dept. Agr. Tech. Bull.,* 1046:1–60 (1951).

Verrall, A. F. and T. C. Scheffer: Control of Stain, Mold, and Decay in Green Lumber and Other Wood Products. *Proc. Forest Products Research Soc.,* 3:480–489 (1949).

CHAPTER 21

Principles of Forest-disease Control

The forest-disease situation in the United States has been outlined by Hartley, Boyce, et al. (1933), and part of the discussion in the following pages has been taken verbatim from that paper. There is such a difference in the effect on our forests between diseases caused by native fungi and by introduced fungi that they must be considered separately for procedure in prevention and control.

NATIVE DISEASES

Diseases caused by native fungi, or fungi that have been introduced in the remote past and to which our forests have become adjusted, are unlikely to cause catastrophic epidemics in native tree species or to threaten the commercial extermination of any forest tree, so that control measures against them can proceed indirectly as an integral part of routine silvicultural practice.

The prevention and control of forest-tree diseases is circumscribed because, in general, forest stands have a relatively low value per unit of area so that expenditures on them must be relatively small. This means that intensive methods of disease control such as cultivation, irrigation, fertilization, crop rotation, spraying, and dusting are rarely applicable in forestry in this country. It is only in forest nurseries where the crop has a high value on a small area that intensive control measures are economically possible. Consequently the forester is largely limited in controlling disease by what can be accomplished during the various cuttings to which a regulated stand is subjected, and in fact such control, if regularly followed, will be sufficiently effective in time to reduce the incidence of most native diseases in natural stands on suitable sites to the point where they can be tolerated. Diseased trees should be removed during the intermediate and final cuttings because each one, unless it is soon to die, continues to occupy growing space that could be better utilized by a thrifty tree and also increases the likelihood of the spread of infectious diseases to other trees in the stand. Besides removing infected trees, decay of the heartwood can be reduced by lowering the rotation, by protection against wounding particularly by fire, and in addition in hardwood stands by favoring seedlings over sprouts. The forester must be familiar with the symptoms of various diseases so that he can intelligently follow pathologic marking rules as they are drawn up for various regions and types (Hepting 1933; Weir and Hubert 1919). Prevention and control are based on a knowledge of the various diseases, which is still inadequate, and even the incidence of common diseases is not always well known. Obtaining information on the distribution and intensity of diseases should be routine procedure to prevent or reduce losses (Weir and Hubert 1918; Kienholz and Bidwell 1938; Peltier, Schroeder, and Wright 1939).

Conditions in natural stands point strongly to the fact that there is no factor more important in relation to disease than tree vigor. Stands on good sites are generally not damaged significantly by native diseases, but those on poor sites often suffer severely. Site selection is all-important because no amount of care can save a stand in an unsuitable environment. With natural regeneration of native species the stand is adapted to the land it occupies, but it also must be accepted that on marginal and submarginal forest land satisfactory timber stands cannot be grown with assurance of success because of diseases, insects, or other factors and that attempts at control on such areas are not economically justifiable. However, it is with artificially reproduced stands that the full significance of site quality is most readily apparent, and in Europe with its extensive planted stands, commonly of trees outside their natural range, there are many examples of an unsuitable environment increasing the incidence of disease (Day 1934). The relation of disease to lack of vigor may be so constant that the activity of a pathogen can be used as a sure indicator of unfavorable growing conditions for the host, as for instance the presence of *Diplodia pinea* on planted pines in New Zealand (Birch 1936:18).

But even stands on good sites must be properly cared for, which particularly applies to planted stands. Too often plantations are established at considerable cost only to be neglected, so that serious losses occur from disease or insects. Timely thinning is essential to maintain stand vigor, and, if a stand stagnates, it becomes susceptible to disease. In Great Britain where most stands are reproduced by planting, the statement has been made that fully 75 per cent of the troubles caused by fungi and insects is primarily chargeable to neglected thinnings (St.

Clair-Thompson 1928:157). In the Northeast, Tympanis canker is known to appear in stagnant stands of red pine.

Where forest conditions have been largely destroyed over extensive areas by fire or by unregulated cutting, the new stand, particularly in its juvenile stages, is likely to suffer more severely from disease than would normally be the case, more especially from noninfectious diseases caused by agencies such as drought, frost, winter injury, and unfavorable soil conditions. The destruction of normal forest conditions can lead to such profound changes in site that on the less favorable locations it may be impossible to establish again a satisfactory stand of the climax species for decades, even by planting (Clark 1954). Forest land can be so changed by heavy cutting followed by repeated fires that it is necessary to restock it first with species less exacting in their requirements in order to restore and improve the site sufficiently so that the more valuable species that formerly occupied the area can again be grown successfully. With this principle in mind, it is evident that in cutting timber the selection method or one of its modifications should be practiced, where possible, instead of clear-cutting, unless clear-cutting has been nature's method. In general the sudden and complete exposure of the soil following clear-cutting cannot be other than less favorable to the new stand and on steep slopes may lead to an actual loss of top soil by erosion. As a rule clear-cutting with natural regeneration is to be preferred to clear-cutting with artificial reproduction.

In general naturally regenerated stands are less susceptible to disease than those artificially reproduced. Seedlings are preferable to sprouts because sprout growth is usually more susceptible to disease. In California the powdery mildew on oak sprouts, caused by *Sphaerotheca lanestris*, which dwarfs them and even causes witches'-broom formation, is of no consequence on seedlings. Hardwoods of sprout origin are often affected by butt rot that enters from the parent stump. Whether stands established by direct seeding are more susceptible to disease than naturally regenerated stands is problematical, but certainly planted stands are (Boyce 1954). At best trees are more or less injured by planting. Roots are particularly subject to injury, thus increasing the incidence of root rots. In Corsica, chestnuts occurring naturally in forest stands on mountain slopes are free from the ink disease, which is devastating those planted in the valleys. On the Island of Bornholm in the Baltic Sea, root rot caused by *Fomes annosus* was more prevalent in planted than in naturally reproduced silver fir (Boyce 1927:14). *Polyporus schweinitzii*, which ordinarily decays the heartwood in the butt of large trees, has damaged planted eastern white and red pines in New York by killing roots (page 114). The same fungus has damaged an eastern white pine plantation in Michigan with the root rot not appearing until nearly 30 years after the plantation was established. Baxter (1937:34) warns that great care must be taken in selecting planting sites, and that eastern white pine should not be planted on areas with water tables that tend to rise during wet seasons into the root zones and saturate the soil so as to drown out the roots, or on areas with soils that are subject to alternate saturation and drying out due to an underlying layer of impervious clay. A layer of hardpan can cause the same unfavorable conditions.

Provided that plantations are established on suitable sites, of species adapted to the locality, with stock grown from seed from the proper locality and from thrifty mother trees, and further provided that stand composition is correct and that the stock has been properly grown in the nursery and is carefully planted, then such plantations should develop almost as well as naturally reproduced stands. Of course there is a weakening from the shock of planting and from the curtailed and unnaturally placed root system characteristic of planted trees. The latter handicap is never entirely overcome, and therefore planted trees are generally less disease resistant.

There is always a tendency to establish plantations of one species, rather than mixtures, without regard to how the planted species occurs naturally, because both silviculture and and utilization are simplified in pure stands and because of the economic pressure to favor a more valuable species over its less valuable associates. Pure stands are more susceptible to diseases, particularly those caused by introduced parasites, than mixed stands. Mixtures of conifers and hardwoods are especially desirable because these two classes of trees in general have their own groups of fungous parasites. A pure stand forms an ideal situation for a pathogen to build up to epidemic proportions. Infection is direct and rapid from tree to tree, and if the one species is destroyed there is nothing left. The most hazardous pure stands are even-aged because fungous parasites are often virulent during only one stage in the development of a tree. Pure stands of trees outside their natural range are particularly liable to difficulty.

The establishment of pure stands by planting, or even by modifications in the reproduction method to favor one species, should be approached with caution unless the species naturally forms pure stands. Champion (1933*a*) maintains that in Europe failures of pure stands cannot be ascribed to the purity of the crop but rather to mistakes in management, to planting up extensive areas with a species outside its natural distribution as regards either climate or soil, to planting on soils perhaps potentially suitable but rendered unsuitable by undue exposure or other faulty treatment, to using unsuitable strains of seed or planting stock, to failure from fungous and insect attack, and so forth. However, the author agrees with Oliphant (1932) that the idea that purity as such is safe should be limited to species that naturally grow in pure stands. It also seems that growing in pure stands species that naturally occur in mixture may ultimately lead to soil deterioration. This will be a slow and insidious process, not culminating for several rotations or more, but once such deterioration is accomplished, it will likely be difficult, if not impossible, to restore the soil to its original state. In western Europe where Norway spruce has been established outside its range in extensive pure stands, commonly on natural hardwood land, and where pure stands of other species have also been long established, reduction in rate of growth accompanied by fungus and insect attacks has become so serious that conversion of these stands back to mixtures approximating their original composition is considered essential (Fourchy 1954; Köstler 1952; Leibundgut 1947; Péter-Contesse 1956).

From the economic standpoint one species may be so much more valuable than its natural associates that it may seem and may prove to be financially sound to grow it in pure stands, although with changing demand there is no assurance that any species will retain its relative value. In countries with adequate forests, forestry should be based on the care and improvement of existing natural forests. It has been pointed out that natural forests are not necessarily healthy forests (Peace 1957), yet they are likely to be healthier than artificial forests. If pure stands are established to replace mixtures, this should be done as it is now done for orchards, only with the full realization that additional expenditures must ultimately be expected for protection against parasites. In the Northeast, eastern white pine, which reaches its best development in mixture, is now growing extensively in pure stands, both planted and naturally reproduced where it has reclaimed abandoned farm lands. Yet, to grow pine of quality and reduce damage by the white pine weevil (*Pissodes strobi* Peck), it is necessary to reestablish the old mixtures or to make expenditures for artificial pruning and insect control. Red pine in pure plantations on sites where it does not occur naturally is severely attacked by *Fomes annosus* in southern New England.

Of course, it is not advisable to change natural pure stands into mixtures in order to lessen the potential danger from virulent introduced parasites. Pure even-aged Douglas fir, pure even-aged lodgepole pine, and pure uneven-aged ponderosa pine have always thriven in nature. From the standpoint of resistance to disease, silvicultural practice should maintain as closely as possible natural-stand composition. It should emulate nature and improve on her methods but not change them radically, because the result of any radical change may be difficulties and ills that cannot be foreseen. Under intensive management there is frequently an urge to modify drastically natural-stand composition, but as Oliphant (1932:186) points out, in the tropics even the limited change in composition involved in converting natural rain forests to regulated stands by natural regeneration accentuates the risk from diseases and insects; therefore it is necessary to avoid changes so drastic that the risk outweighs the economic advantage of an increased proportion of valuable species. According to Dimitroff (1935) the natural forests of Bulgaria were resistant to attacks by insects and fungi, but with the introduction of silvicultural methods, aided by unwise exploitation and considerable injury by fire, the health of the forests declined, and the damage by pathogenic fungi increased. Of course, in the management of any forest for wood production, it is impossible to avoid at least temporarily changing natural conditions every time cutting occurs, but it is important to foresee as far as possible just what the result of such changes will be and thus avoid practices that lead to unfavorable conditions. In placing natural forests under management in this country, the forester should be guided by forest conditions that nature has evolved over the centuries.

In general except on an experimental basis there is little need for introducing exotics into the United States, which has a great variety of conifers and hardwoods admirably adapted to the various forest regions, because it is unlikely that a tree will be found which in the end will be silviculturally or economically more desirable than a native species. Exotics should be introduced only if they fulfill some definite need not met by native species. Furthermore the introduction of exotics is always attended by difficulties in securing seed from the proper source, in selecting suitable sites, and in the reaction of the foreigner to hitherto innocuous native parasites (Boyce 1941). Exotics are usually more liable to injury when weather conditions become critical. Into Europe with its paucity of native trees, especially conifers, many exotics have been introduced since the sixteenth century, but not one of them has attained the commercial importance of native species, and only a few are used to any extent in forest planting. The outstanding example of a successful introduction is Monterey pine in Australia, New Zealand, and South Africa (Moulds 1957), but whether this tree will ultimately justify its initial promise, time alone will tell.

In the United States limited experience with exotics so far indicates that most of them will be attacked by pathogens in some degree, and that much research will be necessary to determine how and where to grow them to obtain satisfactory stands. However, as native species are eliminated by introduced diseases or insects, we will have increased need for trial of exotics. One case of introduction of exotics to replace a lost species has already appeared in the Asiatic chestnuts, which are being introduced in quantity in the hope of securing a satisfactory replacement of the American chestnut; these, although resistant to chestnut blight, are already attacked by several native fungi. It is true that in migrating to a new home a tree species may escape from parasites that have reduced its usefulness in its native range and may make better growth in its new habitat than it did in the country of origin (both these statements so far hold for Monterey pine); but most exotics are not so fortunate. An introduced tree is likely to find among our fungi at least one parasite to which it will lack resistance. Such a fungus may be rare or geographically limited at first and cause no serious trouble to isolated test plantings, but if the exotic tree becomes widely used, the troublesome fungus can multiply and extend, slowly at first and with gradually increasing speed, until at the end of one or two generations of extensive use of the new tree its early promise may entirely disappear. This is what happened to eastern white pine when it was used in Europe. Rapid growth and high quality in early trials led to extensive planting. Now, after more than a half century of additional experience, eastern white pine has been abandoned over a large part of the countries in which it was once popular. The ultimate failure is due to the white pine blister rust fungus, previously rare in Europe, which found a congenial host in the introduced species and gradually became more widespread and abundant until it became the controlling factor. Scotch pine, which at first seemed to be free from enemies in the Northeast, has been attacked by several insects and fungi including western gall rust and sweetfern blister rust. Plantations in Pennsylvania grew vigorously for 20 years, then disintegrated rapidly from a combination of factors including attacks by fungi (Aughanbaugh 1935). Exotics in Michigan have likewise been disappointing (Baxter 1937).

Generally the same difficulties arise in extending the distribution of native trees as in introducing exotics, although not to the same degree, unless the indigenous tree is carried so far outside its natural range that it becomes essentially an exotic. A noteworthy example is European larch on which larch canker is of no importance in the native habitat of the tree but is frequently severe where the range of the tree has been widely extended in Europe. Extension of the range of Norway spruce in Europe has also been accompanied by an increasing incidence of fungous and insect parasites. A survey of silver fir plantations in Germany outside the natural habitat of the tree showed that only 25 per cent of those over 20 years old had been successful (Oldberg and

Röhrig 1955). In the Northeastern states, Tympanis canker of red pine is known only outside the natural range of this conifer. Artificial extension of the range of slash pine in the South will probably result in the development of new diseases that may prove to be limiting factors in production. The range of a tree can generally be extended somewhat with reasonable safety, but as a tree moves farther and farther from its native habitat, climatic and soil conditions become increasingly less favorable, thus reducing vigor of the species and making it more susceptible to attacks by parasites. After all, when there is no topographic or apparent climatic barrier to the natural spread of a species, it is not reasonable to expect that its range can be greatly extended artificially with any permanent success.

It must be emphasized that for an introduced species, whether it is an exotic or a native tree outside its natural habitat, successful establishment and rapid early growth are no assurance of satisfactory development in later years. Early promise must not be accepted as a guaranty for the future to be followed by extensive planting, because this way lies the possibility of severe losses. Not until the end of the first rotation can a reasonable prediction be made as to the success of an introduction; even this is not conclusive. Introductions are not all foredoomed to failure, but the chance of failure is much greater than the chance of success.

The origin of seed, which is important in relation to disease, must be considered from two viewpoints, i.e., the actual condition of the mother trees from which the seed comes as regards health and vigor, and the relation of the climatic environment of the mother trees to climatic conditions of the locality where the seed is to be used. The second consideration is of no consequence with natural reproduction, since the seedlings grow in the immediate locality of seed production. In selecting seed trees, however, those individuals which appear to be less vigorous or actually show disease should be avoided. On the other hand it is unlikely that decay of the heartwood only will have any effect on the quality of seed, although this demands investigation to determine whether undesirable hereditary characters may be associated with liability to decay. Repeated intermediate cuttings should consistently improve the seed quality of a stand, except that selection thinning that takes out the best trees must result in somewhat reducing the hereditary vigor of the remaining stand.

It is in selecting seed for artificial regeneration that great care must be exercised (Baldwin 1942:29). Even though the mother trees are vigorous and healthy, success cannot be expected if the progeny are established in an unsuitable climate or on an otherwise unsuitable site. In Sweden it is recommended that Scotch pine seed provenances be moved not more than 25 miles northward or 20 miles southward, permissible distances varying with the change in altitude (Langlet 1957). It is in Europe particularly that the importance of the source of seed can readily be seen, because there the range of native species has been greatly extended and many exotics introduced. Among other examples the writer saw two adjoining plantations of lodgepole pine in Scotland, one of which was growing vigorously and was free from any parasites, whereas the other was stunted, had unhealthy yellow-green foliage, and was severely attacked by a needle cast fungus. The plantations were the same age, the stock came from the same nursery, but the seedlings were grown from two different lots of seed of unknown origin. The trees in the better plantation resembled high-quality lodgepole pine as it occurs in the Rocky Mountains; those in the poorer plantation, the scrubby shore form of the tree found close to sea level on the Pacific Coast. In 15-year-old loblolly pine plantations in the South, trees grown from seed collected locally were larger in height and diameter and had less rust canker than trees grown from seed collected several hundred miles away (Anonymous 1941).

It is well known that races of trees within a species and individuals within a race vary in their susceptibility to both noninfectious and fungous diseases (Baxter 1933:11; Champion 1933*b*:21; Hartley 1927). The dieback of European larch in maritime climates of Germany is attributed to the use of seed from unfavorable localities such as the Tirol, and the dieback of alder is charged to seed from unfavorable localities in Belgium (Münch 1935, 1936). Münch concludes that, with everything else favorable, seed from an improper source is alone certain to cause failure of artificial reproduction. Individual and racial resistance of Scotch pine to needle cast caused by *Lophodermium pinastri* was demonstrated long ago by Zederbauer (1912), and Baxter (1933:13) cites an example in Bohemia where Scotch pine seed from an unknown source produced worthless scrubby trees severely attacked by *L. pinastri*, whereas seed from local trees produced a satisfactory stand on an adjacent area. Individual resistance within a race is well illustrated in the West by the marked variation to *Rhabdocline pseudotsugae* in natural stands of Douglas fir and to various needle cast fungi in natural stands of pines and balsam firs. Trees of the same age, in the same stand, and with their branches interlaced will range from practically immune to highly susceptible.

Another factor must be considered in the disease susceptibility of future plantations. Forest geneticists are beginning to select particularly desirable types from our timber species and even to breed them. In poplars planted for pulp, vegetatively propagated varieties may be in common use in the not distant future. This means that strains of fungi specially adapted to the particular strain of the tree species that is being used will have an unprecedented opportunity to spread. Although the forest-tree breeder will select his types from those that appear resistant to the more important diseases, it will be difficult to ensure resistance to all the species and strains of fungi concerned, and this will be particularly true for the soil-inhabiting fungi. Unfortunately, hybrid vigor can go hand-in-hand with susceptibility to disease. This has been demonstrated already by canker stain on the hybrid London plane (page 271) and by Septoria canker on hybrid poplars (page 264). There is reason to fear that progress in forest genetics, if or where it goes far enough to give planted forests of selected strains, may conceivably put our forest plantations into somewhat the same condition of susceptibility already apparent in orchards. Stands from a single clone will be especially vulnerable (Hartley 1939). This is particularly likely unless study of diseases and development of methods for testing resistance proceed more rapidly for the species to which the geneticists give their attention (Clapper 1952).

INTRODUCED DISEASES

Diseases caused by introduced fungi are potentially a menace to every one of our commercially important timber species. When an introduced disease is discovered, its capacity for damage must be speedily determined, and, if that capacity is high, the disease must be promptly eradicated or brought under control by direct measures, or else a valuable species may be destroyed. Agriculture and horticulture meet attacks of virulent introduced diseases by quick substitution of resistant varieties, by a shift of the center of cultivation from one part of the country to another, by changes in methods of cultivation, or by expensive methods of direct control such as spraying or dusting. These procedures are not possible for forest trees. The result in agriculture has been increased costs of production, and in forests virulent introduced diseases are likely to end in the loss of valuable tree species and permanent reduction in forest values, as has already happened with the American chestnut.

A foreign fungus will not necessarily do more damage than a native fungus of the same type, since most foreign fungi when transferred to a new habitat find that the climate or lack of suitable food materials militate against them. However, a foreign fungus that does find a congenial host may do to it exactly what the chestnut blight fungus has done to American chestnut, or it may injure it more than any native fungus could, because the tree lacks the specific resistance to the foreign parasite that it has acquired to native fungi in past ages. An introduced fungus may attack less conspicuously and develop more slowly than the chestnut blight fungus but nevertheless to a sufficient degree ultimately to make the management of the host species unprofitable. Furthermore the introduced disease may not be recognized as such at first, thus permitting it to become entrenched beyond hope of eradication. We are yet so little acquainted with the native diseases of our timber trees that when a previously unobserved diseased condition is found it is often difficult to decide whether it is caused by a native or an introduced pathogen. A virulent introduced parasite is infinitely more destructive to pure than to mixed stands, and no more potentially dangerous situation for disaster can be imagined than the extensive pure Douglas fir forests of the Pacific Northwest or the far-flung pure stands of ponderosa pine and lodgepole pine in other parts of the West.

Even though a lost timber species is replaced promptly and completely by others of equal commercial and aesthetic value, there is a chance for indirect and deferred damage to the forest. Most forests are balanced associations of a number of species of trees and shrubs, and the soil organic matter is largely controlled by the species mixture. There are various ways in which these species can affect each other, and the complete removal of any of the commoner species may result in soil changes unfavorable for the entire association, or in otherwise so unbalancing the forest-tree community as to reduce its productiveness seriously. The wholesale killing of American chestnut by blight on thin soils on upper slopes in places in Pennsylvania has in some instances resulted in soil deterioration through opening up of the stand to such an extent that the humus layer has disappeared, and centuries may be required to restore the original value of the watershed.

Introduced diseases, when allowed to run unchecked, can affect the aesthetic and thus the recreational value of the forest much more than most native diseases. With chestnut blight or even white pine blister rust, so many trees are killed at once that it takes years to replace the skeletons of the killed trees with enough trees of other species to restore the beauty of the forest. To many people the species that replace them are aesthetically inferior to the chestnut or the white pine.

The outstanding example of a destructive foreign disease is chestnut blight, an importation from Asia. Since its discovery in 1904, the blight has eliminated the valuable American chestnut as a commercial timber tree. Fortunately chestnut is not gregarious so that other, though less valuable, hardwoods have promptly replaced it, but forest land on which chestnut predominated has been permanently reduced in value.

Another serious introduced disease that eliminates white pines from commercial timber production if allowed to run unchecked is white pine blister rust, an importation from Europe. This disease can be controlled, but of course the cost of control is an additional charge against white pines. However, it is likely that few if any species can be intensively managed without carrying charges for protection against diseases and insects in addition to fire.

The other introduced diseases known at present appear thus far to be less serious because they are less active, because they are attacking trees of less economic importance, or because they have been introduced in localities where the susceptible native trees are not common. Willow blight, which occurs in the Northeast, where willow is of minor importance, may influence flood conditions when it spreads to the Middle West, where willow is the tree best adapted to holding the soil on the banks of streams. Larch canker was introduced into a locality where there was little eastern larch. The Dutch elm disease is so far damaging shade and ornamental trees, but it may become of consequence in those forest stands where elms have some value as timber trees. Beech bark blight, caused by an introduced insect and a fungus of uncertain origin, is destructive in northern Maine, but American beech for various reasons including the high incidence of decay in living trees has never been highly regarded as a timber tree.

The introduction of dangerous organisms is not limited to the bringing in of new species. It is known that many important fungous species consist of numerous strains, some of which clearly differ from others in their virulence. The well-known root rot fungus *Fomes annosus* is widespread in both Europe and North America. In the Douglas fir forests of the Pacific Northwest it is found frequently but appears to be unimportant so far as parasitic activity is concerned. In Europe, on the other hand, it is a serious parasite in planted coniferous forests and causes considerable damage to European plantations of Douglas fir. Although the pathogenic activities of this fungus in Europe may probably be explained by the prevalence of coniferous species planted outside their natural range or on unsatisfactory sites, there is a possibility that the European strain or strains of this root rot fungus are more dangerous than any of those present in the United States; consequently their importation might prove serious, particularly for reforestation projects.

Troublesome diseases may be introduced from other parts of our own country as well as from abroad. Although it is possible that most of the dangerous fungi native to this country are already present in every part of the continental United States in which there is any common native host susceptible to them, this is not certain. A fungus (*Dasyscypha ellisiana*) living harmlessly on the outer bark of eastern hard pines has been associated with severe injury of Douglas fir in the East. This tree in its natural range would probably be more resistant to the fungus; but if the fungus should reach the western forests and should there prove able to attack this species as it seems to in the East, serious losses would result. An example of danger from the movement of native diseases in the opposite direction is afforded by the dwarf mistletoes of pines. Probably the most important of the growth-reducing parasites, they are now limited to the West and are separated from the eastern pine forests by a practically treeless belt hundreds of miles wide along the hundredth meridian. Dwarf mistletoes spread naturally only for short distances and cannot establish themselves in a new place unless both sexes are introduced. It is probable that the eastern pines have never been exposed to them. Artificial inoculation experiments in neutral territory have shown that some of these mistletoes can attack eastern species, and it is possible that if introduced to the East they would be found to be serious enemies of eastern pines. One fungus that was brought from the West to the East is the western gall rust fungus occurring on Scotch pine. Injury by this fungus to loblolly and slash pines is feared if it spreads to the South.

Serious as some of the introduced diseases are, there is reason to think that danger from foreign parasites has just begun. To lose a single species from among our numerous forest trees is not completely disastrous, but to lose several from the same region might seriously cripple forestry from both the timber-production and the recreational standpoints. Of the great number of parasitic fungi on trees, of which some are known (Spaulding 1956, 1958), it may be that the United States already has a large proportion of the dangerous pathogens that are native to Europe, although conspicuous diseases there, but still unknown here, include a bacterial canker of poplars caused by *Pseudomonas syringae* f. sp. *populea* (Sabet and Dawson 1952), a cankerous gall of ash (page 300), and the watermark disease of willow (page 317). Up to 1912 the United States government had no legal authority to exclude diseased propagating stock, and it was only after the quarantine regulations of 1919 that the policy of free trade in plant diseases really ended. However, considering the number of forest trees related to ours that are found in Asia and the relatively small amount of commerce with Asia, Africa, and South America until recent years, it is probable that the United States has so far been exposed to relatively few of the potential tree-disease organisms of those continents. With the increasing tendency of man to bridge natural barriers by more varied and rapid means of transportation, it can be expected that tree diseases will become more and more cosmopolitan.

The first safeguard for our forests against introduced diseases is quarantines (McCubbin 1946). The maintenance of quarantine regulations is beset with difficulties; therefore in the final analysis quarantines must be regarded as measures of delay rather than measures of exclusion. The reasons for considering them as delay measures only are (1) some diseases in the early stage of development cannot be detected by visual inspection, (2) with modern facilities of transport it is practically impossible to supervise adequately all methods of importation, and (3) diseases may be carried on a variety of plant products that if excluded would reduce such imports to the point where public opinion would rebel.

Granted that quarantines in general must be considered as measures of delay, although some diseases may be kept out indefinitely, yet every delay in introduction is important not only because it defers damage or costly readjustment to the intruder but also because the new disease can likely be met in the future with a more organized and efficient effort than is possible now. Quarantines are more essential to forestry than to agriculture or horticulture because silviculture, dealing with timber crops that require decades to a century or more to reach maturity, cannot be quickly readjusted to combat new diseases, nor can disease-resistant strains be developed in reasonable time. Actually foreign trees and shrubs should be introduced only as seeds. Although it seems apparent that some fungi parasitic on trees can be carried by seeds (Collins 1915; Cunningham 1934; Pethybridge 1919), yet seeds can usually be successfully disinfected, whereas living plants usually cannot. In addition the exclusion of forest products such as logs, pulpwood, packing cases, and similar material is necessary for efficient protection, since it is known that the Dutch elm disease reached this country on elm burl logs from Europe and may have reached Europe on forest products brought in with Asiatic labor troops during the First World War. Furthermore the importation of products can bring fungi that may add to deterioration problems. Wood-destroying fungi have been introduced from other continents (Weir 1919), although as yet there is no evidence that they have become established. A heteroecious needle rust native to Europe, where it occurs on Scotch pine, was introduced into Wisconsin from Europe on packing material around a shipment of Norway spruce, from which it spread to a Scotch pine plantation (Davis 1913). Jack and red pines are potential hosts for the rust. Although not a serious disease, the manner of its introduction is significant.

Unfortunately, desirable quarantines cannot be applied to forest products without excessive interference with commerce, and therefore we can look forward to further difficulties from this source with now unknown foreign diseases. Furthermore the movement across the United States of native diseases now limited to the East or West, though less dangerous than the introduction of foreign diseases, is also a basis for concern.

Upon the discovery of a new and potentially dangerous disease, an extensive survey should be made immediately to ascertain its distribution, its capacity for damage, and whether or not it is caused by a native or by a recently introduced parasite; and, if the latter, whether eradication or direct control is necessary, feasible, or justifiable. In fighting an introduced pest it is rarely possible to eradicate it, but it is probable that it can be reduced to the point of tolerable damage. The forest values at stake both present and future must be known or quickly determined. For example, an expensive campaign would not be justified to save a species that would be replaced naturally and promptly by others of satisfactory commercial, silvicultural, and aesthetic value. A campaign against an introduced pest must be carefully planned and soundly organized to avoid undue costs. The principles governing control of introduced parasites have been expressed by Detwiler (1929:20) and later by Fracker (1937). Detwiler points out that (1) scientific research on the pest whose control is being considered or attempted is primary and basic, and in this research should be included determination of threatened values, their location, and the economic limit of protection costs; (2) further spread of the pest by artificial means should be prevented by quarantines based on a quick investigation of the mode of entry of the pest, its nature, and area of establishment; (3) the scope and character of the control plan must be based on a systematic survey to determine the extent of spread and degree of establishment of the invader; (4) the decision as to the necessity for control and the general policy to be followed should be decided in open conference by representatives of all interests affected, since a control program once under way cannot be changed readily; (5) leadership in a general control campaign is a function of government, and control work should be based on legal authority already existent or that must be provided for, although legal authority should be invoked only when persuasive measures fail; and (6) in an extensive control program, particularly one involving sanitation or destruction of host plants, the public must be fully apprised by service and educational work of the values threatened, the seriousness of the pest, and the necessity for cooperation.

The present procedure in this country against introduced diseases is unsound. In general nothing is done until the foreign organism has been introduced and become more or less well established; then an eradication or control campaign is begun. Eradication or control must be predicated on a thorough knowledge of the disease and its causal organism, which can be obtained only by careful research. But this research now largely goes along with control measures against the disease, whereas, if the control work is to be done effectively and economically, it should be preceded by considerable investigative work. However, after a dangerous foreign disease is once established, it is impossible to wait until the fundamental facts are ascertained before attempting to control it; consequently unavoidable mistakes are made and unnecessarily high expenditures for control work result. Virulent diseases known to exist abroad and to be able to attack trees related to our forest species, but which have not yet gained entry into North America, should be studied in their native countries to get information that will enable us to prevent or delay as long as possible their introduction here and to combat them effectively when they do arrive. For example, it was not known that the Dutch elm disease could be transported on logs until after it reached this country, but an early investigation of the disease in Europe by American pathologists would probably have brought this out, resulting in a quarantine on logs and the exclusion of the disease. In addition studies should be made of the pathology of North American trees growing abroad and of foreign trees closely related to ours. For example, there are species of *Pseudotsuga* in Asia that may be harboring a disease dangerous to Douglas fir. Again, although incense cedar is the only representative of its genus in the United States, other species of *Libocedrus* are found in South America, Asia, and Australasia. Many of our other valuable trees have parallel species in other continents.

Although it may be considered that the expense of controlling a virulent introduced pest can be avoided by sacrificing the native host and by crossbreeding or selection developing a resistant variety to replace it, such work is time consuming and costly and carries no certainty of success. It is unlikely that a resistant variety developed by crossbreeding will also have satisfactory silvicultural characteristics and technical properties. Even if a variety is finally produced that fills all requirements, it can be established over only a small portion of the natural range of the species where planting is economically practical. Meanwhile the tree will have lost its position in the competitive timber market.

The replacement of a susceptible American tree by a resistant foreign species has little to recommend it except as a last resort, because any exotic tree is an uncertain quantity and if introduced should be grown only experimentally for a long time, for it is not until at least one rotation has passed that reasonable judgment can be rendered as to the result of an introduction no matter how initially successful it may seem to be. First, there is the problem of securing seed from the optimum native range of the species; second, the difficulty of proper site selection; third, the question of whether the exotic will be silviculturally and commercially desirable even on the best sites in its new home; and, finally, the possibility that the foreign species at some time in its life may be attacked by an indigenous parasite hitherto innocuous, just as Asiatic chestnuts in the United States are attacked by several fungi causing dieback (Bedwell 1937). Any one of the foregoing factors can spell failure.

However, experimental plantations of foreign species should be more generally established in the United States. This activity should yield a relatively high reward for the expenditure involved. In the future, circumstances that cannot be foreseen or controlled may necessitate the widespread establishment of a foreign tree species, and it is only with knowledge gained from experimental plantings that such a situation can be adequately met.

Biological control, i.e., the control or extermination of a parasitic fungus by another parasite, has a strong appeal to the imagination, particularly when considered for an introduced parasite, because it would so simplify procedure and reduce costs, but experience so far offers little hope in this field (Krstic 1956). A careful investigation of any specific organism in its native home would be necessary to determine if a biological agent was keeping it in check. The chances are against this because in general it is host resistance or climatic factors that limit the virulence of parasitic fungi in their native habitats. If another organism were found to be responsible, it might not function effectively in a different environment, and it could not be introduced unless there was practical certainty that it would not attack other valuable plants.

Every effort should be made to protect native species against introduced diseases. Direct control, i.e., operations for the sole purpose of eradicating or controlling a disease, are essential against virulent introduced parasites that may eliminate a native tree as a factor in commercial timber production. Against native parasites direct control is rarely necessary or justifiable.

REFERENCES

Anonymous: Artificial Regeneration. *Southern Forest Expt. Sta. Ann. Rept.,* 21:7 (1941).

Aughanbaugh, J. E.: Scotch Pine: An Enigma. *Penna. Dept. Forests and Waters Serv. Letter,* 6(53):1, 3 (1935).

Baldwin, H. I.: "Forest Tree Seed." Waltham, Mass.: Chronica Botanica Company (1942).

Baxter, D. V.: Observations on Forest Pathology as a part of Forestry in Europe. *Univ. Mich. School Forestry and Conserv. Bull.*, 2:1–39 (1933).

Baxter, D. V.: Development and succession of forest fungi and diseases in forest plantations. *Univ. Mich. School Forestry and Conserv. Circ.*, 1:1–45 (1937).

Bedwell, J. L.: Factors Affecting Asiatic Chestnuts in Forest Plantations. *J. Forestry*, 35:258–262 (1937).

Birch, T. T. C.: *Diplodia pinea* in New Zealand. *New Zealand State Forest Serv. Bull.*, 8:1–32 (1936).

Boyce, J. S.: Observations on Forest Pathology in Great Britain and Denmark. *Phytopathology*, 17:1–18 (1927).

Boyce, J. S.: Exotic Trees and Disease. *J. Forestry*, 39:907–913 (1941).

Boyce, J. S.: Forest Plantation Protection Against Diseases and Insect Pests. *FAO Forestry Development Paper*, 3:1–41 (1954).

Champion, H. G.: European Silvicultural Research—Part IV—Mixtures. *Indian Forester*, 59:22–28 (1933*a*).

Champion, H. G.: The Importance of the Origin of Seed Used in Forestry. *Indian Forest Records*, 17:I-VII, 1–76 (1933*b*). (Illustrated.)

Clapper, R. B.: Breeding and Establishing New Trees Resistant to Disease. *Econ. Botany*, 6:271–293 (1952).

Clark, T. G.: Survival and Growth of 1940–41 Experimental Plantings in the Spruce Type of West Virginia. *J. Forestry*, 52:427–431 (1954).

Collins, J. F.: The Chestnut Bark Disease on Freshly Fallen Nuts. *Phytopathology*, 5:233–235 (1915).

Cunningham, G. H.: Mycology Section. *New Zeal. Dept. Agr. Ann. Rept.*, 1933–1934:39–42 (1934).

Davis, J. J.: The Introduction of a European Pine Rust into Wisconsin. *Phytopathology*, 3:306–307 (1913).

Day, W. R.: The Relation Between Disease and the Constitution and Environment of the Tree. *J. Roy. Agr. Soc. Engl.*, 95:54–72 (1934).

Detwiler, S. B.: "Insect and Disease Control as a Branch of Forest Protection." New Haven, Conn.: Yale University, School of Forestry (1929).

Dimitroff, T.: Contribution to the Study of the Insect Pests and Fungal Diseases of Our Natural and Cultivated Forests. [Translated title. French summary.] *Sofiisk. Univ., Agron. Lesov. Fakult. God.*, 13:220–252 (1935). [*Rev. Appl. Mycol.*, 15:411–412 (1936).]

Fourchy, P.: Some Aspects of Present-day Silviculture in Switzerland. *Quart. J. Forestry*, 48:85–104 (1954).

Fracker, S. B.: Technique of Large-scale Operations in Pest Control. *Wash. Entomol. Soc. Proc.*, 39:41–58 (1937).

Hartley, C.: Forest Genetics with Particular Reference to Disease Resistance. *J. Forestry*, 25:667–686 (1927).

Hartley, C.: The Clonal Variety for Tree Planting: Asset or Liability? *Arborist's News*, 4:25–27 (1939).

Hartley, C., J. S. Boyce, et al.: The Progress of Forest Pathology. 73d Cong. 1st Sess., Senate Doc., 12:695–722 (1933).

Hepting, G. H.: Eastern Forest Tree Diseases in Relation to Stand Improvement. *Emergency Conserv. Work Forestry Publ.*, 2:1–28 (1933).

Kienholz, R., and C. B. Bidwell: A Survey of Diseases and Defects in Connecticut Forests. *Conn. (State) Agr. Expt. Sta. Bull.*, 412:493–559 (1938).

Köstler, J. N.: Gemischte Wälder. *Forstwiss. Centr.*, 71:1–16 (1952).

Krstic, M.: Prospects of Application of Biological Control in Forest Pathology. *Botan. Rev.*, 22:38–44 (1956).

Langlet, O.: Vidgade gränser för förflytning av tallprovenienser till skogsodlingplatser i norra Sverige. [Wider Limits for Moving Pine Provenances to Forest Sites in North Sweden.] *Skogen* 44:319 (1957). [*Forestry Abstr.*, 18:475 (1957).]

Leibundgut, H.: Über die Planung von Bestandesumwandlungen. *Schweiz. Z. Forstw.*, 98:372–389 (1947).

McCubbin, W. A.: Preventing Plant Disease Introduction. *Botan. Rev.*, 12:101–139 (1946).

Moulds, F. R.: Exotics Can Succeed in Forestry as in Agriculture. *J. Forestry*, 55:563–566 (1957).

Münch, E.: Das Larchenrätsel als Rassenfrage. Zweite Mitteilung: Die Lärche im Seeklima. *Z. Forst u. Jagdw.*, 67:421–442, 483–500 (1935).

Münch, E.: Das Larchensterben. *Forstwiss. Centr.*, 58:469–494, 537–562, 581–590, 643–671 (1936*a*).

Münch, E.: Das Erlensterben. *Forstwiss. Centr.*, 58:173–194, 230–248 (1936*b*).

Oldberg, A., and E. Röhrig: Waldbauliche Untersuchungen über die Weisstanne im nördlichen und mittleren Westdeutschland. *Schriftenreihe forstl. Fak. Univ. Göttingen*, 12:1–103 (1955).

Oliphant, J. N.: Artificial v. Natural Regeneration. *Malayan Forester*, 1:186–192 (1932).

Peace, T. R.: Approach and Perspective in Forest Pathology. *Forestry*, 30:47–56 (1957).

Peltier, G. L., F. R. Schroeder, and E. Wright: Distribution and Prevalence of Ozonium Root Rot in the Shelterbelt Planting Area of Oklahoma. *Phytopathology*, 29:485–490 (1939).

Péter-Contesse, J.: Un problème nouveau. *Schweiz. Z. Forstw.*, 107:587–592 (1956).

Pethybridge, G. H. A.: Destructive Disease of Seedling Trees of *Thuja gigantea* Nutt. *Quart. J. Forestry*, 13:93–97 (1919).

Sabet, K. A., and W. J. Dowson: Studies in the Bacterial Die-back and Canker Disease of Poplar I. The Disease and its Cause. *Ann. Appl. Biol.*, 39:609–616 (1952).

St. Clair–Thompson, G. W.: "The Protection of Woodlands by Natural as Opposed to Artificial Methods." London: H. F. & G. Witherby (1928).

Spaulding, P.: Diseases of North American Trees Planted Abroad. *U.S. Dept. Agr., Agr. Handbook*, 100:1–144 (1956).

Spaulding, P.: Diseases of Foreign Forest Trees Growing in the United States. *U.S. Dept. Agr., Agr. Handbook*, 139:1–118 (1958).

Weir, J. R.: Concerning the Introduction into the United States of Extra-limital Wood-destroying Fungi. *Mycologia*, 11:58–65 (1919).

Weir, J. R., and E. E. Hubert: Forest Disease Surveys. *U.S. Dept. Agr. Bull.*, 658:1–23 (1918).

Weir, J. R., and E. E. Hubert: Pathological Marking Rules for Idaho and Montana. *J. Forestry*, 17:666–681 (1919).

Zederbauer, E.: Versuche über individuelle Auslese bei Waldbäumen I. *Pinus silvestris*. *Centr. ges. Forstw.*, 38:201–212 (1912).

APPENDIX A

Fungicides

Although fungicides are commonly used in the control of plant diseases in general, their application to the control of forest-tree diseases is limited, except in nurseries, because of the cost. They are applied either as dusts or as sprays. In recent years a large number of new fungicides have been developed and marketed under many trade names, thus creating problems of identity and considerable confusion (Frear 1958). Relatively few of these new chemicals have yet been tried against pathogens of forest trees (Carter and English 1957). Inasmuch as the results of tests of fungicides against tree diseases are appearing frequently, it is well to have the latest information for a particular disease if a spray program is planned. The best sources of such information are the state and Federal agricultural experiment stations.

Bordeaux Mixture. This is an old fungicide, still valuable for trees, particularly because its effectiveness has been proved for many years and for a number of pathogens.

The spray is a mixture of copper sulfate, lime, and water in varying proportions, resulting in the mixture being termed 4–4–50, 5–5–50, and so forth. The 4–4–50 formula is made by dissolving 4 pounds of copper sulfate, preferably of "snow" grade, in 50 gallons of water. Stir 4 pounds of hydrated lime into a bucket of water, then add this suspension to the copper sulfate solution in the spray tank while the agitator is running. As this spray deteriorates rapidly, it should be used within a few hours.

When spraying conifers, or other foliage with a slick waxy surface, it is usually necessary to add a sticker and spreader. Any glyceride oil, such as cottonseed or corn oil, at the rate of 1 quart to each 50 gallons of spray, can be used. The Bordeaux mixture itself acts as an emulsifier.

If only a small quantity of Bordeax mixture is needed, it can be made from prepared Bordeaux powder now widely sold. The 5–5–50 formula is composed of 1 pound of commercial prepared Bordeaux powder in 5 gallons of water. For a sticker and spreader, if needed, 3 ounces (about ⅕ pint) of cottonseed or corn oil can be added. However, homemade Bordeaux is more effective and cheaper in any but small quantities. If Bordeaux is difficult to obtain, any of the new copper fungicides will be adequate substitutes.

Soil Fumigants. Fumigants are chemical toxicants applied to the soil in a volatile form, and most of them have to be temporarily sealed in. Some organic chemicals have been successful in controlling various pests in the soil, although the ideal fumigant has not yet been found (Parris 1958). In forestry practice their use is restricted to nurseries.

In the southern pine region methyl bromide kills fungi, nematodes, insects, other soil organisms, and weed seeds (Foster 1959). Furthermore, it improves seedling growth. Most of the beneficial effects last 3 years at least. According to Foster the fumigant is applied before planting as follows: "The soil is harrowed or rototilled and watered to a moisture content suitable for planting. The chemical is applied at the rate of 1 pound per 150 square feet under polyethylene plastic sheets when the soil temperature is above 60 degrees F at a depth of 6 inches. Sacks of pine straw are placed in the center of the area under the sheets to raise the cover so as to allow the gas fumes to spread uniformly over the area. Twenty-four hours after the chemical has been applied, the plastic sheets are removed."

Wound Dressings. The various materials applied to trees to disinfect wounds caused by pruning or in other ways may be classed as (1) the air-tight type as represented by asphaltum, coal tar, grafting wax, and rubber latex and (2) the air-porous type as represented by Bordeaux paste.

Bordeaux paste is made by dissolving 1½ pounds of copper sulfate in 1 gallon of water, slaking 3 pounds of stone lime in 1 gallon of water, and then mixing the two solutions. This makes a dressing about the consistency of whitewash that can be applied with a brush. Commercial prepared Bordeaux powder can also be used by mixing it with water until the consistency of whitewash is attained. Bordeaux paint, which is more permanent, is made by slowly adding linseed oil to Bordeaux powder while stirring constantly until a thick paint that can be applied with a brush is obtained. One application of Bordeaux paint may remain effective for several seasons.

REFERENCES

Carter, J. C., and L. L. English: New Insecticides and Fungicides and Their Use. *Arborists' News,* 22:25–32 (1957).

Foster, A. A.: Nursery Diseases of Southern Pines. *U.S. Dept. Agr., Forest Serv., Forest Pest Leaflet,* 32 (1959).

Frear, D. E. H.: "Pesticide Handbook," 10th ed. State College, Pennsylvania: College Science Publishers (1958).

Hartley, C.: Fungicides for Forest Trees, Shade Trees and Forest Products. *Botan. Rev.*, 16:33–50 (1950).

Hough, W. S., and A. F. Mason: "Spraying, Dusting and Fumigation of Plants." New York: The Macmillan Company (1951).

Parris, G. K.: Soil Fumigants and Their Use: A Summary. *U.S. Dept. Agr., Plant Disease Reptr.*, 42:273–278 (1958).

Westcott, Cynthia: "Plant Disease Handbook." Princeton, N.J.: D. Van Nostrand Company, Inc. (1950).

APPENDIX B

List of Common Names of Plants Used with Scientific Equivalents

Acacia—*Acacia* spp.
Ailanthus—*Ailanthus altissima* (Mill.) Swing.
Alder, mountain—*Alnus tenuifolia* Nutt.
Alder, red—*Alnus rubra* Bong.
Alder, Sitka—*Alnus sinuata* (Reg.) Rydb.
Alder, smooth—*Alnus rugosa* (Du Roi) Spreng.
Alder, speckled—*Alnus incana* (L.) Moench.
Alder, white—*Alnus rhombifolia* Nutt.
Apple, Oregon crab—*Malus rivularis* (Dougl.) Roem.
Apple, wild—*Malus* spp.
Apricot—*Prunus armeniaca* L.
Ash, English—*Fraxinus excelsior* L.
Ash, green—*Fraxinus pennsylvanica lanceolata* (Borkh.) Sarg.
Ash, mountain—*Sorbus* spp.
Ash, red—*Fraxinus lanceolata* Marsh.
Ash, velvet—*Fraxinus velutina* Torr.
Ash, white—*Fraxinus americana* L.
Aspen—*Populus tremuloides* Michx.
Aspen, largetooth—*Populus grandidentata* Michx.
Basswood—*Tilia glabra* Vent.
Bay, sweet—*Magnolia virginiana* L.
Beech—*Fagus grandifolia* Ehrh.
Beech, American—*Fagus grandifolia* Ehrh.
Beech, European—*Fagus sylvatica* L.
Beech, red—*Fagus sylvatica* L.
Birch, dwarf—*Betula glandulosa* Michx.
Birch, gray—*Betula populifolia* Marsh.
Birch, paper—*Betula papyrifera* Marsh.
Birch, sweet—*Betula lenta* L.
Birch, water—*Betula fontinalis* Sarg.
Birch, western paper—*Betula papyrifera* var. *occidentalis* Sarg.
Birch, yellow—*Betula alleghaniensis* Britt.
Boxelder—*Acer negundo* L.
Buckeye, California—*Aesculus californica* (Spach) Nutt.

Buckeye, Ohio—*Aesculus glabra* Willd.
Butternut—*Juglans cinerea* L.
California laurel—*Umbellularia californica* (Hook. & Arn.) Nutt.
Catalpa, hardy—*Catalpa speciosa* Ward.
Catalpa, northern—*Catalpa speciosa* Ward.
Cedar, Alaska—*Chamaecyparis nootkatensis* (Lam.) Sudw.
Cedar, eastern red—*Juniperus virginiana* L.
Cedar, incense—*Libocedrus decurrens* Torr.
Cedar, northern white—*Thuja occidentalis* L.
Cedar, Port Orford—*Chamaecyparis lawsoniana* (A. Murr.) Parl.
Cedar, Rocky Mountain red—*Juniperus scopulorum* Sarg.
Cedar, southern red—*Juniperus lucayana* Britt.
Cedar, southern white—*Chamaecyparis thyoides* (L.) B.S.P.
Cedar, western red—*Thuja plicata* D. Don.
Cherry, black—*Prunus serotina* Ehrh.
Cherry, choke—*Prunus virginiana* L.
Chestnut, American—*Castanea dentata* (Marsh.) Borkh.
Chestnut, Chinese—*Castanea mollissima* Blume
Chestnut, European—*Castanea sativa* Mill.
Chestnut, horse—*Aesculus hippocastanum* L.
Chestnut, Japanese—*Castanea japonica* Blume
Chinquapin—*Castanea pumila* (L.) Mill.
Chinquapin, golden—*Castanopsis chrysophylla* (Hook.) A. DC.
Chokeberry—*Aronia* spp.
Coffee-tree, Kentucky—*Gymnocladus dioicus* (L.) Koch.
Cottonwood, black—*Populus trichocarpa* Hook.
Cottonwood, eastern—*Populus deltoides* Marsh.
Cottonwood, valley—*Populus wislizenii* (Wats.) Sarg.
Crab, wild—*Peraphyllum ramosissimum* Nutt.
Cypress, Arizona—*Cupressus arizonica* Greene
Cypress, bald—*Taxodium distichum* (L.) Rich.
Cypress, Lawson—*Chamaecyparis lawsoniana* (A. Murr.) Parl.
Cypress, Monterey—*Cupressus macrocarpa* Gord.
Cypress, smooth—*Cupressus glabra* Sud.
Cypress, southern—*Taxodium distichum* (L.) Rich.
Dogwood, flowering—*Cornus florida* L.
Douglas fir—*Pseudotsuga taxifolia* (Lam.) Br. = *P. menziesii* (Mirb.) Franco
Elder—*Sambucus* spp.
Elm, American—*Ulmus americana* L.
Elm, Chinese—*Ulmus parvifòlia* Jacq.
Elm, English—*Ulmus campestris* L.
Elm, Siberian—*Ulmus pùmila* L.
Elm, slippery—*Ulmus fulva* Michx.
Elm, smoothleaf—*Ulmus foliacea* Gilib.
Elm, winged—*Ulmus alata* Michx.
Fernbush—*Chamaebatiaria millefolium* (Torr.) Maxim.
Fir, alpine—*Abies lasiocarpa* (Hook.) Nutt.
Fir, balsam—*Abies balsamea* (L.) Mill.
Fir, bristlecone—*Abies venusta* (Dougl.) Koch.
Fir, corkbark—*Abies lasiocarpa* var. *arizonica* (Merriam) Lemm.

Fir, Douglas—*Pseudotsuga taxifolia* (Lam.) Br. = *P. menziesii* (Mirb.) Franco
Fir, lowland white—*Abies grandis* Lind.
Fir, noble—*Abies nobilis* Lind.
Fir, Pacific silver—*Abies amabilis* (Lind.) Forbes
Fir, red—*Abies magnifica* A. Murr.
Fir, silver—*Abies alba* Mill.
Fir, southern balsam—*Abies fraseri* (Pursh) Poir.
Fir, subalpine—*Abies lasiocarpa* (Hook.) Nutt.
Fir, white—*Abies concolor* Lind. & Gord.
Firs, balsam—*Abies* spp.
Gum, red—*Liquidambar styraciflua* L.
Gum, sweet—*Liquidambar styraciflua* L.
Gum, tupelo—*Nyssa aquatica* L.
Gums—*Nyssa* spp.
Hawthorn—*Crataegus* spp.
Hazel—*Corylus* spp.
Hemlock, Carolina—*Tsuga caroliniana* Engelm.
Hemlock, eastern—*Tsuga canadensis* (L.) Carr.
Hemlock, mountain—*Tsuga mertensiana* (Bong.) Sarg.
Hemlock, western—*Tsuga heterophylla* (Raf.) Sarg.
Hickory, pignut—*Carya glabra* (Mill.) Sweet
Hickory, shagbark—*Carya ovata* (Mill.) K. Koch
Honeylocust—*Gleditsia triacanthos* L.
Hop-hornbeam—*Ostrya virginiana* (Mill.) Koch
Hornbeam, European—*Carpinus betulus* L.
Juniper, alligator—*Juniperus pachyphloea* Torr.
Juniper, dwarf—*Juniperus communis* L.
Juniper, European—*Juniperus sabina* L.
Juniper, Mexican—*Juniperus mexicana* Spreng.
Juniper, mountain—*Juniperus sibirica* Burgsd.
Juniper, one-seed—*Juniperus monosperma* (Engelm.) Sarg.
Juniper, prostrate—*Juniperus horizontalis* Moench.
Juniper, Rocky Mountain—*Juniperus scopulorum* Sarg.
Juniper, Utah—*Juniperus utahensis* (Engelm.) Lemm.
Juniper, western—*Juniperus occidentalis* Hook.
Larch, alpine—*Larix lyallii* Parl.
Larch, Dahurian—*Larix dahurica* Turcz.
Larch, eastern—*Larix laricina* (Du Roi) Koch
Larch, European—*Larix decidua* Mill.
Larch, golden—*Pseudolarix amabilis* (Nels.) Rehd.
Larch, Japanese—*Larix leptolepis* Gord.
Larch, Siberian—*Larix sibirica* Led.
Larch, western—*Larix occidentalis* Nutt.
Locust, black—*Robinia pseudoacacia* L.
Locust, honey—*Gleditsia triacanthos* L.
Madroño—*Arbutus menziesii* Pursh.
Magnolia, evergreen—*Magnolia grandiflora* L.
Magnolia, southern—*Magnolia grandiflora* L.
Manzanita—*Arctostaphylos* spp.
Maple, bigleaf—*Acer macrophyllum* Pursh.

Maple, dwarf—*Acer glabrum* Torr.
Maple, mountain—*Acer spicatum* Lam.
Maple, Norway—*Acer platanoides* L.
Maple, red—*Acer rubrum* L.
Maple, silver—*Acer saccharinum* L.
Maple, striped—*Acer pennsylvanicum* L.
Maple, sugar—*Acer saccharum* Marsh.
Maple, sycamore—*Acer pseudoplatanus* L.
Maple, vine—*Acer circinatum* Pursh.
Mesquite—*Prosopis* spp.
Mimosa tree—*Albizzia julibrissin* Durazz.
Mulberry, red—*Morus rubra* L.
Oak, black—*Quercus velutina* Lam.
Oak, California black—*Quercus kelloggii* Newb.
Oak, chestnut—*Quercus montana* Willd.
Oak, coast live—*Quercus agrifolia* Née
Oak, Emory—*Quercus emoryi* Torr.
Oak, English—*Quercus robur* L.
Oak, live—*Quercus virginiana* Mill.
Oak, northern red—*Quercus rubra* L.
Oak, Oregon white—*Quercus garryana* Dougl.
Oak, pin—*Quercus palustris* Muench.
Oak, post—*Quercus stellata* Wangh.
Oak, Utah white—*Quercus utahensis* (DC.) Rydb
Oak, water—*Quercus nigra* L.
Oak, white—*Quercus alba* L.
Oak, whiteleaf—*Quercus hypoleuca* Engelm.
Osage orange—*Toxylon pomiferum* Raf.
Paintbrush—*Castilleja* spp.
Pear—*Pyrus* spp.
Pea-tree—*Caragana arboréscens* Lam.
Pecan—*Hicoria pecan* (Marsh.) Britt.
Persimmon—*Diospyros virginiana* L.
Pine, Apache—*Pinus apacheca* Lemm.
Pine, Arizona—*Pinus arizonica* Engelm.
Pine, Austrian—*Pinus nigra* Arn.
Pine, Balkan—*Pinus peuce* Griseb.
Pine, bristlecone—*Pinus aristata* Engelm.
Pine, Chihuahua—*Pinus leiophylla* Schlecht. & Cham
Pine, Corsican—*Pinus laricio* Poir.
Pine, Coulter—*Pinus coulteri* D. Don.
Pine, digger—*Pinus sabiniana* Dougl.
Pine, eastern white—*Pinus strobus* L.
Pine, foxtail—*Pinus balfouriana* Murr.
Pine, Himalayan—*Pinus excelsa* Wall.
Pine, jack—*Pinus banksiana* Lam.
Pine, Japanese black—*Pinus thunbérgii* Parl.
Pine, Japanese red—*Pinus densiflora* Sieb. & Zucc.
Pine, Jeffrey—*Pinus jeffreyi* "Oreg. Com."
Pine, knobcone—*Pinus attenuata* Lemm.

Pine, limber—*Pinus flexilis* James
Pine, loblolly—*Pinus taeda* L.
Pine, lodgepole—*Pinus contorta* Loud.
Pine, longleaf—*Pinus palustris* Mill.
Pine, maritime—*Pinus pinaster* Sol.
Pine, Monterey—*Pinus radiata* D. Don.
Pine, Montezuma—*Pinus montezumae* Lamb.
Pine, mountain—*Pinus pungens* Lam.
Pine, northern white—*Pinus strobus* L.
Pine, Norway—*Pinus resinosa* Sol.
Pine, piñon—see Piñon
Pine, pitch—*Pinus rigida* Mill.
Pine, pond—*Pinus rigida serotina* (Michx.) Loud.
Pine, ponderosa—*Pinus ponderosa* Laws.
Pine, red—*Pinus resinosa* Sol.
Pine, Rocky Mountain ponderosa—*Pinus ponderosa* Laws. var. *scopulorum* Engelm.
Pine, sand—*Pinus clausa* (Engelm.) Sarg.
Pine, Scotch—*Pinus sylvestris* L.
Pine, scrub—*Pinus virginiana* Mill.
Pine, shortleaf—*Pinus echinata* Mill.
Pine, slash—*Pinus elliottii* Engelm. ex Vasey
Pine, spruce—*Pinus glabra* Walt.
Pine, sugar—*Pinus lambertiana* Dougl.
Pine, Swiss mountain—*Pinus montana* Mill.
Pine, Swiss stone—*Pinus cembra* L.
Pine, Virginia—*Pinus virginiana* Mill.
Pine, western white—*Pinus monticola* D. Don.
Pine, western yellow—*Pinus ponderosa* Laws.
Pine, whitebark—*Pinus albicaulis* Engelm.
Pines, southern yellow—Loblolly, longleaf, pitch, pond, shortleaf, slash, spruce, and Virginia pines.
Piñon—*Pinus edulis* Engelm.
Piñon, Mexican—*Pinus cembroides* Zucc.
Piñon, Parry—*Pinus parryana* Engelm.
Piñon, singleleaf—*Pinus monophylla* Torr. & Frem.
Plane, American—*Platanus occidentalis* L.
Plane, California—*Platanus racemosa* Nutt.
Plane, London—*Platanus acerifolia* (Ait.) Willd.
Plane, oriental—*Platanus orientalis* L.
Plum, Pacific—*Prunus subcordata* Benth.
Poplar, balsam—*Populus balsamifera* L.
Poplar, black—*Populus nigra* L.
Poplar, Japan—*Populus maximowiczii* Henry
Poplar, Norway—*Populus eugenii* Simon-Louis
Poplar, white—*Populus alba* L.
Poplar, yellow—*Liriodendron tulipifera* L.
Quince—*Cydonia* spp.
Redbud—*Cercis canadensis* L.
Redwood—*Sequoia sempervirens* Endl.
Sassafras—*Sassafras albidum* (Nutt.) Nees

Serviceberry—*Amelanchier* spp.
Soapberry, western—*Sapindus drummondi* Hook. & Arn.
Sourwood—*Oxydendrum arboreum* (L.) DC.
Spruce, bigcone—*Pseudotsuga macrocarpa* (Torr.) Mayr.
Spruce, black—*Picea mariana* (Mill.) B.S.P.
Spruce, blue—*Picea pungens* Engelm.
Spruce, Engelmann—*Picea engelmannii* Engelm.
Spruce, Norway—*Picea abies* Karst.
Spruce, red—*Picea rubens* Sarg.
Spruce, Sitka—*Picea sitchensis* (Bong.) Carr.
Spruce, weeping—*Picea breweriana* S. Wats.
Spruce, western white—*Picea glauca albertiana* (Br.) Rehd.
Spruce, white—*Picea glauca* (Moench.) Voss.
Sugarberry—*Celtis laevigata* Willd.
Sumach, staghorn—*Rhus typhina* L.
Sweetfern—*Comptonia peregrina* (L.) Coult.
Sweet gale—*Myrica gale* L.
Sweet-gum—*Liquidambar styraciflua* L.
Sycamore, American—*Platanus occidentalis* L.
Sycamore, California—*Platanus racemosa* Nutt.
Syringa—*Philadelphus* spp.
Tamarack—*Larix laricina* (Du Roi) Koch.
Torreya, California—*Torreya californica* Torr.
Tupelo, black—*Nyssa sylvatica* Marsh.
Tupelo, swamp—*Nyssa biflora* Walt.
Walnut, black—*Juglans nigra* L.
Wax-myrtle, southern—*Myrica cerifera* L.
Willow, bay-leaved—*Salix pentandra* L.
Willow, crack—*Salix fragilis* L.
Willow, golden—*Salix alba vitellina* (L.) Stokes
Willow, heart-leaved—*Salix cordata* Muhl.
Willow, osier—*Salix viminalis* L.
Willow, purple—*Salix purpurea* L.
Willow, weeping—*Salix babylonica* L.
Witchhazel—*Hamamelis* spp.
Yellowpoplar—*Liriodendron tulipifera* L.
Yew, Pacific—*Taxus brevifolia* Nutt.

Index

When several page references follow a topic, the principal discussion is indicated by **boldface** numbers.